Lecture Notes in Artificial Intelligence 5223

Edited by R. Goebel, J. Siekmann, and W. Wahlster

Subseries of Lecture No [illegible] Science

Gem Stapleton   John Howse   John Lee (Eds.)

Diagrammatic Representation and Inference

5th International Conference, Diagrams 2008
Herrsching, Germany, September 19-21, 2008
Proceedings

With 13 Color Figures

Series Editors

Randy Goebel, University of Alberta, Edmonton, Canada
Jörg Siekmann, University of Saarland, Saarbrücken, Germany
Wolfgang Wahlster, DFKI and University of Saarland, Saarbrücken, Germany

Volume Editors

Gem Stapleton
Computing, Mathematical and Information Sciences
University of Brighton
Brighton, UK
E-mail: G.E.Stapleton@brighton.ac.uk

John Howse
Computing, Mathematical and Information Sciences
University of Brighton
Brighton, UK
E-mail: John.Howse@brighton.ac.uk

John Lee
Human Communication Research Centre
University of Edinburgh
Informatics Forum
Edinburgh, Scotland, UK
E-mail: J.Lee@ed.ac.uk

Library of Congress Control Number: 2008934909

CR Subject Classification (1998): I.2, D.1.7, G.2, H.5, J.4, J.5

LNCS Sublibrary: SL 7 – Artificial Intelligence

ISSN 0302-9743
ISBN-10 3-540-87729-0 Springer Berlin Heidelberg New York
ISBN-13 978-3-540-87729-5 Springer Berlin Heidelberg New York

Springer is a part of Springer Science+Business Media

springer.com

Printed in Germany

Typesetting: Camera-ready by author, data conversion by Scientific Publishing Services, Chennai, India
Printed on acid-free paper SPIN: 12525874 06/3180 5 4 3 2 1 0

Preface

Diagrams is an international and interdisciplinary conference series, covering all aspects of research on the theory and application of diagrams.

Recent technological advances have enabled the large-scale adoption of diagrams in a diverse range of areas. Increasingly sophisticated visual representations are emerging and, to enable effective communication, insight is required into how diagrams are used and when they are appropriate for use. The pervasive, everyday use of diagrams for communicating information and ideas serves to illustrate the importance of providing a sound understanding of the role that diagrams can, and do, play. Research in the field of diagrams aims to improve our understanding of the role of diagrams, sketches and other visualizations in communication, computation, cognition, creative thought, and problem solving. These concerns have triggered a surge of interest in the study of diagrams.

The study of diagrammatic communication as a whole must be pursued as an interdisciplinary endeavour. Diagrams 2008 was the fifth event in this conference series, which was launched in Edinburgh during September 2000. Diagrams attracts a large number of researchers from virtually all related fields, placing the conference as a major international event in the area.

Diagrams is the only conference that provides a united forum for all areas that are concerned with the study of diagrams: for example, architecture, artificial intelligence, cartography, cognitive science, computer science, education, graphic design, history of science, human-computer interaction, linguistics, logic, mathematics, philosophy, psychology, and software modelling. We see issues from all of these fields discussed in the papers collected in the present volume.

For the 2008 conference, no preferred theme was specified, with the result that the interdisciplinary range was perhaps broader than ever. Contributions were solicited in the categories long paper, short paper and poster. Submissions were received representing both academia and industry, from 24 countries. Of 70 papers submitted, 25 were accepted for either long or short presentations, an acceptance rate of 36%, demonstrating that the conference continues to offer an outlet as high in quality as it is unique in breadth. This year, a substantial number of posters contributed to the range of work presented. All submissions were reviewed by members of the large and distinguished International Program Committee, or by reviewers that they nominated. The papers and posters were augmented by three highly distinguished keynote presentations—perfectly representing both the quality and breadth of the conference as a whole—and two excellent tutorials.

For the first time in its history, Diagrams was co-located, running in conjunction with the IEEE Symposium on Visual Languages and Human-Centric Computing and the ACM Symposium on Software Visualization, as part of Visual Week. This co-location provided a lively and stimulating environment, enabling researchers from related communities to exchange ideas and more widely disseminate research results. The program featured joint keynote speakers (including Wilhelm Schäfer), and joint sessions (including the papers under "Diagram Aesthetics and Layout" in this volume).

Any successful conference depends on the efforts of a large team of people, and Diagrams is no exception. The hard work of the Program Committee is paramount in ensuring the quality and substance of the conference, and we are greatly indebted to their dedication and generosity with their time. This volume is evidence of the important support of our publishers, Springer, and especially Ursula Barth. Barbara Wirtz of the *Haus der bayerischen Landwirtschaft* provided invaluable assistance with the conference venue. As noted, Diagrams 2008 was co-located with Visual Week, but also very nearly with the celebrated Munich Oktoberfest! There was therefore an exceptional effort by Local Chair Mark Minas and his team, including Steffen Mazanek, Sonja Maier (who oversaw Oktoberfest arrangements), and Florian Brieler (who created the registration Web page). Alan Blackwell helped greatly with finding the Diagrams tutorials.

Sponsors are vital to any successful conference, and we are delighted to acknowledge the generous support of the Cognitive Science Society, for the Best Student Paper award, and Nokia for providing two N810s as Best Paper prizes as well as financial support.

Finally, we would like to thank all of the organizers mentioned on the following pages, and our own institutions, the University of Brighton and the University of Edinburgh. We are especially grateful for the indispensable assistance of our administrator, Carol Suwala in Brighton.

July 2008

Gem Stapleton
John Howse
John Lee

Conference Organization

General Chair

Gem Stapleton	University of Brighton, UK

Program Chairs

John Howse	University of Brighton, UK
John Lee	University of Edinburgh, UK

Local Chair

Mark Minas	Universität der Bundeswehr, Germany

Publicity Chair

Andrew Fish	University of Brighton, UK

Website and Technical Support

Aidan Delaney	University of Brighton, UK

Sponsoring Organizations

Nokia
Cognitive Science Society

Program Committee

Gerard Allwein	Naval Research Laboratory, USA
Michael Anderson	University of Hartford, USA
Dave Barker-Plummer	Stanford University, USA
Alan Blackwell	Cambridge University, UK
Dorothea Blostein	Queen's University, Canada
B. Chandrasekaran	Ohio State University, USA
Peter Cheng	University of Sussex, UK
Phil Cox	Dalhousie University, Canada
Richard Cox	University of Sussex, UK
Frithjof Dau	University of Wollongong, Australia

Max J. Egenhofer	University of Maine, USA
Stephanie Elzer	Millersville University, USA
Yuri Engelhardt	University of Amsterdam, The Netherlands
Jacques Fleuriot	University of Edinburgh, UK
Jean Flower	Autodesk, UK
David Gooding	Bath University, UK
Corin Gurr	University of Reading, UK
Mary Hegarty	University of California, Santa Barbara, USA
Mateja Jamnik	Cambridge University, UK
Yasuhiro Katagiri	Future University, Japan
Hans Kestler	University of Ulm, Germany
Zenon Kulpa	Institute of Fundamental Technological Research, Poland
Oliver Lemon	University of Edinburgh, UK
Stefano Levialdi	University of Rome – "La Sapienza", Italy
Richard Lowe	Curtin University of Technology, Australia
Grant Malcolm	University of Liverpool, UK
Kim Marriott	Monash University, Australia
Bernd Meyer	Monash University, Australia
Nathaniel Miller	University of Northern Colorado, USA
N. Hari Narayanan	Auburn University, USA
James Noble	Victoria University of Wellington, New Zealand
Jesse Norman	University College London, UK
Jon Oberlander	University of Edinburgh, UK
Luis Pineda	Universidad Nacional Autónoma de México, Mexico City
Helen Purchase	Glasgow University, UK
Thomas Rist	Fachhochschule Augsburg, Germany
Peter Rodgers	University of Kent, UK
Frank Ruskey	University of Victoria, Canada
Atsushi Shimojima	Doshisha University, Japan
Sun-Joo Shin	Yale University, USA
John Sowa	VivoMind Intelligence Inc., USA
Keith Stenning	University of Edinburgh, UK
Nik Swoboda	Universidad Politécnica de Madrid, Spain
Gabi Taentzer	Technical University of Berlin, Germany
Susan Trickett	Naval Research Laboratory, USA
Barbara Tversky	Stanford University and Columbia University, USA

Additional Referees

Bonny Banerjee
Richard Bosworth
Andrew Fish
Unmesh Kurup

Laura Meikle
Matthew Ridsdale
Andrew Seniuk
John Taylor
Sean Wilson
Graham Winstanley
Michael Wybrow
Nissa Yestness

Table of Contents

Keynote Reflections

Tutorials

Diagram Aesthetics and Layout

Psychological and Cognitive Issues

Applications of Diagrams

Theoretical Aspects

Diagrams in Education

Understanding and Comprehension

Posters

Heterogeneous Reasoning

John Etchemendy

Stanford University
Stanford, CA, 94305-4101, USA

During the 1990s, we developed the theory of *formal heterogeneous deduction*: logically valid inference that involves information expressed using multiple different representations. The Hyperproof program was developed as an implementation of that theory, and permitted deductions using sentences of first-order logic and blocks world diagrams. We have since been generalizing both the theory and the implementation to allow applications in a wide variety of domains.

On the theoretical side, we have identified features of heterogeneous reasoning which occur in everyday domains, but which do not correspond to logical deductions. Examples here include applications in design, where a desired design may not be a logical consequence of some initial information, but rather is preferred based on notions of aesthetics, cost or other non-logical features of the design domain. The architecture that we developed for Hyperproof generalizes to these situations and permits the modeling of this kind of reasoning. The notion of a logically valid inference is generalized to a *justified modification*, where the justification may be a piece of text written by the user, a calculation of the cost of materials in a design, or any other rationale that the user finds appropriate to the domain at hand. The general principles of information flow through a structured argument remain the same in both contexts.

On the implementation side, we have designed the Openproof architecture, an application framework for building heterogeneous reasoning systems which permits "plug-ins". Individual developers can provide components to the architecture which are guaranteed to work with other existing components to create heterogeneous reasoning environments in a flexible manner. A researcher interested in Venn diagrams, for example, might implement components for this representation, and use them together with our first-order logic components to construct a proof environment for those representations. This frees researchers and developers from the tasks of developing proof management facilities, and permits a focus on the properties of the representations themselves.

In this talk we will describe our generalized theory of heterogeneous reasoning motivated by examples from design and by problems taken from GRE and SAT tests. We will demonstrate a variety of applications based on the Openproof framework to show the range of heterogeneous reasoning applications which may be built using the system.

G. Stapleton, J. Howse, and J. Lee (Eds.): Diagrams 2008, LNAI 5223, p. 1, 2008.

Rich Data Representation: Sophisticated Visual Techniques for Ease and Clarity

W. Bradford Paley

Digital Image Design Incorporated and Columbia University
brad@didi.com

Within the fields of Interaction Design and especially Information Visualization diagram-like constructions are used fairly often. I will leave it up to this community to define the exact characteristics that make a diagram, but expect that definition to include visual representations of abstract data entities and relationships; I also expect that clarity and ease of interpretation are among the goals for which many diagrams are created.

I have devoted my efforts toward making clear, comfortable representations of abstract data with a concentration on supporting very constrained domains of practice. As a result I believe I have found (and developed) techniques for both structuring and rendering certain kinds of abstract data, and that such techniques are sadly underutilized.

The rendering techniques largely focus on putting visual richness into the representations. In most cases I now advocate adding visual richness to complex representations—beyond just color—to texture, varying line widths, curving lines; even pseudo-3D edges, drop shadows, and node shapes that suggest their meaning. This practice is saved from being decorative, from being what might be called "chart junk," by direct ties in two directions: to the data and to the mind. The extra dimensions of visual variation some techniques offer can better show the variability within the data. And if the techniques take into account how the human eye and mind extract knowledge from natural scenes, we can ease the transformation from photons into ideas considerably.

The structuring techniques likewise tie both into data structure and the mental mechanisms of Vision and Cognitive Science, and they do so by paying attention to a mapping technique as questionable in the domain of layout as chart junk is in the domain of rendering: the use of metaphors. But well-developed domains of practice often have mappings from their abstract concepts onto more physical grounds. It helps people to understand the basics more easily, even automatize this understanding, so that work can proceed on higher levels, with bigger chunks of information.

I believe the areas of rich visual representation and metaphorically-driven layout are under researched for two reasons: difficulty and fear. It is easy, given modern tools, to change the color of a node or line; much harder to change its texture or to develop a carefully consistent and expressive variation in shape—people with the right skill set may not only be on another floor in the building, but in another school altogether. I believe fear drives some people from expressive graphics to abstract graphics: fear that embellishment of the graphics is unnecessary or even suggests embellishment of the results. Abstraction seems to be mistaken for objectivity: Bacon

G. Stapleton, J. Howse, and J. Lee (Eds.): Diagrams 2008, LNAI 5223, pp. 2–3, 2008.

forbid that the reports on findings show any kind of subjectivity—perhaps the experiments themselves were tainted by the same sloppy flaw.

But with more data to hand, more easily processed than ever before; and with the increasingly complex models we can use to operate upon and understand the data, we will need stronger tools to help people conceptualize the models and understand the results. We need to understand when it is misleading or limiting (in a Whorfian way) to use a metaphor, and when one simply makes it easy for practitioners to absorb the structure. We need to understand when visual richness is there just to excite the eye, and when it helps the visual parsing of the image, the layering of the information, and when it acts as a visual mnemonic: easing the mental connection of glyph and idea. I propose we carefully distinguish objectivity in the scientific method from paucity in making the results clear. And I propose that we extend Tufte's parsimonious data/ink ratio to a more process-grounded "data/think" ratio (pardon the rhyme).

I will show many examples of work that tries to be rich in order to be clear: not distorting data to be a distracting illustration, but folding more data into the image by respecting the visual processes we engage to decode it. This work is my own, that of my students at Columbia University, and some that is over a century old: charts, graphs and maps—perhaps diagrams—that take advantage of a more ecologically-valid mode of presentation than many stripped-bare data presentations do. And along the way I will try to give some structure to how we might rigorously and repeatedly design such work, verify its readability and correctness, and figure out exactly where and what kinds of techniques can support easier thinking in the quickly rising sea of things we want to understand.

Model Driven Development with Mechatronic UML

Wilhelm Schäfer

University of Paderborn

Abstract. Model Driven Development with Mechatronic UML Visual languages form a constituent part of a well-established software development paradigm, namely model driven development. The structure and functionality of the software is precisely specified by a model which can be formally analyzed concerning important (safety and liveness) properties of the system under construction. Executable code is automatically generated from the model.

Although model-driven development has been recognized as a potential to improve significantly the productivity and quality of software. success stories are restricted to particular domains, mainly in business applications. Other domains, especially embedded systems employ model-driven development only in very special cases and on a limited scale, namely in the continuous world of control theory. This is due to the complex nature of the software of advanced mechatronic (or embedded) systems which includes complex coordination between system components under hard real-time constraints and reconfiguration of the control algorithms at runtime to adjust the behavior to the changing system goals. In addition, code generation must obey very limited and very specific resources of the target platform, i.e. the underlying hardware or operating system. Finally, techniques for modeling and verifying this kind of systems have to address the interplay between the discrete world of complex computation and communication and the "traditional" world of continuous controllers. The safety-critical nature of these systems demands support for the rigorous verification of crucial safety properties.

Our approach, called Mechatronic UML addresses the above sketched domain by proposing a coherent and integrated model-driven development approach. Modeling is based on a syntactically and semantically rigorously defined and partially refined subset of UML. It uses a slightly refined version of component diagrams, coordination patterns, and a refined version of statecharts including the notion of time. Code generation obeys the resource limits of a target platform. Verification of safety properties is based on a special kind of compositional model checking to make it scalable. The approach is illustrated using an existing cooperative project with the engineering department, namely the Railcab project (www.railcab.de). This project develops a new type of demand-driven public transport system based on existing and new rail technology.

G. Stapleton, J. Howse, and J. Lee (Eds.): Diagrams 2008, LNAI 5223, p. 4, 2008.

Cognitive Dimensions of Notations: Understanding the Ergonomics of Diagram Use

Alan F. Blackwell

University of Cambridge Computer Laboratory
Cambridge CB3 0FD UK
alan.blackwell@cam.ac.uk

Abstract. The Cognitive Dimensions of Notations framework provides an analytic approach to understanding the way that diagrams are used in real tasks. It is intended as a tool for people who need to invent new diagrams or notational conventions. It offers such practitioners a vocabulary with which to discuss the properties of the notation that will be cognitively relevant, and that are likely to have an impact on the usability of notational systems. This tutorial presents the original motivation for the framework, an illustrated overview of the vocabulary, and a survey of the tools that have been developed for applying the framework in practical design and research contexts.

1 Introduction

Many novel kinds of diagram have been invented in the pursuit of computer systems that can be operated more easily, or by a wider range of people. Ever since Sutherland's Sketchpad (1963/2003), many designers of programming languages and user interfaces have expected that diagrammatic representations would be more natural or intuitive than textual or algebraic languages (Blackwell 1996, 2006). However, both empirical user studies and market sales repeatedly demonstrate that, while a new diagrammatic representation might make some tasks easier, other tasks become more difficult. The work of Green and others described this as a challenge to notational "superlativism" (Green et. al. 1991) – the belief that some notation might be superior for all purposes (presumably motivated by an unstated hypothesis of structural correspondence to an internal human cognitive representation or language of thought).

2 Objectives

The Cognitive Dimensions of Notations (CDs) framework was specifically developed to provide better design guidance for those hoping to create notations with improved usability. Rather than simply collecting results from the many empirical studies of perception and diagrammatic reasoning, it aimed to provide "broad brush" design guidance. One essential contribution was the reminder that computer systems provide not only a notation (the diagram or other visual structure as statically rendered), but an environment for manipulating that notation. Even notations that are highly

G. Stapleton, J. Howse, and J. Lee (Eds.): Diagrams 2008, LNAI 5223, pp. 5–8, 2008.

appropriate to a particular task can be made unusable by a poor set of manipulation tools. Secondly, it was important to highlight likely design trade-offs, where improvements in one aspect of system usability (e.g. ease of understanding a descriptive command name) might be accompanied by deterioration in another (e.g. the verbose command name takes longer to type).

A further contribution was to draw attention to the way that "user interface" operation had become distinct from "programming" in the understanding of interaction design. Early texts on the ergonomics of computing had regarded many varieties of computer use as kinds of programming (Weinberg 1971, Shneiderman 1980). But mass-market computing encouraged the view that a well-designed display must inform naïve users about which buttons to push, in a correct sequence defined by the predetermined goal structure of their tasks. In contrast, the CDs framework emphasised that people use visual representations in unpredictable ways, often changing or refining their goals in response to what they see.

3 Illustrative Examples

The core of the CDs framework is a set of descriptive terms (the dimensions) that emphasise the consequences for a user when the combination of notation and environment enforces or allows certain kinds of interaction. For example, the dimension of Viscosity (Green 1990) describes the kind of notation and environment where a relatively small change in the conceptual structure of the user's goals requires many actions to produce the corresponding change in the visual representation. Viscosity is particularly problematic in activities where the user is likely to change his or her goals frequently, such as exploratory design. A common cause of viscosity is the fact that, after an initial change has been made, there can be "knock-on" or "domino" effects requiring further changes elsewhere in the diagram. The syntax rules for the diagram could potentially be modified in a way that would reduce such domino effects. However the tradeoff would be that this would result in the creation of Hidden Dependencies that make it difficult to anticipate the consequences of change. Alternatively, the notation designer might allow users themselves to define changes to the syntax, in order to better support the kinds of goal change that the user knows to be likely. However the tradeoff would be that support for explicit modification of diagram syntax increases the Abstraction Gradient for the user.

4 Application to Practice

The CDs are intended to provide a discussion vocabulary for the designers of new notations, such that these trade-offs and design manoeuvres can be identified and compared. The names and definitions have been refined over the past 20 years, and the total number of dimensions has been gradually increased to incorporate findings from Diagrams research and other fields. However, there is a tension between completeness of the framework and its utility to designers. CDs researchers feel that, in order to meet the original goal of improving professional practice in the design of new notations, the ideal number of dimensions would probably be 6 or less. In practice, individual dimensions are passing into professional knowledge one at a time, without the intended rigor of analysis, to the extent even senior HCI researchers seem unaware that terms like viscosity arose from rigorous analytic research (Olsen 2008).

5 Research Applications

The use of the word "Cognitive" in the framework name does not refer to a coherent cognitive theory, but only to the fact that these dimensions are intended to be cognitively relevant (as opposed to social or affective dimensions). This relatively a-theoretic stance was intentional, because cognitive theories had previously been associated with a "death by detail" in HCI, where close analysis of cognitive models frustrates or distracts people needing to make holistic design decisions. Nevertheless, the CDs framework does offer a useful analytic stance for further diagrams research (and indeed, citation analysis suggests that the framework has often had more impact among researchers than among visual design practitioners).

The essence of this analytic stance is the understanding that the syntax and semantics of visual formalisms, with their associated manipulation environment, impose structure on the user's task. At the same time, the task itself has inherent structure that is not derived from the visual formalism, but from the user's mental representation and the task context. These structures are not static, but fluid, such that mis-fits between them result not only from problems with diagrammatic syntax, but from failures to account for the temporal structure of human cognitive activities such as exploration, understanding, communication and design. There are many opportunities for the Cognitive Dimensions of Notations not only to support the professional application of Diagrams research in graphic and interaction design, but also to suggest future agendas for Diagrams research itself.

6 Further Reading

There is a fairly large literature related to the CDs framework. Many of the earlier publications, together with significant recent developments, are indexed from the cognitive dimensions resource site[1]. The most-cited reference is a journal article by Green & Petre (1996). The first conference tutorial on CDs was presented in 1998, and is still available online (linked from the CDs site). There is a well-known textbook chapter by Blackwell and Green (2003) that includes more recent developments. A special edition of the Journal of Visual Languages and Computing was published to mark the 10 years of active research since the publication of the Green and Petre paper. That issue includes reflections by Green, Petre and the present author, as well as a survey of recent CDs research, including application of the framework to tangible user interfaces (which are considered as 3-D diagrams (Edge & Blackwell 2006)) in addition to subsequent applications and theoretical work by Green and others.

[1] http://www.cl.cam.ac.uk/~afb21/CognitiveDimensions/

Acknowledgements

The CDs tutorial presentation at Diagrams 2008 is prepared in collaboration with Thomas Green and Luke Church.

References

Blackwell, A.F.: Metacognitive theories of visual programming: What do we think we are doing? In: Proceedings IEEE Symposium on Visual Languages, pp. 240–246 (1996)

Blackwell, A.F., Green, T.R.G.: Notational systems - the Cognitive Dimensions of Notations framework. In: Carroll, J.M. (ed.) HCI Models, Theories and Frameworks: Toward a multidisciplinary science, pp. 103–134. Morgan Kaufmann, San Francisco (2003)

Blackwell, A.F.: The reification of metaphor as a design tool. ACM Transactions on Computer-Human Interaction (TOCHI) 13(4), 490–530 (2006)

Edge, D., Blackwell, A.F.: Correlates of the cognitive dimensions for tangible user interface. Journal of Visual Languages and Computing 17(4), 366–394 (2006)

Green, T.R.G.: The cognitive dimension of viscosity: a sticky problem for HCI. In: Diaper, D., Gilmore, D., Cockton, G., Shackel, B. (eds.) Human-Computer Interaction – INTERACT 1990. Elsevier, Amsterdam (1990)

Green, T.R.G., Petre, M., Bellamy, R.K.E.: Comprehensibility of visual and textual programs: a test of 'superlativism' against the 'match-mismatch' conjecture. In: Koenemann-Belliveau, J., Moher, T., Robertson, S. (eds.) Empirical Studies of Programmers: Fourth Workshop, pp. 121–146. Ablex, Norwood (1991)

Green, T.R.G., Petre, M.: Usability analysis of visual programming environments: a 'cognitive dimensions' framework. J. Visual Languages and Computing 7, 131–174 (1996)

Olsen Jr., D.R.: Evaluating user interface systems research. In: Symposium on User Interface Software and Technology (UIST 2007), pp. 251–258 (2007)

Shneiderman, B.: Software psychology. Winthorp, Cambridge (1980)

Sutherland, I.E.: Sketchpad, A man-machine graphical communication system. Ph.D Thesis at Massachusetts Institute of Technology, online version and editors' introduction by A.F. Blackwell & K. Rodden. Technical Report 574. Cambridge University Computer Laboratory (1963/2003)

Weinberg, G.M.: The psychology of computer programming. Van Nostrand Reinhold (1971)

Getting Started with Sketch Tools
A Tutorial on Sketch Recognition Tools

Beryl Plimmer[1] and Tracy Hammond[2]

[1] University of Auckland
[2] Texas A&M University, College Station

1 Introduction

Diagrams are an important, if not pivotal, part in both education and design. In an effort to understand abstract concepts, students play an active role in their education by specifying visual and abstract concepts in hand-sketched diagrams. While students are understanding abstract concepts through hand-drawn diagrams, designers are creating those abstract concepts. Just as the act of hand-drawing a diagram (as opposed to using a mouse-and-palette CAD tool) better engages the student in the learning process, the act of hand-drawing a diagram also improves the design process by freeing the designer of constraints that may otherwise impede creativity and innovation.

Diagrams, and thus hand-sketched diagrams, are fundamental to many domains including circuit diagrams, mechanical engineering, concept maps, flow charts, finite state machines, civil engineering, and foreign language learning. Hand-sketched diagrams offer many benefits not yet present in a traditional CAD tool, such as intuitive input methods. Hand-sketched diagrams are pervasive, but currently most computer programs are unable to understand this natural and informal form of input. Diagrams are commonly sketched on paper, and then re-entered into a computer-understood program using a mouse and palette tool for processing, wasting time and effort, as well as interrupting the design and learning flow.

Tablet-PCs have attempted to bridge that gap, allowing people to use a digital pen to draw directly on a computer screen and store the information for later examination. The tablet-PC technology on its own leaves the sketch mostly unrecognizable by a CAD tool. Sketch recognition is a subfield in computer science that develops algorithms using artificial intelligence to recognize and understand the hand-drawn strokes on a screen just as a human would; these understood sketches can then be sent to a CAD or other computer program for simulation or other intelligent interaction.

Sketch recognition technologies exist for a limited set of domains but would benefit a myriad of currently unsupported domains. Two tools, InkKit and LADDER, currently exist to allow users to create their own sketch recognition systems without expertise in sketch recognition. The goal of this tutorial is to provide each participant with an overview of the two currently existing sketch tools with which they can design domain specific sketch recognition technology.

G. Stapleton, J. Howse, and J. Lee (Eds.): Diagrams 2008, LNAI 5223, pp. 9–12, 2008.

Each toolkit provides 1) recognition engines to recognize hand-drawn sketches, 2) innovative ways to define new shapes and provide details that can help recognition of those shapes, and 3) methods for interaction, including display and editing settings, to help make the created user interface as natural and intuitive as possible. Interaction settings can have a profound effect on the usability of a system. Changes in display can act as an important method of recognition feedback, whereas self-modifying objects (such as instant object beautification) on the screen may be distracting for certain domains.

2 InkKit

InkKit is a fully-featured sketch toolkit. The fundamental goal is to minimize the effort required to support recognition of a specific type of diagram (for example entity-relationship diagrams). There are three main parts of InkKit: the user interface, the core recognition engine, and extensibility via plug-ins. It has been developed for the Microsoft Vista OS in Visual Studio .Net using C#. This has the advantages of providing a good hardware platform and the Ink SDK for basic ink data support and character recognition. The InkKit user interface contains two main interaction spaces, sketch pages and a portfolio manager (Figure 1). Sketch pages belong to a portfolio and pages can be linked together the meaning of the links is dependent on the plug-in. There are supplementary interfaces for defining recognition components and other options.

The recognition engine is able to recognize diagrams containing shapes and text. The core InkKit recognizer consists of a divider, which separates writing and drawing; a basic shape recognizer for rectangles, circles, etc.; and a component recognizer, which combines basic shapes and text into diagram components.

To add another diagram type to InkKit the plug-in developer writes a small interpreter, provides some example components, and writes appropriate output

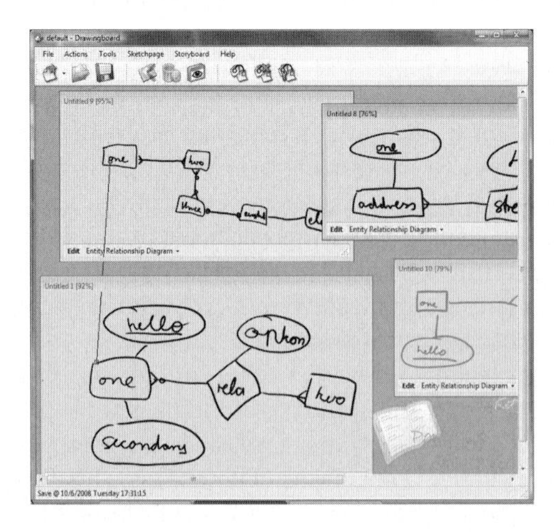

Fig. 1. InkKit User Interface Showing ER Diagrams

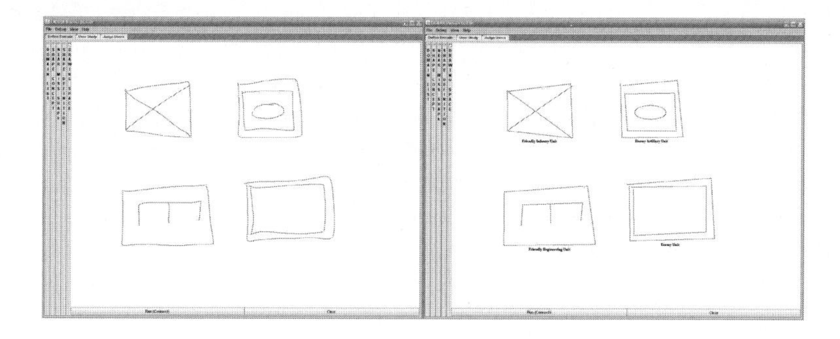

Fig. 2. Course of action symbols in LADDER before and after beautification

modules. The developer can exclude unused basic shapes and add new basic shapes from the user interface. Also from the UI the developer provides a few examples of each component the core recognition engine analyses these examples to derive the recognition rules. The plug-in must define the component names and can be used to supplement the recognizer with additional rules. InkKit does not support any beautification of the digital ink. Formalization is achieved by writing an output module that converts the recognized diagrams into a format for other tools. We have written plug-ins for a range of diagrams and output formats including UI designs to Java and html; directed and undirected graphs and organization charts to text, picture formats or MS Word diagrams; and ER diagrams to Java. A typical plug-in consists of less than 500 lines of simple code.

3 LADDER/GUILD

LADDER (initially an abbreviation for a Language for Describing Drawing, Display, and Editing for use in sketch Recognition) has grown to be a multi-tier recognition system which generates a user interface based on a LADDER description. A developer can use LADDER to recognize shapes and gestures in a particular domain. In order to recognize a shape, a user first draws an example of the shape or gesture. If the user specifies that he or she is drawing a shape, the system will automatically generate a geometric description of the shape composed of geometric primitives: lines, arcs, curves, ellipses, circles, spirals, or helixes. The user is then shown the shape's generated geometric constraint list and provided the option (if the description needs improvement) to either modify the list by hand, draw more examples, or choose from a number of automatically generated near-miss examples. Contrarily, if the user specifies that he or she is drawing a gesture, a list of features specific to the gesture are recognized and the user is asked to draw several more examples. Since context plays a significant part in recognition the user can also specify contextual information by hand that may help to better recognize the object, preventing ambiguity.

As editing and display are important in creating a usable sketch recognition system, in LADDER, a developer has the option to specify how each shape should be displayed and edited either by using some simple preset selections (such as translating an object by pressing and holding the pen briefly over the object to 'grab' it before moving it and/or displaying the beautified, icon-replaced, or original strokes) or by specifying it more specifically using the LADDER sketch language. LADDER supplies a Java API to connect to an external program. Thus, the developer can create his or her own display panel, monitor and control recognized objects, as well as connect to an existing program to provide simulation or other intelligent interaction.

4 Conclusion

The two sketch recognition toolkits described above provide complimentary advantages. InkKit is available for download at `https://www.cs.auckland.ac.nz/~beryl/inkkit.html`, whereas LADDER is available for download at `http://srl.cs.tamu.edu`. We are always looking for collaborators who would like to be trial users for our software, and we encourage you to try our toolkits in unprecedented domains. These tools are still in beta versions, and we look forward to your comments as to how we can make them better.

5 Contributors

Dr. Beryl Plimmer, Ph.D.: Dr Plimmer is a Senior Lecturer in Computer Science at the University of Auckland, New Zealand. She has many years experience in industry and education. She has a BBus, MSc in Science Maths Education and a PhD in Computer Science. Her passion is to make computers more usable and useful sketch tools is one part of this endeavour. Her other research interests include digital ink annotation, using multi-modal interaction to support visually impaired students and user centred design.

Dr. Tracy Hammond, Ph.D.: Dr. Hammond is the Director of the Sketch Recognition Lab and Assistant Professor in the Computer Science department at Texas A&M University. Previously, she taught computer science courses for many years at Columbia University. Additionally, she worked for four years full time at Goldman Sachs as a telecom analyst developing speech recognition, and other intelligent user interface applications. She earned her B.A. in Mathematics, B.S. in Applied Mathematics, M.S. in Computer Science, and M.A. in Anthropology all from Columbia University in the City of New York. She earned an F.T.O in Finance and a Ph.D. in Computer Science from the Design Rationale Group (DRG) in the Computer Science and Artificial Intelligence Laboratory (CSAIL) at the Massachusetts Institute of Technology.

Acknowledgements

The authors would like to thank NSF IIS grants 0757557 & 0744150 as well as members of the Sketch Recognition Lab.

General Euler Diagram Generation

Peter Rodgers[1], Leishi Zhang[1], and Andrew Fish[2]

[1] Computing Laboratory, University of Kent, UK
[2] Computing Mathematical & Information Sciences, Brighton University of Brighton, UK
P.J.Rodgers@kent.ac.uk, L.Zhang@kent.ac.uk,
Andrew.Fish@brighton.ac.uk

Abstract. Euler diagrams are a natural method of representing set-theoretic data and have been employed in diverse areas such as visualizing statistical data, as a basis for diagrammatic logics and for displaying the results of database search queries. For effective use of Euler diagrams in practical computer based applications, the generation of a diagram as a set of curves from an abstract description is necessary. Various practical methods for Euler diagram generation have been proposed, but in all of these methods the diagrams that can be produced are only for a restricted subset of all possible abstract descriptions.

We describe a method for Euler diagram generation, demonstrated by implemented software, and illustrate the advances in methodology via the production of diagrams which were difficult or impossible to draw using previous approaches. To allow the generation of all abstract descriptions we may be required to have some properties of the final diagram that are not considered nice. In particular we permit more than two curves to pass though a single point, permit some curve segments to be drawn concurrently, and permit duplication of curve labels. However, our method attempts to minimize these bad properties according to a chosen prioritization.

Keywords: Euler Diagrams, Venn Diagrams.

1 Introduction

Euler diagrams are sets of (possibly interlinking) labelled closed curves and are popular and intuitive notation for representing information about sets and their relationships. They generalize Venn diagrams [16], which represent all possible set intersections for a given collection of sets. Euler diagrams allow the omission of some of these set intersections in the diagram, enabling them to make good use of the spatial properties of containment and disjointness of curves. Euler diagrams are said to be effective since the relationships of the curves matches the set theoretic relationships of containment and disjointness [10]; they provide 'free rides' [17] where one obtains deductions with little cognitive cost due to the representation. For example, if A is contained in B which is contained in C then we get the information that A is contained in C for free.

The motivation for this work comes from the use of Euler diagrams in a wide variety of applications, including the visualization of statistical data [2,13], displaying the results of database queries [19] and representing non-hierarchical computer file systems [3]. They have been used in a visual semantic web editing environment [18] and for viewing clusters which contain concepts from multiple ontologies [11]. Another

G. Stapleton, J. Howse, and J. Lee (Eds.): Diagrams 2008, LNAI 5223, pp. 13–27, 2008.

major application area is that of logical reasoning [12] and such logics are used for formal object oriented specification [14].

A major requirement for many application areas is that they can automatically produce an Euler diagram from an abstract description of the regions that should be present in the diagram. This is called the Euler diagram generation problem.

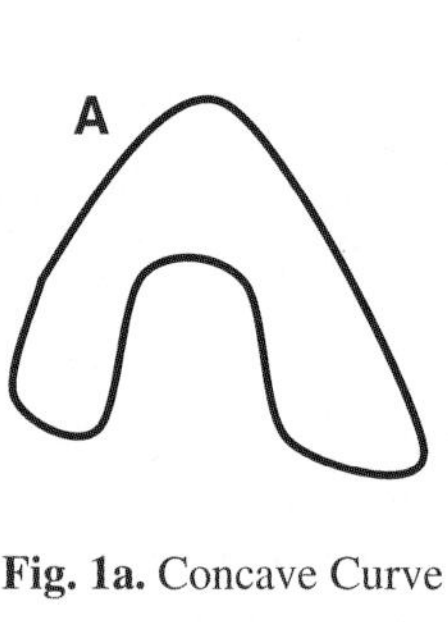

Fig. 1a. Concave Curve

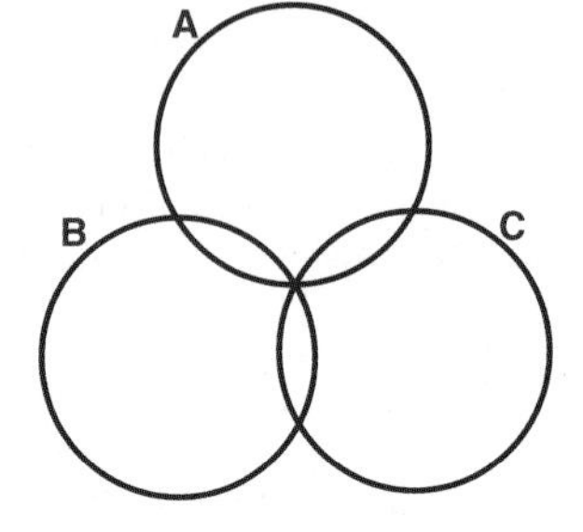

Fig. 1b. A triple point

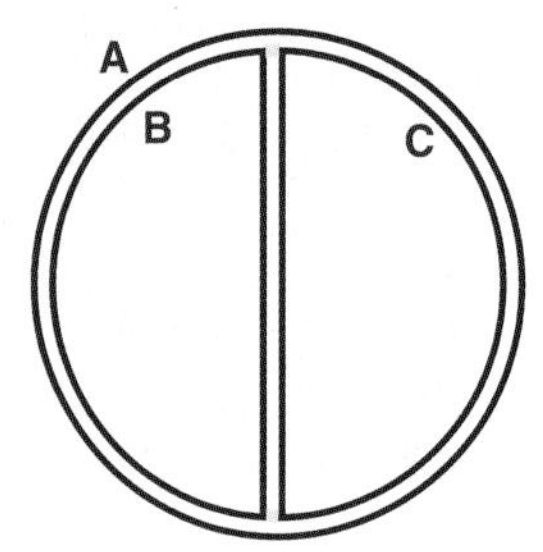

Fig. 1c. Concurrent curves

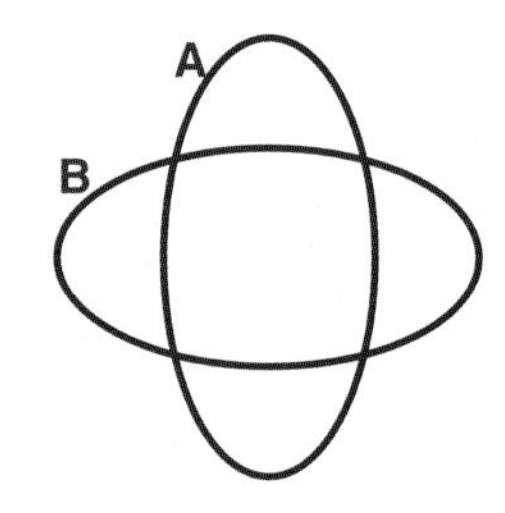

Fig. 1d. Disconnected zones

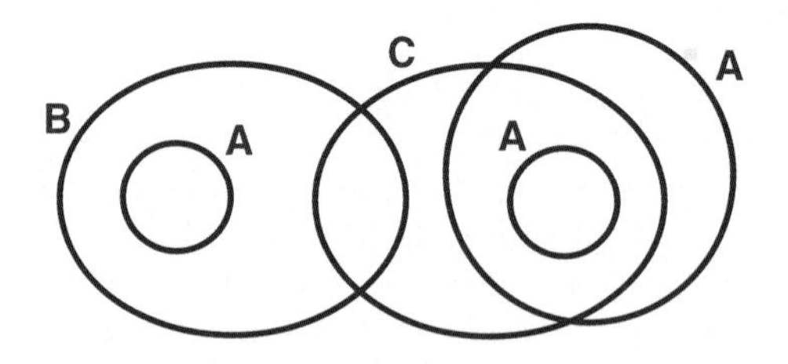

Fig. 1e. Duplicated curve label

Fig. 1f. Non simple curve

A range of diagram properties, called wellformedness conditions, which are topological or geometric constraints on the diagrams, have been suggested with the idea of helping to reduce human errors of comprehension; these properties can also be used as a classification system. Some of the most common properties are shown in Fig. 1. Note that the term *zone* in Fig. 1d refers to the region enclosed by a particular set of curve labels, and excluded by the rest of the curve labels. For example the region that is inside the curve labelled A but outside the curve labelled B is disconnected here.

The definition of what constitutes an Euler diagram varies in the literature, and can usually be expressed in terms of these wellformedness conditions. Although Euler himself [5] did not formally define the diagrams he was using, his illustrations do not break any of wellformedness conditions given in Figure 1. In [7] the first Euler diagram generation algorithm was presented and further formalized in [6]. This work guaranteed the production of an Euler diagram that meets all of the wellformedness conditions in Fig. 1 except Fig 1a, from an abstract descriptions whenever it was possible to do so. An implementation of the algorithm was also provided which had a limited guarantee of being able to draw any such diagram with up to four contours in any one connected component. In [1] the relaxation of the wellformedness conditions to allow multiple points and concurrent contours was adopted, and although no conversion from theory to practise was provided, it was shown that the Euler diagram generation problem is NP-Complete in this case. Since imposing some wellformedness conditions implies that some abstract descriptions are not realisable as Euler diagrams, in [19] the notion of an Euler diagram was extended so that any abstract description with at most nine sets could be drawn: they used Euler diagrams that had holes, which are a restricted version of allowing duplicate curve labels.

In this paper, we integrate and significantly extend the work of these three major attempts at the Euler diagram generation problem and provide a complete solution to general Euler diagram generation in the sense that any abstract description is drawable using our method.

We define an *Euler diagram* to be a set of labelled closed curves in the plane. We call the set of all of the labelled curves with the same label a *contour*. A *zone* of an Euler diagram is a region of the plane enclosed by a set of contours, and excluded by the rest of the contours. The diagrams obtained via our generation process can have curves of any geometric shape and they may have duplicate contour labels, multiple points, and concurrent curves. However, we guarantee not to generate any diagrams with duplicate zones or non-simple curves.

Utilising this broad definition of Euler diagrams makes the general generation problem of any abstract description possible, but typically, the "more non-wellformed" a diagram is the less desirable it is from a usability perspective. Therefore, we adopt a strategy which guides the output towards being as wellformed as possible, according to a chosen prioritisation of the wellformedness conditions, whilst ensuring that we generate a diagram with the correct set of zones (i.e. it complies with the abstract description). However, we give no guarantee that the diagrams generated are the most wellformed diagrams possible since some of the problems that need to be solved to ensure this are NP-Complete.

In this paper we adopt the convention of using single letters to label contours. Each zone can be described by the contour labels in which the zone is contained, since the excluding set of contour labels can be deduced from this set. An abstract description is a description of precisely which zones are required to be present. For example, the abstract description for the Euler diagram in Figure 2b is ∅ **b c ab ac abc**, where indicates the zone which is contained by no contours, called the *outside zone*, which must be present in every abstract description.

In Section 2, we give details of the generation method. Section 3 gives our conclusions and future directions.

2 The Generation Process

First we give a high level outline of the methodology used, with details and explanation of the terminology appearing in later sections. Given an abstract description of an Euler Diagram, we produce an embedded Euler diagram using the following steps:

1. Generate the superdual graph for the abstract description.
2. Using edge removal, find a planar subgraph that is either wellconnected or close to wellconnected.
3. If the graph is not wellconnected, add concurrent edges to increase the closeness of the graph to being wellconnected whilst maintaining planarity.
4. Find a plane layout for the graph.
5. Add edges to reduce unnecessary tangential intersections, forming the dual of the Euler graph.
6. Find subgraphs where duplicate curves will be required.
7. Construct the Euler diagram from the dual of the Euler graph using a triangulation based method.

Since we are constructing the dual of the Euler graph, planarity is clearly essential. If the dual graph constructed is not wellconnected then the Euler diagram will have either duplicate curves or concurrency. Steps 2 and 3 try to reduce the instances of either. However, step 3 may add concurrent edges (i.e. those with multiple contour labels) which can reduce the number of duplicate curves used at the expense of causing concurrency. Step 5 removes unnecessary tangential intersections (those that can be removed without introducing concurrent curve segments). Checking the face conditions, as in [7] would identify if multiple points will appear, but since attempting to search for an Euler dual which passed the face conditions (and so has no multiple points) is so time consuming, we omit this step.

2.1 Generating a Super Dual

Recall that an abstract description of a diagram is list of zone descriptions (which are just the sets of contour labels that will contain the zones). As described in [7] we can construct the superdual by taking one node for each required zone, and labelling each one with its zone description. When an Euler diagram is drawn, each contour's curves will enclose the nodes whose label set contains the contour label. Two nodes in the superdual are connected by an edge precisely when the labels differ by a single contour label. The edges are labelled with the difference between their incident node labels. Fig. 2 shows an example of a superdual, and resultant Euler diagram generated for the abstract description ∅ **b c ab ac abc**. In this case, and for other small examples that can be drawn without concurrent contours or duplicate curve labels, the superdual can be embedded without requiring steps 2,3,5,6 of our process. However, many superduals are not planar, and so methods to find a planar dual need to be applied, as described in the next section.

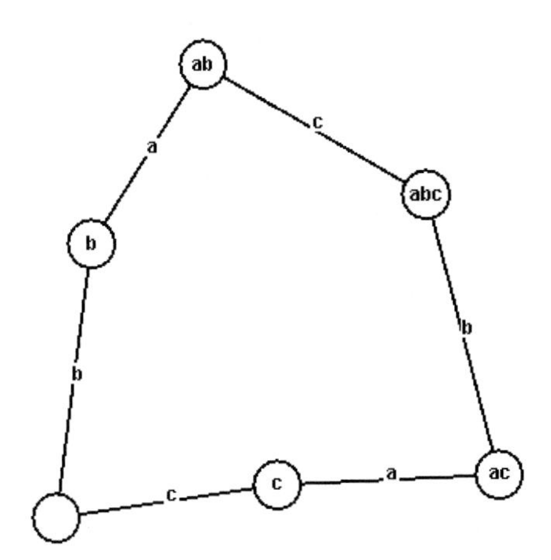

Fig. 2a. Superdual for ∅ **b c ab ac abc**

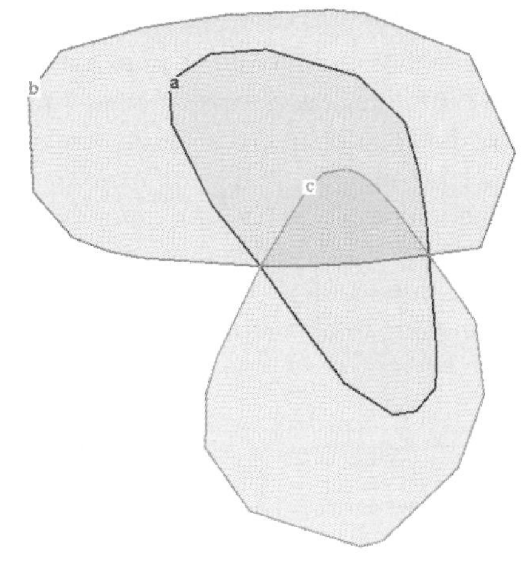

Fig. 2b. Embedded diagram

2.2 Edge Removal to Achieve Planarity

Given a superdual that is non-planar, we try to find a planar subgraph of the superdual that can be used to generate a general Euler diagram that has no concurrency or duplicated curve labels; i.e. it must be wellconnected, which means that it must pass the connectivity conditions below. Even if such properties are necessary, the amount of concurrency and the number of curves in a contour may be reducible (by finding a subgraph that is "close" to passing the connectivity conditions). The connectivity conditions state that

a. the graph is connected
and for each contour label in the abstract description:

b. if the nodes with that contour label present are removed (recall, a node is labelled by a collection of contour labels) then the graph remains connected and,

c. if the nodes without that contour label present are removed then the graph must also remain connected.

If condition *a* does not hold in the superdual, then concurrency is required in the Euler diagram, and Step 3 of our method will be applied to add a multiply labelled edge to the dual (corresponding to concurrency of edges in the Euler diagram). If conditions *b* or c do not hold and they cannot be fixed by the addition of edges without breaking planarity, then duplicate curve labels will be used for that contour – in the case of condition c failing, curves are placed "inside" another curve of that contour, forming holes in the contour.

In Step 2 we attempt to find a wellconnected planar dual by removing edges from the superdual. If this cannot be done, our edge removal strategy attempts to find a planar dual that has as much connectivity as possible, that is the occurrences of conditions *b*, or *c* are as few as can be achieved.

To guarantee to find a wellconnected planar dual where one exists is an NP-Complete problem [1]. Therefore we resort to heuristics to do as good a job as possible. Our current technique is to take a fairly lightweight approach of discovering potentially removable edges, checking those that may be removed from the dual and exploring the effects of removing combinations of these. We first layout the superdual graph using a spring embedder [4] and remove highly crossed edges, preferring the potentially removable edges. Once a planar layout has been found we then attempt to add back any unnecessarily deleted edges that improve the wellconnectedness. This paper does not focus on heuristics, and we give only a simple demonstration of a possible technique. As with other NP-Complete problems we expect there to be a number of alternative heuristics.

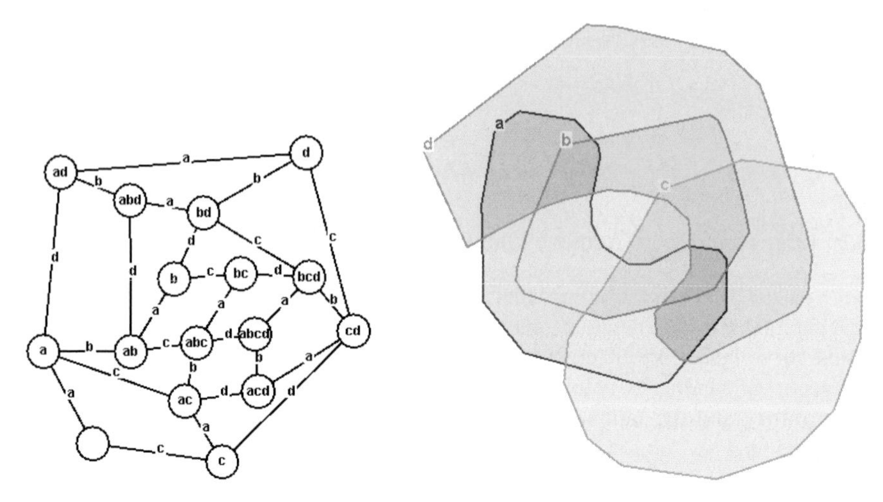

Fig. 3a. Planar dual for 4Venn

Fig. 3b. 4Venn embedded

Fig. 3a shows a wellconnected plane dual for 4Venn (the Venn diagram on 4 sets), and the corresponding diagram generated from this dual is shown in Fig 3b. Various edges have been removed from the superdual to achieve planarity, including the edge between ∅ and **b**, and the edge between **c** and **bc**. Depending on the starting conditions, and on how much time is given to the search, it is also possible that versions of 4Venn which have triple points and duplicate curve labels might be created (see Fig. 8 for example).

2.3 Concurrent Edge Addition

If, after Step 2, the graph is disconnected then, in Step 3, we attempt to make it connected by adding edges whilst maintaining planarity. In addition, for each contour label, if removal of nodes without that contour label present would result in multiple disconnected subgraphs, then we attempt to add edges which would connect those subgraphs.. Similarly, we attempt add edges which connect any multiple disconnected subgraphs formed by the removal of nodes with that contour label present. Recall that edges in the dual graph are labelled with the difference between the labels of the nodes they are incident with. Edges that are labelled by more than one contour label are called *concurrent edges* since they correspond to the use of concurrency in the

Euler diagram. Adding edges in this manner can reduce the number of duplicate curve labels but can also add extra concurrency.

Given that there appears to be a combinatorially explosive number of possible ways of connecting up the various subgraphs of the dual graph, and only one of which might be optimal, we expect that the problem of finding a planar dual which is as close to wellconnected as possible by adding edges to be at least NP-Complete. Hence, we take a heuristic approach to deciding how to add edges. Again we adopt a simple method, taking a greedy approach, but with the small examples we are currently exploring (less than 10 sets) we find that relatively few disconnected components appear. We take one disconnected component and attempt to connect it to another by an edge that is labelled with the least number of contours. This process continues until the dual is connected or no more improvements can be made.

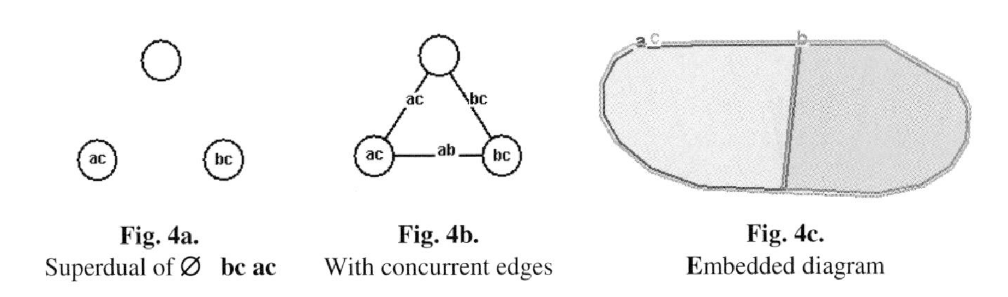

Fig. 4a. Superdual of ∅ **bc ac**

Fig. 4b. With concurrent edges

Fig. 4c. Embedded diagram

An example of the process is shown in Fig 4 for the abstract description ∅ **ac bc**. Fig 4a shows the superdual which is disconnected; no two nodes have label sets differing by one label. Fig 4b shows the dual graph with concurrent edges added as a result of Step 3. We note that adding any two edges to the superdual does not make the graph wellconnected. For example, if we only added one edge between nodes labelled "∅" and "ac", and another edge between nodes labelled "∅" and "bc" then for the contour labelled "c" condition *c* of the connectivity conditions does not hold since the nodes "ac" and "bc" would not be not adjacent. Similarly, leaving either of the pair of nodes labelled "∅" and "ac" or the pair of nodes labelled "∅" and "bc" not adjacent breaks condition *b* of the connectivity conditions. The Euler diagram created (using the dual graph in Fig 4b) is shown in Fig. 4c, where we slightly separate the concurrent curve segments to ease comprehension.

2.4 Planar Layout

In this step we embed the dual graph in the plane. There are various standard approaches to planar layout. At the moment we use a method provided by the ODGF software library. We make one adjustment to ensure the node labelled with "∅" is in the outer face of the drawn graph, as this node represents the part of the diagram enclosed by no contour. The layout of the dual has a significant impact on the drawing of the diagram, and Section 3 includes some discussion of methods to layout planar graphs to improve the usability of the final diagram.

2.5 Edge Addition to Remove Tangential Intersections

For the purposes of embedding we treat the dual as the dual of an Euler graph [1]. An Euler graph can be formed from an Euler diagram by placing a node at each point where curves meet or cross, and connecting them up with edges that parallel the curve segments. Using the dual of an Euler graph means that, unlike the treatment of the dual in [7], each face in the dual has at most one point where contours meet or concurrent edges separate. However, it can lead to the introduction of unnecessary tangential intersections and so we apply an edge addition process to remove them (subdividing the faces of the dual separates the tangential meetings of the curves).

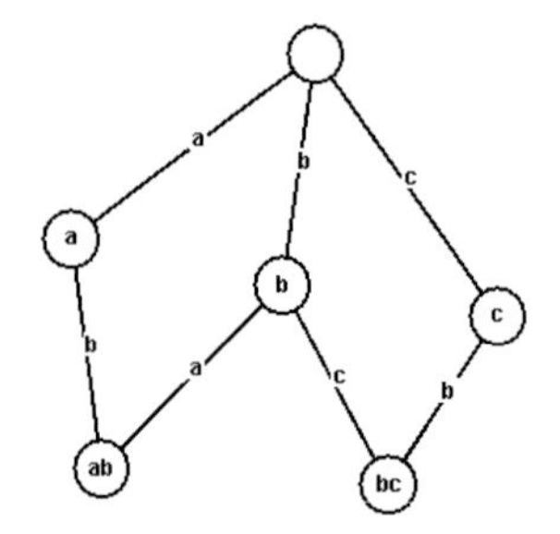

Fig. 5a. Dual graph for ∅ **a ab b bc c**

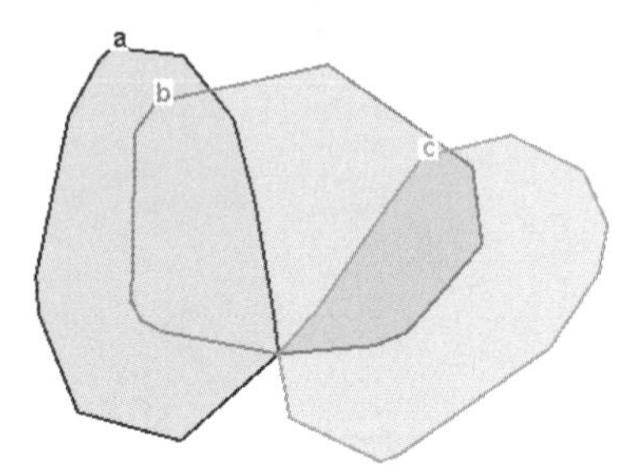

Fig. 5b. Without edge addition

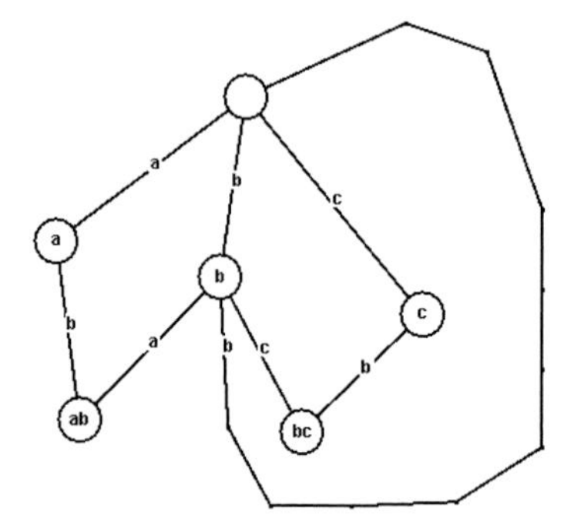

Fig. 5c. Additional edge between ∅ and **b**

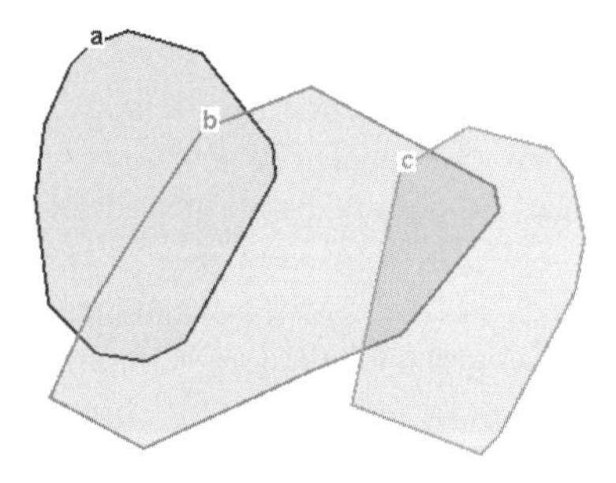

Fig. 5d. With edge addition

We detect the need for extra edges by testing each face in the dual graph. If it is possible to add an edge between two non adjacent nodes in the face and the new edge will be labelled with one of the edge labels of the face, then that edge is added (recall that edges are labelled with the difference in their incident node labels). An example is shown in Fig. 5, where Fig. 5a shows a dual graph and Fig. 5b shows the corresponding Euler diagram which contains an unnecessary tangential intersection (the point where all of the three curves meet). The graph in Fig. 5a has an outside face that has an edge labelled "b", but it can also have another edge labelled "b" added to it between nodes labelled ∅ and **b**, as shown in Fig. 5c. The Euler diagram constructed from the dual with the additional edge is shown in Fig. 5d. We route this edge without bends if possible, but often it is not possible, as is the case in Fig. 5c. In this case, a triangulation of the face is made, and the edge is routed through appropriate triangulated edges.

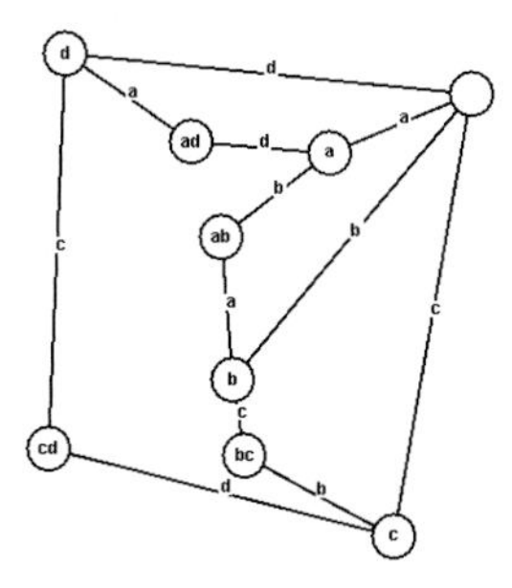

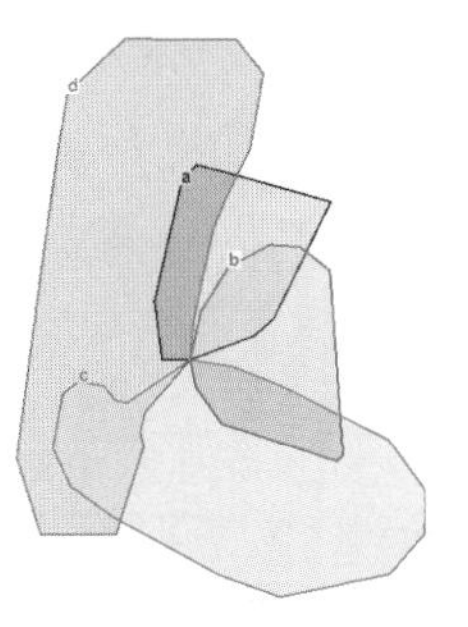

Fig. 6a. Dual of the Euler graph for
∅ a b c d ab ad bc cd

Fig. 6b. Embedded diagram

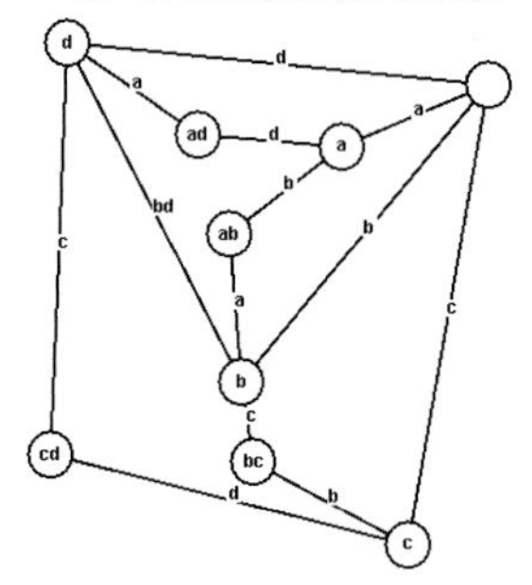

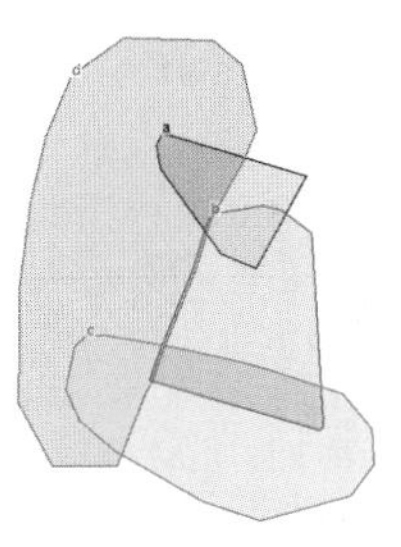

Fig. 6d. Diagram with reduced tangentiality but extra concurrency

Fig. 6c. Additional edge between **b** and **d**

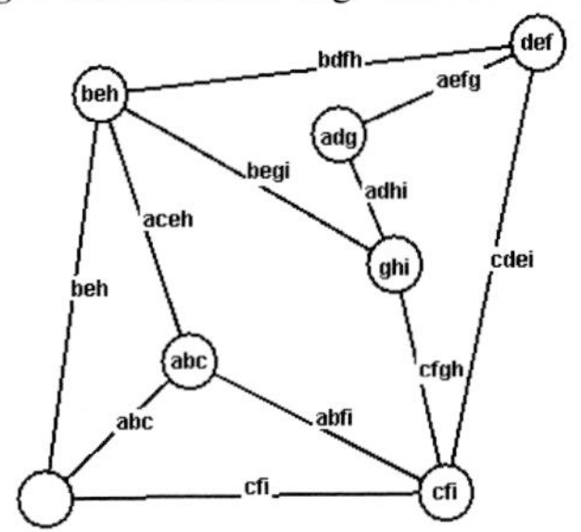

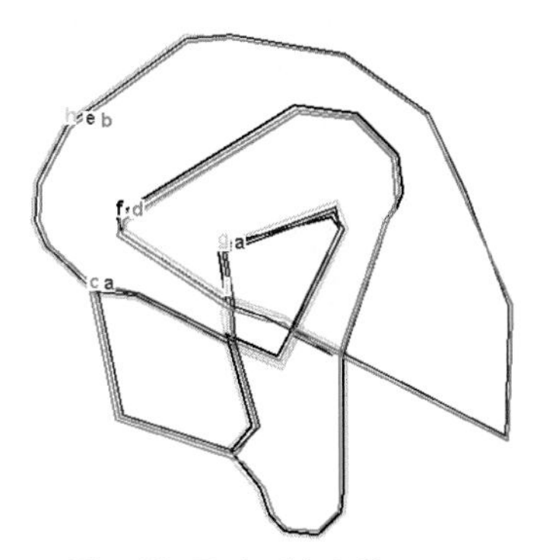

Fig. 7a. Dual of the Euler graph for
∅ abc def ghi adg beh cfi

Fig. 7b. Embedded diagram

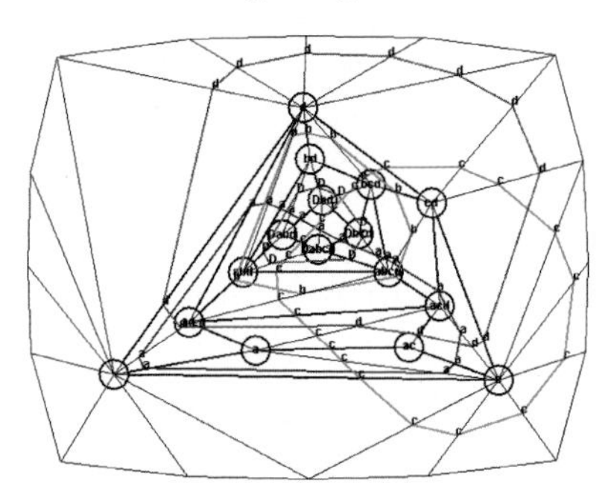

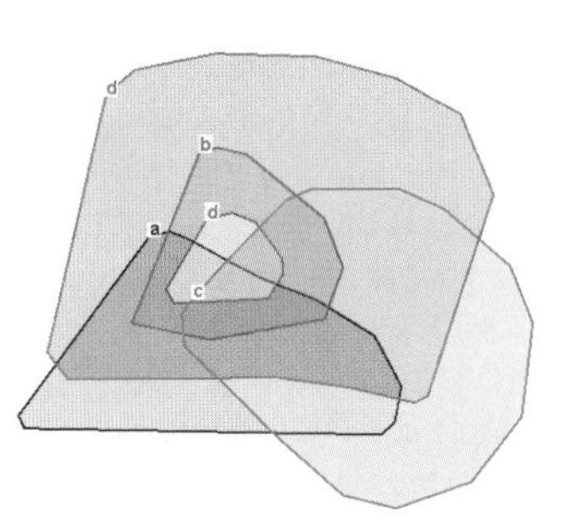

Fig. 8a. Contour Routing

Fig. 8b. Embedded diagram

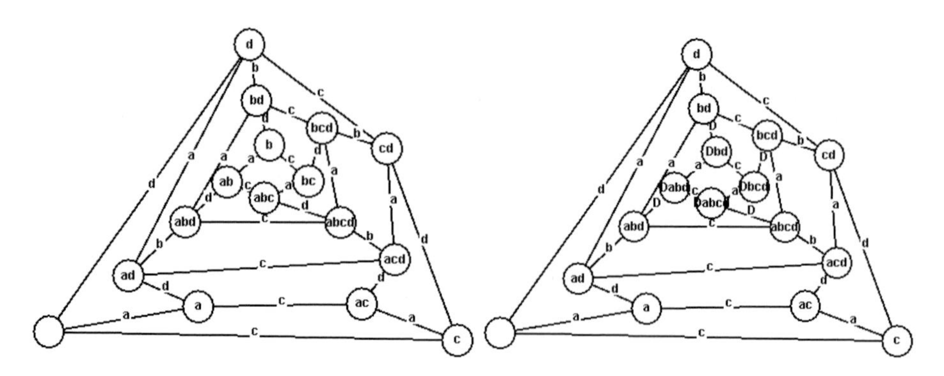

Fig. 8c. Plane dual for 4Venn **Fig. 8d.** Hole has labels "D" and "d" added

Not all tangential intersections are removed by this method because some can only be removed at a cost of adding extra concurrency. For example Fig. 6a shows the dual of the Euler diagram in Fig 6b. Addition of an edge labelled "bd" between nodes labelled "b" and "d" would result in the removal of the tangential connection between the nodes labelled "a" and "c", however this would also result in a concurrent segment "bd" shown in figures 6c and 6d.

2.6 Duplicate Curve Labels

As shown in [19], not all Euler diagrams can be embedded in the plane with simple, uniquely labelled curves. The reason is that there are abstract descriptions for which any wellconnected dual graph is non-planar. The example in the paper is ∅ **abc def ghi adg beh cfi**, for which any wellconnected graph with the corresponding nodes contains a subgraph that is isomorphic with the non-planar $K_{3,3}$. This limits the method of [19] to guarantee the existence of a drawing only if there are 8 sets or less. We demonstrate that this example can be drawn by our method in Fig. 7b; the red contour "a" has two curves, since the subgraph of the dual with nodes containing the label a consists of the two nodes **abc** and **adg** which not being adjacent in Fig. 7a, breaking condition *c* of the connectivity conditions. Given the need for large amounts of concurrency when drawing this diagram, it is not likely to have a particularly usable embedding, but this example demonstrates the ability of our method to embed a diagram from any abstract description.

The example in Fig.7 uses duplicate curves for the same contour. Given a dual graph (obtained from Step 3 of our method), we discover the duplicate contours required by looking at the connected components of the subgraphs of the dual that include the contour label present (corresponding to the wellconnected condition *c*, Section 2.2) or removed, corresponding to the use of holes (wellconnected condition *b*). To enable us to draw the Euler diagram, we re-label the nodes of the graph that contain the contour label which requires the use of duplicate curves, being careful to distinguish the case of holes. Essentially, we keep the label of the contour the same for one of the components (in the label present case) and change it for the other components (thereby assigning new curve labels; here we adopt the convention of using capital letters for the duplicates to help distinguish from the usual lowercase). Then we alter the labels of the nodes on the components in the label removed case, so that they indicate the new curve labels assigned as well as the fact that these nodes are within a hole in that contour. This relabeling procedure allows us to draw the curves correctly, but when labelling the contours of the final diagram, we revert to the original labels for the curves.

Fig. 8 shows an example for 4Venn drawn with a hole. Here there is a duplicate label "d" required for two curves, because, when nodes including "d" are removed from the dual graph in Fig 8c the subgraph with nodes labelled "b" "ab" "bc" and "abc" is not connected to the rest of the graph. Therefore, these four nodes have the label "d" added, together with the label "D", indicating a hole, as shown in Fig 8d. "D" will be mapped back to "d" when the diagram is embedded, as in Fig 8b.

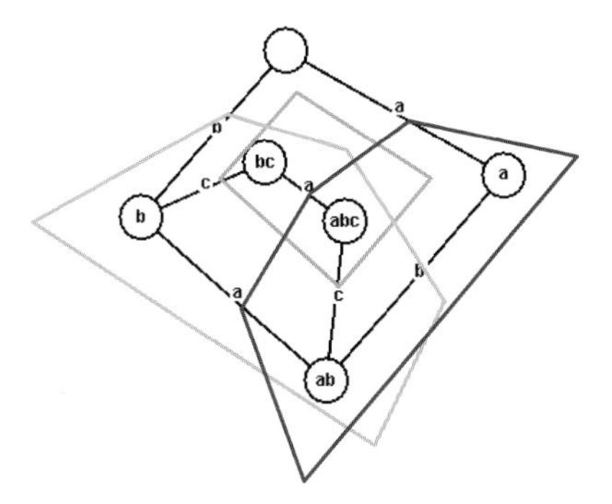

Fig. 9. An incorrect embedding for ∅ a b ab ac abc

2.7 Constructing the Euler Diagram from the Dual of the Euler Graph

In general, straight lines cannot simply be drawn between edges of the dual to indicate where the contours pass through the faces, because a face may not be convex. This could cause the lines to intersect edges of the dual graph whose labels do not include the same contour label, possibly introducing incorrect contour intersections that cause the diagram generated to not have the required zone set. If an arbitrary polyline routing through the face is taken, incorrect intersections can again occur, also possibly failing to form a diagram with the required zone set. For example, see Figure 9 where the zone **c** appears but does not exist in the abstract description, and the zone **a** is disconnected, appearing both at the bottom and top right of Figure 9. The difficulty of routing contours motivates the use of a triangulation. The convex nature of the triangles means that the above problems can be avoided, but we must establish how to route contours through the triangles.

First, we triangulate the bounded faces of the plane dual graph, and for the outer face we form a border of nodes with empty labels around the graph and triangulate the polygon that is formed (see Fig. 10d, where the border nodes have been hidden). As with the dual graph, each triangulation edge is labelled with the difference between the labels present in its incident nodes, see Figure 10a. Again, as with the dual, the labels on the triangulation edges indicate which contours will cross them when we produce an embedding.

We choose one triangle in each face to be the *meeting triangle* in which all contours in that face will cross or meet. In the current implementation, this is taken to be the triangle that contains the centriod of the polygon formed from the face (or is the triangle closest to the centriod, if none contain it). We mark a point called the *meeting point* in the centre of the meeting triangle, and all contours in the face must pass through that point.

Next we assign an ordering of the contours which must pass through each triangulated edge in the face. This will enable us to assign points on the triangulation edges where the contours cross them. For the purposes of this method we add triangulation edges to parallel dual graph edges. Concurrent contours that are drawn across the face maintain concurrency until at least the meeting point, where they may separate if that concurrency is not maintained in the face.

The ordering of contours that pass through a triangulation edge that parallels a dual edge is trivial because there is either only one contour or group of concurrent contours. Also, any triangulation edge with no contours passing through it can be trivially assigned an empty order. It is then necessary to assign a contour ordering to the other triangulation edges of the face. Fig. 10 shows an example of this process. Once the trivial above triangulation edge orderings have been performed there will be at least two triangles with two triangulation edges assigned, see Fig. 10a. If the face is not a meeting triangle (shown as the triangle containing a green circle as the marked point) then the order of the third triangulation edge of such triangle can be assigned; Fig. 10b shows the assignment of one of these. This third triangulation edge will have contours ordered to avoid any contour crossings in the face by reading the order of the two assigned triangulation edges in sequence and using a similar order for the edge, as shown in Fig. 10c where both triangulation edges now have an assigned order. In addition, we enforce the condition that all contours on the other two triangulation edges must also be present in the third triangulation edge, to ensure that all contours reach the meeting point.

The assignment of a contour ordering on a triangulation edge means that another triangle has an additional triangulation edge with contour ordering assigned, as each triangulation edge (that is not a dual graph edge) is shared between two triangles in the face. Hence the process continues until all triangulation edges are assigned an order. At this point the meeting triangle should also have all three of its triangulation edges with assigned order, as the triangles that surround it should all have triangulation edges with assigned order.

This method can be shown to terminate due to the restricted nature of the triangulation, where any triangles without any triangulation edges parallel to face edges imply that there is an extra triangulation face with two triangulation edges parallel to face edges.Once the triangulation edges have been assigned an ordering, the curves can be routed around the face by linking up the appropriate triangulation edge points, except where the triangulation face is the meeting triangle, where they must first pass through the meeting point in the triangle (shown as a filled circle inside a triangle in figures 10a, 10b and 10c).

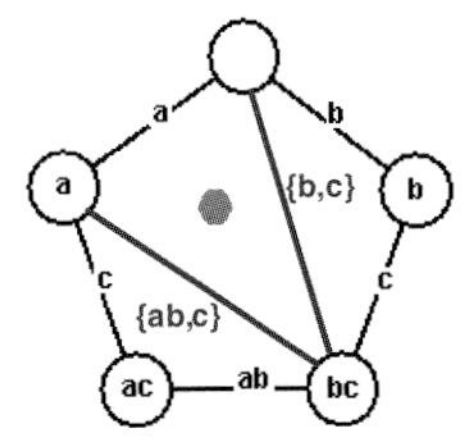

Fig. 10a. Unassigned Triangulation Edges

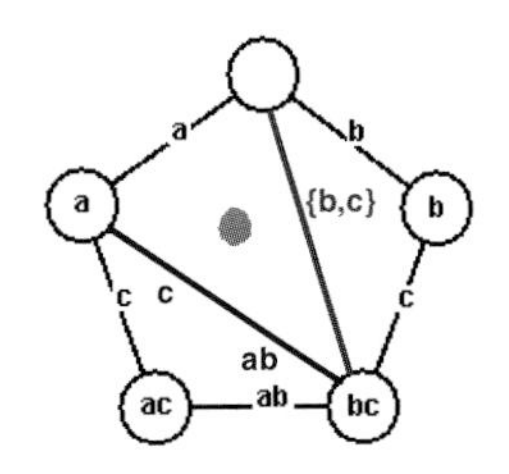

Fig. 10b. One Triangulation Edge Assigned

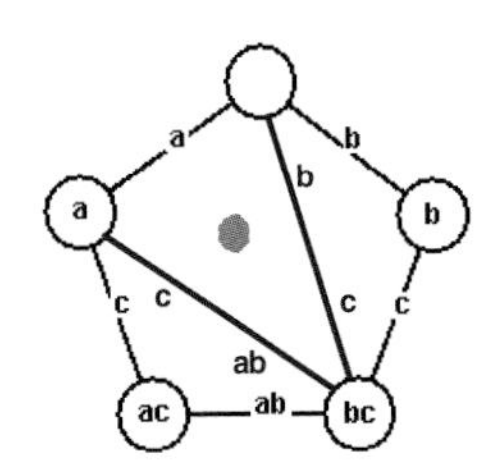

Fig. 10c. Both Triangulation Edges Assigned

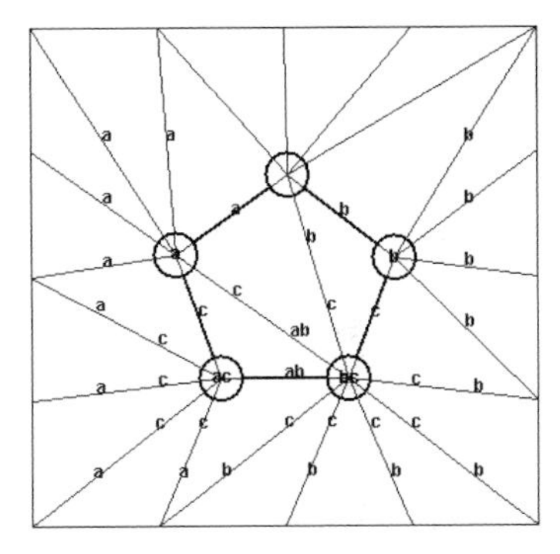

Fig. 10d. Every Triangulation Edge

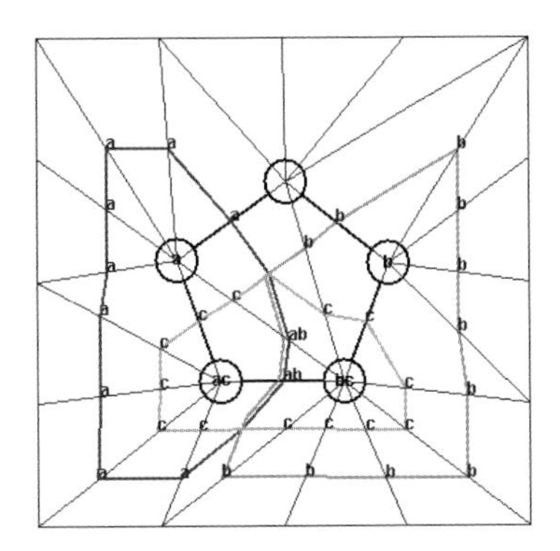

Fig. 10e. Contours Routed Through Cut Points

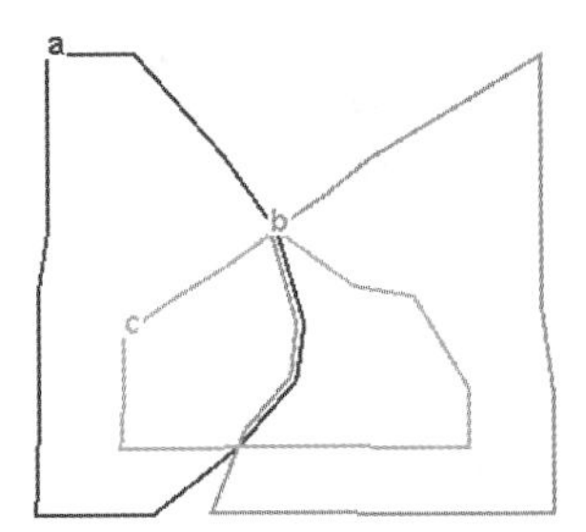

Fig. 10f. Final Diagram

2.8 Non-atomic Diagrams

Up to this point we have only shown examples of atomic diagrams, which are diagrams that can be drawn with disconnected contours. [8].. The above method can be used to draw both atomic and non-atomic diagrams, with the atomic components tangentially connected. However, for reasons of algorithmic efficiency, as well as improved layout, it is desirable to lay these diagrams out as separate components, which are joined at a later date. Figure 11 shows a non-atomic diagram that has nested components: ∅ **a b ab ac ad ae acd.**

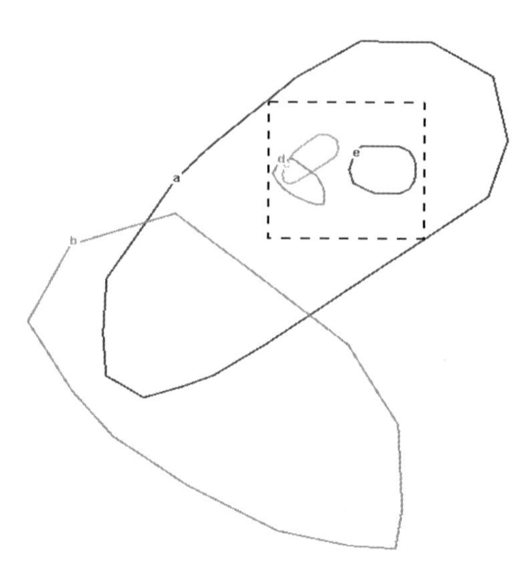

Fig. 11. A nested diagram, showing the rectangle in which the nested components can appear

Non-atomic components can be identified from the dual graph from the abstract description [1] and placed in a maximal rectangle that can found in the appropriate zone, as shown in Figure 11. Nested components can also be created when there is more than one curve in a contour and the additional curves do not intersect with any other curves in the diagram. Where multiple nested components are to be embedded within a single zone, the rectangle is simply split into the required number of sub rectangles. Any nested component may have further nested components inserted by simply repeating the process.

3 Conclusions and Further Work

We have presented the first generation method for generating an Euler diagram for any abstract description.. To do this we have brought together and extended various approaches in the literature, and developed new mechanisms for the embedding process. We have demonstrated these ideas with output from a working software system the implements the method. In terms of the methodology adopted, Step 2 - edge removal to find a plane dual, and Step 3 - adding concurrent edges, are computationally intractable problems to solve exactly in the general case, so improved heuristics and optimizations are a rich area of further work. Initially, utilizing effective search techniques such as constraint satisfaction and adapting well known heuristics such as insertion methods from the Travelling Salesman Problem are likely to improve current performance.

A further avenue of research is in improvements of the final layout which is an essential feature in usability terms. Methods, such as those discussed in [9,15] have been applied to the some of the diagrams shown in this paper, and further heuristics that more accurately measure contour smoothness, and as well as measuring other aesthetic features of the diagram not currently considered could be introduced. Also, the plane embedding of the dual has significant impact on the usability of the drawing, and methods to control the layout at Step 4 could impact on the number of triple points generated, which is currently not restricted, for example.

Acknowledgments. This work has been funded by the EPSRC under grant refs EP/E010393/1 and EP/E011160/1.

References

1. Chow, S.: Generating and Drawing Area-Proportional Venn and Euler Diagrams. Ph.D Thesis. University of Victoria, Canada (2008)
2. Chow, S., Ruskey, F.: Drawing area-proportional Venn and Euler diagrams. In: Liotta, G. (ed.) GD 2003. LNCS, vol. 2912, pp. 466–477. Springer, Heidelberg (2004)
3. DeChiara, R., Erra, U., Scarano, V.: VennFS: A Venn diagram file manager. In: Proc. IV 2003, pp. 120–126. IEEE Computer Society, Los Alamitos (2003)
4. Eades, P.: A Heuristic for Graph Drawing. Congressus Numerantium 22, 149–160 (1984)
5. Euler, L.: Lettres à une Princesse d'Allemagne, vol 2, Letters No. 102–108 (1761)
6. Flower, J., Fish, A., Howse, J.: Euler Diagram Generation. Journal of Visual Languages and Computing (2008)
7. Flower, J., Howse, J.: Generating Euler Diagrams. In: Hegarty, M., Meyer, B., Narayanan, N.H. (eds.) Diagrams 2002. LNCS (LNAI), vol. 2317, pp. 61–75. Springer, Heidelberg (2002)
8. Flower, J., Howse, J., Taylor, J.: Nesting in Euler diagrams: syntax, semantics and construction. Journal of Software and Systems Modeling, 55–67 (2003)
9. Flower, J., Rodgers, P., Mutton, P.: Layout Metrics for Euler Diagrams. In: Proc. IEEE Information Visualization (IV 2003), pp. 272–280 (2003)
10. Gurr, C.: Effective diagrammatic communication: Syntactic, semantic and pragmatic issues. Journal of Visual Languages and Computing 10(4), 317–342 (1999)
11. Hayes, P., Eskridge, T., Saavedra, R., Reichherzer, T., Mehrotra, M., Bobrovnikoff, D.: Collaborative knowledge capture in ontologies. In: Proc. of 3rd International Conference on Knowledge Capture, pp. 99–106 (2005)
12. Howse, J., Stapleton, G., Taylor, J.: Spider diagrams. LMS J. Computation and Mathematics 8, 145–194 (2005)
13. Kestler, H.A., Müller, A., Gress, T.M., Buchholz, M.: Generalized Venn diagrams: a new method of visualizing complex genetic set relations. Bioinformatics 21(8) (2005)
14. Kim, S.-K., Carrington, D.: Visualization of formal specifications. In: Proc. APSEC, pp. 102–109 (1999)
15. Rodgers, P.J., Zhang, L., Fish, A.: Embedding Wellformed Euler Diagrams. In: Proc. Information Visualization (IV 2008) (to appear, 2008)
16. Ruskey, F.: A Survey of Venn Diagrams. The Electronic Journal of Combinatorics (March 2001)
17. Shimojima, A.: Inferential and expressive capacities of graphical representations: Survey and some generalizations. In: Blackwell, A.F., Marriott, K., Shimojima, A. (eds.) Diagrams 2004. LNCS (LNAI), vol. 2980, pp. 18–21. Springer, Heidelberg (2004)
18. Tavel, P.: Modeling and Simulation Design. AK Peters Ltd. (2007)
19. Verroust, A., Viaud, M.-L.: Ensuring the drawability of Euler diagrams for up to eight sets. In: Blackwell, A.F., Marriott, K., Shimojima, A. (eds.) Diagrams 2004. LNCS (LNAI), vol. 2980, pp. 128–141. Springer, Heidelberg (2004)

Euler Diagram Decomposition

Andrew Fish* and Jean Flower

Visual Modelling Group,
School of Computing, Mathematical and Information Sciences,
University of Brighton, Brighton, UK
Andrew.Fish@brighton.ac.uk
www.cmis.brighton.ac.uk/research/vmg/

Abstract. Euler diagrams are a common visual representation of set-theoretic statements, and they have been used for visualising the results of database search queries or as the basis of diagrammatic logical constraint languages for use in software specification. Such applications rely upon the ability to automatically generate diagrams from an abstract description. However, this problem is difficult and is known to be NP-complete under certain wellformedness conditions. Therefore methods to identify when and how one can decompose abstract Euler diagrams into simpler components provide a vital step in improving the efficiency of tools which implement a generation process. One such decomposition, called diagram nesting, has previously been identified and exploited. In this paper, we make substantial progress, defining the notion of a disconnecting contour and identifying the conditions on an abstract Euler diagram that allow us to identify disconnecting contours. If a diagram has a disconnecting contour, we can draw it more easily, by combining the results of drawing smaller diagrams. The drawing problem is just one context which benefits from such diagram decomposition - we can also use the disconnecting contour to provide a more natural semantic interpretation of the Euler diagram.

1 Introduction

Euler diagrams are a diagrammatic method for representing information about the relationships between sets. They have been used in various forms since Euler [5] first introduced them, and they generalize Venn diagrams [19,24] which represent all set intersections. Many notations have been based on Euler diagrams, such as Higraphs [13], Spider diagrams [16] and Constraint Diagrams [17]. Currently Euler diagram variants are being applied in a multitude of areas including: file-information systems [3,4], library systems [25], statistical data representation [1,2,18] and as the basis for logical specification and reasoning systems [7,12,16,21,22,23].

Whilst humans using Euler diagram applications will be interacting with concrete diagrams (i.e. diagrams drawn in the plane), frequently the underlying

* Funded by UK EPSRC grant EP/E011160: Visualisation with Euler Diagrams.

G. Stapleton, J. Howse, and J. Lee (Eds.): Diagrams 2008, LNAI 5223, pp. 28–44, 2008.

system interacts with abstract diagrams (i.e. some abstract information encapsulating the recorded semantic information). This separation of concrete and abstract levels is commonplace for visual languages, but was first applied to Euler diagrams in [14]. Whether considering applications for data representation or diagrammatic logical reasoning systems, the system may be performing transformations as the abstract level (e.g. computation of a proof in theorem proving), but displaying concrete diagrams for the user to view and interact with.

The problem of converting an abstract Euler diagram into a concrete representation, called the generation problem, was first addressed in [9] and then addressed in more depth in [8]. These approaches tend to be based upon the creation of a dual graph from the abstract diagram which is then used to construct the required Euler diagram. Since the generation problem is inherently difficult, identifying ways to reduce the size of the problem is clearly beneficial.

In [11] the authors identify *nested* diagrams - those concrete diagrams with a separating curve. The wellformedness condition of non-concurrency of contours was removed in [2]. There, the notion of a nested Euler diagram is extended to that of a composite Euler diagram, which essentially means that either the diagram is nested (can be written as d_1 embedded in a zone z of d_2) or it is "almost nested" (can be viewed as d_1 embedded in a zone z of d_2, followed by the removal of the region that is "in z but outside d_2"). This allows the decomposition of concrete Euler diagrams into its prime factors. Furthermore, translations of such a decomposition to the abstract level are provided, thereby allowing the application of the techniques to the Euler diagram generation problem.

While the property of nesting is a powerful way to decompose some large diagrams into smaller, more easily drawable, pieces, there remain classes of large diagram which pose problems for drawing algorithms. In this paper we introduce the concept of a *disconnecting contour* which can be used to split many more large diagrams into smaller diagrams, adding value to all the various existing drawing algorithms.

Section 2 begins with the necessary background notation and definitions for the rest of the paper, including the notion of a concrete Euler diagram, the Euler dual graph and the intersection graph of a concrete Euler diagram. In Section 3 we review work on nested diagrams. The key concepts of disconnecting contours and 1-separating curves are introduced in Section 4, together with theorems relating these concepts with the Euler dual graph.

In Section 5 we recall the Euler diagram generation problem and describe an algorithm which uses of disconnecting contours within an existing generation algorithm. Although the main advantages of this approach are likely to be for examples with large numbers of contours, it can be useful for small cases too and we provide an indication of the proportion of atomic diagrams that have disconnecting contours. Furthermore, it is not only the generation problems that benefit from the work here, and in Section 6, we give examples of the beneficial effects diagram decomposition can have on the expression of diagram semantics.

2 Concrete Euler Diagrams

We (informally) recall the basic concepts of concrete Euler diagrams and their dual graphs that will be needed later. Detailed formal definitions about Euler diagrams and their well-formedness conditions can be found in [8], and elaboration on the property of nestedness can be found in [11].

A *concrete Euler diagram* is a set of labeled *contours* (closed curves) in the plane. A *minimal region* is a connected component of the complement of the contour set, whilst a *concrete zone* is a union of minimal regions determined by being inside a set of contours and outside the rest of the contours of the diagram. [1] The usual semantic interpretation is that each of the contours represents a set; the spatial relationships between the curves (containment, overlap and disjointness) correspond to the set-theoretic relationships (subset, intersection and disjointness). They are thought to be an effective representation since the properties of the spatial and domain relationships match, and this gives rise to free rides [20], which are extra inferences that the reader obtains without additional cognitive overhead, due to the effectiveness of the representation.

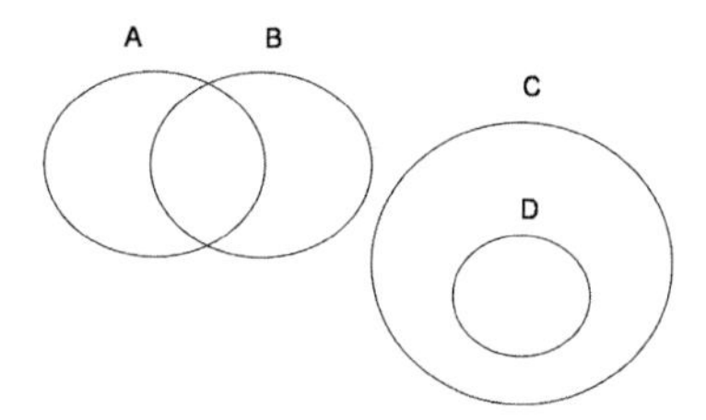

Fig. 1. A well-formed concrete Euler diagram

For example, in Figure 1 we have 4 contours representing the sets A, B, C and D together with the properties that the set C is disjoint from the sets A and B, and the set D is a subset of the set C. There are 6 zones in this diagram, which would be formally written as $(\{\}, \{A, B, C, D\}), (\{A\}, \{B, C, D\}), (\{A, B\}, \{C, D\}), (\{B\}, \{A, C, D\}), (\{C\}, \{A, B, D\}), (\{C, D\}, \{A, B\})$. The first set of contours labels is the set of labels that "contain" the zone, whilst the second set is the set that "exclude" it. This notation, first introduced in [15] and given in more detail in [16], is rather cumbersome to read, we will adopt the convention of using single letters as the contour labels, and concatenating contour labels to give a zone descriptor, so the above set of zones is listed as follows: $\{o, a, ab, b, c, cd\}$. Here we have adopted the conventions of using lower case letters for the zone descriptors and o for the outside zone descriptor to help to distinguish them from contour labels in the figures.

[1] The notions of minimal region and concrete zone have often been conflated, but they are are only the same thing if we impose this as a condition on the diagrams under consideration (an example of a well-formedness condition).

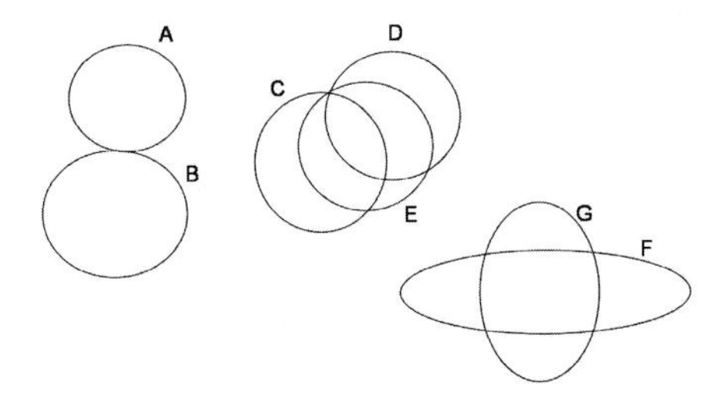

Fig. 2. Non well-formed Euler diagrams

As mentioned earlier, topological or geometric conditions, called wellformedness conditions are often imposed on Euler diagrams in order to try to prevent user misunderstanding. Throughout this paper, we assume the following common set of wellformedness conditions:

1. The contours are simple closed curves.
2. Each contour has a single, unique label.
3. Contours meet transversely (so no tangential meetings or concurrency).
4. At most two contours meet at a single point.
5. Each concrete zone is a minimal region.

Example 1. Figure 2 also shows three non well-formed concrete diagrams. The left one shows A and B with a tangential intersection; the middle one shows C, D and E with a triple point (called a multiple point in general); the right one shows a diagram with F and G which has disconnected zones: for example the zone f (which is inside F but outside G) is disconnected.

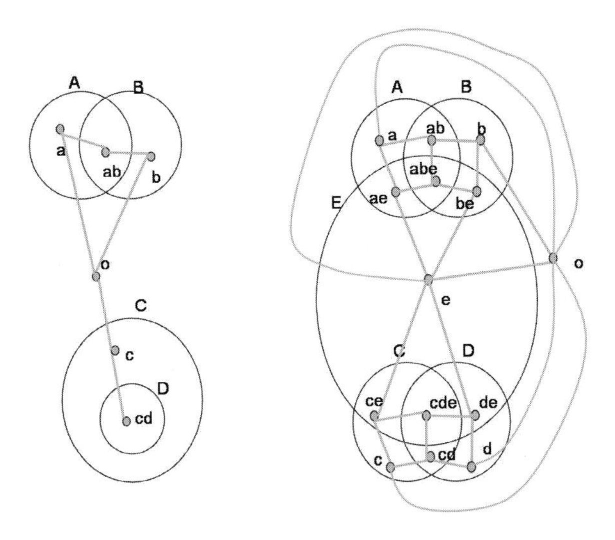

Fig. 3. Two concrete Euler diagrams and their Euler dual graphs

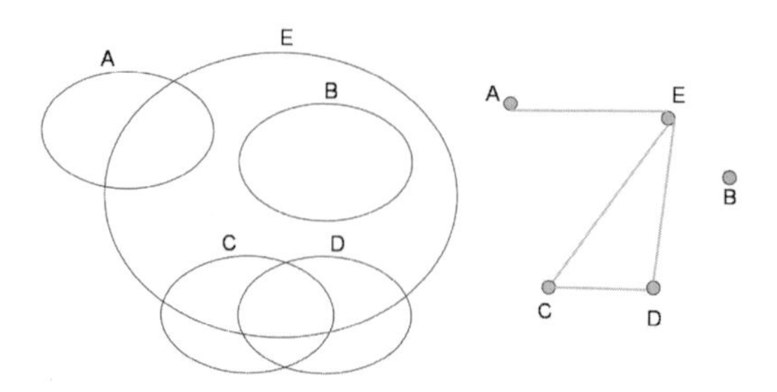

Fig. 4. A concrete Euler diagram and its intersection graph

Recall the well-known notion of the (Euler) dual graph [2] of an Euler diagram.

Definition 1. *Let d be a concrete Euler diagram. The Euler dual graph of d is the plane labelled graph obtained by:*

1. *placing exactly one vertex in each minimal region of d;*
2. *labelling each vertex by the set of contour labels that contain that region (i.e. by the zone descriptor);*
3. *placing exactly one edge between two topologically adjacent regions in the plane for every maximally connected segment of the curves separating those regions (and the edge crosses the corresponding segment).*

Figure 3 shows Euler diagrams with the Euler dual graphs overlaid. Each vertex is labelled by the corresponding zone descriptor. In the second example, there is more than one edge between the vertices o and e (so the dual graph has multiple edges).

Another useful graph that can be derived from a concrete Euler diagram d is the intersection graph of d which has one node for each contour, and nodes are joined by an edge if the contours cross in d. An example of an intersection graph can be seen in Figure 4; note that this only captures the pairwise intersection properties of the diagram and semantically different diagrams can have the same intersection graph.

3 Nested Euler Diagrams

The idea of a *nested* Euler diagram was introduced in [10] and studied in more depth in [11]. The property of nested-ness can be easily stated in terms of the intersection graph: a diagram is called nested if and only if its intersection graph is disconnected; otherwise it is called *atomic*. This property has been exploited to assist with diagram generation, diagram interpretation and reasoning.

A nested diagram d has a partition of its contours into d_1 and d_2, and we can think of d as having been constructed by inserting d_2 into a zone of d_1. The

[2] This is equivalent to the dual used in [2], but differs slightly from the duals used in [8,9] where multiple edges were not allowed.

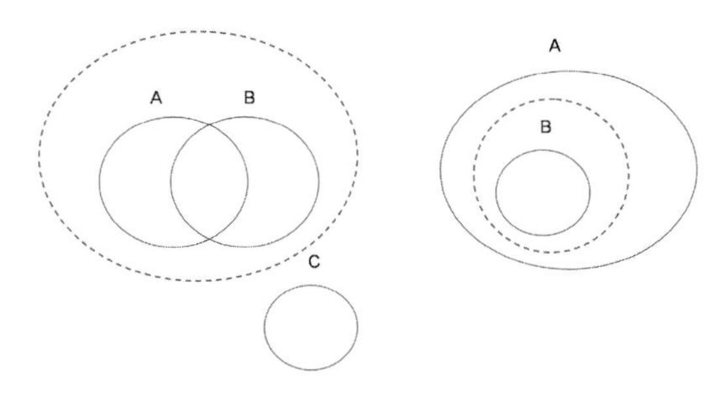

Fig. 5. Two nested Euler diagrams and separating curves

left-hand diagram[3] of Figure 5 can be split into d_1 with contours $\{A, B\}$ and d_2 with contour set $\{C\}$. The composite diagram d can be built by inserting d_2 into the outside zone o of d_1, or by inserting d_1 into the outside zone of d_2. The second nested example on the right of the figure has just two contours. The diagram with contour set $\{B\}$ can be inserted into zone a of the diagram with contour set $\{A\}$. The figure also shows some curves which separate the sub-diagrams described above.

Definition 2. *A closed curve embedded in the plane is a* separating curve *of a concrete Euler diagram if the curve meets no contour of the Euler diagram and splits the plane into two pieces, each of which contains at least one contour of the Euler diagram.*

Finally we observe that in Figure 3, the left-hand diagram is nested, falls into three atomic components, and the nodes labelled o and c are cut vertices of the Euler dual graph (recall that a cut vertex is a vertex which disconnects the graph if it, together with its adjacent edges, is deleted). Now we can recall the following theorem.

Theorem 1 (from [10]). *Let d be a concrete Euler diagram. Then the following statements are equivalent:*

1. *d is nested.*
2. *d has a separating curve.*
3. *The Euler dual graph of d has a cut vertex.*

4 Disconnecting Contours

In this section we extend the idea of nested-ness of an Euler diagram. We wish to decompose diagrams, even when they are atomic. A key concept which will aid us is that of a *disconnecting contour*. Here we define this concept for the first time, together with a "dual" concept of a *1-separating curve*, and show that a diagram has a disconnecting contourif and only if it has a 1-separating

[3] The dotted curves are not part of the Euler diagram but are examples of separating curves of the diagram (see Definition 2).

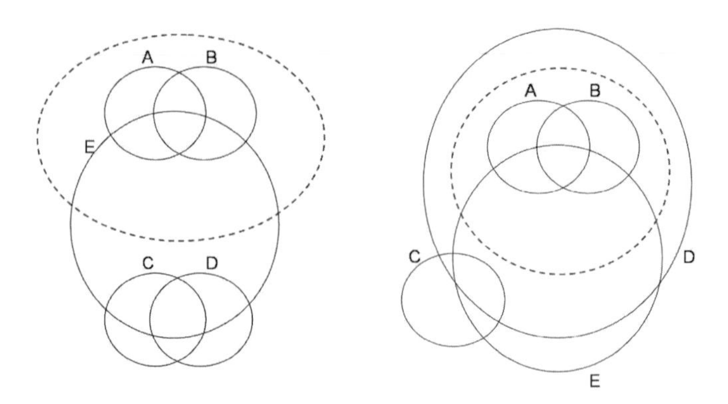

Fig. 6. Two Euler diagrams and 1-separating curves

curve (but note that although these are strongly related ideas, a disconnecting contour itself is not a 1-separating curve). Furthermore, we relate these notions to the dual graph, a step which is necessary to support the automatic generation applications of this work.

Definition 3. *Let d be an atomic concrete Euler diagram with contour set $C(d)$. Then $C \in C(d)$ is a* disconnecting contour *for d if the removal of C from d leaves a nested diagram.*

Definition 4. *Let d be an atomic concrete Euler diagram with contour set $C(d)$. Then a* 1-separating curve *for a contour $C \in C(d)$ in d is a curve that meets only the contour C, and splits the plane into two components, each of which contains at least one contour of $C(d) - C$.*

Each 1-separating curve of an Euler diagram meets just one contour, and that contour is a disconnecting contour of the diagram. In the examples in Figure 6, the contour E is a disconnecting contour, and the dotted curves are not part of the Euler diagrams but are 1-separating curves for E in the diagram. Both of the diagrams in Figure 7 are shown with 1-separating contours. The left hand diagram has two disconnecting contours, E and F, each of which has a single 1-separating curve (which is unique up to isotopy on the 2-sphere, S^2). The right hand diagram has only one disconnecting contour, G, but the figure shows non-isotopic 1-separating curves for G.

Theorem 2. *Let d be a concrete Euler diagram with contour set $C(d)$. Then the following are equivalent:*

1. *$C \in C(d)$ is a disconnecting contour of d.*
2. *There is a 1-separating curve for C in d.*
3. *C corresponds to a cut vertex in the intersection graph of d.*

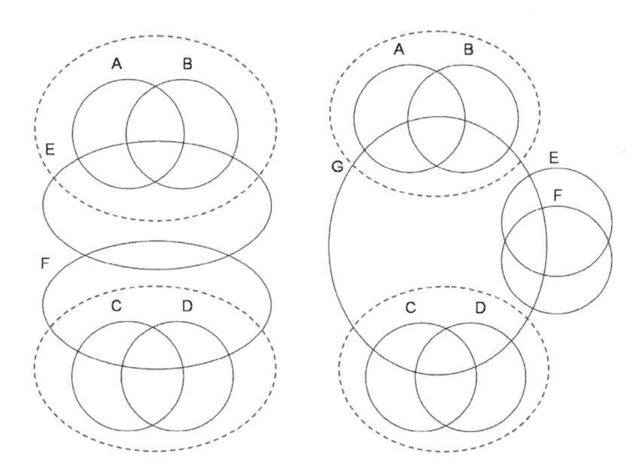

Fig. 7. Euler diagrams with multiple 1-separating curves

In order to identify disconnecting contours in abstract Euler diagrams, just from the set of zones, we will use the following theorem:

Theorem 3. *Let d be an atomic concrete Euler diagram. Then $C \in C(d)$ is a disconnecting contour if and only if there exists a cut pair* [4] $\{v_1, v_2\}$ *in the Euler dual graph where the label set of v_1 differs from the label set of v_2 by precisely the label of C.*

Proof. Firstly, suppose that C is a disconnecting contour of d. Then there exists p which is a 1-separating curve for C by Theorem 2. The curve p induces a cycle in the Euler dual graph. Since C is a disconnecting contour, the cycle induced by p is a cut-cycle. Furthermore, since p is a 1-separating curve it also only meets contour C in d. Therefore, the cycle induced in the Euler dual graph by traversing once around p must be of the form $v_1, v_2, v_1, v_2, \ldots v_1$, where the label set of v_1 differs from the label set of v_2 by precisely c (if any other vertex appeared then p must have crossed another contour), where c is the label of C. Thus, $\{v_1, v_2\}$ is a cut-pair of the dual graph [5].

Conversely, suppose that the removal of vertices v_1 and v_2 whose label sets differ by exactly the label of C splits G into n components (with $n > 1$) having vertex sets $V_1, \ldots, V_n$. For each i, let S_i denote the connected subgraph of G induced by the vertices V_i; the S_i's are disjoint by construction. Since $\{v_1, v_2\}$ is a cut pair, we have $n \geq 2$ and the only paths in the Euler dual graph between S_i and S_j (with $i \neq j$) must pass through either v_1 or v_2. Therefore there exists a path p in $\mathbb{R}^2$ consisting of the edge from the vertex v_1 to the vertex v_2 together with an arc joining v_2 to v_1 which: encloses at least one S_i; excludes at least one S_j; and meets only edges of the dual graph whose incident vertex labels differ by exactly c (and so it meets only contours labelled by c). Since the regions that are contained and those that are excluded by the curve p each contain at least

[4] We use the term cut-pair for a cut-set of size 2.

[5] If disconnected zones and hence nodes of the dual with the same label were permitted then we would need to consider a cut-path of vertices.

one vertex (and these are labelled by distinct labels that are not the same as the label of c), p is a 1-separating curve for d. □

Another approach to this proof is to consider the effect on the dual graph of the removal of a disconnecting contour C from an atomic diagram d. Contour removal merges the pairs of vertices of the graph whose labels differ by exactly c (the label of C), and it may also alter some vertex labels (but here we are primarily interested in connectivity of the dual graph). Since C is a disconnecting contour, we know that the transformed dual graph with merged vertices has a cut vertex. The original dual did not have a cut vertex because the diagram was atomic. So the cut vertex must have been obtained by merging a cut-pair of vertices v_1 and v_2 whose labels differ by exactly c. And conversely, if we began with a cut-pair of vertices whose labels differ by exactly c, then we can immediately see that removing contour C will merge the cut pair into a cut vertex, so C must have been a disconnecting contour.

The main benefits of using disconnecting contours are likely to be seen for large diagrams where Euler diagram algorithms (e.g. drawing algorithms) run into computational difficulties as the number of contours increases. If we choose an arbitrary diagram with many contours but few zones then it is likely to be nested. As the number of zones increases, relative to the number of contours, the diagram is more likely to be atomic, but there are many examples with a disconnecting contour. As the number of zones increases further, the likelihood of the presence of a disconnecting contour decreases, and so one might want to consider disconnecting sets of contours (see Section 7).

We have implemented the check for a disconnecting contour on an abstract Euler diagram and have been able to collect some data to assess how common diagrams with disconnecting contours are. For diagrams on 3 contours (ignoring wellformedness conditions): if there are only 4 or 5 zones, then the diagram is nested; for 6 zones, there are 12 diagrams, 75% of which have disconnecting contours; but for diagrams with 7 and 8 zones, there are no disconnecting contours. For diagrams on 4 contours (ignoring wellformedness conditions): if there are only 5, 6 or 7 zones, then the diagram is nested; for 8 zones, there are 564 diagrams, 40% of which are nested, and a further 45% have disconnecting contours; there are 765 diagrams with 9 zones, only 4% of which are nested, but a further 42% have disconnecting contours.

5 The Application to the Diagram Generation Problem

Until now we have been considering concrete diagrams and their properties. However, the process of automatically generating diagrams requires that we start from an abstract diagram. So we must lift the properties and relationships to the abstract level. The identification of disconnecting contours, for example, at the abstract level will enable the generation of smaller diagrams which can then be pieced together to form the required larger diagram. We continue with our slightly informal style for recalling the basic definitions (again, see [8] for more formal details).

An *abstract Euler diagram* comprises a set of *contour labels* and a set of *zones* which are subsets of the set of contour labels (for discussion on Euler diagram abstraction see [6]). We impose the constraint that the union of the sets of zones is the contour label set to mimic the same relationship between contours and zones of a concrete Euler diagram.

Example 2. An example of an abstract diagram is given by $d = \{o, a, b, c, d, f, af, bc, bf, cf, de, df, bcf, def\}$. Later, on the right of Figure 9, we will show a concrete Euler diagram generated from this abstract diagram.

Providing an abstract diagram of a concrete diagram just requires the computation of the set of all of the zone descriptors for the diagram. However, constructing concrete representations of abstract diagrams is a very challenging task. Given an abstract diagram, what we would ideally like to do is construct an Euler dual graph for that diagram, enabling the construction of the required concrete Euler diagram. One thing that we can construct more easily is the "combinatorial dual graph" for an abstract diagram.

Definition 5. *The* (combinatorial) dual [6] *of an abstract Euler diagram d, called $dual(d)$, is the abstract labeled graph whose: labels are the contour labels of d, whose vertex set is in 1-1 correspondence with the the zone set of d (with each zone descriptor labelling a vertex), which has a single edge between two vertices if and only if the vertex label sets differ by a single label.*

We follow the generation process of [8,9], although the variations in [2,25] follow a similar theme and the ideas outlined here could equally well be applied in these cases. The generation process takes the combinatorial dual and tries to embed it (or a spanning subgraph of it) in the plane under certain conditions called well-connectedness and the face conditions. Well-connectedness states that the graph should be connected and the sets of vertices that contain and exclude each contour label are connected, thereby ensuring that simple closed curves with unique labels can be used for the concrete Euler diagram. The face conditions place a combinatorial constraint on the faces of the embedded plane graph to ensure that no multiple points are necessary for the Euler diagram. The process uses this generated graph as a dual for the Euler diagram. However, since edges might have been removed from the combinatorial dual $dual(d)$ in order to planarise, we can have adjacent vertices in $dual(d)$ corresponding to zones which are not topologically adjacent in the plane. We also note that the plane graph constructed might not be quite the Euler dual graph since it does not have multiple edges by construction, but it is still sufficient to construct an Euler diagram that complies with d.

For space reasons, we will just give an example which demonstrates the new generation technique employed when disconnecting contours occur to give an outline of the improved generation procedure which takes advantage of both nesting and disconnecting curves.

---

[6] Sometimes called the superdual [8,9].

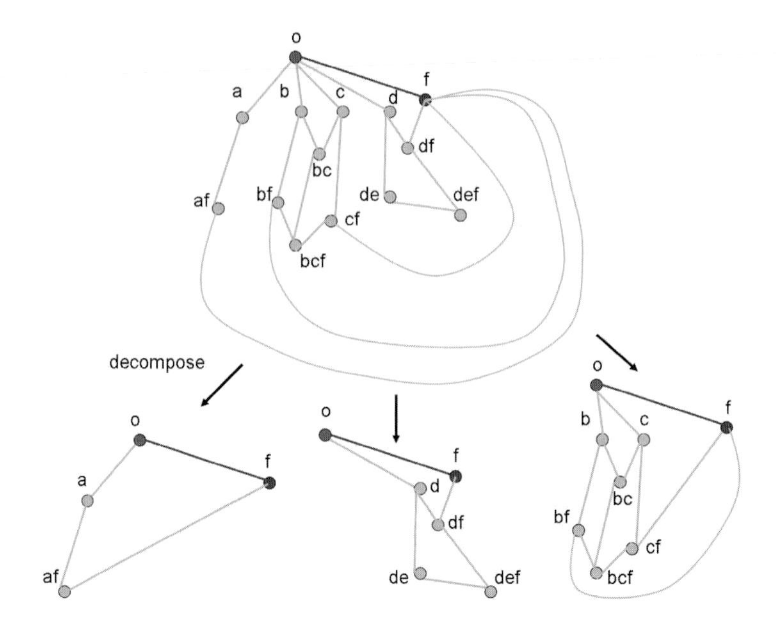

Fig. 8. The combinatorial dual of an abstract diagram d, a minimal cut pair $\{o, f\}$ and the graph decomposition

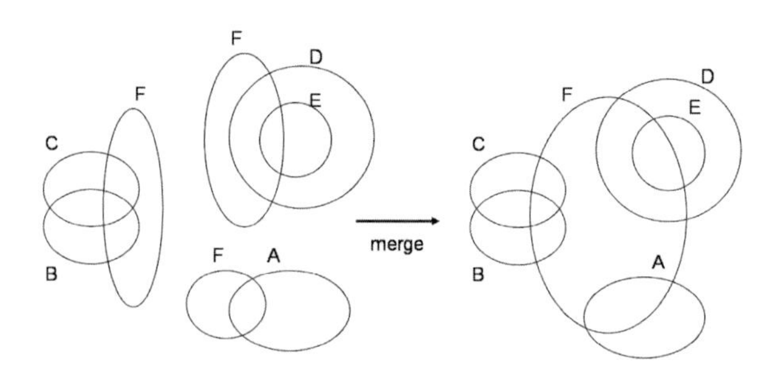

Fig. 9. Merging together the generated diagram pieces

Example 3. We describe an example of the whole process, using Figures 8 and 9. Starting with an abstract diagram description $d = \{o, a, b, c, d, f, af, bc, bf, cf, de, df, bcf, def\}$, we construct the combinatorial dual of d with a minimal cut pair $\{o, f\}$. The decomposition of this graph into pieces using the disconnecting curve F is shown in Figure 8. This enables the construction of diagrams from the three pieces; all three of the diagrams corresponding to the three pieces contain a contour labelled F, but the other contours appear in only one piece. These three diagrams are combined into the final diagram by merging the three occurrences of F in each diagram into one contour F, as shown in Figure 9.

Remark 1. We could have defined a notion of "component diagrams" by simply removing the disconnecting contour, and then drawing these components individually, but then it is not easy to describe how to overlay the disconnecting contour over the existing component diagrams, in general. We have instead used a disconnecting contour to decompose a diagram by splitting the disconnecting contour into multiple contours, one in each "component diagram". The "split and recombine" approach we have adopted makes the transformation of recombining the diagram components local to one zone, rather than global across entire diagram components.

Remark 2. There is a subtle issue regarding the dual edges of the combinatorial dual graph of an abstract diagram and the Euler dual graph of a concrete diagram. The diagram components created in the above process may have Euler dual graphs where some edges from the corresponding induced subgraphs of the combinatorial dual graph are missing. In order to be able to recombine the diagram components, we need the edge of the combinatorial dual graph which is incident with the cut pair to persist in the Euler dual graph. In the example shown, the edge between vertices labelled by o and f is present in the graph pieces. This kind of requirement comes for free if the diagram components are small (three contours or fewer), but places a requirement on the drawing process for more complex diagrams components.

A brief outline of the basic algorithm is presented. Given an abstract diagram d we construct G, the combinatorial dual graph of d. Then we create a new graph G', by splitting G at every cut vertex v, creating duplicate copies of v in every component that its removal would create. Then we create a new graph G'', by splitting G' at a cut pair $\{v_1, v_2\}$, creating duplicate copies of $\{v_1, v_2\}$ in every component that its removal would create. The individual pieces can then be generated and merged together. Further decompositions before generation as well as the use of disconnecting sets of contours (see Section 7) could also be used.

6 Semantics Predicate

The semantics of Euler diagrams are formally defined in a model theoretic manner (details can be found in [16] for example). Briefly, an *interpretation* for a diagram d is a universal set U and a function which assigns contours (or contour labels) to subsets of U. It is said to be a *model* of d if the natural extension of this function to zones satisfies the *semantics predicate* which can be stated as the *plane tiling condition* or the *missing zones condition.* The semantics predicate can be viewed as a conjunction of set theoretic statements that constrains the model appropriately. The plane tiling condition essentially says that the union of the images of all of the zones in the diagram is the universal set, whilst the missing zones condition says that zones that are missing represent the empty set. Although such semantic statements are useful, the difficulty with them is that they are not very readable by humans since they both involve unions of lists of zones and every contour label is used to describe each zone. For diagrams with many contours, the zone descriptors become cumbersome. We can use disconnecting contours to provide a more natural semantic statement automatically from a diagram than was previously possible; it is possible to convert back to standard zone-based description if required.

Consider the example given in Figure 1. The diagram is nested and can be viewed as the component comprising A and B, with the component of C inserted, and then the component of D inserted. The semantics predicate of each component is just $True$. The semantics of the whole diagram expresses how the components are combined: D is "inside" C, so $D \subseteq C$ and the component with contours $\{A, B\}$ is "outside" the contour C, so $C \subseteq \overline{A \cup B}$. Combine these to give

$$D \subseteq C \wedge C \subseteq \overline{A \cup B}.$$

Now consider examples with disconnecting contours. A particularly nice case of the semantics is when a disconnecting contour C of a diagram d "splits every zone of $d - C$ in two". For example, in the right hand side of Figure 6 the contour E is a disconnecting contour and it splits every zone of $d - E$ in two. The effect of splitting every zone of $d - E$ in two is that we can use the same associated semantic statement for d as for the nested diagram $d - E$, that is $A \cup B \subseteq D \cap \overline{C}$.

The first diagram of Figure 7 has two disconnecting contours, E and F. First we will write the semantics of d using the fact that E is a disconnecting contour. Reading the semantics of the nested diagram $d - E$ gives $(A \cup B) \cap (C \cup D \cup F) = \emptyset$ and the placement of the disconnecting contour E gives $E \subseteq \overline{C \cup D}$. Therefore, the semantics of d can be written as:

$$E \subseteq \overline{C \cup D} \wedge (A \cup B) \cap (C \cup D \cup F) = \emptyset.$$

Equivalently, treating F as the disconnecting contour gives

$$F \subseteq \overline{A \cup B} \wedge (C \cup D) \cap (A \cup B \cup E) = \emptyset.$$

Note that when reading the semantics of a diagram with a disconnecting contour, we have transformed the diagram into a nested diagram by contour removal. This allows us to rewrite the semantics predicate as the semantics of the nested diagram with contour removed conjoined with the semantics associated to the contour removed. This contrasts with the technique of splitting the disconnecting contour up which was more useful in diagram generation.

Although the output in the examples above might need some post processing, one approach for the basic step in automatically producing the semantics is to use a normal form for the semantics predicate of a nested diagram (see [11]), as follows.

Theorem 4. *Let d be an atomic Euler diagram with a disconnecting contour C. Let $N(d - C)$ denote the semantics predicate of the nested diagram $d - C$,* *written in normal form as in [11], and let $P(C)$ denote the semantics predicate* [7] *of the contour C. Then we can write the semantics predicate of d as: $P(d) = N(d - C) \wedge P(C)$.*

7 Higher Orders of Disconnectedness

We briefly mention some natural extensions of the ideas so far. Removal of a disconnecting contour turns an atomic diagram into a nested diagram. The idea of a disconnecting contour generalises to an n-disconnecting contour set. Removal of an n-disconnecting contour set turns an atomic diagram into a nested diagram, and n is minimal. A separating curve meets exactly one contour of the diagram, and separates the remainder of the diagram appropriately. This generalises to an n-separating curve, which meets exactly n contours and separates the remainder of the diagram appropriately. Then, a nested diagram has a disconnected intersection graph, an atomic diagram with a disconnecting contour has a cut vertex in its intersection graph, and an n-disconnecting contour set corresponds to a minimal cut set of n vertices in the intersection graph.

A nested diagram has a cut vertex in the dual graph, a diagram with a disconnected contour has a cut pair (differing by a single label) in its dual graph, and in general an n-disconnecting contour set has a connected subgraph of the dual graph of d whose vertices differ by the labels of X, which is a maximal such subgraph, and whose removal disconnects the dual graph.

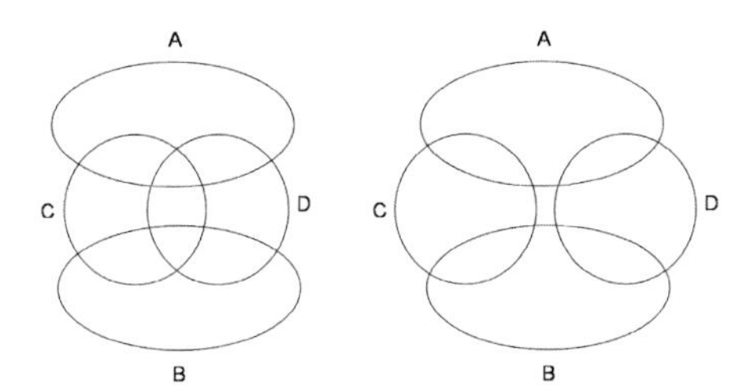

Fig. 10. Two diagrams with a disconnecting contour-pair

In both examples of Figure 10, the contour set $\{C, D\}$ is a disconnecting contour set. We can visualise a separating curve passing through C and D but separating A from B. The intersection graph would have a cut pair of vertices labelled C and D. The dual graph would have a subgraph with vertices from $\{o, c, d, cd\}$ whose removal would disconnect the graph. The second diagram is

[7] Here we mean a semantic statement which encapsulates the relationships of C with the other contours in the diagram, which could be a phrase such as "$R_{in}(C) \subseteq C \subseteq R_{out}(C)$" where $R_{in}(C)$ and $R_{out}(C)$ are descriptions of the regions of d that are maximally contained within C and minimally containing C, without referencing C, respectively (see [6] for details).

not well-formed because the zone o is disconnected, but we have included it because it is indicative of a more general kind of behaviour where the disconnecting contours do not form a Venn diagram.

The generalisations of Theorem 2 can be proved by induction:

Theorem 5. *Let d be a concrete Euler diagram. Then the following statements are equivalent:*

1. *d has an n-disconnecting contour set $X \subseteq C(d)$.*
2. *X corresponds to a cut vertex set of the intersection graph of d.*
3. *d has an n-separating curve which passes through only X.*
4. *There is a connected subgraph of the dual graph of d whose vertices differ by the labels of X, which is a maximal such subgraph, and whose removal disconnects the dual graph.*

8 Conclusion

Euler diagrams are a popular diagrammatic system for representing set-theoretic based information. They are used as the basis of logical systems for formal software specification purposes and in various application areas for information presentation, such as for displaying the results of queries on library system databases. Two areas of research related to them can help improve the user-experience. Firstly, the automatic generation of Euler diagrams from an abstract description is essential in order for the system to be able to display the correct diagrams to users. This causes serious computational problems as the number of contours in the diagram increases. Secondly, if users want to be presented with a more traditional sentential version of the information contained in a diagram in set-theory for instance, then presenting this in a human readable form is desirable.

We have developed the new idea of disconnecting contours for Euler diagrams, which are contours whose removal disconnects the diagram, and provided several equivalent conditions, the most useful, from an implementation point of view, being that of cut pairs of the dual graph. As well as being useful in understanding the structure of these diagrams in their own right, this enables us to decompose a diagram generation problem into smaller problems if the diagram has a disconnecting contour. We provide an indication of the algorithms that can used as refinements of existing generation algorithms. Furthermore, this work enhances out ability to automatically produce more natural human-readable semantics than the traditional zone-based semantic descriptions.

References

1. Chow, S., Ruskey, F.: Drawing area-proportional Venn and Euler diagrams. In: Liotta, G. (ed.) GD 2003. LNCS, vol. 2912, pp. 466–477. Springer, Heidelberg (2004)
2. Chow, S.C.: Generating and Drawing Area-Proportional Euler and Venn Diagrams. Ph.D thesis, University of Victoria (2007)

3. DeChiara, R., Erra, U., Scarano, V.: VennFS: A Venn diagram file manager. In: Proceedings of Information Visualisation, pp. 120–126. IEEE Computer Society, Los Alamitos (2003)
4. DeChiara, R., Erra, U., Scarano, V.: A system for virtual directories using Euler diagrams. In: Proceedings of Euler Diagrams 2004. Electronic Notes in Theoretical Computer Science, vol. 134, pp. 33–53 (2005)
5. Euler, L.: Letters a une princesse dallemagne sur divers sujets de physique et de philosophie. Letters 2, 102–108 (1775)
6. Fish, A., Flower, J.: Abstractions of Euler diagrams. In: Proceedings of Euler Diagrams 2004, Brighton, UK. ENTCS, vol. 134, pp. 77–101 (2005)
7. Fish, A., Flower, J.: Investigating reasoning with constraint diagrams. In: Visual Language and Formal Methods 2004, Rome, Italy. ENTCS, vol. 127, pp. 53–69. Elsevier, Amsterdam (2005)
8. Flower, J., Fish, A., Howse, J.: Euler diagram generation. Journal of Visual Languages and Computing (2008), `http://dx.doi.org/10.1016/j.jvlc.2008.01.004`
9. Flower, J., Howse, J.: Generating Euler diagrams. In: Proceedings of 2nd International Conference on the Theory and Application of Diagrams, Georgia, USA, April 2002, pp. 61–75. Springer, Heidelberg (2002)
10. Flower, J., Howse, J., Taylor, J.: Nesting in Euler diagrams. In: International Workshop on Graph Transformation and Visual Modeling Techniques, pp. 99–108 (2002)
11. Flower, J., Howse, J., Taylor, J.: Nesting in Euler diagrams: syntax, semantics and construction. Software and Systems Modelling 3, 55–67 (2004)
12. Hammer, E., Shin, S.J.: Euler's visual logic. History and Philosophy of Logic, 1–29 (1998)
13. Harel, D.: On visual formalisms. In: Glasgow, J., Narayan, N.H., Chandrasekaran, B. (eds.) Diagrammatic Reasoning, pp. 235–271. MIT Press, Cambridge (1998)
14. Howse, J., Molina, F., Shin, S.-J., Taylor, J.: Type-syntax and token-syntax in diagrammatic systems. In: Proceedings FOIS-2001: 2nd International Conference on Formal Ontology in Information Systems, Maine, USA, pp. 174–185. ACM Press, New York (2001)
15. Howse, J., Stapleton, G., Flower, J., Taylor, J.: Corresponding regions in Euler diagrams. In: Proceedings of 2nd International Conference on the Theory and Application of Diagrams, Georgia, USA, April 2002, pp. 76–90. Springer, Heidelberg (2002)
16. Howse, J., Stapleton, G., Taylor, J.: Spider diagrams. LMS Journal of Computation and Mathematics 8, 145–194 (2005)
17. Kent, S.: Constraint diagrams: Visualizing invariants in object oriented modelling. In: Proceedings of OOPSLA 1997, October 1997, pp. 327–341. ACM Press, New York (1997)
18. Kestler, H., Muller, A., Gress, T., Buchholz, M.: Generalized Venn diagrams: A new method for visualizing complex genetic set relations. Journal of Bioinformatics 21(8), 1592–1595 (2005)
19. Ruskey, F.: A survey of Venn diagrams. Electronic Journal of Combinatorics (1997), `www.combinatorics.org/Surveys/ds5/VennEJC.html`
20. Shimojima, A.: Inferential and expressive capacities of graphical representations: Survey and some generalizations. In: Blackwell, A.F., Marriott, K., Shimojima, A. (eds.) Diagrams 2004. LNCS (LNAI), vol. 2980, pp. 18–21. Springer, Heidelberg (2004)
21. Shin, S.-J.: The Logical Status of Diagrams. Cambridge University Press, Cambridge (1994)

22. Stapleton, G., Thompson, S., Howse, J., Taylor, J.: The expressiveness of spider diagrams. Journal of Logic and Computation 14(6), 857–880 (2004)
23. Swoboda, N., Allwein, G.: Using DAG transformations to verify Euler/Venn homogeneous and Euler/Venn FOL heterogeneous rules of inference. Journal on Software and System Modeling 3(2), 136–149 (2004)
24. Venn, J.: On the diagrammatic and mechanical representation of propositions and reasonings. Phil. Mag. (1880)
25. Verroust, A., Viaud, M.-L.: Ensuring the drawability of Euler diagrams for up to eight sets. In: Blackwell, A.F., Marriott, K., Shimojima, A. (eds.) Diagrams 2004. LNCS (LNAI), vol. 2980, pp. 128–141. Springer, Heidelberg (2004)

Smooth Linear Approximation of Non-overlap Constraints

Graeme Gange[1], Kim Marriott[2], and Peter J. Stuckey[1,3]

[1] Department of Comp Sci and Soft Eng
University of Melbourne, 3010, Australia
ggange@csse.unimelb.edu.au
[2] Clayton School of IT Monash University, 3800, Australia
Kim.Marriott@infotech.monash.edu.au
[3] National ICT Australia, Victoria Laboratory
pjs@csse.unimelb.edu.au

Abstract. Constraint-based placement tools and their use in diagramming tools has been investigated for decades. One of the most important and natural placement constraints in diagrams is that their graphic elements do not overlap. However, non-overlap of objects, especially non-convex objects, is difficult to solve and, in particular, to solve sufficiently rapidly for direct manipulation. Here we present the first practical approach for solving non-overlap of possibly non-convex objects in conjunction with other placement constraints such as alignment and distribution. Our methods are based on approximating the non-overlap constraint by a smoothly changing linear approximation. We have found that this in combination with techniques for lazy addition of constraints, is rapid enough to support direct manipulation in reasonably sized diagrams.

1 Introduction

Diagram editors were one of the earliest applications areas for constraint-solving techniques in computing [22]. Constraint solving allows the editor to preserve design aesthetics, such as alignment and distribution, and structural constraints, such as non-overlap between objects, during manipulation of the graphic objects. The desire for real-time updating of the layout during user interaction means that fast incremental constraint solving techniques are required. This, and the difficult, non-linear, combinatorial nature of some geometric constraints, has motivated the development of many specialized constraint solving algorithms.

Geometric constraint solving techniques have evolved considerably and modern diagram editors, such as MicrosSoft Visio and ConceptDraw, provide placement tools that impose persistent geometric constraints on the diagram elements, such as alignment and distribution. However, the kind of constraints provided is quite limited and, in particular, does not include automatic preservation of non-overlap between objects. This is, perhaps, unsurprising since solving non-overlap constraints is NP-hard. Here we address the problem of how to solve non-overlap constraints in conjunction with alignment and distribution constraints

G. Stapleton, J. Howse, and J. Lee (Eds.): Diagrams 2008, LNAI 5223, pp. 45–59, 2008.

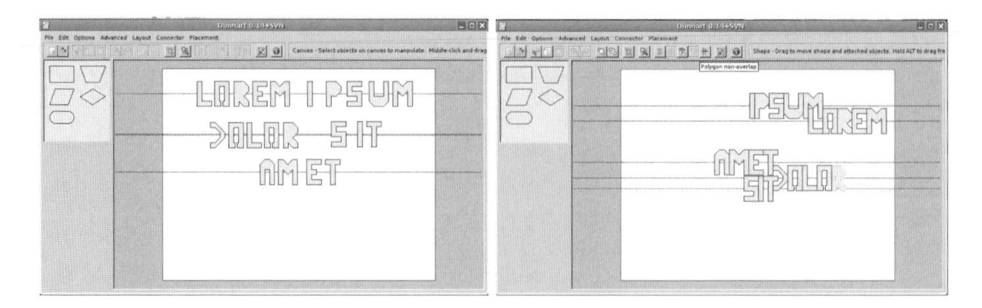

Fig. 1. An example of non-overlap interacting with alignment. Each word is aligned horizontally, and no letter can overlap another. Starting from the position on the left, direct manipulation of the letters updates the position of all objects in real time as letters are dragged around to reach the right position.

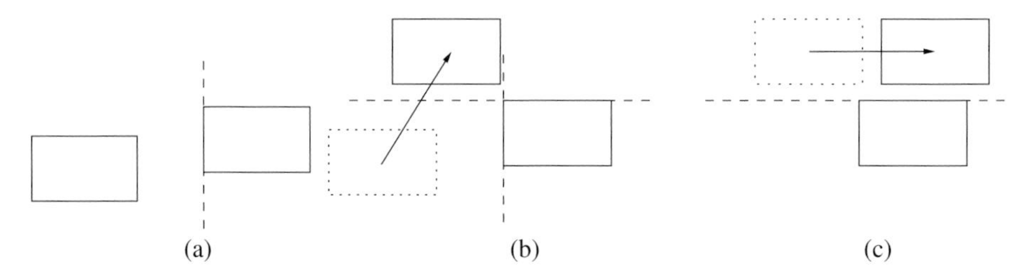

Fig. 2. Dynamic linear approximation of non-overlap between two boxes. Satisfaction of any of the constraints: *left-of*, *above*, *below* and *right-of* is sufficient to ensure non-overlap. Initially (a) the *left-of* constraint is satisfied. As the left rectangle is moved (b), it passes through a state where both the *left-of* and *above* constraint are satisfied. When the *left-of* constraint stops the movement right, the approximation is updated to *above* and (c) motion can continue.

sufficiently quickly to support real-time updating of layout during user interaction. We have integrated our non-overlap constraints into the constraint based diagramming tool Dunnart. An example of using the tool is shown in Figure 1.

The key to our approach is the use of dynamic linear approximation (DLA) [12,16]. While many geometric constraints, such as alignment and distribution are linear, non-overlap is inherently non-linear. In DLA, non-linear constraints are approximated by linear constraints. In a specialization of this technique, smooth linear approximation (SLA), as the solution changes the linear approximation is *smoothly* modified. The approach is exemplified in Figure 2. Efficient incremental solving techniques for linear constraints [1] mean that this approach is potentially fast enough for interactive graphical applications. It is worth pointing out that SLA is not designed to find a new solution from scratch, rather it takes an existing solution and continuously updates this to find a new locally optimal solution. This is why the approach is tractable and also well suited to interaction since, if the user does not like the local optimum the system has found, then they can use direct manipulation to escape the local optimum.

Previous research has described a number of proof-of-concept toys that demonstrate the usefulness of SLA for non-overlap of two simple shapes, axis-aligned rectangles and circles. In this paper we extend this to provide the first systematic investigation of how to use SLA to handle non-overlap between arbitrary polygons of fixed sized and orientation. Handling convex polygons is a relatively straightforward extension from axis aligned rectangles. A single linear constraint between each pair of objects suffices to ensure non-overlap. Handling non-convex polygons is considerably more difficult and the main focus of this paper.

The first approach (described in Section 5.1) is to *decompose* each non-convex polygon into a collection of convex polygons joined by equality constraints. Unfortunately, this leads to a large number of linear constraints between each pair of non-convex polygons since if these are decomposed into m and n convex polygons, respectively, then this gives rise to to mn constraints. In our second approach (described in Section 5.2), we *invert* the problem and, in essence, model non-overlap of polygons A and B by the constraint that A is contained in the region that is the complement of B.

As we have described them, the above two methods are *conservative* in the sense that for each pair of objects there is always a linear constraint in the solver ensuring that the objects will not overlap. However, if the objects are not near each other this incurs the overhead of keeping an inactive inequality in the linear solver. We also investigate non-conservative, *lazy*, variants of the above two methods in which the linear constraints are added only if the objects are "sufficiently" close and removed once they are sufficiently far apart. These utilize fast collision-detection algorithms [14][1] and are described in Section 6.

We provide a detailed empirical evaluation of these different approaches and their variants in Section 7. We find that SLA using the inverse approach with lazy addition of constraints is the fastest technique. Furthermore, it is very fast and certainly fast enough to support immediate updating of object positions during direct manipulation for practically sized diagrams. We believe the algorithms described here provide the first viable approach to handling non-overlap of (possibly non-convex) polygons in combination with other constraints in constraint-based diagramming tools.

2 Related Work

Starting with Sutherland [22], there has been considerable work on developing constraint solving algorithms for supporting direct manipulation in interactive graphical applications. These approaches fall into four main classes: propagation based (e.g. [23]); linear arithmetic solver based (e.g. [3]); geometric solver-based (e.g. [13]); and general non-linear optimisation methods such as Newton-Raphson iteration (e.g. [17]). However, none of these techniques support non-overlap.

[1] It is perhaps worth emphasizing that collision-detection algorithms by themselves are not enough to solve our problem. We are not just interested in detecting overlap: rather, we must ensure that objects do not overlap while still satisfying other design and structural constraints and placing objects as close as possible to the user's desired location.

Non-overlap constraints have been considered by Baraff [2] and Harada, Witkin, and Baraff [10], who use a specialised force based approach, modelling the non-overlap constraint between objects by a repulsion between them if they touch. However, they do not consider non-overlap constraints in conjunction with other linear constraints. Hosobe [11] describes a general purpose constraint solving architecture that handles non-overlap constraints and other non-linear constraints. The system uses variable elimination to handle linear equalities and a combination of non-linear optimisation and genetic algorithms to handle the other constraints. Our approach addresses the same issue but is technically quite different, and we believe much faster.

The most closely related work are our earlier papers introducing DLA [12,16]. The present paper extends these by providing a more detailed investigation of non-overlap and, in particular, of non-overlap of non-convex polygons. The algorithms given here to compute the linear approximation are all new.

3 Background: Smooth Linear Approximation

In constraint-based diagram editors the author can place geometric constraints on the objects in the diagram. During subsequent editing these geometric constraints will be maintained until the user explicitly removes them. Standard geometric constraints are:

- horizontal and vertical alignment
- horizontal and vertical distribution
- horizontal and vertical ordering that keeps objects a minimum distance apart horizontally or vertically while preserving their relative ordering
- an "anchor" tool that allows the user to fix the current position of a selected object or set of objects.

Each of the above geometric relationships can be modelled as a linear constraint over variables representing the position of the objects in the diagram. For this reason, a standard approach in constraint-based graphics editors is to use a constraint solver that can support arbitrary linear constraints.

Direct manipulation is handled as follows. Assume that the variables $y_1, ..., y_m$ correspond to objects which are the subject of the direct manipulation and that the variables $x_1, ..., x_n$ correspond to the remaining objects, and let C be the set of linear constraints representing geometry constraints. During directly manipulation the system successively solves a sequence of linear optimization problems of the form:

$$\text{Minimize } w_s \sum_{i=1}^{n} |x_i - c_i| + w_e \sum_{i=j}^{m} |y_j - d_j| \text{ subject to } C$$

where c_i is the current value of x_i and d_j is the desired value of y_j. The weighting constants w_s and w_e specify how important it is to move they_j's to their desired position as opposed to leaving the x_i's at their current position. Typically w_e is much greater than w_s.

The effect of the optimization is to move the objects being directly manipulated to their desired value, i.e. where the user wishes to place them, and leave the other objects where they are unless they are connected by constraints to the objects being moved. This optimization problem is repeatedly resolved during direct manipulation for different values of the d_i's. Fast incremental algorithms have been developed to do this so that the object positions can be updated in real-time [1].

However, not all geometric constraints are linear. Dynamic linear approximation (DLA) [12,16] is a recent technique for handling non-linear constraints that builds upon the aforementioned efficient linear constraint solving algorithms. In DLA, non-linear constraints are approximated by linear constraints.

Consider a complex constraint C. A *linear approximation* of a complex constraint is a (possibly infinite) disjunctive set of linear configurations $\{F_0, F_1, \dots\}$ where each *configuration* F_i is a conjunction of linear constraints. For example for the non-overlap constraint of the two boxes in Figure 2 there are four configurations left-of, above, below and right-of each consisting of a single linear constraint. We require that the linear approximation is *safe* in the sense that each linear configuration implies the complex constraint and *complete* in the sense that each solution of C is a solution of one of the linear configurations. For the purposes of this paper we will consider the complex constraint C to be a disjunctive set of its linear configurations.

In a specialization of the DLA technique, *smooth linear approximation (SLA)*, as the solution changes the linear approximation is smoothly modified. SLA works by moving from one configuration for a constraint to another, requiring that both configurations are satisfied at the point of change. This *smoothness* criteria reduces the difficulty of the problem substantially since we need to consider only configurations that are satisfied with the present state of the diagram. It also fits well with continuous updating of the diagram during direct manipulation.

The initial approach to smooth linear approxiation [16] involved keeping all disjunctive configurations in the underlying linear solver, and relying on the solver to switch between disjunctions. This has a number of problems: the number of configurations must be finite, and there is significant overhead in keeping all configurations in the solver. This was improved in [12] by storing only the current configuration in the solver and dynamically generating and checking satisfiability of other configurations when required.

The basic generic algorithm for solving a set of smooth linear approximations is given in Figure 3. Given a set of complex constraints $\mathbf{C}$ and their current configurations $\mathbf{F}$ as well as an objective function o to be mininimized, the algorithm uses a linear constraint solver to find an minimal solution θ with the current configuration. It then searches for alternative configurations that are satisfied currently, but if replaced would allow the objectivefunction to decrease further.

```
sla(C, F, o)
Let C = {C_0, C_1, ..., C_n} be the set of complex constraints.
Let F = {F_0, F_1, ..., F_n} be the current set of configurations.
repeat
  θ := minimize o subject to ⋀_{j=1}^{n} F_j
  finished := true
  for i ∈ 1..n
    F'_i = update(θ, C_i, F_i, ⋀_{j=1,j≠i}^{n} F_j, o)
    if F_i ≠ F'_i then
      F_i := F'_i
      finished := false
until finished

update(θ, C, F, F, o)
if ∃F' ∈ C where F' ≠ F and θ satisfies F'
  and F' may improve the solution
    return F'
else return F
```

Fig. 3. General algorithm for solving sets of smooth linear approximations

The algorithm repeatedly replaces configurations until no further improvement is possible. The algorithm is generic in:

- The choice and technique for generating the linear configurations for the complex constraint.
- How to determine if an alternative linear configuration might improve the solution.

In the next two sections we describe various choices for these operations for modelling non-overlap of two polygons.

4 Non-overlap of Convex Polygons

Smooth Linear Approximation (SLA) is well suited to modelling non-overlap of convex polygons. The basis for our approach is the Minkowski difference, denoted by $P \oplus -Q$, of the two polygons P and Q we wish to ensure do not overlap. Given some fixed point p_Q in Q and p_P in P the *Minkowski difference* is the polygon M s.t. the point $p_Q - p_P$ (henceforth referred to as the *query point*) is inside M iff P and Q intersect.

For convex polygons, it is possible to "walk" one polygon around the boundary of the second; the vertices of the Minkowski difference consist of the offsets of the second polygon at the extreme points of the walk. It follows that the Minkowski difference of two convex polygons is also convex. An example of the Minkowski difference of two convex polygons is given in Figure 4 while an example of a non-convex Minkowski sum is shown in Figure 5.

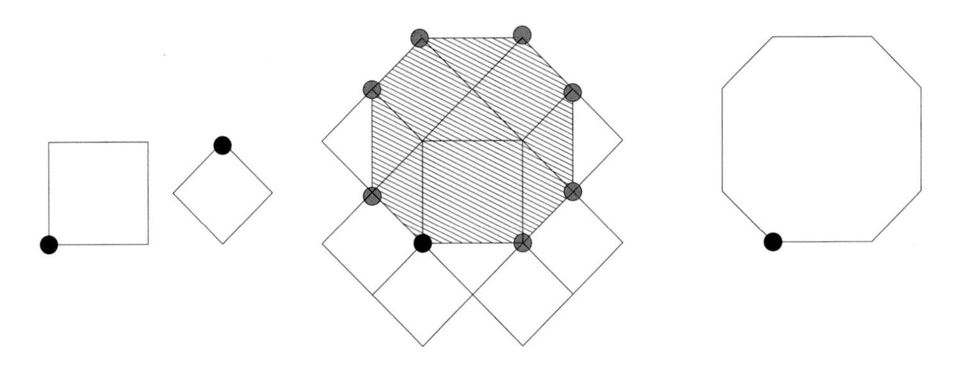

Fig. 4. A unit square S and unit diamond D and their Minkowski difference $S \oplus -D$. The local origin points for each shape are shown as circles.

There has been considerable research into how to compute the Minkowski difference of two polygons efficiently.[2] Optimal $O(n+m)$ algorithms for computing the Minkowski difference of two convex polygons with n and m vertices have been known for some time [8,18]. Until recently calculation of the Minkowski difference of non-convex polygons decomposed the polygons into convex components, constructed the convex Minkowski difference of each pair, and took the union of the resulting differences. Recently direct algorithms have appeared based on convolutions of a pair of polygons [19,6]. We make use of the implementations in the CGAL (`www.cgal.org`) computational geometry library.

We can model non-overlap of convex polygons P and Q by the constraint that the query point is *not* inside their Minkowski difference, M. As the Minkoskwi difference of two convex polygons is a convex polygon, it is straightforward to model non-containment in M: it is a disjunction of single linear constraints, one for each side of M, specifying that the query point lies on the outside of that edge.

Consider the unit square S and unit diamond D shown in the left of Figure 4. Their Minkowski difference is shown in the right of Figure 4. The linear approximation of the non-overlap constraint between S and D is the disjunctive configuration: $\{y \leq 0, x+y \leq 0, x \leq -0.5, y-x \geq 2, y \geq 2, x+y \geq 3, x \geq 1.5, y-x \leq -1\}$ which represent the 8 sides of the Minkowski difference shown at the right of the figure, from the bottom in clockwise order, where $(x, y) = (x_D, y_D) - (x_S, y_S)$ represents the relative position of the diamond to the square.

The approximation is conservative and accurate. It is also relatively simple and efficient to update the approximation as the shapes are moved. Note that the Minkowski difference only needs to be computed once. However before getting into details of how to update the approximation, we need to introduce Lagrange multipliers. These are a fundamental notion in constrained optimization, but

[2] More precisely, research has focussed on the computation of their Minkowski sum since the Minkowski difference of A and B is simply the Minkowski sum of A and a reflection of B.

describing their properties is beyond the scope of this paper; the interested reader is referred to [7]. Here, it suffices to understand that the value of the Lagrange multiplier λ_c for a linear inequality $c \equiv \sum_{i=1}^{n} a_i x_i \leq b$ provides a measure of how "binding" a constraint is; it gives the rate of increase of the objective function as a function of the rate of increase of b. That is, it gives the cost that imposing the constraint will have on the objective function, or conversely, how much the objective can be increased if the constraint is relaxed.[3]

Thus, intuitively, a constraint with a small Lagrange multiplier is preferable to one with a large Lagrange multiplier since it has less effect on the objective. In particular, removing a constraint with a Lagrange multiplier of 0 will not allow the objective to be improved and so the Lagrange multiplier is defined to be 0 for an inequality that is not active, i.e. if $\sum_{i=1}^{n} a_i x_i < b$. Simplex-based linear constraint solvers, as a byproduct of optimization, compute the Lagrange multiplier of all constraints in the solver.

Updating of the linear approximation to non-overlap with a convex polygon works as follows. Assume that the current linear approximation is the linear constraint c corresponding to boundary edge e and that the current solution is θ. If $\lambda_c = 0$ we need do nothing since changing the constraint could not improve the solution. Otherwise assume that $\lambda_c > 0$. We consider the two edges e_1 and e_2 adjacent to e and their corresponding constraints c_1 and c_2. If θ does not satisfy c_1 and does not satisfy c_2, then we do not change the choice of linear approximation since it cannot be done smoothly. Otherwise, it will satisfy at most one of the two, say c_1. We must now determine if it will improve the solution if we swap c for c_1. We do this by tentatively adding c_1 to the solver and computing its Lagrange multiplier λ_{c_1}. (Note that this is very efficient since the current solution will still be the optimum.) If $\lambda_{c_1} < \lambda_c$ then we are better off swapping to use c_1. This is done by removing c from the solver, leaving c_1 in the solver. Otherwise, we are better off staying with the current approximation, so we remove c_1 from the solver.

5 Non-convex Polygons

We now give two extensions of our technique for handling non-overlap of convex polygons to the case when one or both of the polygons are non-convex. We only consider simple polygons and do not allow holes in the polygons.

5.1 Decomposition into Convex Polygons

Probably the most obvious approach is to simply decompose the non-convex polygons into a union of convex polygons which are constrained to be joined together (either using equality constraints, or simply using the same variables to denote their position), and add a non-overlap constraint for each pair of polygons.

[3] It follows that at an optimal solution the Lagrange multiplier λ_c for an inequality cannot be negative.

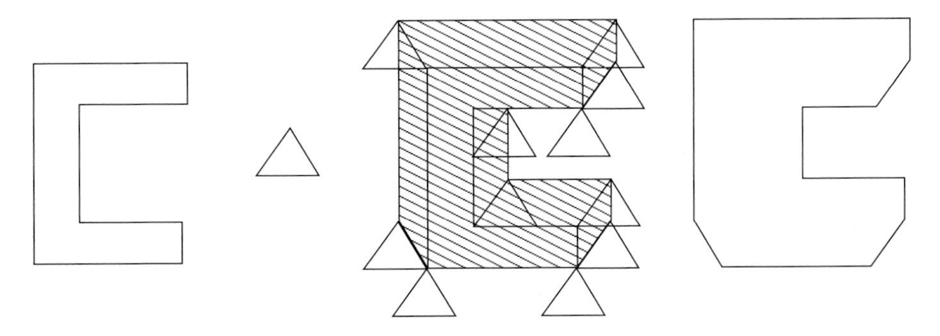

Fig. 5. The Minkowski difference of a non-convex and a convex polygon. From left, A, B, extreme positions of A and B, $A \oplus -B$.

This decomposition based method is relatively simple to implement since there are a number of well explored methods for convex partitioning of polygons, including Greene's [9] dynamic programming method for optimal partitioning. However, it has a potentially serious drawback: in the worst case, even the optimal decomposition of a non-convex polygon will have a number of convex components that is linear in the number of vertices in the polygon. This means that in the worst case the non-overlap constraint for a pair of non-convex polygons with n and m vertices will lead to $\Omega(nm)$ non-overlap constraints between the convex polygons.

5.2 Inverse Approach

However, in reality most of these $\Omega(nm)$ non-overlap constraints are redundant and unnecessary. An alternative approach is to use our earlier observation that we can model non-overlap of convex polygons P and Q by the constraint that the query point is *not* inside their Minkowski difference, M. This remains true for non-convex polygons, although the Minkowski difference may now be non-convex. An example Minkowski difference for a non-convex polygon is shown in Figure 5.

In our second approach we pre-compute the Minkowski difference M of the two, possibly non-convex, polygons P and Q and then decompose the space not occupied by the Minkowski polygon into a union of convex regions, $R_1, .., R_m$. We have that the query point is not inside M iff it is inside one of these convex regions. Thus, we can model non-overlap by a disjunction of linear constraints, with one for each region R_i, specifying that the query point lies inside the region. We call this the *inverse* approach.

These regions cover the region outside the polygon's convex hull, the non-convex *pockets* where the boundary of the polygon deviates from the convex hull, and the *holes* inside the polygon. The key to the inverse approach is that whenever the query point is not overlapping with the polygon, it must be either outside the convex hull of the polygon (as in the convex case), or inside one of the pockets or holes. If each pocket and hole is then partitioned into convex

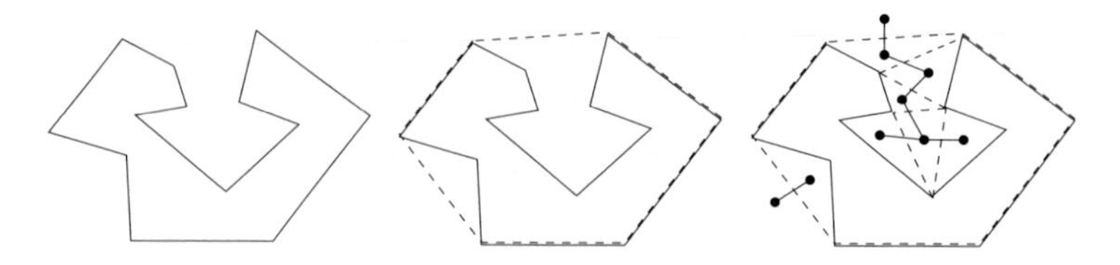

Fig. 6. A non-convex polygon, together with convex hull, decomposed pockets and adjacency graphs. From a given region, the next configuration must be one of those adjacent to the current region.

regions, it is possible to approximate the non-overlap of two polygons with either a single linear constraint (for the convex hull) or a convex containment (for a pocket or hole). An example is shown in Figure 6.[4]

As the desired goal is to reduce the number of constraints in the linear constraint solver, it seems reasonable to partition the pockets into triangular regions, resulting in at most 3 constraints per pair of polygons. The use of a triangulation has the added benefit of defining an adjacency graph which allows selection of adjacent regions in $O(1)$ time (Figure 6). There are a number of algorithms for partitioning simple polygons into triangles [4,21], of varying complexity; the present implementation uses a simple approach described by O'Rourke [18].

One of the advantages of the inverse approach is that, in most cases, particularly when the pairs of polygons are distant, the two polygons are treated as convex. It is only when the polygons are touching, and the query point lies upon the opening to a *pocket* that anything more complex occurs.

Updating of the approximation extends that given earlier for convex polygons. The new case is that the current linear approximation is one of the triangular regions, R, in a pocket or hole. In this case it has three constraints c_1, c_2, c_3 corresponding to each boundary edge. We have pre-computed the subset of those constraints that correspond to boundary constraints that are permeable in the sense that the boundary is shared with the convex hull or another triangular region. We compare the Lagrange multipliers of these constraints and determine which constraint has the largest Lagrange multiplier, say c. If $\lambda_c = 0$ we need do nothing since changing the region cannot improve the solution. Otherwise the current solution is on the boundary corresponding to c and so we move to the adjacent region sharing the same boundary.

6 Lazy Addition of Constraints

The preceding methods are all *conservative* in the sense that for each pair of objects there is always a linear constraint in the solver ensuring that the objects will not overlap. However, if the objects are not near each other this incurs the

[4] Our current implementation does not properly handle holes in the Minkowski difference, currently they are discarded yielding a simple polygon. This is a limitation of our implementation, not of the basic algorithm.

overhead of keeping an inactive inequality in the linear solver. A potentially more efficient approach is to lazily add the linear constraints only if the objects are "sufficiently" close and remove them once they are sufficiently far apart.

We have investigated two variants of this idea which differ in the meaning of sufficiently close. The first method measures closeness by using the bounding boxes of the polygons. If these overlap, a linear approximation for the complex constraint is added and once they stop overlapping it is removed. We also investigated a more precise form of closeness, based on the intersection of the actual polygons, rather than the bounding box. However, we found the overhead involved in detecting intersection and the instability introduced by repeatedly adding and removing the non-overlap constraint made this approach infeasible. Thus we focus on the first variant.

Implementation relies on an efficient method for determining if the bounding boxes of the polygons overlap. Determining if n 2-D bodies overlap is a well studied problem and numerous algorithms and data structures devised including Quad/Oct-trees [20], and dynamic versions of structures such as range, segment and interval-trees [5]. The method we have chosen to use is an adaptation of that presented in [14].

The algorithm is based, as with most efficient rectangle-intersection solutions, on the observation that two rectangles in some number of dimensions will intersect if and only if the span of the rectangles intersect in every dimension. Thus, maintaining a set of intersecting rectangles is equivalent to maintaining (in two dimensions) two sets of intersecting intervals.

The algorithm acts by first building a sorted list of rectangle endpoints, and marking corresponding pairs to denote whether or not they are intersecting in either dimension. While this step takes, in the worst case $O(n^2)$ time for n rectangles, it is in general significantly faster. As shapes are moved, the list must be maintained in sorted order, and intersecting pairs updated. This is done by using insertion sort at each time-step, which will sort an almost sorted list in $O(n)$ time.

Note that it is undesirable to remove the linear constraint enforcing non-overlap between two polygons as soon as the solver moves them apart and their bounding boxes no longer intersect; instead, such pairs of polygons are added to a removal buffer, and then removed only if their bounding boxes are still not intersecting after the solver has reached a stable solution.

A change in intersection is registered only when a left and right endpoint of different bounding boxes swap positions. If a left endpoint is shifted to the left of a right endpoint, an intersection is added if and only if the boxes are already intersecting in all other dimensions. If a left endpoint is shifted to the right of a right endpoint, the pair cannot intersect, so the pair is added to the removel buffer (if currently intersecting). (See Figure 7)

Unfortunately, we have found that this simple approach to lazy (or late) addition of constraints has the significant drawback of violating the conservativeness of the approximation and somewhat undermines the smoothness of the approximation since objects can momentarily overlap during direct manipulation. This

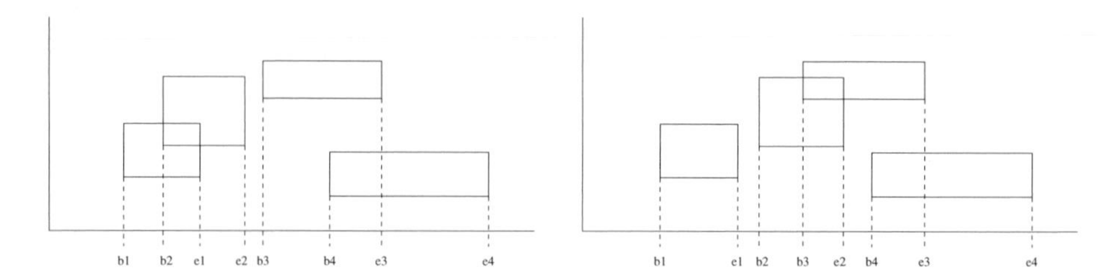

Fig. 7. The sorted list of endpoints is kept to facilitate detection of changes in intersection. As the second box moves right, b_2 moves to the right of e_1, which means that boxes 1 and 2 can no longer intersect. Conversely, endpoint e_2 moves to the right of b_3, which means that boxes 2 and 3 may now intersect.

can cause problems when the objects are being moved rapidly; that is, the distance moved between solves is large compared to the size of the objects. This is not very noticeable with the inverse approach but is quite noticeable with the decomposition method, as the convex components (and hence bounding boxes) are often rather small; if two shapes are moved sufficiently far between solves, the local selection of configurations may be unsatisfiable.

One possible solution would be to approximate the shapes by a larger rectangle with some "padding" around each of the objects. Another possible approach is to use rollback to recover from overlapping objects. When overlap is detected using collision detection, we roll back to the previous desired values (which did not lead to overlap), add the non-overlap constraint and re-solve and, finally, solve for the current desired value. This should maintain most of the speed benefits of the current lazy addition approaches, while maintaining conservativeness of approximation; and using a separate layer for the late addition avoids adding additional complexity to the linear constraint solver.

7 Evaluation

The algorithms were implemented using the Cassowary linear inequality solver, included with the QOCA constraint solving toolkit [15]. Non-convex Minkowski difference calculation was implemented using the Minkowski_sum_2 CGAL package produced by Wein [24]. We implemented both the decomposition and inverse approach for handling non-overlap of non-convex polygons. The decomposition was handled using Greene's dynamic programming algorithm [9]. For each version, both the conservative and lazy versions were implemented for comparison, and (of course) executed with exactly the same input sequence of interaction.

Two experiments were conducted. Both involved direct manipulation of diagrams containing a large number of non-convex polygons some of which were linked by alignment constraints. We focussed on non-convex polygons because all of our algorithms will be faster with convex polygons than non-convex. The experimental comparison of the approaches were run on an Intel Core2 Duo E6700 with 2GB RAM.

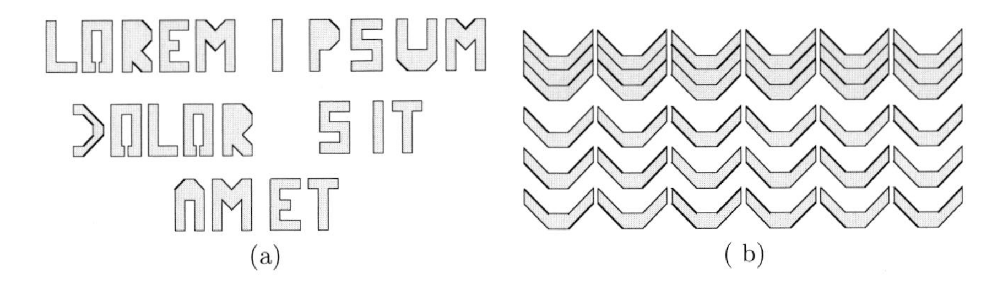

Fig. 8. Diagrams for testing. (a) Non-overlap of polygons representing text. The actual diagram is constructed of either 2 or 3 repetitions of this phrase. (b) The top row of shapes is constrained to align, and pushed down until each successive row of shapes is in contact.

Table 1. Experimental results. For the text diagram, we show the average and maximum time to reach a stable solution, and the average and maximum number of solving cycles to stabilize. For the U-shaped polygon test we show average time to reach a stable solution as the number of rows increases. All times are in milliseconds.

	Direct		Inverse	
	Cons	Lazy	Cons	Lazy
Ave Time	–	15.49	4.77	3.15
Max Time	–	26.03	31.54	11.26
Ave Cycle	–	2.58	1.89	2.55
Max Cycle	–	18	16	10

(a) Text diagram

	Direct		Inverse	
Rows	Cons	Lazy	Cons	Lazy
5	16.99	1.86	2.79	0.87
6	34.95	2.21	2.93	1.23
7	23.12	4.13	4.99	1.49
8	34.35	4.05	7.01	2.02
9	49.02	7.40	10.99	2.51
10	61.90	7.68	12.07	5.39

(b) U-shaped polygons

The first experiment measured the time taken to solve the constraint system for a particular set of desired values during direct manipulation of the diagram containing non-convex polygons representing letters, shown in Figure 8(a). Individual letters were selected and moved into the center of the diagram in turn. The results are given in Table 1(a). Note that the conservative decomposition method was unusable on the full diagram with 3 copies of the text, this is indicated with "–" in the table entry. On a smaller version of the diagram with 2 copies of the text it took on average 107 milliseconds.

In order to further explore scalability, a diagram of tightly fitting U-shapes, Figure 8(b), of a varying number of rows was constructed, and the top row pushed through the lower layers. Results are given in Figure 1(b).

The results clearly demonstrate that the inverse approach is significantly faster than the decomposition approach. They also show that for both approaches the lazy versions are significantly faster than the conservative versions. Interestingly, the conservative version of the inverse approach appears to outperform even the lazy version of the decomposition approach in cases where the convex hulls do not penetrate (such as the text example).

Most importantly the results also demonstrate that the inverse approach (with or without laziness) can solve non-overlap constraints sufficiently rapidly to facilitate immediate update of polygon positions during direct manipulation of realistically sized diagrams.

8 Conclusions

We have explored the use of smooth linear approximation (SLA) to handle solving of non-overlap constraints between polygons. We presented two possible approaches for handling non-overlap of non-convex polygons and have shown that the inverse method (which models non-overlap of polygons A and B by the constraint that A is contained in the region that is the complement of B) is significantly faster than decomposing each non-convex polygons into a collection of adjoining convex polygons.

We have also shown that the inverse method can be sped up by combining it with traditional collision-detection techniques in order to lazily add the non-overlap constraint only when the bounding boxes of the polygons overlap. This is capable of solving non-overlap of large numbers of complex, non-convex polygons rapidly enough to allow direct manipulation, even when combined with other types of linear constraints, such as alignment constraints.

Acknowledgements

We thank Michael Wybrow for his help in integrating the non-overlap constraint solving code into the Dunnart diagramming tool and Peter Moulder and Nathan Hurst for insightful comments and criticisms.

References

1. Badros, G.J., Borning, A., Stuckey, P.J.: The Cassowary linear arithmetic constraint solving algorithm. ACM Trans. CHI. 8(4), 267–306 (2001)
2. Baraff, D.: Fast contact force computation for nonpenetrating rigid bodies. In: Proceedings of the 21st Annual Conference on Computer Graphics and Interactive Techniques, pp. 23–34 (1994)
3. Borning, A., Marriott, K., Stuckey, P., Xiao, Y.: Solving linear arithmetic constraints for user interface applications. In: Proceedings of the 10th Annual ACM Symposium on User Interface Software and Technology, pp. 87–96. ACM Press, New York (1997)
4. Chazelle, B.: Triangulating a simple polygon in linear time. Discrete and Computational Geometry 6(1), 485–524 (1991)
5. Chiang, Y., Tamassia, R.: Dynamic algorithms in computational geometry. Proceedings of the IEEE 80(9), 1412–1434 (1992)
6. Flato, E.: Robust and efficient construction of planar Minkowski sums. Master's thesis, School of Computer Science, Tel-Aviv University (2000)
7. Fletcher, R.: Practical methods of optimization. Wiley-Interscience, New York (1987)

8. Ghosh, P.K.: A solution of polygon containment, spatial planning, and other related problems using minkowski operations. Comput. Vision Graph. Image Process. 49(1), 1–35 (1990)
9. Greene, D.H.: The decomposition of polygons into convex parts. In: Computational Geometry, pp. 235–259 (1983)
10. Harada, M., Witkin, A., Baraff, D.: Interactive physically-based manipulation of discrete/continuous models. In: Proceedings of the 22nd Annual Conference on Computer Graphics and Interactive Techniques, pp. 199–208 (1995)
11. Hosobe, H.: A modular geometric constraint solver for user interface applications. In: Proceedings of the 14th annual ACM symposium on User Interface Software and Technology, pp. 91–100 (2001)
12. Hurst, N., Marriott, K., Moulder, P.: Dynamic approximation of complex graphical constraints by linear constraints. In: Proceedings of the 15th annual ACM symposium on User Interface Software and Technology, pp. 27–30 (2002)
13. Kramer, G.A.: A geometric constraint engine. Artificial Intelligence 58(1-3), 327–360 (1992)
14. Lin, M., Manocha, D., Cohen, J.: Collision detection: Algorithms and applications (1996)
15. Marriott, K., Chok, S.C., Finlay, A.: A tableau based constraint solving toolkit for interactive graphical applications. In: Proceedings of the 4th International Conference on Principles and Practice of Constraint Programming, London, UK, pp. 340–354. Springer, London (1998)
16. Marriott, K., Moulder, P., Stuckey, P.J., Borning, A.: Solving disjunctive constraints for interactive graphical applications. In: Proceedings of the 7th International Conference on Principles and Practice of Constraint Programming, London, UK, pp. 361–376. Springer, Heidelberg (2001)
17. Nelson, G.: Juno, a constraint-based graphics system. In: Proceedings of the 12th annual conference on Computer Graphics and Interactive Techniques, pp. 235–243. ACM Press, New York (1985)
18. O'Rourke, J.: Computational Geometry in C, 2nd edn. (1998)
19. Ramkumar, G.D.: An algorithm to compute the Minkowski sum outer-face of two simple polygons. In: Proceedings of the Twelfth Annual Symposium on Computational geometry, pp. 234–241 (1996)
20. Samet, H.: The Design and Analysis of Spatial Data Structures (1990)
21. Seidel, R.: A simple and fast incremental randomized algorithm for computing trapezoidal decompositions and for triangulating polygons. Comput. Geom. Theory Appl. 1(1), 51–64 (1991)
22. Sutherland, I.E.: Sketch pad a man-machine graphical communication system. In: DAC 1964: Proceedings of the SHARE design automation workshop, pp. 6.329–6.346. ACM Press, New York (1964)
23. Vander Zanden, B.: An incremental algorithm for satisfying hierarchies of multiway dataflow constraints. ACM Transactions on Programming Languages and Systems (TOPLAS) 18(1), 30–72 (1996)
24. Wein, R.: Exact and efficient construction of planar Minkowski sums using the convolution method. In: Proceedings of the 14th Annual European Symposium on Algorithms, September 2006, pp. 829–840 (2006)

Extremes Are Better: Investigating Mental Map Preservation in Dynamic Graphs

Helen C. Purchase and Amanjit Samra

Department of Computing Science, Univerity of Glasgow, G12 8QQ, UK
hcp@dcs.gla.ac.uk

Abstract. Research on effective algorithms for efficient graph layout continues apace, and faster technology has led to increasing research on algorithms for the depiction of dynamic graphs which represent changing relational information over time. Like the static layout algorithms that preceded these, empirical work lags behind their design, and assumptions are made about how users' comprehension may be enhanced without human data to support them. This paper presents an experiment investigating an existing dynamic layout algorithm, focusing on the effect of the preservation of the mental map on comprehension. The results indicate that extremes produce better performance, suggesting that individual preference may be important.

Keywords: Dynamic graph layout, mental map, empirical study.

1 Introduction

Static graph layout algorithms are typically valued for their computational efficiency, and the extent to which they conform to well-known layout criteria (e.g. minimizing the number of crossings, maximizing the display of symmetry). Similarly, designers of dynamic graph layout algorithms typically advertise efficiency and the method by which the algorithm embodies mental map preservation [8,9]. The mental map criterion attempts to allow the user to maintain an evolving cognitive model of the graph as it changes over time: it is often defined as limiting the distance that nodes move over time [2].

Like their static counterparts, these dynamic layout algorithms need to be subject to empirical studies. As the design of these algorithms often focuses on the need to maintain the users' mental map, this particular feature of dynamic graph layout algorithms requires investigation [2].

While several dynamic graph layout systems exist, many of them are domain-specific [1,4] and do not allow for the easy manipulation of a mental map parameter that is necessary for conducting comparative empirical tests. We have identified GraphAnimation [9] and Graphael [7] as potential systems to investigate, and have developed our own system GDG [5] which embodies eight dynamic graph layout algorithms with an easy to use interface for graph creation. Our previous investigations considered hierarchical layout

G. Stapleton, J. Howse, and J. Lee (Eds.): Diagrams 2008, LNAI 5223, pp. 60–73, 2008.

(GraphAnimation [9]) and trivial movement limiting algorithms (GDG [13]) and have been inconclusive. The implementation of the mental map in Gaphael takes a very different approach, and is therefore the one considered here.

1.1 The User's Mental Map

As reported before [11] the creation of dynamic graph drawings from graph data that varies over time poses a problem: the dynamic layout algorithm has to either ensure that each individual timeslice conforms to well-known static layout criteria (e.g.: minimising crossing, no node overlap etc), or it must attempt to assist the user's understanding of the evolution of the graph over time.

Assisting understanding over time is typically done by attempting to maintain the users' 'mental map'; in most cases this is defined as minimising the movement of nodes between timeslices [10].

The notion of 'maintaining the user's mental map' is often used in the design of automatic graph layout algorithms to justify the algorithm's design: it is assumed that minimising the movement of nodes will assist with comprehension of this graph, and a 'mental map' parameter is included which allows the extent of mental map preservation to be adjusted.

It is likely that the two requirements (good layout of timeslices, and minimal node movement between timeslices) may be in conflict [2]. Limiting the movement of nodes from one timeslice to another may result in a poor layout for the second timeslice.

Our previous study [11] evaluated the effect of the mental map parameter in a hierarchical layout algorithm, GraphAnimation. We discovered that maintaining the mental map is important for the comprehension of an evolving graph for tasks that require that nodes be identified by name, but that it is less important for tasks that focus on edges rather than nodes or which do not require that nodes be nominally differentiated from each other. Since then, we have also investigated two dynamic layout algorithms that limit the extent of permitted movement of nodes in a spring algorithm by geometric restriction or proportional restriction with respect to desired movement [13]. The results suggested that dynamic algorithms that resulted in key static aesthetic criteria being violated in the individual time slices performed poorly in terms of human comprehension.

All three of the algorithms so far investigated are "on-line"; thus, the layout of each timeslice depends on its predecessor, and the mental map criterion is applied from timeslice-to-timeslice, with no reference to the subsequent timeslices. The experiment reported here investigates an off-line algorithm, Graphael, where the layout of each timeslicse takes into account the structure of the whole graph, over all timeslices, and is determined before the first timeslice is displayed. Using a spring-based offline experiment allows us to address the node-overlap situation discovered in Saffrey [13], and our choice of questions addresses the results found in Hoggan [11], where the mental map only appeared to be relevant to particular types of task.

Our overall aim is the same in all cases: to investigate the influence of the preservation of the mental map in a dynamic graph drawing on human understanding, where several existing algorithms implement the notion of 'mental map' in different ways.

1.2 The Graphael Algorithm

Graphel [6,8] is based on a spring algorithm, and determines the layout of all the individual timeslices in advance. While each timeslice is obviously a single graph of nodes and edges, Graphael creates new 'inter-timeslice edges' between the same node in two subsequent timeslices, thus creating one single large graph which includes all timeslices as well as all inter-timeslice edges. This large graph is then laid out using the spring algorithm, the inter-timeslice edges removed, and the layout of each individual timeslice extracted from this larger graph drawing.

The presence of the inter-timeslice edges prevents individual nodes from changing geometric position radically from timeslice to timeslice. The strength of these inter-timeslice edges determines the extent to which such movement is restricted. Thus, the strength of the inter-timeslice edges determines the extent of mental map preservation: strong edges mean a high mental map (as the nodes cannot move far); weak edges mean a low mental mode (as their restrictive effect is minimal).

By way of example, Figure 1 shows the seven time slices in the same dynamic graph, firstly with no mental map applied (the inter-time slice edges have no strength at all) and secondly with a high mental map parameter (the inter-time slice edges are very strong).

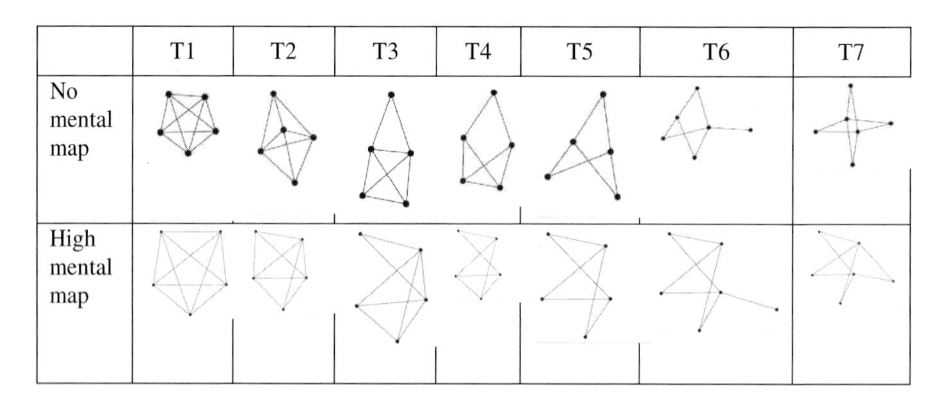

Fig. 1. The seven timeslices for the same example graph, under No and High mental map conditions

This mode of maintaining the mental map is known as 'rods', and is the weak version. Gaphael also allows a stronger mental map mode, that of 'cliques'. In this case, inter-timeslice edges do not only connect instances of the same node between subsequent timeslices, but connect them between all timeslices in the entire dynamic graph. This results in an even stronger restriction on node movement, and is also dependent on the strength of the intertimeslice edges. For the first user study using Graphael, we consider the weaker version (rods) as confirmation of our assumptions in the weak version would extrapolate to the strong version (but not vice versa).

2 The Experiment

2.1 Methodology

The experimental methodology used was based the extensive experience gained from our former static graph layout experiments [12], and our previous hierarchical dynamic graph experiments [11]. We used an online system to present the graphs, asking the participants to enter answers to multiple choice questions on the graphs, and collecting error and response time data. As before, tutorials and worked example material was given at the beginning to introduce the participants to the terminology and context, and to familiarize the participants with the experimental tasks. A qualitative and preference questionnaire was given at the end. A within-subjects methodology was used to reduce any subject variability, with the inclusion of practice tasks and randomization of tasks controlling for the learning effect. User-controlled rest breaks were included at regular periods throughout the duration of the experiment to address any problems of fatigue. The dependent variables were errors and response time.

2.2 The Mental Map

Graphael allows for an Inter-Timeslice Edge Weight parameter to be set. The authors claim that by changing the weight of the inter-timeslice edges, a balance can be found between individual timeslice graph readability and mental map preservation [8].The parameter determines the strength of these inter-timeslice edges, with a strong weight resulting in less movement of nodes over time, and a weak weight allowing the nodes to move more freely.

The parameters for *high, medium* and *low* mental map preservation were chosen after extensive visual exploration of graphs of the size used in the experiment.

The choices were made by perception, ensuring that there was a clear visual mental map distinction between the three conditions. There was a clear difference between graphs with a parameter of 30 and those of 40, but for values greater than 40, the difference was less clear. 0.1 was the lowest value allowed by Graphael. A parameter value of 5 was chosen for the medium value, as in this case it appeared that there was equal emphasis on preservation of layout as on preserving the mental map, while this was not the case with values of 2 or 10. These values were validated by observation by six other people not involved in the design of the experiment.

The Inter-Timeslice Edge Weight used in Graphael for the three mental map conditions were therefore:

- High mental map: 40
- Medium mental map: 5
- Low mental map: 0.1

The Graphael default setting of 'rods' was used; thus, node equivalents were linked between consecutive time slices, but not between all timeslices.

Figure 2 shows the difference between the mental map setting for graphs A and B. The second timeslice for the low mental map condition in both cases is noticeably different in shape when compared with the first timeslice, as the nodes have been free to move large distances. This is not the case for the high mental map condition when the nodes are more restricted in their movement.[1]

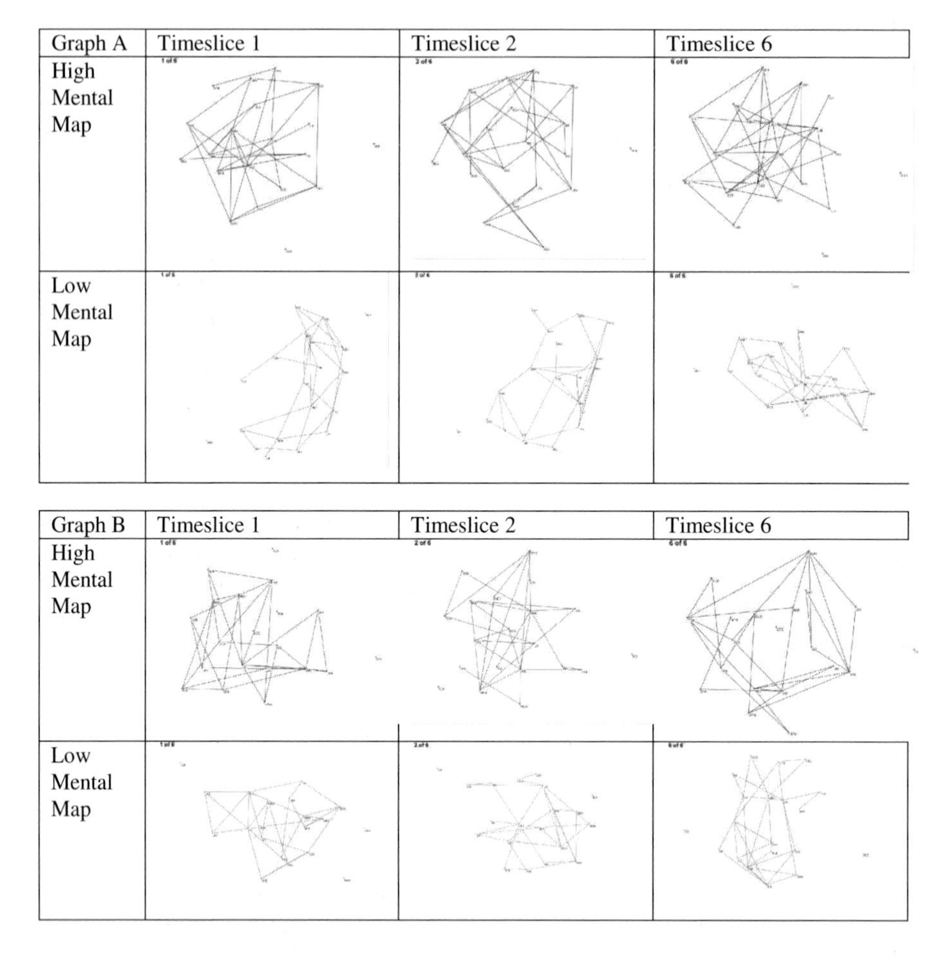

Fig. 2. The first two timeslices and the final timeslice for graphs A and B, each under both High and Low mental map conditions

2.3 Experimental Design

Three different dynamic graphs were created, each with six timeslices. So as to make the tasks easy for participants to understand, and to provide some simple context, these abstract graphs were represented as telephone communications between

[1] The experiment graphs are all available at http://www.dcs.gla.ac.uk/~hcp/edge/diag08/.

students. The static graph in each time slice represented a week, with students depicted as nodes, and edges indicating the presence of at least one phone call between two students during that week. These graphs were based on an imaginary scenario where students phone each other to discuss an upcoming assignment. Sometime during the six weeks, one student works out the solution to the problem, and hence subsequently takes part in more phone calls. This scenario was not intended to be representative of real phone call activity; it was merely used to make the experiment easier to conduct and explain.

The three graphs were of comparable size (20 nodes, 30 edges), with a comparable number of edge insertions and edge deletions per time-slice (approx 22). No new nodes were inserted, and no existing nodes deleted. Despite these controls, we aimed to make the graphs distinctive so that the answers to the questions for each graph were clearly different.

The graphs were animated using Macromedia Fireworks and the .png files produced by Graphael. The animation parameters were decided after running several pilot versions of the experiment:

- 3 seconds for timeslice 1, which included time to read the question
- 1 second for each of the other timeslices
- 1 second animation time between the six timeslices[2]
- 3 additional seconds for timeslice 6, during which time the participant was expected to answer the question, and after which the next task was displayed. A bell was sounded one second before the next question was displayed.

Thus, each task took a total of 16 seconds.

For each of these three graphs, three versions were created, one for each of our three experimental mental map conditions, resulting in nine graph animations in total. These are referred to by their graph identifier (A, B, C) and their mental map condition (H, M, L). For example, the high mental map version of graph A is referred to as GA-H.

Each of the nine animations used an entirely different set of names on the nodes used in their timeslices, thus ensuring that they were all perceived by the participants as being nine different scenarios, even though they were actually the same three scenarios presented three times each. These names were all of length three (and occasionally four) (e.g.: Meg, Ann, Jai, Jill), and all in upper case, so as not to introduce any confounding factor related to relative reading difficulty of the nodes.

The three questions used were:

- Q1: Which student was called the least (overall) during the 6 week term?
  This is the node with the lowest overall degree over all timeslices, excluding those students who were not called at all.
- Q2: Which student's phone was disconnected during 1 of the 6 weeks?
  In each case, there was only one node that had a degree 0 in only one of the timeslices.
- Q3: Which student was called the most (overall) during the 6 week term?
  This was the node with the highest overall degree over all timeslices.

[2] In practise, Graphael 'fades' inserted edges in and deleted edges out. This means that the 1 second for displaying a timeslice and the 1 second for animating between timeslices are blurred, and the animation has the appearance of always moving.

These questions were selected after several pilot experiments from an initial list of 16 possible questions. Questions that were abandoned for being too difficult or too easy included “How many more people did Jan call than Jon” and “Who was the least popular student in week 1?” The chosen questions did not reveal any floor or ceiling effects in the extensive pilot testing.[3]

The questions were explained and demonstrated to the participants carefully at the start of the experiment, to ensure that they knew exactly what they meant, and what answer was required.

Each of the nine animations was presented twice for each of three questions, making a total of 54 tasks. Each task was therefore a combination of graph, mental map, and question (for example, GA-H-3 is question 3 being asked of the high mental map versions of graph A). The tasks were presented in random order, with no task was repeated consecutively.

Before commencing the experimental tasks, the participants completed a practice set of 16 tasks randomly selected from the bank of 54 tasks. The participants were not aware that these practice tasks did not form part of the data collection.

2.4 Experimental Process

30 students were recruited from undergraduate Science courses in the University of Glasgow. As this is a within-subject experiment, and participants’ performance in one condition is compared with their own performance in another condition, any variation or similarity in the nature of participants does not affect the data analysis.

Each experiment, including time spent at the beginning on the tutorial and the worked example, and on the questionnaire at the end, took approximately one hour. No problems were experienced during the experiments and all participants appeared to engage in the task seriously.

3 Results and Analysis

3.1 Data Filtering

The response time data was filtered to exclude those data points when the participant answered the question before the end of the animation, as all questions required that the whole dynamic graph be viewed before the correct answer could be identified. The sixth timeslice was first revealed in the animation after 11.2 seconds, so all datapoints where the response time was recorded as being before this time were ignored when the averages were calculated.

The response time data is measured from this 11.2s point, when the sixth timeslice is revealed. Any tasks which were timed-out (ie: the participant did not answer before the next task was presented) were recorded as the maximum response time, 4.8s.

The error data was recorded as a 0 or 1, where 1 represents getting the incorrect answer for the task. Thus, a high value for both data measures (time and errors) implies poor performance.

[3] A floor/ceiling effect arises when the task is so difficult/easy that the manipulation of experimental condition has no effect on the data variation.

3.2 Results by Mental Map

Our hypothesis was that the extent to which the mental map was maintained between time-slices would affect performance. Intuitively, we thought that an animation which attempts to maintain the mental map would produce a better performance than one which does not consider the mental map at all.

The average response time and the average number of errors for the three mental map conditions over all three graphs and all three questions are shown in Figure 3.

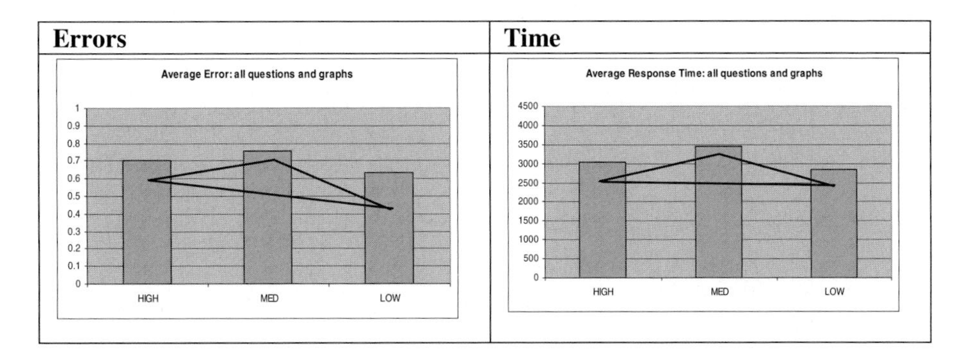

Fig. 3. The average number of errors and average response time for the three mental map conditions, over all graphs, and over all questions. Lines indicate statistical significance between conditions.

For both time and errors, there is statistical significance in performance over all conditions (errors: F=15.66> F(2,58)=3.156; time: F=42.98> F(2,58)=3.156)[4]. Post-hoc tests between conditions lead to the following conclusions[5].

- Low mental map produces fewer errors than high mental map, which in turn produces fewer errors than medium mental map,
- Low mental map has a faster response time than high mental map, which in turn has a faster response time than medium mental map,

These are strong conclusions indeed, revealing that the two extremes produce better performance than the medium value, and the low mental map produces the best results. There is also no inverse correlation between time and errors; thus, it is not the case that a longer response time produced fewer errors (and vice versa).

---

[4] The statistical analysis used here is the standard two-tailed ANOVA analysis, based on the critical values of the F distribution. Identifying a statistical significance between results collected from different conditions indicates that the difference between the results can be attributed to the differing nature of the conditions, rather than being due to mere chance. The F distribution allows us to test whether the probability of the difference being due to chance is less that the value of α (here we use $\alpha = 0.05$).

[5] The post-hoc test used here is the Tukey test, based on the critical values of the HSD (Honestly Significant Difference); it enables pair-wise comparison between the results of the conditions. As before, $\alpha = 0.05$.

3.3 Results by Question

Our experience in prior studies [11] is that, despite our best efforts to use questions of equal difficulty this is difficult to achieve, and that it is often useful to consider the effect of the mental map for each question separately.

We therefore performed further analysis to investigate whether there were any differences in the data with respect to questions, to see whether any were inherently more difficult than the others.

The average errors and average response time for each of the three questions (aggregated over all mental map conditions and all graphs) is shown in Figure 4.

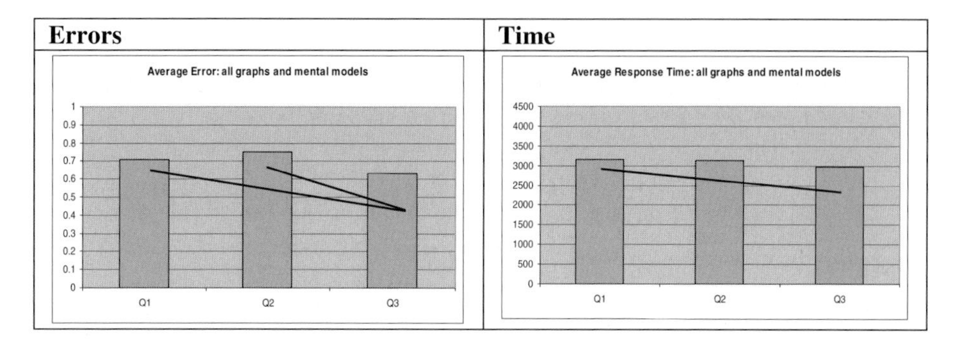

Fig. 4. The average errors and response time for the questions, over all mental map conditions,and graphs. Lines indicate statistical significance between conditions.

There are significant differences in performance between the three questions (errors: $F=8.05>F(2,58)=3.156$; time: $F=3.58>F(2,58)=3.156$). This is not surprising, as each question requires a very different visual process:

- Q1 requires keeping track of nodes with low degree, which will typically be on the edge of the drawings. This task also includes keeping track of these nodes even when they have 0 degree.
- Q2 requires looking out for 0 degree nodes, counting in how many time slices they have 0 degree, and identifying the node that has 0 degree in only one timeslice. The other nodes are irrelevant.
- Q3 requires keeping track of nodes with high degree, which will typically be central to the drawings in an area with many edges and connected nodes.

Figure 5 depicts the difference in performance according to mental map, separated by question.

There are significant differences between the mental map conditions for errors in Question 3 ($F=6.67>F(2,58)=3.934$)[6]. There are significant differences between the mental map conditions for response time in all three questions (Q1: $F=17.457>F(2,58)=3.934$; Q2: $F=21.5>F(2,58)=3.934$; Q3: $F=14.0>F(2,58)=3.934$). There are no other significant differences.

---

[6] In this case, $\alpha=0.025$, an appropriate statistical adjustment for second-level analysis according to the Bonferonni correction.

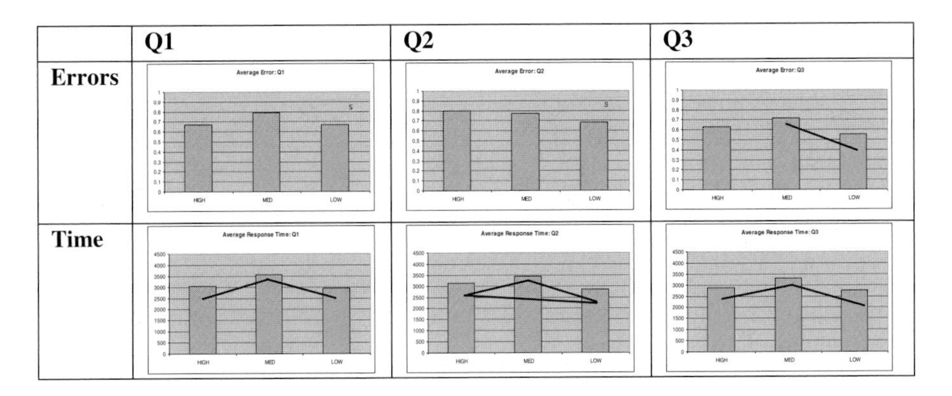

Fig. 5. The average errors and response time for the mental map conditions, separated according to question. Lines indicate statistical significance between conditions.

Even when considering the effect of the mental map with respect to the different questions, we find no support for the high mental map, and, for all questions, the medium mental map produces the worst performance.

3.4 Questionnaire Data

For each question, participants were asked to identify which mental map condition they thought helped best in answering that question. Figure 6 shows the frequency of choice of the mental map conditions.

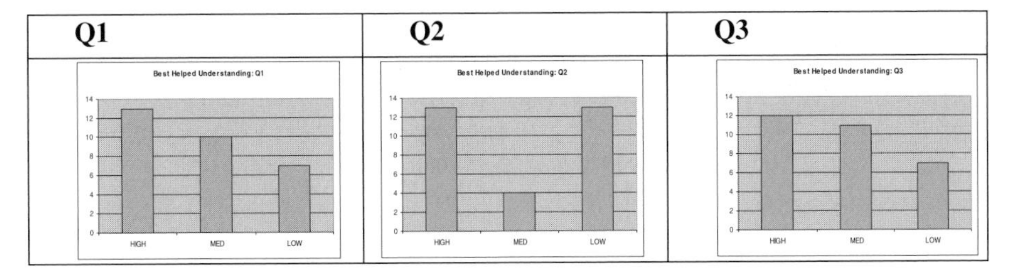

Fig. 6. Bar charts showing the frequency of participant choice for the three conditions for each question

Table 1 shows representative comments made about these conditions in the context of each question when the participants were asked to compare the three conditions both positively and negatively. The negative comments are in italics.

In their open questionnaire responses, most participants indicated that they found the tasks difficult, that they were typically unsure of their answers, and that they would have liked to have had more time to answer the questions.

The participants' responses reveal that although all graphs were depicted at exactly the same speed, those using a low mental map condition appeared faster, as the visual

Table 1. Representative comments from the participants for each condition. Comments given negatively are in italics.

High	Med	Low
more spread out, slower, more spacious, the outer nodes have fewest edges, least movement, easier to keep track of nodes, nodes kept apart, clustered and therefore easier to spot disconnected nodes, looks nice, easier to understand, more time to look for answers, the central nodes were the high degree ones, was able to keep track of the degree of the nodes, nodes didn't move too much.	right balance, easier to spot least connected nodes, least messy, most visually pleasing, slow enough node movement to prevent overlap but not so fast as to be untraceable, can look at nodes on [the] edge while animation is not too fast, disconnected people at edges and right speed, sensible speed, most enjoyable, lines seemed clearer, high degree nodes were centred, slow movement, easier to keep track.	loners at edge, lines moved around more, straightforward, more fun to watch, faster and easier, loners moved away and therefore easier to spot, more fluid and free-flowing, good shape, high degree nodes moved to the centre, fast movement made high degree nodes more obvious, high movement prevented unconnected nodes appearing connected, nodes spread out.
lines more tangled, most messy, more cluttered, nodes didn't move enough, seems like too many lines, less clear, too slow, lines all clustered together, too cluttered.	*large mess in the middle, too many lines close together, does not have the attributes of High nor the extreme movements of Low, central nodes often not connected.*	*too fast, erratic node movement, too much node movement, hard to keep track of nodes, too fast, more spread out, nodes move in and out a lot – confusing.*

changes between the timeslices are more extreme and therefore more eye-catching. Those with the high mental map condition appeared slower as the visual changes are less dramatic.

4 Discussion

The layout of dynamic graphs requires a compromise between maintaining the mental map between timeslices and conforming to static layout criteria within each timeslice. Our expectation was that the high mental map would result in the best performance, or, if this was not the case, the compromise 'medium' condition would be best as it would encompass the best features of both the high and low conditions. This has been shown not to be the case.

What the data reveals is that the medium mental map produces the worst performance, and the low mental map produces the best performance.

In analyzing the data according to the three different questions, we investigated whether the effect of the mental map differed according to the nature of the interpretation task, but still found that the medium condition produced the worst results for all three questions, and the low mental map condition produced the best.

We considered whether these results could be due to participant difference, that is, that some participants just perform better with a high mental map while some participants just perform better with a low one. We therefore performed a correlation analysis between average response time and average errors for each student in each condition. No significance was found ($r=0.088<0.25$, $df=60$, $\alpha=0.05$).

The qualitative data shows that the participants liked the high mental map graphs because they were more spread out and the nodes didn't move much, and the low mental map graphs because the movement helped in answering the questions. But it is the comments about the medium graphs that are most revealing: several participants recognized the compromise being made ("slow enough node movement to prevent overlap, but not so fast as to be untraceable", "does not have the attributes of High nor the extreme movements of Low"). In recording their preferences, some students found this compromise a good thing, others not.

It is clear that the stated preferences of the participants are not directly related to performance, apart from in question 2 where the medium condition was deemed to be least useful. The comments made about the extreme conditions (e.g. for the high condition: "more spread out", "lines more tangled") are frequently contradictory. What these comments do reveal, however, are the perceived negative features of the extremes (e.g. high: "more cluttered", "lines more tangled"; low: "too much node movement", "too fast").

We can attempt to explain the quantitative results in the context of these comments: it appears that the medium condition, rather than being a good compromise between the benefits of the high and low conditions, has adopted the negative features of them both, producing a bad compromise and poorer performance than either. Thus, the medium condition is not fast enough for the movement of the nodes to help with identifying the between timeslice changes, and was not spread out enough for easy identification of the relationships between the nodes within the timeslices. Observation of the graphs confirms this: the low mental map graphs tended to have clusters of nodes and edges near the centre of the drawing (a feature that is retained in the medium mental map graphs), while the slow movement of nodes in the high mental map graphs did not enable easy identification of nodes whose relationships were changing significantly between timeslices.

The results also indicate that the low mental map condition performed better than the high mental map overall, thus indicating that the extreme movement of nodes which indicates significant changes was more crucial for good performance than being able to track the nodes by similar position. In the medium and high mental map conditions, nodes could become connected or disconnected without a strong or immediate visual change to the graph layout, so while the nodes may have been easier to track individually, the changes in their connectivity were more difficult to identify.

5 Conclusion

Much as we would like to, we cannot argue with objective data! We have attempted to confirm the hypothesis that maintaining a high mental map assists with understanding a dynamic graph, but have shown that not only is this not the case, a compromise condition performs even worse. We have conducted our experiment in the light of our previous experience through our careful choice of questions (addressing the results reported in Hoggan [11]) and though our choice of an off-line algorithm that avoids node-overlapping (addressing the results reported in Saffrey [13]). And, within the context of this experiment, we conclude that there is no benefit in a dynamic graph layout algorithm maintaining a mental map between timeslices when mental map preservation is defined as minimising node movement.

These results need, of course, to be interpreted within the context of this experiment and its limitations and parameters. The experiment used three graphs of a particular size, three particular questions and one particular algorithm. Using more than one graph and more than one question assists in producing generalisable results, but these are still constrained by the necessary limitations of the experimental method.

In particular, the graphs used in this experiment change radically over time, with a large proportion of the edges added or deleted between timeslices (approx 73%). This may be why the low mental map condition performed best, and mental map preservation may therefore be most appropriate for dynamic graphs that change slowly over time. Further experimentation will tell.

As always, there is still much work to be done in this area. In particular, there are other definitions of the mental map whose effectiveness needs to be assessed (e.g. [10]): these may give more predictable results. In addition, experimenting with real world data is necessary; we are currently devising an experiment based on metabolic pathways which investigates the relative benefit of the mental map when observing the effect of different treatments.

Acknowledgments. We are grateful to Eve Hoggan for developing the initial version of the experimental software, and for her advice throughout the experimental process. David Forrester gave crucial and timely advice on Graphael, and discussions with Stephen Kobourov were instrumental in interpreting the data. We thank the participant volunteers without whom there would be no data. Ethical approval was given by the University of Glasgow (ref FIMS00177).

References

1. Brandes, U., Corman, S.R.: Visual unrolling of network evolution and the analysis of dynamic discourse. Information Visualization 2(1), 40–50 (2003)
2. Branke, J.: Dynamic Graph Drawing. In: Kaufmann, M., Wagner, D. (eds.) Drawing Graphs. LNCS, vol. 2025, pp. 228–246. Springer, Heidelberg (2001)
3. Coleman, M.K., Stott Parker, D.: Aesthetics-based graph layout for human consumption. Software Practice and Experience 26(12), 1415–1438 (1996)

4. Collberg, C., Kobourov, S., Nagra, J., Pitts, J., Wampler, K.: A system for graph-based visualization of the evolution of software. In: Proceedings of the ACM symposium on Software Visualization, pp. 77–86. ACM Press, New York (2003)
5. Dempster, L.M.: Information Visualisation: Dynamic graph layout with GDG. Unpublished Computing Science Honours project, Department of Computing Science, University of Glasgow (2006)
6. Diehl, S., Görg, C.: Graphs, they are changing. In: Proceedings of Graph Drawing 2002. LNCS, vol. 2538, pp. 23–30. Springer, Heidelberg (2002)
7. Erten, C., Harding, P.J., Kobourov, S.G., Wampler, K., Yee, G.V.: Graphael: Graph animations with evolving layouts. In: Liotta, G. (ed.) GD 2003. LNCS, vol. 2912, pp. 98–110. Springer, Heidelberg (2004)
8. Forrester, D., Kobourov, S.G., Navabi, A., Wampler, K., Yee, G.V.: Graphael: A System for Generalized Force-Directed Layouts. In: Pach, J. (ed.) GD 2004. LNCS, vol. 3383, pp. 454–464. Springer, Heidelberg (2005)
9. Görg, C., Birke, P., Pohl, M., Diehl, S.: Dynamic graph drawing of sequences of orthogonal and hierarchical graphs. In: Pach, J. (ed.) GD 2004. LNCS, vol. 3383, pp. 228–238. Springer, Heidelberg (2005)
10. Lee, Y., Lin, C., Yen, H.: Mental Map Preserving Graph Drawing Using Simulated Annealing. In: Proceedings of the 2006 Asia-Pacific Symposium on Information Visualisation, pp. 179–188. Australian Computer Society (2006)
11. Purchase, H.C., Görg, C., Hoggan, E.: How Important is the "Mental Map"? - an Empirical Investigation of a Dynamic Graph Layout Algorithm. In: Kaufmann, M., Wagner, D. (eds.) GD 2006. LNCS, vol. 4372, pp. 184–195. Springer, Heidelberg (2007)
12. Purchase, H.C.: Effective information visualisation: a study of graph drawing aesthetics and algorithms. Interacting with Computers 13(2), 147–162 (2000)
13. Saffrey, P., Purchase, H.C.: The "Mental Map" versus "Static Aesthetic" Compromise in Dynamic Graphs: A User Study. In: Proceedings of 9th Australasian User Interface Conference, pp. 85-93 (2008)

An Eye-Tracking Study of Exploitations of Spatial Constraints in Diagrammatic Reasoning

Atsushi Shimojima[1] and Yasuhiro Katagiri[2]

[1] Faculty of Culture and Information Science, Doshisha University, Japan
ashimoji@mail.doshisha.ac.jp
[2] Department of Media Architecture, Future University - Hakodate, Japan
katagiri@fun.ac.jp

Abstract. The semantic studies of diagrammatic notations [1,2,3] have revealed that so-called "perceptual," "non-deductive," or "emergent" effects of diagrams [4,5,6,7] are all rooted in a common inferential process, namely, the exploitation of spatial constraints on graphical structures. Thus, theoretically, this process is a key factor in inference with diagrams, explaining the oft-observed unburdening of the inferential load. In the present study, we inspect the empirical basis of this theoretical suggestion. Eye-movements were recorded while the participants were engaged in three-term transitive inference problems. They were provided with simple positions diagrams, on which we can define positions that should be fixated if the hypothesized inferential process occurs. Our analysis has revealed that the participants could exploit spatial constraints on graphical structures even when (1) they were not in the position of actually manipulating diagrams, and (2) the semantic rule of the provided diagrams did not match their preference. These findings indicate that the hypothesized practice is in fact robust, with a potential to broadly account for the inferential advantage of diagrams.

1 Introduction

Many, perhaps all, systems of diagrams have the function of letting the user to exploit spatial constraints on their graphical structure, and thus lightening the load of inferences. Consider a system of simple position diagrams, where letter symbols are arranged vertically to express a certain transitive relation holding between the symbolized objects. Figure 1-a shows a sample diagram in this system, which expresses that the object *A* is lighter than the object *B*. Now, modify this diagram to express another piece of information, that the object *C* is lighter than the object *A*. We obtain the new position diagram shown in Figure 1-b.

This diagram is the result of expressing two pieces of information in the current system of position diagrams. Yet, it expresses a *third* piece of information, namely, the information that *C* is lighter than *B*. Furthermore, given the transitivity of the relation *lighter*, this additional piece of information is a logical consequence of the original two pieces of information. Thus, just by expressing the two premises in this system of position diagrams, the user obtains a diagram that expresses a particular logical consequence of them *automatically*. As aptly put by Barwise and Etchemendy [1], the user "never need infer" this consequence from the premises, but "can simply read [it] off from the diagram as needed."

G. Stapleton, J. Howse, and J. Lee (Eds.): Diagrams 2008, LNAI 5223, pp. 74–88, 2008.

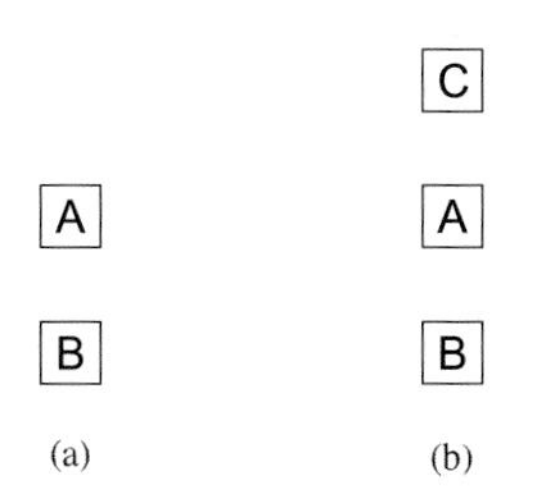

Fig. 1. (a) A position diagram expressing that A is lighter than B. (b) The result of adding the information that C is lighter than A.

Note that this "automaticity" of expression is largely due to a spatial constraint on the arrangement of letter symbols in a position diagrams: if a letter symbol x is placed above another letter symbol y, which is placed above still another symbol z, then the symbol x necessarily comes above the symbol z. The system of position diagrams is designed so as to exploit this spatial constraint for the "automatic" expression of certain logical consequences. The user can ride on this function of the system and significantly reduce his or her inferential task, replacing it with a reading-off task.

This inferential advantage of diagrams has been noted by many researchers, conceptualized variously as "perceptual inference" [4], "non-deductive representation" [5], and "emergent" effect [6,7]. The semantic studies of diagrams [1,2,3] clarified that the inferential process is a form of exploiting spatial constraints on graphical structures in the relevant diagrams.

Characterized this way, the advantage can be easily seen to exist in a variety of diagrammatic representation systems. For example, expressing the information that all As are Bs in a Venn diagram (Figure 2-a) and adding the information that no Bs are Cs (Figure 2-b) result in the expression of the information that no Cs are As, due to constraints on the shading of sub-regions. Expressing the information that $A \subset B$ in an Euler diagram (Figure 2-c) and then expressing the information that $C \cap B = \emptyset$ (Figure 2-d) result in the expression of the information that $C \cap A = \emptyset$, due to constraints on the inclusion-exclusion relation between regions. Geographical maps have the same function, and to a much larger extent. Adding an icon of a house to a particular position in a map results in the expression of various new pieces of information, concerning

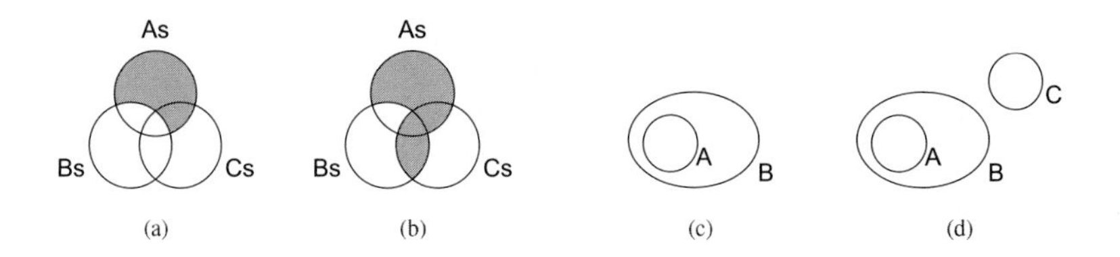

Fig. 2. (a) A Venn diagram expressing that all As are Bs. (b) The result of adding the information that no Bs are Cs. (c) An Euler diagram expressing that $A \subset B$. (d) The result of adding the information that $C \cap B = \emptyset$.

the spatial relationship of the house to many other objects already mapped. This is due to spatial constraints governing map symbols, which are pretty isomorphic to spatial constraints governing mapped objects.

Main Question

Thus, from the viewpoint of semantic theory, the exploitation of spatial constraints on graphical structures seems to be a good explanation for why diagrams are so useful for inference in certain contexts. From an empirical point of view, the question remains how *generally* this account applies to the real inferential practices with diagrams.

In the present study, we are particularly interested in inferences with diagrams that take place when the user is just looking at them. To illustrate the problem, consider the following situation. You are given the position diagram in Figure 1-a, which expresses the first premise that *A* is lighter than *B*. Then the second premise that *C* is lighter than *A* is given, but you are not in the position of changing the first diagram to add this premise to it. You are then asked, "Is *C* lighter than *B*?"

How would you answer this question? Because you cannot manipulate the diagram in the first place, you might entirely ignore the diagram and just think about the given premises solely in your head. Or you might make a mental copy of the diagram, and add a mental symbol [C] to this internal diagram to express the second premise. Or you might directly "draw" on the external diagram, where the "drawing" is not physical but sharing an inferential function with a physical drawing. In this case, it would amount to placing a non-physical symbol [C] in an appropriate place on the external diagram.

This paper reports on an experiment that tested this third possibility. Testing this possibility in a focused and systematic manner is important in at least two respects. First, as already suggested, it amounts to testing the general applicability of one promising explanation of the inferential advantage of diagrams. If a non-physical drawing in the above sense is possible and actually in practice, then it would mean that exploitations of spatial constraints on diagrams can occur even when they are not in the position of actually manipulating the diagrams. The advantage of diagrams in these cases is then explained from the (partial) sustainment of the inferential load by external spatial constraints, just as it is explained in the cases involving physical manipulations of diagrams.

Secondly, as we will see later, various works in diagrammatic reasoning presuppose or suggest the existence of non-physical drawing sharing an inferential role with a physical drawing. The concepts such as "envisaging" [8], "perceptual operation" [4], "visualization process" [9], and "spatial transformation" [10] all seem to point to, or include, such non-physical operations on the external diagrams. Although some empirical evidence has been obtained for its existence [11,12,13], the phenomenon has never undergone a sufficiently focused and systematic test. The experiment to be reported here aimed at providing such a test, by formulating an explicit experimental hypothesis with the help of the recent research on visual indexing (sections 2.1 and 2.2), systematically varying semantic rules of diagrams used as stimuli (section 2.3), detecting the participants' preference for semantic rules (section 2.4), and evaluating the experimental results in relation to the participants' semantic preference thus detected (section 3).

2 Methods

2.1 General Hypothesis

A non-physical "drawing" may sound too mysterious or vague to be subjected to an experimental test. However, given the recent research on the function of deictic indices in visual routines [14,15,16,17,18], it is not so farfetched to hypothesize that people use such an operation to take advantage of spatial constraints on external diagrams.

Deictic indices are pointers to particular objects in the external space. Through the attentional mechanism, we can maintain a small pool of such indices at once, and can easily direct our focal attention to any of these indices. Usually postulated are the operations of *placing* an index in the external space, returning to a single index to *identify* its visual property, and returning to multiple indices to *check* their spatial relation.

Pylyshyn [18] summarizes a series of experiments showing our ability to maintain indices to multiple moving objects. Ballard et al. [17] reports on an interesting experiment where such indices help to serialize a perception-motor coordination task and thus to lighten the working memory load. Specifically, by placing indices to objects in the external space, we seem to replace the task of remembering their visual properties with the task of identifying their visual properties through frequent returns to them. Spivey et al. [19] gives a survey of other possible roles of deictic indices.

On the footing of these studies of deictic indices, we propose the following general hypothesis. Whenever one wants to add a piece of information into a diagram, one could actually place an object with an appropriate visual feature in an appropriate location of the diagram, *or instead*, one could place a deictic index to the appropriate location, tagging it with the appropriate visual feature stored in the internal memory.

Then in the first case, one could observe what spatial relationship or pattern is produced as the result of the addition of the new object, and use this observation to draw a conclusion of the inference. This is the usual process of constraint exploitation, based by an actual drawing. In the second case, one could also observe the resulting spatial relationship or pattern, but this time the one produced by the addition of the new *index*. For example, the newly indexed location may have a certain spatial relationship with a particular object in the diagram; it may produce a particular shape (say, a triangular region) in combination with multiple objects in the diagram, and so on. This observation could be used to draw a conclusion for the inferential task, and when this happens, it is a case of exploiting a spatial constraint on graphical structure of the diagram. The only difference from the usual process is that the observed spatial relationship or pattern is partly formed by an indexed location, whose visual feature is stored not *in situ* but in the internal memory.

2.2 Predictions of Eye-Movements

To test this general hypothesis, we designed a number of of three-term transitive inference problems, to be solved with simple position diagrams. Each problem was presented in three steps. The following is a sample procedure:

First Step. An audio-recording, "A is lighter than B," is played, while the diagram in Figure 3-a is presented on a computer display at the same time.

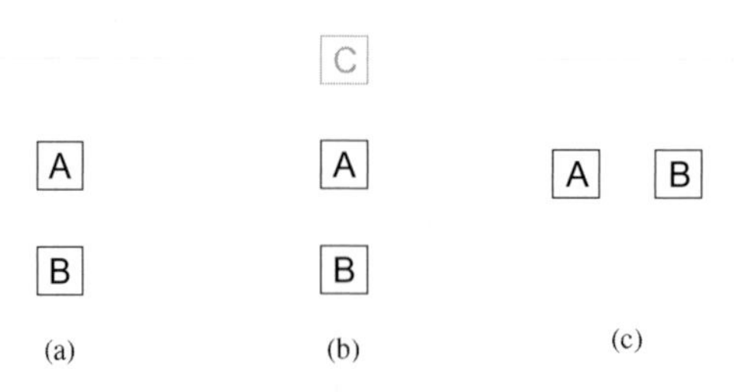

Fig. 3. (a) A sample diagram used in the experiment. (b) A rough location of hypothetical indexing in the sample problem. (c) A horizontal diagram used in the experiment.

Second Step. Another audio-recording, "C is lighter than A," is played, while the diagram on the display is unchanged.

Third Step. An audio recording, "Is C lighter than B?" is played, while the diagram is unchanged.

From the premise that A is lighter than B (first step) and the premise that C is lighter than A (second step), it follows that C is lighter than B. Thus, the correct answer to the question in the third is "yes" in this particular problem.

In this procedure, the participant is instructed to interpret the diagram in the first step as expressing the same information expressed by the audio recording played at the same time. Thus, in this problem, the participant should interpret Figure 3-a to mean that A is lighter than B, assuming the semantic rule that a symbol's being above another symbol means that the referent of the first symbol is lighter than the referent of the second symbol. (For brevity, we will employ the symbol [*above* ⇒ *lighter*] to indicate this semantic rule; similar symbols will be used for other semantic rules.)

Throughout the problem, the diagram only expresses the first premise presented in the first step. The second premise supplied in the second step remains unexpressed in the diagram. However, if the participant exploits spatial constraints on position diagrams in solving this problem, he or she would "draw" a symbol in the blank area above the square symbol [A]. According to our general hypothesis, this amounts to placing a deictic index. Thus, the area above the square symbol [A] is the *hypothetical indexing position* for this problem. The greyed symbol [C] in Figure 3-b indicates this fact. Since the operation of placing an index requires a focal attention, we hypothesize that eyes would move to this area at this point, namely, in the second step of this problem.

In the similar vein, we predict that eyes would move to this same area in the third step. For, if the participant uses the constraint-exploitation strategy, he or she would need to check the spatial relation between this indexed position and the position of the symbol [B], and a relation-check requires focal attentions to positions to be compared.

2.3 Semantic Variation

However, just verifying eye-movements in this particular problem is not sufficient to establish the occurrence of the constraint-exploitation practice. For eyes may be move to this area for a reason *unrelated to* the presence of the external diagram. Perhaps the participant simply had the habit to look at the upper area of the display to process the information verbally given in the second and the third step. Or perhaps, the participant was "looking" at his or her own, independently constructed internal image, and upper-directed eye-movements were only epi-phenomenal on this internal operation.

In order to filter out these possibilities, we systematically varied the semantic rules of the diagrams shown to the participants. For example, in one type of problems, we presented the diagram in Figure 3-a in the first step while giving the same premises and question verbally. This would lead the participants to assume a semantic rule opposite to the one assumed in the previous sample problem, namely, the semantic rule [*below* ⇒ *lighter*]. It would also shift the hypothetical indexing position to the area below the square symbol [B], and this is where we predict eyes would move to in this type of problems. In another type of problems, we presented a "horizontal" diagram such as Figure 3-c while verbally providing the same set of premises and a question. This would lead the participants to assume the semantic rule [to_the_left ⇒ *lighter*] and the hypothetical indexing position would be the area to the left of the symbol [A]. In yet another type of problems, the semantic rule was [*to_the_right* ⇒ *lighter*] and the indexing point was the area to the right of the symbol [B].

The idea is that by adjusting the diagram presented in the first step and the premises in the first and the second steps, we can systematically vary the semantic rules of the diagram given in the problem and hence the indexing position for the problem. If we could observe the participants "track" this variance, moving their eyes to the particular indexing points differing from problems to problems, we can take it to be a good evidence that the participant does operate on the externally given diagram, following the particular semantic rule associated with it. Such a tracking would be unlikely if the participant's eye-movements result from his or her personal habit or from the operations on internal images constructed independently of the given diagram.

In creating problems used in the experiment, we adopted a total of 6 pairs of a relational predicate and its inverse: *heavier* and *lighter*, *clearner* and *dirtier*, $infront$ and *behind*, to_the_east and to_the_west, *bigger* and *smaller*, and *brighter* and *darker*.[1] A half of the problems came with vertical position diagrams such as Figure 3-a, while the other half came with horizontal position diagrams such as Figure 3-c.

We created a total of 32 types of problems for each pair of transitive relations. The 32 types were derived from 2 variations in diagram direction × 4 variations in semantic rule × 2 variations in predicate-pairing × 2 variations in logical consequence. A half of the 16 vertical types of problems had the hypothetical indexing position above the upper square symbol in the diagram (see Figure 3-b), so they will be called *higher-predictive* problems. The other half had the HIP below the lower square symbol, so they are *lower-predictive* problems. The horizontal types of problems are divided into *left-predictive* and *right-predictive* in an analogous fashion.

[1] These are the same pairs of relational predicates adopted in the experiment by Shimojima and Fukaya [11], who checked that these relations fairly diverse in visualizability and spatializability assessments [20].

2.4 Semantic Preference Test

After the main session of the experiment (measuring the participants' eye-movements during transitive inference tasks), we administered an additional test on each participant, in order to detect the participant's *semantic preference*. For cultural, social, and personal reasons, people may have developed certain preference over the way the given transitive relation is represented by a spatial relation. For example, most readers, accustomed to a standard convention of geographical mapping, would prefer *to_the_east* to be represented by the *to_the_right* relation rather than by the *to_the_left* relation. On the other hand, the same people may prefer *in_front* to be represented by *to_the_left* rather than by *to_the_right*. With the analogy to gravity, some people may prefer *heavier* to be represented by the *below* relation rather than by the *above* relation.

As explained earlier, the problems in the main session of the experiment involve a variety of semantic rules for each transitive relation. For example, *to_the_east* can be alternatively represented by the vertical spatial relation *below* and its inverse *above*, and by the horizontal spatial relation *to_the_right* and its inverse *to_the_left*. It is likely that these different semantic rules are preferred in different degrees by an individual participant, making the diagrams used in the relevant problems familiar or unfamiliar to the participant.

One may then suspect that this difference in familiarity with the presented diagrams may affect the participant's inferential process, particularly the occurrence of a non-physical drawing and a constraint exploitation. One the one hand, people may conduct non-physical drawing *only when* the semantic rule associated with the given diagram matches onefs semantic preference. One the other hand, people may be rather adaptive to unfamiliar diagrams, engaged in the constraint-exploitation process even when it requires the use of externally given, non-preferred semantic rules. In either way, detecting the participants' semantic preference is important in evaluating *how persistent* the constraint exploitation process is, and hence *how central* it can be to human inference processes with diagrams.

Our semantic preference test involves a number of binary-choice questions presented in the following way:

Step 1. An audio-recording, "A is cleaner than B," is played, while the two alternative diagrams such as Figure 4 are presented on a computer display at the same time.
Step 2. The participant is to choose which diagram is intuitively more appropriate for expressing the information presented in the audio-recording.

If the participant answers that the left diagram is more appropriate, it counts as a partial evidencethat he or she prefers the *cleaner* relation to be represented by the *above*

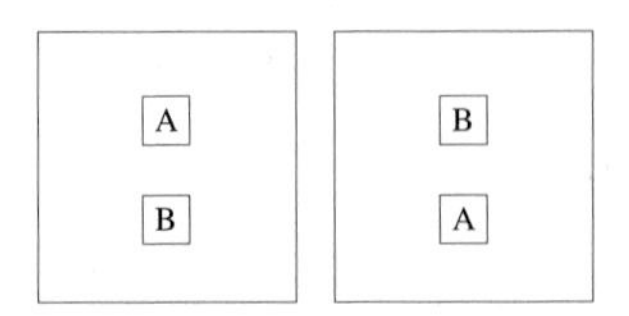

Fig. 4. A sample stimulus used in the semantic preference test

relation rather than by the *below* relation. Yet, one answer is not sufficient for determining the participant's preference about the way the *cleaner* relation is represented, so we created 4 questions about the way this relation is represented in vertical diagrams, with modifications on the positions of the two diagrams and the orders of A and B mentioned in the audio-recordings. Since 12 transitive relations was used in the main experiment, a total of 4×12 questions were created for the test on vertical diagrams. These questions were asked to a participant in a random order. After these questions on vertical diagrams, the participant proceeded to the questions on horizontal diagrams, which were designed and administered in a similar manner.

If, for example, the participant was very consistent in his or her answers to the four questions about the *cleaner* relation, so that all four answers indicate the preference for the use of the *above* relation rather than the *below* relation, the preference score of the semantic rule [*above* $\Rightarrow$ *cleaner*] is assessed at 4 for this participant, whereas the opposite semantic rule [*below* $\Rightarrow$ *cleaner*] is assessed at 0. Generally, the preference score of a semantic rule for a participant is the number of times the participant issued the answer indicating the preference for the rule in the four relevant questions. This procedure let us score a total of 48 semantic rules (12 transitive relations $\times$ 4 spatial relations) on the basis of each participant's semantic preference.

Based on the scores thus obtained, the problems used in the main session of the experiment were classified into three categories: (1) a problem was classified as "Match" if the semantic rule of the diagram used in the problem had the preference socre 4, (2) "OK" if the semantic rule had a preference score in the range from 1 to 3, and (3) "Mismatch" if the semantic rule had the score 0. These categories of semantic preference were then used to evaluate how much a participant's semantic preference affects the process of constraint exploitation on the relevant diagrams.

2.5 Subjects and Procedure

Our experiment consists of the main session engaged in transitive inference problems (described in section 2.2 and section 2.3) and the follow-up test of the participant's semantic preference (described in section 2.4). The presentation of the problem materials, including instructions, was controlled by SuperLab 4.0 (Windows version). A total of 31 undergraduate students (17 females and 14 males) participated for monetary reward.

In the main session of the experiment, the participant first solved 6 exercise problems, and then proceeded to solve a total of 96 transitive inference problems (3 problems involving different transitive relations $\times$ 32 types). These problems were presented in a random order, with a break in every 24 problems. The participant used a response pad (Cedrus RB-530) to issue an answer, being instructed to answer as quickly and accurately as possible.

Throughout the main session, eye-movements of the participant were recorded by NAC eye-tracker EMR-AT. Although the eye-trackin system is fairly tolerant of head movements, the participant's chin and forehead were placed to a fixed support for steady recording. The diagrams were presented on a 17-inch LCD display, at the resolution of 1024×768 pixels. The viewing angles were approximately 34 degrees (horizontal) and 28 degrees (vertical). The eye-tracker samples eye-movement at 60 Hz, recording their momentary positions as coordinate values on the display, using the pixel unit.

The follow-up test of the participant semantic preference was administered immediately after the main session. After 3 practice questions, the participant answered 48 questions (4 questions × 12 transitive relations) involving vertical diagrams, and then 48 more questions involving horizontal diagrams. The maximum of 7 seconds were allowed for answering each question.

2.6 Analysis of Eye-Tracking Data

In analyzing the eye-tracking data, we focused on the "reaches" of eye-movements in crucial steps of the problem. That is, for the problems involving vertical diagrams, we analyzed the highest and the lowest positions on the display to which eyes moved in the second and the third steps, and for the problems involving horizontal diagrams, we analyzed their left-most and right-most positions.

Remember that the hypothetical indexing position on a vertical diagram is either the area above the top letter symbol of the diagram or the area below the bottom letter symbol (section 2.3). These positions are determined by the types of problems (higher-predictive or lower-predictive types), so comparing the highest and the lowest positions of eyes in different types of problems is sufficient to test our experimental hypothesis.

The same goes for the problems involving horizontal diagrams. The hypothetical indexing position on a horizontal diagram are either the area to the left of the left-hand square symbol or the area to the right of the right-hand square symbol (section 2.3). Thus, just comparing the left-most positions and the right-most positions of eyes in different types of problems (left-predictive or right-predictive problems) is sufficient to test the experimental hypothesis.

Positions of eyes analyzed in this study are momentary positions of eye-movements sampled at a fixed rate, and not necessarily within areas of eye-fixations. However, assuming that any fixation lasts longer than the sampling interval (i.e., 1/60 second), the highest, lowest, left-most, or right-most position sampled in a given period should be within the highest, lowest, left-most, or right-most area of fixation in that period.

To wit, suppose, for example, p be the highest position sampled in this period. Let A be the highest fixation area in this period. Since a fixation is assumed to last longer than our sampling interval, at least one momentary position within A is sampled. Let q be the highest position of all sampled positions in this area. Since a saccade is ballistic and to be directed to the next fixation position, no other momentary position sampled in this period is higher than q. So, q is the highest position sampled in this period, namely, p. Thus, p is within the highest fixation area in this period. The same observation applies to the cases of the lowest, the left-most, and the right-most sampled position.

3 Results

The data of 6 participants were excluded from the analysis, due to serious misunderstanding of instructions (1 participant), errors in problem randomization (2 participants), failures in calibration (2 participants), and excessive blinking (1 participant). Eye-movements were analyzed only for those trials in which the participant returned a correct answer, which accounts for 87.1% (2090 trials) of the total trials in our data.

Table 1. Vertical coordinate values of the highest and the lowest positions of eye-movements in the second steps of the problems with vertical diagramsiunit: pixels, N = 24

	Problem catagory	Match	s.d.	OK	s.d.	Mismatch	s.d.
highest position	higher-predictive	554.6	81.3	537.9	70.4	533.9	64.6
	lower-predictive	492.1	52.9	501.3	52.7	509.0	78.2
lowest position	higher-predictive	245.3	79.9	226.9	86.3	251.0	86.2
	lower-predictive	181.9	85.3	167.8	86.7	167.8	82.2

Table 2. Horizontal coordinate values of the left-most and the right-most positions of eye-movements in the second steps of the problems with horizontal diagramsiunit: pixels, N = 21

	Problem catagory	Match	s.d.	OK	s.d.	Mismatch	s.d.
left-most position	left-predictive	351.6	115.5	330.3	100.4	314.0	120.8
	right-predictive	389.9	104.2	410.0	80.5	400.8	108.8
right-most position	left-predictive	657.0	103.1	650.6	69.6	627.5	65.6
	right-predictive	714.5	93.2	722.0	98.4	726.7	107.2

3.1 Eye-Movements in the Second Steps

As the sample problem in section 2.2 illustrates, the second step of our transitive inference problem is when the participant is predicted to place an index in a particular position on the diagram.

Table 1 shows the highest and the lowest positions to which eyes moved on vertical diagrams, grouped by problem categories and semantic preference categories. The numbers are vertical coordinate values in pixels. The origin of the coordinates is set to the bottom-left corner of the display, so the greater the number of the highest positions, the higher on the display eyes reached, and the smaller the number of the lowest positions, the lower eyes reached.

We used a 2 × 3 repeated-measure ANOVA to see if eye-movements really depend on the indexing positions prescribed by problems categories (main effects), and if the participants' semantic preference affects the dependency (interactions). We found a highly significant main effect of problem categories on highest positions ($F(1,24) = 29.00, p < .001$), but the interaction with semantic preference was only marginal ($F(2,48) = 2.75, p < .1$). Problem categories had a highly significant main effect on lowest positions too ($F(1,24) = 44.62, p < .001$), but the interaction with semantic preference was not significant ($F(2,48) = 1.25, n.s.$).

The results for the cases involving horizontal diagrams were similar. Table 2 shows the left-most and the right-most positions to which eyes moved on horizontal diagrams, grouped by problem categories and semantic preference categories. The numbers are horizontal coordinate values in pixels. The smaller the number of the left-most positions, the more left on the display eyes reached, and the greater the number of the right-most positions, the more right eyes reached.

An 2 × 3 repeated-measure ANOVA indicated a highly significant main effect of problem categories on left-most positions ($F(1,21) = 36.93, p < .001$), and a

Table 3. Vertical coordinate values of the highest and the lowest positions of eye-movements in the third steps of the problems with vertical diagramsiunit: pixels, N = 21

	Problem catagory	Match	s.d.	OK	s.d.	Mismatch	s.d.
highest position	higher-predictive	550.5	91.2	534.7	64.9	539.8	85.9
	lower-predictive	490.9	51.6	497.4	55.6	508.3	74.1
lowest position	higher-predictive	242.1	78.9	229.2	88.3	206.2	87.8
	lower-predictive	179.3	70.2	193.8	88.5	180.0	97.8

Table 4. Horizontal coordinate values of the left-most and the right-most positions of eye-movements in the third steps of the problems with horizontal diagramsiunit: pixels, N = 21

	Problem catagory	Match	s.d.	OK	s.d.	Mismatch	s.d.
left-most position	left-predictive	306.6	129.3	308.0	115.8	327.1	113.1
	right-predictive	392.5	100.2	435.9	98.9	388.6	156.3
right-most position	left-predictive	676.4	102.6	670.0	111.8	670.3	106.5
	right-predictive	712.6	94.8	737.3	106.7	727.8	98.4

significant interaction with semantic preference ($F(2, 42) = 3.42, p < .05$). Problem categories had a significant main effect on the right-most positions too ($F(1, 21) = 43.28, p < .001$), but the interaction with semantic preference was not significant ($F(2, 42) = 0.33, n.s.$).

3.2 Eye-Movements in the Third Steps

As the sample problem in section 2.2 illustrates, the third step of our transitive inference problem is when the participant is predicted to return to the index placed in the second step in order to to check the spatial relation between this index and another index created in the first step. We want to see if eyes really return to the indexing positions in the second step, and if the participants' semantic preference affects this behavior.

Table 3 shows the highest and the lowest positions of eye-movements over vertical diagrams, grouped by problem categories and semantic preference categories.

On the 2×3 repeated-measure ANOVA, we found a highly significant main effect of problem categories on highest positions ($F(1, 24) = 12.84, p < .01$), but no significant interaction with semantic preference ($F(2, 48) = 1.69, n.s.$). Problem categories had a highly significant main effect on lowest positions too ($F(1, 24) = 23.59, p < .001$), but the interaction with semantic preference was not significant ($F(2, 48) = 2.19, n.s.$).

The results for the cases of horizontal diagrams are shown in Table 4. On the 2×3 repeated-measure ANOVA, the main effect of problems categories was highly significant on the left-most positions ($F(1, 21) = 38.79, p < .001$) and significant on the right-most positions ($F(1, 21) = 13.88, p < .01$). The interactions with semantic preference were marginally significant for the left-most positions ($F(2, 42) = 2.88, p < .1$) but non-significant for the right-most positions ($F(2, 42) = 0.83, n.s.$).

4 Discussion

Thus, overall, problem categories had a very strong effect on the outward limit of eye-movements everywhere. Moreover, the way eye-movements were affected by problem-categories was exactly as we predicted.

Figure 5 plots the data in Tables 1 and 3 to show where the highest and the lowest positions of eye-movements are located relative to the letter symbols in vertical diagrams. As the left figure shows, the reach of eye-movements in the second steps tended to be higher in higher-predictive problems than in lower-predictive problems, and tended to be lower in lower-predictive problems than in higher-predictive problems. This indicates that eyes tended to move toward the hypothetical indexing positions determined by the types of problems, suggesting that the participants actually moved focal attention to those areas to place deictic indices.

The average reach of eye-movements in the third steps (the right figure) had the same tendency, going toward the hypothetical indexing positions again. This suggests that, in the third steps, the participants moved focal attention to the locations indexed in the second steps, in order to check the spatial relation between these locations and other locations in the diagrams.

Note that these tendencies did not so much depend on the participants' semantic preference. Both in the second and the third steps, the arrows in higher-predictive problems *equally* go higher than those in lower-predictive problems, and the arrows in lower-predictive problems *equally* go lower than those in higher-predictive problems. This reflects our results of interaction analysis, where the main effects of problems categories had no or only marginal interactions with semantic preference.

Thus, the participants were fairly adaptive to vertical diagrams drawn with unfamiliar semantic rules; their eyes seem to have tracked the hypothetical indexing positions, complying with externally given semantic rules, whether or not they were familiar ones to them. This suggests that the exploitation of spatial constraints is a robust process, applicable to externally given diagrams with unfamiliar semantic rules.

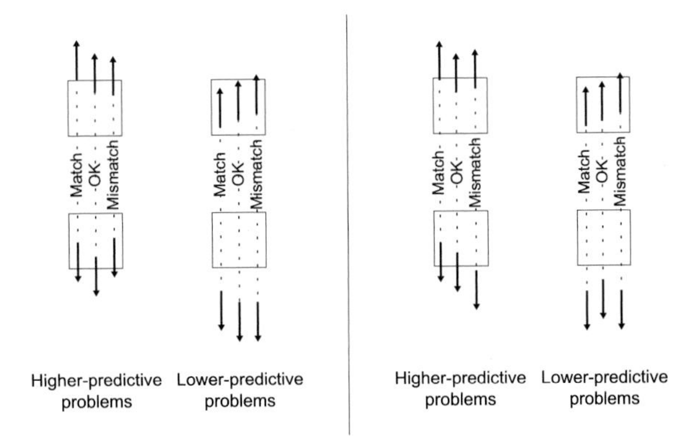

Fig. 5. The higher and the lower reaches of eye-movements on vertical diagrams. Left: average reach in the second steps. Right: average reach in the third steps.

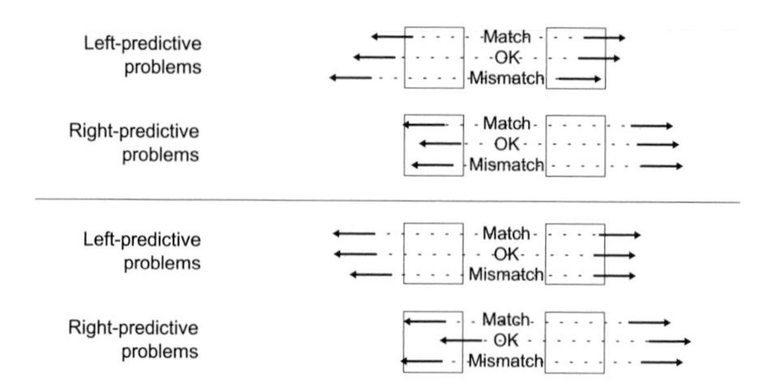

Fig. 6. The leftward and the rightward reaches of eye-movements on horizontal diagrams. Top: average reach in the second steps. Bottom: average reach in the third steps.

Figure 6 shows similar tendencies for the case of horizontal diagrams. In both second steps (the top figure) and third steps (the bottom figure), eyes tracked the hypothetical indexing positions prescribed by the problem categories, and this tracking did not depend on the difference in semantic preference categories.

The only notable difference is that a significant interaction with semantic preference was found for the left-most positions of eye-movements in the second steps. On a closer look at the data, this result reflects the fact that problem categories had a *greater* effect on the leftward reach of eye-movements when the semantic rules did *not* exactly match the participants' preference. Although this effect is statistically significant ($F(2,42) = 3.42, p < .05$), we found no comparable interactional effect in any other combination of diagram types and problem steps. Thus, the effect is rather isolated one. Also, the effect itself is rather weak, in the face of stronger and more consistent main effects of problem categories on eye-movements.[2]

Related Work

In fact, a number of studies in diagrammatic reasoning have suggested the existence of non-physical operations on diagrams of the kind identified in our experiment. Back in

[2] This, however, does not mean that the difference in relational predicates used in the problems makes no difference. In particular, Knauff and Johnson-Laird [20] reported that performance in transitive inference problems was impeded when more visualizable (but less spatializable) relational predicates were used. Our data corroborate their finding. Response time to the premises was longest when predicates in this category were used (the mean value of 9.47 second for "cleaner" and "dirtier"). It is shorter when less visualizable and less spatializable predicates were used (8.90 second for "brighter" and "darker") and shortest when more visualizable and more spatializable predicates were used (8.50 second for "front of" and "back of"). This trend was significant (Page's L = 326 in 3 treatments on 26 subjects, p ¡ .05). Response time to the question had a similar trend (mean value of 4.04 second for the first category, 3.82 second for the seond category, and 3.73 second for the third category). The trend did not reach significance, however. The categorization of the relational predicates assumed in this discussion draws upon the visualizability and spatializability test partly reported in Shimojima and Fukaya [11].

1971, Sloman [8] already cited our ability to "imagine or envisage rotations, stretches and translations of parts" of a diagram as an explanation of the efficacy of inference with diagrams. Beside their well-known analysis of information-indexing functions of diagrams, Larkin and Simon [4] also discussed "simple, direct perceptual operations" on an economy graph that helps the user "read off" the effect of an economic policy from the graph. In characterizing "reasoning about a picture as the referent" in contrast to "reasoning about the picture's referent," Schwartz [21] assumed a mental operation that makes the line representing the upper leg of a hinge swing down to the line representing the lower leg of a hinge. Narayanan et al. [9] postulated the "visualization" process on a schematic diagram of a mechanical device that move, rotate, copy, and delete elements of the diagram. These studies therefore seem to point to a rich field of diagrammatic inferences to which the account from constraint exploitations can be potentially applied.

Recently, several researchers started to investigate non-physical drawing on external diagrams in experimental settings. Trafton and Trickett [10] also used an eye-tracking method to investigate "spatial transformations" applied on an "external image" such as a weather map. Shimojima and Fukaya [11] is the direct predecessor of the present study, which investigated "hypothetical drawing" on position diagrams used in transitive inference tasks. Yoon and Narayanan [12] used an eye-tracking data to investigate imaginative drawing on a near-blank screen in the context of problem-solving.In an eye-tracking study of kinematic problem-solving, Kozhevnikov et al. [13] identified the ability of "spatial transformation" on external diagrams as characteristic to students with a higher spatial-visualization ability.

The present study adds a new evidence for the existence of such a process in a more focused and systematic manner. Specifically, by varying the semantic rules of the stimulus diagrams, we could obtain a strong evidence that the observed pattern of eye-movements were really caused by the particular diagrams presented externally. Furthermore, the follow-up test of the participants' semantic preference also let us verify that the process of non-physical drawing is fairly tolerant to diagrams with unfamiliar semantic rules. This result seems to point to the general applicability of the process to wide varieties of diagrams.

5 Conclusion

In our experiment, the participants' eye-movements conformed to a particular pattern, as predicted from the hypothesis that they used deictic indices on the external diagrams to take advantage of spatial constraints on their graphical structures. This indicates that, for the kind of position diagrams used in our experiment, one can make inferences exploiting their spatial constraints, by just looking at them, without physically drawing on them. Moreover, this process is tolerant to diagrams with unfamiliar semantic rules.

These findings in turn indicate that the explanation of the inferential advantage of diagrams in terms of constraint exploitations can be extended to a larger class of diagrammatic inferences than expected from its face value. At least potentially, inferences involving no physical manipulations of the given diagrams, and inferences involving diagrams with unfamiliar semantic rules are shown to be in the scope of explanation.

References

1. Barwise, J., Etchemendy, J.: Visual information and valid reasoning. In: Allwein, G., Barwise, J. (eds.) Logical Reasoning with Diagrams, pp. 3–25. Oxford University Press, Oxford (1990)
2. Shimojima, A.: On the Efficacy of Representation. PhD thesis, Department of Philosophy, Indiana University (1996b)
3. Stenning, K., Lemon, O.: Aligning logical and psychological perspectives on diagrammatic reasoning. Artificial Intelligence Review 15(1-2), 29–62 (2001)
4. Larkin, J.H., Simon, H.A.: Why a diagram is (sometimes) worth ten thousand words. In: Glasgow, J., Narayanan, N.H., Chanrasekaran, B. (eds.) Diagrammatic Reasoning: Cognitive and Computational Perspectives, pp. 69–109. AAAI Press, Menlo Park (1987)
5. Lindsay, R.K.: Images and inference. In: Glasgow, J.I., Narayanan, N.H., Chandrasekaran, B. (eds.) Diagrammatic Reasoning: Cognitive and Computational Perspectives, pp. 111–135. The MIT Press and the AAAI Press, Cambridge, Menlo Park (1988)
6. Kulpa, Z.: From picture processing to interval diagrams. Technical report, Instytut Podstawowych Problemów Techniki Polskiej Akademii Nauk, Warszawa (2003)
7. Chandrasekaran, B., Kurup, U., Banerjee, B., Josephson, J.R., Winkler, R.: An architecture for problem solving in diagrams. In: Diagrammatic Representation and Inference: Third International Conference, Diagrams 2004, pp. 151–165 (2004)
8. Sloman, A.: Interactions between philosophy and ai: the role of intuition and non-logical reasoning in intelligence. Artificial Intelligence 2, 209–225 (1971)
9. Narayanan, N.H., Suwa, M., Motoda, H.: Hypothesizing behaviors from device diagrams. In: Glasgow, J., Narayanan, N.H., Chanrasekaran, B. (eds.) Diagrammatic Reasoning: Cognitive and Computational Perspectives, pp. 501–534. AAAI Press, Menlo Park (1995)
10. Trafton, J.G., Trickett, S.B.: A new model of graph and visualization usage. In: Proceedings of the Twenty-Third Annual Conference of the Cognitive Science Society, pp. 1048–1053 (2001)
11. Shimojima, A., Fukaya, T.: Do we really reason about the picture as the referent? In: Proceedings of the Twenty-Fifth Annual Conference of the Cognitive Science Society, pp. 1076–1081 (2003)
12. Yoon, D., Narayanan, H.: Mental imagery in problem solving: An eye tracking study. In: Proceedings of the 2004 symposium on Eye tracking research & applications, pp. 77–84 (2004)
13. Kozhevnikov, M., Motes, M.A., Hegarty, M.: Spatial visualization in physics problem solving. Cognitive Science 31(4), 549–579 (2007)
14. Ullman, S.: Visual routines. Cognition 18, 97–159 (1984)
15. Chapman, D.: Penguins can make cake. AI Magazine 10(4), 45–50 (1989)
16. Pylyshyn, Z.W.: The role of location indexes in spatial perception: A sketch of the finst spatial-index model. Cognition 32, 65–97 (1989)
17. Ballard, D.H., Hayhoe, M.M., Pook, P.K., Rao, R.P.N.: Deictic codes for the embodiment of cognition. Behavioral and Brain Sciences 20(4), 723–767 (2001)
18. Pylyshyn, Z.W.: Seeing and Visualizing: It's Not What You Think. The MIT Press, Cambridge (2003)
19. Spivey, M.J., Richardson, D.C., Fitneva, S.A.: Thinking outside the brain: Spatial indices to visual and linguistic information. In: Henderson, J.M., Ferreira, F. (eds.) The Interface of Language, Vision, and Action: Eye Movements and the Visual World, pp. 161–189. Psychology Press, New York (2004)
20. Knauff, M., Johnson-Laird, P.N.: Visual imagery can impede reasoning. Memory and Cognition 30(3), 363–371 (2002)
21. Schwartz, D.L.: Reasoning about the referent of a picture versus reasoning about the picture as the referent: an effect of visual realism. Memory and Cognition 23(6), 709–722 (1995)

What Diagrams Reveal about Representations in Linear Reasoning, and How They Help

Krista E. DeLeeuw and Mary Hegarty

Department of Psychology,
University of California
Santa Barbara, CA 93101, USA
{deleeuw,hegarty}@psych.ucsb.edu

Abstract. Previous research has shown that linear reasoning about visual relations is slower than reasoning about spatial or abstract relations. It has been proposed that this "visual impedance" effect occurs because visual information interferes with creating a spatial mental model of the problem. Here we examine whether the construction of a spatial mental model depends on the content (visual, abstract, and spatial) of the problems and on individual differences. Half of the participants made notes during reasoning, and all were asked about the strategies they used. Accuracy on visual relations was worse than on spatial or abstract relations. When reasoners were allowed to make notes, their external representations were typically spatial (i.e., abstract diagrams), and accuracy was higher and depended less on ability measures than for those who did not make notes; yet the visual impedance effect remained. These data imply reasoners typically externalize their representations of linear reasoning problems as diagrams, and with these diagrams, problem solving is less effortful.

Keywords: reasoning, diagrams, cognitive ability, individual differences, mental representations.

1 Introduction

Identifying the mental representations and strategies that people use to solve reasoning problems is a fundamental issue in cognitive psychology. Some researchers have claimed that reasoning problems are solved by constructing spatial mental models of the contents of the problems [1, 2], whereas others have claimed that the fundamental representations underlying linear reasoning are propositional [3]. In recent years there has been new evidence in support of the spatial mental models view. Dual-task studies examining the role of working memory in reasoning have shown that linear reasoning is more impaired by secondary tasks that depend on visual-spatial working memory suggesting that visual-spatial representations underlie this form of reasoning. In contrast, these studies suggest that verbal working memory is only minimally involved in linear reasoning [4, 5, 6]. Converging evidence comes from functional magnetic resonance imaging (fMRI) studies. For example, Knauff et al. [2] showed that brain areas involved in spatial working memory (specifically, the occipito-parietal pathway, including the precuneus), as well as the visual association cortex,

G. Stapleton, J. Howse, and J. Lee (Eds.): Diagrams 2008, LNAI 5223, pp. 89–102, 2008.

were activated during both conditional and relational reasoning, but areas related to verbal working memory were not.

Furthermore, when reasoning problems contain highly visual content rather than abstract or spatial content, it appears that reasoners solve the problems less efficiently. Linear reasoning problems containing relations between visual properties of objects (that are easy to imagine visually) take more time than those containing abstract or spatial relations [1]. For example, the linear reasoning problem which is about a visual property (cleanliness):

A is cleaner than B
C is dirtier than B
Does it follow that A is cleaner than C?

takes longer to solve, and is sometimes solved less accurately than the following problem, which is about a more abstract property:

A is smarter than B
C is dumber than B
Does it follow that A is smarter than C?

even though the two problems are isomorphic and differ only in the relation to be compared. An example of a spatial problem is as follows:

A is in front of B
C is behind B
Does it follow that A is in front of C?

Note that this problem is classified as spatial, not visual, because it refers to the locations of objects rather than their appearance. Knauff and colleauges refer to these as visuospatial problems, but we prefer the term spatial (although vision may be the dominant modality by which we sense spatial relations, it is not the only one, as one can encode the location of objects from other senses; touch, hearing etc).

Knauff and colleagues have argued that problems with highly visual content are solved less efficiently because of an interference of visual information with the process of creating a spatial mental model of the problem. This is known as the *visual impedance effect*. From a process perspective, this may occur because it is difficult for reasoners to ignore the easily imageable information and so they construct detailed visual representations of the problems, which can be solved more efficiently by creating a more schematic representation of the relations expressed in the problem. The additional visual details are assumed to take time to create, and maintaining these representations uses cognitive resources that could otherwise be used for representing the problem itself. Interestingly, Knauff and May [7] found that congenitally blind individuals do not show the visual impedance effect. They argue that because blind people have never experienced visual detail, they do not represent visual properties in transitive inferences like this, and therefore highly visual problems are not any more difficult than abstract or spatial relations for this population.

Knauff and colleagues assume the same type of representation and reasoning strategy hold for all individuals and problems. That is, all problems are eventually solved by creating a spatial mental model. However, in other research, Bacon, Handley, and Newstead [8] have found individual differences in the representations and processes that people use in reasoning problems. In their research, both written and verbal protocols indicated that people vary with respect to whether they use verbal or spatial strategies on reasoning problems, and these results were consistent with people's retrospective reports. Similarly, Egan and Grimes-Farrow [9] found that participants usually fell into one of two groups regarding the strategies they used during reasoning. People were either "abstract directional thinkers", who constructed mental orderings of the terms in transitive problems, or "concrete properties thinkers", who tended to represent the terms (geometric shapes) as mental images and then attribute physical properties to them. Interestingly, Egan and Grimes-Farrow found that differences in strategy also led to differences in accuracy: abstract directional thinkers were more accurate than concrete properties thinkers.

In related research, on problem solving in mathematics and physics, Kozhevnikov, Hegarty and Mayer [10] have shown that people differ in the way they prefer to process information. While verbalizers tend to use verbal or logical strategies, visualizers use imagery to process the same information. Furthermore, visualizers fall into two categories: those with low spatial ability tend to use iconic imagery and those with high spatial ability tend to use schematic spatial representations. These groups are similar to Egan and Grimes-Farrow's categories of "concrete properties thinkers" and "abstract directional thinkers" suggesting that, preferences such as these may affect how people reason about linear relationships, and may be related to spatial ability.

Other studies have shown that the extent to which people rely on spatial representations depends not just on the individual, but also on the content of the problem. For example, Duyck, Vandierendonck, and De Vooght [11] used a dual-task paradigm to show that conditional reasoning problems containing spatial relations such as "right of" and "left of" depended on visuospatial working memory more than problems containing non-spatial relations such as "likes" and "laughs at". Specifically, a spatial matrix-tapping secondary task (assumed to draw on spatial working memory) interfered more with solving reasoning problems with spatial content than problems with non-spatial content.

In work on the visual impedance effect, Knauff and Johnson-Laird do not address the possibility of individual differences in this effect, or the possibility that people might use different representations depending on the content about which they are reasoning. Furthermore, they base their conclusions on patterns of response time and accuracy across different problems and do not include process measures or ask reasoners to report the type of solution strategies that they used. Here, we consider the possibility that different people may use different strategies depending on their cognitive abilities (spatial, verbal, and reasoning abilities) and whether they are reasoning about visual, abstract or visual-spatial relations.

To provide information on people's strategies, in this study, we asked all participants to retrospectively report on the strategies they used for linear reasoning problems with different content. These data were used to classify reasoners based on their strategies. We also allowed half of the participants to make notes as they solved the problems and analyzed these notes for evidence of visual-spatial vs. propositional representations. We hypothesized that the representations and strategies revealed by the retrospective reports and notes made would depend on the problem content and that they would differ between individuals. We examined whether these differences depended on cognitive abilities. We also predicted that performance would be better when reasoners were able to make notes, since this reduces the load on working memory, and that this facilitation would reduce the visual impedance effect.

2 Method

2.1 Participants

Our participants were 60 undergraduate students at the University of California, Santa Barbara, who received course credit for their participation in the experiment. Participants were randomly assigned to either the Notes (n=30) or No Notes (n=30) condition.

2.2 Design and Materials

Each participant evaluated 24 reasoning problems that were identical to those used by Knauff and Johnson-Laird [1], except that indeterminate problems were changed to be determinate[1]. Of the 24 problems, eight contained the abstract relations (e.g. "the cat is smarter than the dog"), eight contained spatial relations (e.g., "the cat is above the dog"), and eight contained visual relations (e.g. "the cat is dirtier than the dog"). Half of the problems were valid and half were invalid; half of them were three-term series and half were four-term series. An example of a three-term inference with a valid conclusion and a visual relation is:

The dog is cleaner than the cat.
The ape is dirtier than the cat.
[Does it follow:]
The dog is cleaner than the ape?

An example of a four-term inference with an invalid conclusion and an abstract relation is:

The ape is smarter than the cat.
The cat is smarter than the dog.
The bird is dumber than the dog.
[Does it follow:]
The bird is smarter than the ape?

Finally, half of the participants were allowed to make notes while they read the premises and solved the problems; the other half read the premises and conclusions on the computer screen and had to remember them while solving the problems. The resulting design was a 3 (relation type) x 2 (validity) x 2 (number of terms) x 2 (notes) mixed model design, with the first three variables as within-subjects variables and notes as a between-subjects variable.

The order of the problems was pseudo-random. We created three different random orders of problems, to which the participants were randomly assigned. For participants in the Notes condition, paper booklets were provided for them to make notes. These contained one page for each problem, which showed the problem number and the options 'Yes' and 'No'.

Additional materials were presented on paper. A Retrospective Strategies Questionnaire (see Appendix A) provided participants with three example reasoning problems

---

[1] An indeterminate problem is one in which the answer to the final question (e.g., "is the dog cleaner than the ape?") is "it is not possible to tell from the premises". A determinate problem is one in which the answer to the final question is either "yes" or "no".

(one for each type of relation) and asked participants to state which of three strategies (verbal, visual or spatial) they had used to solve the problem. Participants were allowed to check as many of the strategies as they had used and were also given an opportunity to describe a different strategy. Participants also completed a test of spatial ability (the Vandenberg Mental Rotations Test [12]), a test of verbal ability (Vocabulary Version III [13]), and a test of general reasoning (an abstract figural reasoning test [14]).

2.3 Procedure

Participants were tested either alone or in groups of two or three. Each participant was seated in a cubicle containing a desktop computer, which displayed the task instructions and the reasoning problems. The premises and conclusion of each problem were presented on separate screens which participants advanced at their own pace. The premises were presented in black letters, and conclusions in red letters. Participants pressed a 'Yes' key on the keyboard if they judged that the conclusion followed from the premises, and they pressed a 'No' key if they judged that the conclusion did not follow. Four practice problems were administered before the experimental problems to familiarize participants with the procedure.

Participants in the notes condition were instructed to "feel free to take notes as you work through these problems". They wrote their notes in the booklets provided and marked their answer on paper by circling "yes" or "no" as well as pressing the corresponding key on the keyboard. Participants in the No Notes condition answered using the keyboard alone. In both conditions, the computer recorded the response to the conclusion (yes or no) as well as the reading times of the premises and the response latency.

After the reasoning trials, participants completed the Retrospective Strategies questionnaire, the Mental Rotation Test, the Vocabulary test, and the Abstract Reasoning test in that order.

3 Results

Three of the 30 participants in the Notes condition were excluded from the analyses because they did not produce any notes. The scores on the abstract reasoning test for four participants (three from the No Notes condition and one from the Notes condition) were also unavailable for analysis because they were not given enough time to complete the test. However, all other data from these participants was retained since the abstract reasoning test was the final task in the experiment.

3.1 Accuracy

We conducted a 3 (content: visual, spatial, or abstract) x 2 (valid or invalid) x 2 (three or four terms) x 2 (Notes or No Notes) mixed model ANOVA, with content, validity, and number of terms as repeated measures, and notes as a between subjects factor (see Table 1 for means and standard deviations for each condition). A main effect of content showed that accuracy differed depending on the type of relation used in the inference, $F(2, 110)=8.46$, $p<.001$, $\eta^2_{partial} = .133$. Participants were less accurate when reasoning about visual relations (84.2% correct) than when reasoning about spatial

(91.9% correct) or abstract relations (91.4% correct). There was a main effect of notes, $F(1, 55)=8.63$, $p=.005$, $\eta^2_{partial} = .136$, such that participants in the notes condition were significantly more accurate (93.7% correct) than those in the no-notes condition (85.1% correct). We also found a main effect of validity, $F(1, 55)=8.47$, $p=.005$, $\eta^2_{partial} = .133$, which showed that participants were more accurate on invalid (91.5% correct) than on valid problems (86.8% correct). There was no main effect of the number of terms, $F(1, 55)=2.04$, $p=.159$, $\eta^2_{partial} = .036$., and no significant interactions were detected.

Table 1. Means (and standard deviations) of number of problems answered correctly. Maximum correct for each condition = 2.0. The data for No Notes condition are shown in the top left corner of the cells, and the Notes condition in the bottom right corner.

No notes / Notes		*Problem type*	
Validity/ # of Terms	**Visual**	**Abstract**	**Spatial**
Valid/ 3-term	1.60(.56) / 1.70(.47)	1.70(.60) / 1.85(.46)	1.77(.54) / 1.89(.32)
Valid/ 4-term	1.43(.63) / 1.67(.48)	1.77(.50) / 1.85(.36)	1.70(.60) / 1.96(.19)
Invalid/ 3-term	1.77(.43) / 1.93(.27)	1.87(.35) / 2.00(.00)	1.70(.54) / 1.93(.27)
Invalid/ 4-term	1.60(.56) / 1.70(.50)	1.70(.60) / 1.93(.27)	1.83(.38) / 1.96(.19)

The analysis of accuracy therefore showed a visual impedance effect. Participants were less likely to respond correctly to inferences that contained visual relations than to those that contained spatial or abstract relations. Post-hoc tests (LSD) showed a significant difference between visual and abstract relations ($p=.003$) and between visual and spatial relations ($p<.001$), but not between abstract and spatial relations ($p=.81$).

Although participants in the Notes condition scored better than those in the No Notes condition, there was no interaction between content and condition, indicating that the visual impedance effect held for both conditions. In other words, although making notes leads to better performance, it does not eliminate the visual impedance effect.

3.2 Response Latencies

We analyzed the response latencies to the conclusions for only the No Notes condition because any such measurement for the Notes condition would also include the time spent on making notes, and therefore not yield a meaningful comparison. Our No Notes condition was essentially a replication of Knauff and Johnson-Laird's (2002) Experiment 1, so it is perhaps not surprising that we found very similar results.

We computed a 3 (content) x 2 (validity) x 2 (terms) repeated measures ANOVA[1] on the latencies for the correct trials. This yielded a main effect of content, $F(2, 42)=4.44$, $p=.018$, $\eta^2_{partial} = .174$, and no other significant main effect. Post hoc (LSD) tests showed that this effect was driven by the difference between visual and spatial problems ($p = .003$), and there was no significant difference between abstract and visual problems ($p = .25$) or between abstract and spatial problems ($p = .113$). No two-way interaction was found, but there was a significant three-way interaction of content, validity, and number of terms, $F(2, 42)=4.83$, $p=.013$, $\eta^2_{partial} = .187$. This interaction appears to be driven by a two-way interaction between validity and number of terms, for spatial relations only, $F(1, 26)=10.02$, $p=.004$, $\eta^2_{partial} = .278$.

We also analyzed the latencies with a nonparametric test (*Pages's L)*, used in the original Knauff and Johnson-Laird [1] analysis. This test detects ranked differences in a specified order. In the original paper [1] visual relations took the longest, followed by abstract and spatial relations. A test of this ordering was not significant for our data. However, because Knauff and Johnson-Laird did not find a significant difference between the abstract relations and spatial relations, we also tested the ordering visual>spatial>abstract. This analysis did reach significance, $L=2.32$, $p=.011$. In accordance with Knauff and Johnson-Laird's results, we also found a significant difference in latencies between visual relations and abstract relations, $z = -2.77$, $p=.006$, but no significant difference between spatial and abstract relations, $z = -1.14$, $p=.254$ or between visual and spatial relations, $z = 1.62$, $p=.11$.

3.3 Individual Differences

Does accuracy in linear reasoning depend on more general reasoning, verbal, or spatial abilities? To answer this question, we conducted regression analyses to test the hypotheses that relations between reasoning accuracy and these three cognitive abilities is moderated by whether or not participants made notes during reasoning. A separate simultaneous regression was computed for each cognitive ability measure, entering the ability measure, condition (notes or no notes) and the interaction (the product of these two variables) simultaneously.

These analyses suggest that linear reasoning depends on individual differences in verbal and reasoning ability but not spatial ability, and that the dependence on individual abilities is moderated by whether or not one is able to take notes during reasoning. Reasoning accuracy was significantly predicted by performance on the abstract reasoning test for the No Notes condition, $\beta=.696$, $t=4.15$, $p<.001$, but not for the Notes condition, $\beta=.154$, $t=1.01$, $p=.32$, with a significant interaction between the score on the abstract reasoning test and condition, $b=-.157$, $t=-2.38$, $p=.021$. Scores on the vocabulary test showed similar predictive effects, being a significant predictor of accuracy for the No Notes group, $\beta=.591$, $t=3.38$, $p=.001$, but not for the Notes group,

β=.117, t=.79, p=.43, with a significant interaction between the score on the vocabulary test and condition, b= -.36, t=-2.07, p=.043. Performance on the MRT, on the other hand, was not predictive of accuracy for either the No Notes, β=.211, t=1.39, p=.17, or Notes conditions, β=.009, t=.046, p=.96. These results suggest that when reasoners are able to externalize their representations of the problem, accuracy depends far less on more general reasoning and on verbal abilities.

The fact that accuracy is not related to MRT performance is notable, given previous evidence that reasoning depends on visual-spatial working memory and that mental rotation tests are correlated with spatial working memory measures [15]. It is possible that participants did not depend on their spatial capabilities for this task, either because they did not represent the premises spatially, or because creating spatial mental models of these problems does not require exceptional spatial skills. Alternatively, the MRT may not depend on the same spatial skills as those required to building spatial mental models of transitive inferences.

3.4 Retrospective Strategies

The retrospective strategies questionnaire (see Appendix) revealed that participants' representations and strategies depended on whether they were allowed to make notes and on problem content. Table 2 shows the proportion of participants in each group who reported using a verbal, visual or spatial strategy for each type of problem. Multiple between-groups t-tests comparing strategy use for each problem showed that for all types of problems, participants in the No Notes condition reported using verbal and visual strategies significantly more than participants in the Notes condition. In contrast participants in the notes condition reported using spatial strategies (diagrams) significantly more often. These trends were significant (at p<.05) for all but one pairwise comparison; there was no significant difference (p>.10) between the two conditions in how often they reported using a visual strategy on visual problems, with neither group reporting this strategy very much. Instead, the No Notes group preferred a verbal strategy and the Notes group tended to report a spatial strategy on these problems.

We also found that reasoning strategy differed as a function of problem content. Multiple repeated measures ANOVAs indicated that for each strategy reported by the No Notes group, there was a significant main effect of problem type (Fs>3.5, ps<.04). That is, the reported use of each strategy depended on the type of problem. As Table 1 shows, the no-notes group reported using primarily verbal representations for visual problems, primarily visual representations for spatial problems, and used spatial representations on both visual and abstract problems. In contrast, the No Notes group almost always reported a spatial strategy regardless of the type of problem.

It is possible that our participants did not understand the distinction between a visual versus a spatial representation in our questionnaire. In a final analysis, we coded strategies as either visual/spatial (+1 if reported strategy was visual or spatial – we did not distinguish between the two for the purposes of this analysis) or verbal (-1), and summed across problems, resulting in a score that represents a person's tendency to use a visual/spatial or verbal strategy. The Notes condition reported primarily visual/spatial spatial strategies (M=2.30, SD=1.38, maximum score = 3.0) than the No Notes condition (M=.77, SD=2.19). Overall, we found that strategy was not significantly correlated

Table 2. Proportion of reasoners who chose verbal, visual, and spatial strategies on the three types of problems, split by condition. Reasoners could choose more than one strategy. The proportions for No Notes condition are shown in the top left corner of the cells, and the Notes condition in the bottom right corner.

Strategy (No notes / Notes)	*Problem type* Visual	Abstract	Spatial
Verbal	.53 / .07	.33 / .11	.33 / .04
Visual	.17 / .04	.30 / .00	.70 / .07
Spatial	.37 / .78	.40 / .78	.17 / .81

with accuracy for either the Notes condition (r=.18, p=.4) or for the No Notes condition (r=-.10, p=.6). Thus, while strategies differed across conditions, the strategy used was not related to performance within conditions.

3.5 Type of Notes Produced

While the Retroactive Strategies questionnaire provided us with a self-report measure of the representations and strategies that participants used, the external representations of the premises that were produced by the participants in the Notes condition provided a more direct measure of how participants actually represented the premises while solving the problems.

As with strategies, we coded the notes made by each participant while solving each problem as propositional (coded as -1), or spatial (coded as +1). Making no notes or both types of notes on a single problem was coded as 0. Notes were coded as propositional if the participant represented the original premises individually. This included both writing a premise verbatim, shortening it, e.g. "dog smarter cat" or more abstract translations such as "dog > cat". For example, the following notes would be coded as propositional:

Dog > cat
Cat > Bird
Ape > Dog

Notes were coded as spatial if the premises were represented in a single spatial array such as vertical or horizontal ordering. For example, the following notes would be coded as spatial:

Ape > Dog > Cat > Bird

It is important to point out that in no single case did we observe notes that explicitly included visual detail, such as a sparkling clean ape or a very fat dog. Additionally, it is probably the case that "spatialness" of the notes varied on a continuum, with notes containing both spatial and propositional properties falling somewhere in the middle, but for reasons of analysis we opted to code these as discrete categories.

We summed the scores for the notes across all eight problems for each type of content to obtain a measure of where the notes lay on a spatial/propositional continuum. Negative sums indicated that an individual made more propositional than spatial notes (minimum possible = -8), and positive sums indicated that an individual made more spatial than propositional notes (maximum possible = 8).

The means for all relation types were positive (visual, M=3.04, SD=5.82; abstract, M=4.07, SD=5.84; spatial, M=5.41, SD=4.77), indicating that the majority of the notes made for all content types were spatial, providing converging evidence for the strategies that these participants retrospectively reported using on the problems (see Table 2). Additionally, the use of spatial notes as defined by this coding system differed across relation types. A repeated measures ANOVA showed a main effect of content, $F(2, 52) = 8.83$, $p<.001$. All post-hoc pairwise comparisons were significant at the $p < .05$ level.

Importantly, participants' notes were least likely to be spatial if the inference contained a visual relation. This effect was driven by participants making more propositional notes on these problems, rather than making hybrid (both propositional and spatial) notes or leaving the pages blank (see Figure 1, which shows the frequency, out of a maximum of 27, with which participants created each type of notes for each type of problem.). Thus, participants either found it more difficult to create spatial notes for these highly imageable problems, or found it more helpful to make propositional notes for these problems. In either case, we have shown that the representations used for

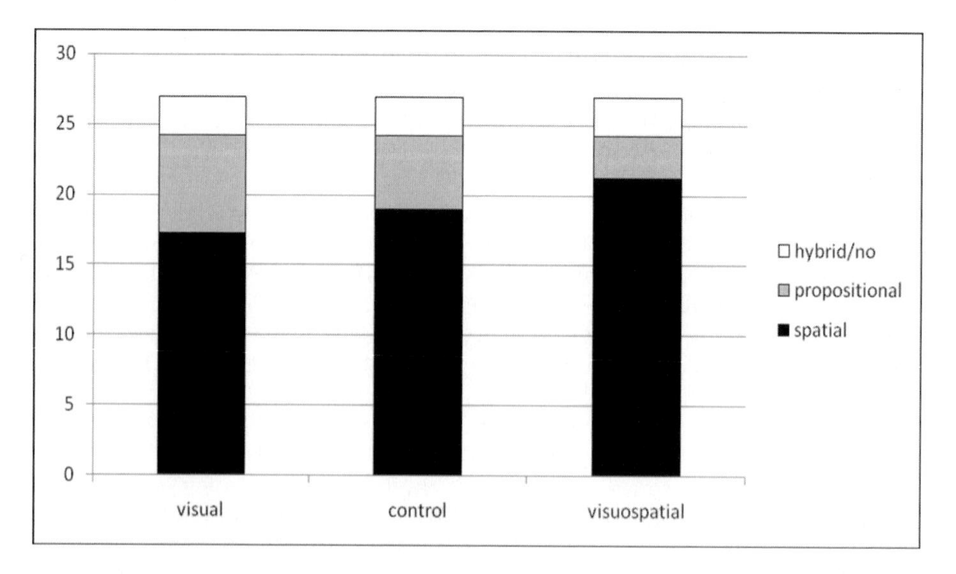

Fig. 1. Mean frequency of each type of notes made for the three problem content types (maximum = 27)

inferences containing visual relations are significantly different from those used for abstract and spatial relations. Finally, when we compare across participants, in no case was the spatial/propositional quality of notes for each type of problem related to the accuracy on that type of problem (*r*s<.31, *p*s>.10 in each case), or overall accuracy, $r(27)=.24$, $p=.20$. However, accuracy on visual problems was significantly lower than for the other two types of problems, and notes on visual problems were significantly less spatial, so this suggests that there is at least a relation between the representation used and solution accuracy when we compare across problems.

4 Discussion

In summary, we found evidence for the visual impedance effect in both accuracy and response times. Our data replicate the finding that when reasoning about visual information, participants (a) take more time to solve the problem, and (b) are less accurate in their conclusions than when reasoning about abstract or visual-spatial relations. Not surprisingly, when reasoners had the opportunity to externalize their representations of the problems, their accuracy increased. This is likely due to a decreased load on working memory. However, allowing participants to make notes does not eliminate the visual impedance effect, demonstrating its robustness.

Knauff and Johson-Laird [1] have proposed that visual content in reasoning problems impedes the construction of a spatial mental model of the problem. In support of this view, we have shown that the notes reasoners make for visual problems are less spatial (and more propositional) than the notes they make for abstract or spatial problems, and that the representations and strategies that reasoners report are less spatial for visual problems than for the other types of problems. However, our research suggests that reasoners do not incorporate visual information to their spatial mental models, simply take more time to create these models, or create less accurate spatial mental models. Instead, it seems that reasoners actually use a different type of mental representation when presented with these challenging visual problems. As revealed by the notes made by those in the Notes condition and the strategies reported by those in the No Notes condition, reasoners appear to represent visual relations propositionally whereas they are more likely to represent abstract and spatial relations spatially.

Reasoning performance in this task was more dependent on verbal and abstract reasoning ability than on spatial ability, a result that is surprising, given evidence for the importance of visual-spatial representations in reasoning [4, 5, 6]. However, it is important to realize that most of the variance in accuracy on these problems came from the visual problems, in which propositional representations were predominant. Accuracy was generally high on the abstract and visual-spatial problems, which were solved with a visual-spatial strategy, and in the notes condition, in which people were more likely to use a visual-spatial strategy. When participants are able to use a spatial strategy, they are very accurate, and therefore there is very little variance in performance. When they are unable to use a spatial strategy, they fall back on verbal-propositional representations that are very close to the problem statements, and this strategy is more successful for those with higher verbal and abstract ability.

Overall, we have shown that the diagrams and other notes that people make when solving reasoning problems can give us important insight into their representations and solution strategies. Furthermore, allowing people to make notes seems to induce a diagrammatic strategy, which in turn enhances performance. In general, we have shown that people differ in terms of the types of strategies and representations they use while solving linear reasoning problems, and while a spatial strategy may be common, it is not the only way to successfully reason about transitive inferences.

These data suggest that teaching individuals to use a spatial strategy to solve reasoning problems and to use external spatial representations of the problems when possible may improve performance in verbal reasoning. Some previous research [16] has indicated that teaching people to create diagrams or other spatial representations does not always enhance reasoning, and that the usefulness of such methods depends on individual differences between reasoners [17]. It is important to note that in those studies, reasoners were instructed on how to create specific types of diagrams for problems (although spontaneous use of diagrams was also studied). In our study, reasoners spontaneously created their own representations. This may be important for a number of reasons. For example, reasoners may be able to better understand their own representations than a type they have just learned, and this in turn may affirm their self-efficacy for being able to solve the problem. In sum, we suggest that creating a situation where students are allowed to make notes while solving reasoning problems may induce them to develop efficient ways of representing the information spatially, which in turn enhance problem solving. While it is expected that training in producing external spatial representations will improve performance in domains that are inherently spatial, such as architecture, this study additionally suggests that training reasoners to use spatial representations can also help in domains that superficially appear to be more linguistic in nature.

References

1. Knauff, M., Johnson-Laird, P.N.: Visual Imagery can Impede Reasoning. Memory & Cognition 30, 363–371 (2002)
2. Knauff, M., Mulack, T., Kassubek, J., Salih, H.R., Greenlee, M.W.: Spatial Imagery in Deductive Reasoning: A Functional MRI Study. Cognitive Brain Research 13, 203–212 (2002)
3. Clark, H.H.: Linguistic Processes in Deductive Reasoning. Psychological Review 76, 387–404 (1969)
4. Klauer, K.C., Stegmaier, R., Meiser, T.: Working Memory Involvement in Propositional and Spatial Reasoning. Thinking and Reasoning 3, 9–47 (1997)
5. Knauff, M., Strube, G., Jola, C., Rauh, R., Schlieder, C.: The Psychological Validity of Quantitative Spatail Reasoning in One Dimension. Spatial Cognition and Computation 4, 167–188 (2004)
6. Vandierendonck, A., De Vooght, G.: Working Memory Constraints on Linear Reasoning with Spatial and Temporal Contents. The Quarterly J. of Exp. Psychology A: Human Experimental Psychology 50A, 803–820 (1997)
7. Knauff, M., May, E.: Mental Imagery, Reasoning, and Blindness. The Quarterly J. of Exp. Psychology 59, 161–177 (2006)
8. Bacon, A.M., Handley, S.J., Newstead, S.E.: Individual Differences in Strategies for Syllogistic Reasoning

9. Egan, D.E., Grimes-Farrow, D.D.: Diffferences in Mental Representations Spontaneously Adopted for Reasoning. Memory and Cognition 10, 297–307 (1982)
10. Kozhevnikov, M., Hegarty, M., Mayer, R.E.: Revising the Visualizer-Verbalizer Dimension: Evidence for Two Types of Visualizers. Cognition and Instruction 20, 47–77 (2002)
11. Duyck, W., Vandierendonck, A., De Vooght, G.: Conditional Reasoning with a Spatial Content Requies Visuo-spatial Working Memory. Thinking and Reasoning 9, 267–287 (2003)
12. Vandenberg, S.G., Kuse, A.R.: Mental Rotations, A Group Test of Three-Dimensional Spatial Visualization. Perceptual and Motor Skills 47 (1978)
13. French, J.W., Ekstrom, R.B., Price, L.A.: Manual for Kit of Reference Tests for Cognitive Factors. Educational Testing Service, Princeton (1969)
14. Bennett, G.K., Seashore, H.G., Wesman, A.G.: Differential Aptitudes Test. The Psychological Corportation, Sidcup, UK (1981)
15. Shah, P., Miyake, A.: The Separability of Working Memory Resources for Spatial Thinking and Language Processing: An Individual Differences Approach. Journal of Experimental Psychology: General 125 (1996)
16. Calvillo, D.P., DeLeeuw, K., Revlin, R.: Deduction with Euler Diagrams: Diagrams that Hurt. In: Barker-Plummer, D., et al. (eds.) Diagrams 2006. LNCS (LNAI), vol. 4045, pp. 199–203. Springer, Heidelberg (2006)
17. Stenning, K., Cox, R., Oberlander, J.: Contrasting the cognitive effects of graphical and sentential logic teaching: Reasoning, representation and individual differences. Language and Cognitive Processes 10, 333–354 (1995)

Appendix: Retrospective Strategies Questionnaire

Instructions. Please indicate which (if any) of the methods described below that you used to solve the following examples of the reasoning problems. If you used several methods, indicate all the methods that you used. If you drew or imagined a diagram please draw your diagram below. If you think that your method(s) for solving the problem was unlike any of those described below, please describe your method, giving as many details as possible.

Example 1

The dog is behind the cat.
The dog is in front of the ape.
The ape is in front of the cat?

__ I solved the problem by remembering the order of the words as I read each premise and the conclusion.
__ I solved the problem by imagining a mental picture the dog, the cat, and the ape standing in a line from front to back.
__ I solved the problem by drawing (or imagining) a diagram representing the problem.
__ Please draw your diagram here:
__ I did not use any of the above methods. I attempted the problem in this way:

Example 2:

The dog is worse than the cat.
The dog is better than the ape.
The ape is better than the cat?

__ I solved the problem by remembering the words of the problem as I read each premise and the conclusion.
__ I solved the problem by making a mental picture of the dog, the cat, and the ape competing and one being better than the other.
__ I solved the problem by drawing (or imagining) a diagram representing the problem.
Please draw your diagram here:
__ I did not use any of the above methods. I attempted the problem in this way:

Example 3:

The dog is cleaner than the cat.
The dog is dirtier than the ape.
The ape is dirtier than the cat?

__ I solved the problem by remembering the order of the words as I read each premise and the conclusion.
__ I solved the problem by imagining a mental picture of the dog, the cat, and the ape with dirt on them.
__ I solved the problem by drawing (or imagining) a diagram representing the problem.
Please draw your diagram here:
__ I did not use any of the above methods. I attempted the problem in this way:

What Can Pictorial Representations Reveal about the Cognitive Characteristics of Autism?

Maithilee Kunda and Ashok Goel

School of Interactive Computing, Georgia Institute of Technology,
85 Fifth Street NW, Atlanta, GA 30318, USA
{mkunda,goel}@cc.gatech.edu

Abstract. In this paper, we develop a cognitive account of autism centered around a reliance on pictorial representations. This Thinking in Pictures hypothesis shows significant potential for explaining many autistic behaviors. We support this hypothesis with empirical evidence from several independent behavioral and neuroimaging studies of individuals with autism, each of which shows strong bias towards visual representations and activity. We also examine three other cognitive theories of autism—Mindblindness, Weak Central Coherence, and Executive Dysfunction—and show how Thinking in Pictures provides a deeper explanation for several results typically cited in support of these theories.

Keywords: Autism; cognition; mental imagery; visual reasoning; visual representation.

1 Introduction

Ever since its discovery in the 1940s by physician Leo Kanner, autism has been defined and diagnosed by the atypical behaviors that it produces. In particular, it is a developmental condition characterized by atypical social interactions, communication skills, and patterns of behavior and interests, as described in the American Psychiatric Association's Diagnostic and Statistical Manual of Mental Disorders [1]. While the specific causes of autism are not known, an etiological framework, shown in Fig. 1, has been traced out that leads from genetic and possibly environmental factors, through neurobiological development and cognitive functioning, and finally to behavioral manifestations (adapted from [2]).

Many theories have attempted to give a cogent account of the changes in cognitive functioning that lead to the behavioral characteristics of autism. Some prominent theories include: Mindblindness, which hypothesizes that individuals with autism lack a "theory of mind," i.e. they cannot ascribe mental beliefs to other people [3]; Weak Central Coherence, which posits a bias towards local instead of global information processing [4]; and Executive Dysfunction, which suggests that individuals with autism have deficits in executive functions such as planning, mental flexibility, and inhibition [5].

However, many individuals on the autism spectrum have given introspective descriptions that are quite different from the above theories. One of the most

G. Stapleton, J. Howse, and J. Lee (Eds.): Diagrams 2008, LNAI 5223, pp. 103–117, 2008.

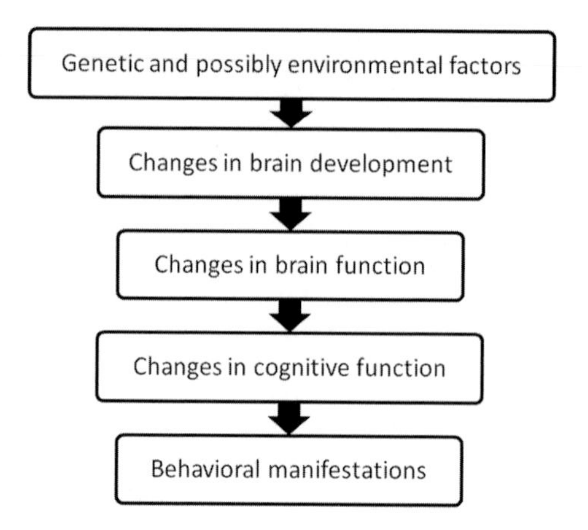

Fig. 1. Hypothesized etiology of autism, adapted from [2]

famous is the account by Temple Grandin in her book Thinking in Pictures [6]. Grandin, a high-functioning adult with autism, states that her mental representations are predominantly visual, i.e. that she thinks in pictures, and that this representational bias affects how she performs a range of cognitive operations, from conceptual categorization to the interpretation of complex social cues. Numerous other individuals with autism have also informally reported being aware of similar biases in mental representation, suggesting that Grandin is not an isolated case.

While Grandin's account of visual thinking has been primarily an introspective study, we aim to show that the Thinking in Pictures hypothesis does, in fact, represent a very powerful way to look at cognition in autism. We begin by considering what it might mean to think in pictures and how this would differ from typical cognition. Second, we present relevant empirical data from a range of literature, including behavioral and neurobiological studies of individuals with autism. Third, we examine how Thinking in Pictures relates to other cognitive theories of autism. Fourth, we examine current cognitive theories of visual thinking and identify several open issues.

2 What Does It Mean to Think in Pictures?

The literature on cognition uses numerous terms to talk about several different kinds of internal representations, e.g. modal, amodal, multimodal, digital, discrete, descriptive, verbal, linguistic, propositional, symbolic, sub-symbolic, analogical, imagistic, pictorial, depictive, visual, spatial, etc. It does not help that different authors often use a single term to mean very different things, and sometimes the same author uses the same term to mean different things. Instead of trying to define all these terms here, we specify what we mean by Thinking in Pictures using a minimal characterization that is sufficient for stating our hypothesis about autistic cognition.

Our characterization of Thinking in Pictures uses Paivio's dual-encoding theory of cognition as its starting point [7]. A knowledge representation can generally be unwound into *content* and *encoding* (i.e. form), where content pertains to *what* knowledge is being represented, and encoding refers to *how* that knowledge is represented. We define pictorial representations as having two key properties, as illustrated in Fig. 2:

1. Encoding is analogical in that it maintains a structural isomorphism between what is represented and how it is represented.
2. Content pertains to the appearance of objects, including both "what" and "where" information.

Verbal representations, in contrast, have the following properties:

1. Encoding is propositional.
2. Content can be arbitrarily assigned based on inferential needs.

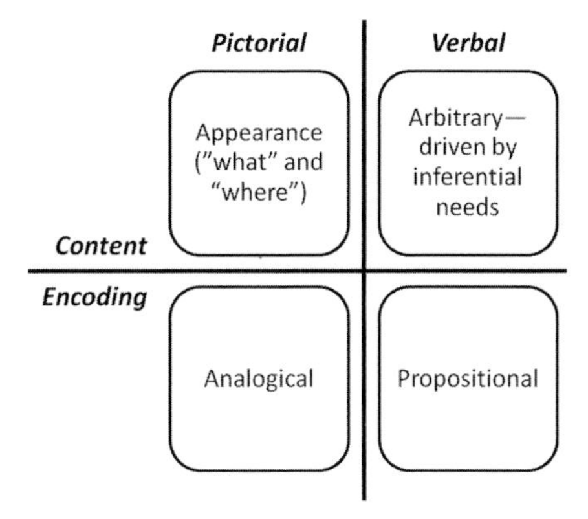

Fig. 2. Properties of content and encoding for pictorial and verbal knowledge representations

This characterization of pictorial and verbal knowledge representations imposes several interesting constraints on the types of knowledge that can be represented in each encoding and the kinds of inferences that can be drawn from each representation. Because pictorial representations are restricted to appearance-related content, it is difficult to explicitly represent abstractions such as causality, intention, or type/token relationships. Causality, for example, is at most implicit in a pictorial representation. However, these types of abstractions can easily be represented with propositions in a verbal representation. On the other hand, because pictorial representations maintain a structural correspondence between representation and content, certain inferences can be made more efficient or effective by exploiting this additional information. For instance, certain spatial inferences can be performed much more quickly using analogical representations than with propositions.

2.1 Our Hypothesis

Given the above characterization of pictorial thinking, we can now state our core hypothesis about autistic cognition more precisely:

1. Typical cognition uses both pictorial and verbal representations for different tasks.
2. Autistic cognition does not use verbal representations for typically verbal tasks.
3. For a subset of these typically verbal tasks, autistic cognition uses pictorial representations as a compensatory mechanism.
4. This difference in representation between typical and autistic cognition leads to observable effects on behavior.

Based on this Thinking in Pictures hypothesis, we expect that the limitations on content imposed by pictorial representations would lead to atypical behaviors and diminished performance on certain tasks. In Sect. 3.1, we describe how these sorts of representational limits can account for many of the characteristic behaviors of autism.

However, we also expect from our hypothesis that tasks or behaviors that can be performed pictorially would not be affected to the same extent. Furthermore, the additional inferential power lent by pictorial representations from their property of structural correspondence would lead to improved performance on tasks that exploited this information. Section 3.2 outlines several independent, empirical studies that support these expectations.

Finally, Sect. 3.3 presents results from neuroimaging studies that show a neurobiological basis for specifying a distinction between the uses of pictorial versus verbal representations in the autistic population.

2.2 Our Methodology

In addition to establishing a general consistency with existing empirical evidence (which we address in this paper), our Thinking in Pictures hypothesis suggests a specific methodology to further investigate tasks that can be performed pictorially. In particular, while current "deficit" accounts of cognition in autism can explain diminished performance on certain tasks, it is harder to explain performance across a wide range of tasks, for instance to account for the so-called "islets of ability" often observed in the autistic population.

However, the four postulates of the basic Thinking in Pictures hypothesis suggest a different methodology for classifying task performance. First, consider the spaces $\boldsymbol{P}$ and $\boldsymbol{V}$ (as illustrated in Fig. 3) of all tasks and behaviors that are typically performed using pictorial and verbal representations, respectively. Some of the tasks in $\boldsymbol{P}$ should straightaway correspond to some of the "islets of ability" seen in autism. Also, for some subset $\boldsymbol{V_1}$ of the tasks that typical cognition performs using verbal representations, individuals with autism can compensate using pictorial representations instead. If we can identify tasks in $\boldsymbol{V_1}$, then the Thinking in Pictures hypothesis should be able to make specific

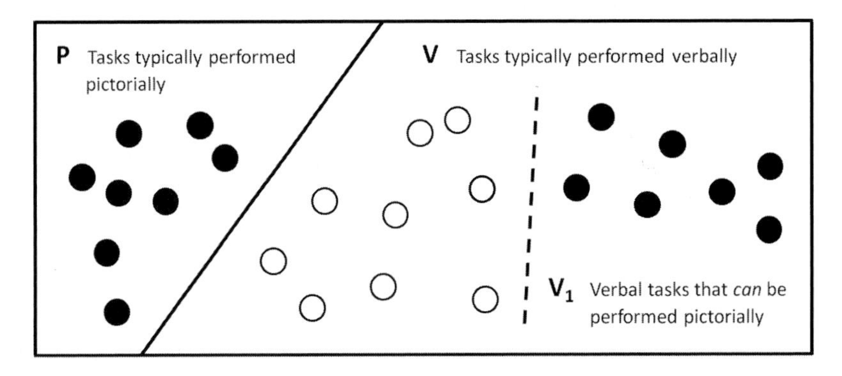

Fig. 3. Tasks typically performed pictorially and verbally, with a subset of verbal tasks that *can* be performed pictorially

predictions about how individuals with autism will perform on these tasks in comparison to typically developing individuals, based solely on the information-processing properties of pictorial versus verbal representations.

In future work, we plan to identify tasks in these subsets and use computational experiments to generate predictions based on our Thinking in Pictures hypothesis. These predictions can then be evaluated in light of real-world performance on the same tasks by individuals with autism.

3 Thinking in Pictures and Autism

The Thinking in Pictures hypothesis outlined in this paper is supported by a significant amount of existing empirical evidence. In the following three subsections, we present the results of several, independent studies that provide evidence from behavior, cognition, and neurobiology, respectively.

3.1 Behavioral Evidence

The definition of autism is centered on three areas: social interaction, communication, and stereotyped patterns of behavior and interests. Each of the following paragraphs addresses a subset of the behaviors in each area, which are listed in the DSM-IV-TR [1].

Atypical social behaviors of autism include a lack of seeking to share enjoyment with others and a lack of social or emotional reciprocity. Both of these types of behavior rely on an ability to infer the mental states of others, which is a highly abstract concept that cannot easily be represented pictorially and has been shown to be impaired in many individuals with autism [8]. Without this concept, individuals with autism would have difficulty in desiring to induce certain mental states in others, like enjoyment, and in reciprocating or even perceiving emotional or social intentions.

Communication issues in autism include the delayed development or inappropriate use of language along with deficits in imaginary play. Thinking in Pictures explicitly allows for problems in verbal language development. Regarding imaginary play, it has been shown that symbolic play in typically developing children evolves from using objects that share perceptual similarity with the target representation to objects that are perceptually dissimilar [9], suggesting a progression from play that is perceptually grounded to play that is free from perceptual constraints. Accordingly, imaginary play deficits in autism could be explained by an atypical adherence to pictorial representations during play.

Finally, autism is also characterized by stereotyped patterns of behavior and interests, such as a preoccupation with parts of objects or an adherence to nonfunctional routines. Function, as mentioned previously, is an abstract concept not well-suited to pictorial representations. Without functional interpretations, object use by children with autism could remain centered on visual features, within the sensorimotor-based frameworks of early developmental play [10]. Several studies have indeed shown less [11] or less complex [12] functional play in children with autism compared to their typically developing peers. Also, as noted previously, pictorial representations do not provide for the explicit representation of causality, which could lead to nonfunctional routines, for instance if routines were structured temporally instead of causally.

3.2 Cognitive Evidence

Memory and Access. We discuss the results of three experiments that examined the uses of pictorial versus verbal representations in individuals with autism as compared to typically developing individuals. First, in studies of word recall tasks in the typical population, the recall of short words is generally better than the recall of longer words, but this effect can be eliminated by articulatory suppression, suggesting that verbal encoding is used to some extent [13]. Furthermore, this effect is still seen if pictures are used instead of words, but only in subjects of a certain age [14]. In one study of this type of picture recall task in children [15], the pictures had either long or short labels, and subjects were asked to either remain silent or verbalize each label. As expected, the control group performed significantly better with the short labels than with the long labels, and whether they verbalized the labels had no effect. In contrast, the autism group exhibited a much smaller word-length effect overall, and the effect was smaller in the silent condition than in the verbalizing condition. These results suggest that the children with autism verbally encoded the pictures to a lesser extent than did the control group, and also that their use of verbal encodings increased when prompted to verbalize the labels.

The second experiment looked at the effects of articulatory suppression on a task-switching test [15]. Children were given a sequence of pairs of numbers to add or subtract alternately, and they either had to remain silent or to repeat "Monday" as a form of articulatory suppression (AS). The control group performed far better when they were silent than under AS. However, the autism group showed no difference between the silent and AS conditions, suggesting that they did not use verbal representations to guide their task-switching.

The third experiment looked at a word-completion task in which semantic priming was provided using either picture or word cues [16]. The control group performed similarly under both conditions, but the autism group performed much better with picture cues than word cues. This suggests that the individuals with autism were better able to retrieve verbal information through pictorial representations than through other verbal representations.

Together, results from these three experiments suggest that individuals with autism use pictorial representations in several tasks that typically recruit verbal representations. They seem to encode pictures pictorially until required to produce a verbal representation and also to rely more on pictorial representations than verbal ones for recall, task-switching, and semantic retrieval.

Classification. As discussed earlier, pictorial representations do not support the explicit formation of type/token (i.e. prototype) categories. In [17], individuals with autism were shown under a variety of tasks to exhibit fairly typical abilities in concept identification but significantly lower abilities in concept formation, which relies on the use of prototypes. Another study showed a preference in individuals with autism for categorizing objects based on physical attributes instead of on more abstract qualities, namely for sorting books by color instead of by genre as typically developing individuals did [18].

Visual Attention and Reasoning. Much empirical evidence has shown that individuals with autism are adept at certain tasks relying on pictorial modes of processing. One such task is the Embedded Figures Task (EFT), in which a small, simple shape must be found within a larger figure. Numerous studies have shown that individuals with autism are often more accurate [19] or more efficient [20] on the EFT than typically developing individuals.

Recent studies have looked at another visual search task in which a target must be found amid a group of distracters that share either shape alone (feature search) or shape or color (conjunctive search) [21][22]. Results showed that individuals with autism had significantly faster search times than typically developing individuals and, unlike the control group, had the same search times under both feature and conjunctive search conditions. Even more unusual were findings that, while typically developing individuals showed a characteristic linear increase in search time as the number of distracters increased in conjunctive search, the increase in search times for the autism group remained fairly flat.

These results suggest that individuals with autism might be using fundamentally different visual search strategies than the typical population. While the view of Thinking in Pictures presented in this paper does not explicitly propose a model for visual attention, these results could be explained by attentional strategies that are not mediated by verbal representations, which might be more efficient for certain tasks.

Finally, while many of these types of studies cast their findings as evidence for isolated skills or "islets of ability" in individuals with autism, recent research has suggested that, given the opportunity to reason pictorially, individuals with autism can exhibit significantly higher measures of general intelligence than shown on standard tests. In particular, a study conducted using groups of both children and adults with autism demonstrated that their performance on Raven's Progressive Matrices fell into dramatically higher percentile ranges than their performance on Wechsler scales, a discrepancy not seen in the typically developing control groups [23]. In fact, whereas a third of the children with autism fell into the mentally retarded range on the Wechsler scales, only 5 percent did so on the Raven's test, with a third of them scoring at the 90th percentile or higher (as compared to none in this range on the Wechsler scales). Raven's Progressive Matrices, while a pictorial test, is cited in this study as requiring inference, planning, control, and other complex abilities. This result is strongly in accord with our Thinking in Pictures hypothesis, which provides for the pictorial execution of all of these high-level reasoning processes.

3.3 Neurobiological Evidence

We discuss two neuroimaging studies using functional MRI that show neurobiological evidence for our Thinking in Pictures hypothesis. The first study looked at differences in brain activation between individuals with autism and typically developing individuals when they had to answer true/false questions about high or low imagery sentences [24]. High imagery sentences included statements like, "The number eight when rotated 90 degrees looks like a pair of eyeglasses," while low imagery sentences included statements like, "Addition, subtraction, and multiplication are all math skills." The control group showed a significant difference between the high and low imagery conditions, with the high imagery condition eliciting more activity from temporal and parietal regions associated with mental imagery as well as from inferior frontal regions associated with verbal rehearsal. In contrast, the autism group showed similar activation in both conditions, with less activity in inferior frontal language regions than the control group in the high imagery condition, and greater activity in occipital and parietal visual regions in the low imagery condition. These results suggest that individuals with autism rely on visuospatial brain regions to process both high and low imagery sentences, unlike typically developing individuals who use these areas more for high imagery sentences and use verbal areas for low imagery sentences.

The second fMRI study looked at the differences in brain activation between individuals with autism and typically developing individuals while they performed the Embedded Figures Task (EFT) which, as described earlier, is a visual search task [25]. While many brain regions showed similar activation between the two groups, the control group showed greater activation than the autism group in prefrontal cortical regions that are associated with working memory and serial search. In contrast, the autism group showed greater activation in occipito-temporal regions that represent low level visual processing and have been linked to mental imagery (and possibly motion). These results suggest a difference in high-level attentional strategy between individuals with autism and typically developing individuals while performing the EFT, with typically developing individuals recruiting a serial search strategy and individuals with autism using an imagery-based strategy.

4 Other Cognitive Theories of Autism

4.1 Mindblindness

Mindblindness hypothesizes that individuals with autism lack a "theory of mind," or ability to ascribe beliefs to other beings [3]. This limitation could lead to atypical social and communicative behaviors. The classic study used in support of this theory is the false-belief task, in which two characters (typically dolls named Sally and Ann) are shown alongside two baskets. Sally places a marble in her basket and exits the room, after which time Ann switches the marble from one basket to another. When Sally returns, the subject is asked in which basket Sally will search for her marble. Responding that Sally will look in the first basket, where she still supposes the marble to be, requires ascribing a false belief to Sally, i.e. a belief that does not match the state of the real world.

The original study [8] looked at three groups for this task: children with autism, children with Down's syndrome, and a control group. Both the Down's syndrome group (who had a lower mean verbal mental age than the autism group) and the control group (who had a lower mean physical age than the autism group) averaged percent-correct scores in the mid-80s, while the autism subjects scored only in the mid-20s.

While Mindblindness holds that theory of mind is a distinct mental mechanism [3], one can also consider representations in general. As mentioned in Sect. 2, the explicit formation of concepts like intentionality or mental states is difficult using pictorial representations. If these concepts were made accessible through pictorial representations, for instance through diagrams or metaphors, then we would expect to see improvements in theory of mind capabilities. A recent study [26] used cartoon-drawn thought-bubbles to teach children with autism about mental states. After this training, most of the children passed standard false-belief tasks that they had previously failed as well as other theory of mind tasks. However, this study did not compare training using pictorial aids versus training using verbal aids, which would have provided a more compelling case for the importance of the pictorial nature of the training.

In addition, a dissociation has been found between children with autism and typically developing children in their abilities to represent different kinds of "false" states. In [27], two groups of children were tested on false belief tasks, in which subjects had to reason about beliefs that did not match the state of the real world, and on false photograph and false map tasks, in which subjects had to reason about external visual representations that did not match the state of the real world. The control group performed well on the false belief tasks but poorly on the false photograph and false map tasks. As expected, the autism group performed poorly on the false belief tasks, but they actually outperformed controls on the photograph and map tasks, suggesting that, while impaired in understanding false beliefs, the children with autism had access to richer pictorial representations or stronger pictorial reasoning skills than the control group.

4.2 Weak Central Coherence

Weak Central Coherence hypothesizes that individuals with autism have a limited ability to integrate detail-level information into higher-level meanings, or are at least biased towards local instead of global processing [4]. This trait is presumed to account for some of the stereotyped patterns of behaviors and interests in individuals with autism. Also, superior performance on certain tasks like the EFT is explained with the rationale that in individuals with autism, reasoning is unhindered by potentially distracting gestalt perceptions.

However, as described above, these results can also be explained under the Thinking in Pictures hypothesis by enhanced visual attentional strategies that exploit the properties of pictorial representations. Other evidence for Weak Central Coherence often includes verbal tests, such as deficits in homograph pronunciation in sentence contexts (as cited in [4]). These tests, while putatively measuring local, word-level versus higher-order, sentence-level processing, can also be interpreted as tests of verbal reasoning skills, which would be impaired under the Thinking in Pictures account.

4.3 Executive Dysfunction

The final cognitive theory of autism that we discuss is Executive Dysfunction, which hypothesizes that individuals with autism have limitations in their executive functions such as planning, mental flexibility, and inhibition, among others [5]. Many studies cited in support of the Executive Dysfunction theory include verbal tests of memory and inhibition and sorting-based tests of mental flexibility. However, in the Thinking in Pictures account, we would expect to see atypical performance in these verbal and category-dependent areas.

Furthermore, as cited earlier, a recent study [23] showed that both children and adults with autism performed considerably better on Raven's Progressive Matrices than on Wechsler scales of intelligence. Raven's Progressive Matrices are deemed to test fluid intelligence, which includes "coordinated executive function, attentional control, and working memory" (as described by that study). Therefore, these results do not seem to indicate a general executive dysfunction in individuals with autism.

One possible explanation, using our Thinking in Pictures hypothesis, is that individuals with autism have deficits in executive functions that are verbally mediated but not in executive functions that are (or *can* be) pictorially mediated. This view is consistent with current models of working memory that propose two distinct storage components—the phonological loop and the visuospatial sketchpad—that operate under a central executive [28]. This model is discussed in more detail in Sect. 5.1.

5 Cognitive Theories of Pictorial Representations

In this section, we present interpretations of our Thinking in Pictures hypothesis using three existing theories of pictorial and verbal representations. In particular, we look at how each theory would formulate 1) pictorial representations, 2) verbal representations, and 3) differences in inferential power between the two.

5.1 Working Memory

In [28], Baddeley summarizes his three-component model of working memory, which consists of a central executive and two memory buffers: the phonological loop and the visuospatial sketchpad.

1. The visuospatial sketchpad is the short-term storage system for pictorial representations. Subsystems for visual versus spatial representations are differentiated, as suggested by behavioral dissociations on psychological tasks on visual (pattern) span and spatial span. This distinction is also linked to dynamic (spatial) versus static (pattern) representations.
2. The phonological loop is the short-term storage system for acoustic and phonological representations. Linguistic representations used in the phonological loop correspond to our definition of verbal representations, although other, purely acoustic representations do not.
3. Because the two buffers are presumed to comprise distinct neural subsystems, one could differentiate 1) operations performed within each representational buffer as well as 2) tasks for which each buffer is typically recruited. In particular, because quantitative properties of each buffer have been identified through numerous behavioral studies, the presence (or absence) of these properties could shed light on whether a particular buffer is, in fact, being recruited. The breakdown of standard predictions (whether for augmented or diminished performance) is generally taken to imply that another, non-standard strategy is being used on a given task.

5.2 Mental Imagery

In [29], Kosslyn describes a model of mental imagery, which is the recreation of an experience that resembles actually perceiving an object or an event.

1. Long-term memory contains two kinds of deep knowledge representations: analogical and propositional. Analogical representations are skeletal, individual, and hierarchically organized. Propositional representations capture information about categories and qualitative spatial locations and relations. Mental images are recreated in the short-term storage system through the retrieval and combination of both analogical and propositional representations from long-term memory.
2. Kosslyn does not directly provide an account of verbal representations in short-term memory. However, since part of long-term memory is formulated as propositions, we interpret using verbal representations as operating directly on these representations.

3. The construction of mental images enables a range of inferences such as finding an object in an image, changing image resolution, rotation, etc. Kosslyn does not, however, directly address the issue of different kinds of inferences being drawn by different kinds of representations.

5.3 Computational Imagery

In [30], Glasgow and Papadias present a computational model of imagery. Although not a cognitive theory per se, their computational model makes more precise and detailed commitments that are generally consistent with Kosslyn's model of mental imagery.

1. Pictorial representations are instantiated using two working-memory buffers: one for visual information contained in a 3D occupancy array, and another for spatial information contained in a 3D symbolic array. While the visual representation is purely pictorial and depictive, the symbolic array uses a combination of pictorial and verbal information and is therefore classed as descriptive, because elements in the array are words linked to semantic long-term memory, which is represented using propositions.
2. Glasgow and Papadias do not directly provide for the use of verbal representations in working memory. However, they do formulate long-term memory using propositions. Within this scheme, we interpret using verbal representations as operating directly on these propositional representations.
3. Although the information in each representation may be derivable from the other, "the representations are not computationally equivalent" because inferencing will operate at different efficiencies within each. However, in this formulation, the only difference between using each type of representation will be efficiency; Glasgow and Papadias do not allow for any differences in the actual inferences that can be made, because they consider only visuospatial content in their verbal representations.

5.4 Perceptual Symbol Systems

In [31], Barsalou presents a case for perceptually grounded representations in cognition. Perceptual symbols are set up in contrast to amodal symbols (e.g. propositions), which are defined as being "nonperceptual" and arbitrary, in that the symbol does not bear any structural relation to what it represents.

1. Pictorial representations are easy to formulate in the perceptual symbol system framework: they are just representations that correspond to the visual perceptual modality. Barsalou does not separate visual from spatial knowledge; we can assume that any such distinctions would match the underlying neural architecture, since perceptual symbols in this scheme are instantiated in the neural correlates of the actual perceptual events that they represent.
2. Barsalou talks about verbal representations in long term memory as consisting of "a schematic memory of a perceived event, where the perceived event is a spoken or a written word." How these word memories function as

verbal representations is through becoming "associated with simulators for the entities and events to which they refer." Presumably these "entities and events" are other perceptual symbols (e.g. visual, olfactory, emotional, etc.) that represent perceptions of things in the world.

3. Barsalou does not directly address the types of inferences that might be carried out in each type of representation. In general, the operations on perceptual symbols use the functionality of a simulator for each symbol. One would have to define classes of simulators for each type of representation in order to make predictions about specific inferences.

5.5 Open Issues

These cognitive theories raise several issues for our account of autistic cognition:

- Where does the change in representation between typical cognition and autistic cognition occur—long term memory or working memory or both?
- Why do individuals with autism use pictorial representations? Is it a deficit in verbal subsystems? A strength in pictorial subsystems? Or something else altogether?
- Can individuals with autism recruit verbal subsystems at all for these tasks? I.e., is it an actual deficit or just a bias that can be overcome by increased attention and/or training?
- Is autistic pictorial cognition the same as typical pictorial cognition, or is it fundamentally different?

6 Conclusion

In this paper, we have developed the Thinking in Pictures hypothesis about cognition in autism. At the minimum, behavioral, cognitive, and neuroimaging evidence suggests that Thinking in Pictures is a cogent alternative explanation of autistic cognition, alongside current accounts. We posit further that Thinking in Pictures has significant strengths not found in existing theories, both in terms of its explanatory breadth regarding the behaviors of autism as well as the depth to which it can account for many different pieces of empirical data.

Of course, the range of autistic behaviors that any of these theories, including Thinking in Pictures, can explain of and by itself remains an open issue, as does the question of whether a theory might apply more to one particular subset of the autistic population than another. It is possible that a full cognitive account of autism requires a combination of theories or the identification of specific subgroups of individuals on the autism spectrum beyond what has already been established in the literature.

Acknowledgments. Kunda was supported by the Office of Naval Research through the NDSEG fellowship program, and Goel's work was partially supported by an NSF IIS Grant (#0534622) on Multimodal Case-Based Reasoning in Modeling and Design. The authors thank Gregory Abowd and the anonymous reviewers for their helpful comments on an earlier draft of this paper.

References

1. American Psychiatric Association: Diagnostic and Statistical Manual of Mental Disorders. American Psychiatric Association, Washington, DC (2000)
2. Minshew, N.J., Goldstein, G.: Autism as a disorder of complex information processing. Mental Retardation & Developmental Disabilities Res. Rev. 4, 129–136 (1998)
3. Baron-Cohen, S.: Mindblindness. MIT Press, Cambridge (1995)
4. Happe, F., Frith, U.: The weak coherence account: Detail-focused cognitive style in autism spectrum disorders. J. Autism and Developmental Disorders 36, 5–25 (2006)
5. Russell, J.: Autism as an executive disorder. Oxford University Press, NY (1998)
6. Grandin, T.: Thinking in pictures, expanded edition. Vintage Press, New York (2006)
7. Paivio, A.: Dual coding theory—Retrospect and current status. Canadian Journal of Psychology 45, 255–287 (1991)
8. Baron-Cohen, S., Leslie, A.M., Frith, U.: Does the autistic child have a "theory of mind"? Cognition 21, 37–46 (1985)
9. Ungerer, J.A., Zelazo, P.R., Kearsley, R.B., Oleary, K.: Developmental changes in the representation of objects in symbolic play from 18 to 34 months of age. Child Development 52, 186–195 (1981)
10. Fenson, L., Kagan, J., Kearsley, R.B., Zelazo, P.R.: Developmental progression of manipulative play in 1st 2 years. Child Development 47, 232–236 (1976)
11. Stone, W., Lemanek, K., Fishel, P., Fernandez, M., Altemeier, W.: Play & imitation skills in the diagnosis of autism in young children. Pediatrics 86, 267–272 (1990)
12. Williams, E., Reddy, V., Costall, A.: Taking a closer look at functional play in children with autism. J. Autism and Developmental Disorders 31, 67–77 (2001)
13. Cowan, N., Baddeley, A.D., Elliott, E.M., Norris, J.: List composition and the word length effect in immediate recall: a comparison of localist and globalist assumptions. Psychonomic Bulletin and Review 10, 74–79 (2003)
14. Hitch, G.J., Halliday, M.S., Dodd, A., Littler, J.E.: Development of rehearsal in short-term memory—Differences between pictorial and spoken stimuli. British Journal of Developmental Psychology 7, 347–362 (1989)
15. Whitehouse, A.J.O., Maybery, M.T., Durkin, K.: Inner speech impairments in autism. Journal of Child Psychology and Psychiatry 47, 857–865 (2006)
16. Kamio, Y., Toichi, M.: Dual access to semantics in autism: Is pictorial access superior to verbal access? Journal of Child Psychology and Psychiatry and Allied Disciplines 41, 859–867 (2000)
17. Minshew, N., Meyer, J., Goldstein, G.: Abstract reasoning in autism. Neuropsychology 16, 327–334 (2002)
18. Ropar, D., Peebles, D.: Sorting preference in children with autism: the dominance of concrete features. J. Autism and Developmental Disorders 37, 270–280 (2007)
19. Shah, A., Frith, U.: An islet of ability in autistic children—A research note. Journal of Child Psychology and Psychiatry and Allied Disciplines 24, 613–620 (1983)
20. Jolliffe, T., BaronCohen, S.: Are people with autism and Asperger syndrome faster than normal on the Embedded Figures Test? Journal of Child Psychology and Psychiatry and Allied Disciplines 38, 527–534 (1997)
21. Plaisted, K., O'Riordan, M., Baron-Cohen, S.: Enhanced visual search for a conjunctive target in autism: a research note. Journal of Child Psychology and Psychiatry and Allied Disciplines 39, 777–783 (1998)

22. O'Riordan, M.A., Plaisted, K.C., Driver, J., Baron-Cohen, S.: Superior visual search in autism. Journal of Experimental Psychology—Human Perception and Performance 27, 719–730 (2001)
23. Dawson, M., Soulieres, I., Gernsbacher, M.A., Mottron, L.: The level and nature of autistic intelligence. Psychological Science 18, 657–662 (2007)
24. Kana, R.K., Keller, T.A., Cherkassky, V.L., Minshew, N.J., Just, M.A.: Sentence comprehension in autism: thinking in pictures with decreased functional connectivity. Brain 129, 2484–2493 (2006)
25. Ring, H., Baron-Cohen, S., Wheelwright, S., Williams, S., Brammer, M., Andrew, C., Bullmore, E.: Cerebral correlates of preserved cognitive skills in autism—A functional MRI study of EFT performance. Brain 122, 1305–1315 (1999)
26. Wellman, H.M., Baron-Cohen, S., Caswell, R., Gomez, J.C., Swettenham, J., Toye, E., Lagattuta, K.: Thought-bubbles help children with autism acquire an alternative to a theory of mind. Autism 6, 343–363 (2002)
27. Leslie, A.M., Thaiss, L.: Domain specificity in conceptual development—neuropsychological evidence from autism. Cognition 43, 225–251 (1992)
28. Baddeley, A.: Working memory: Looking back and looking forward. Nature Reviews Neuroscience 4, 829–839 (2003)
29. Kosslyn, S.M.: Image and brain. MIT Press, Cambridge (1994)
30. Glasgow, J., Papadias, D.: Computational imagery. Cog. Science 16, 355–394 (1992)
31. Barsalou, L.: Perceptual symbol systems. Behavioral and Brain Sciences 22, 577–660 (1999)

Visual Thinking with an Interactive Diagram

Colin Ware[1], Anne T. Gilman[1], and Robert J. Bobrow[2]

[1] University of New Hampshire
anne.gilman@unh.edu
[2] BBN Technologies Corporation

Abstract. To investigate the process of reasoning with an interactive diagram, we recorded eye movements and mouse clicks of 28 users as they investigated social relationships in a 313-node network diagram. The MEgraph application used to display this network combines techniques such as topological range searching and motion highlighting to enable interactive exploration of complex network diagrams. Long-term memory encoding was assessed with a surprise recall protocol one week later, with and without lightweight visual history traces. Frequent video-game players relied more on peripheral vision, moving their gaze less often. History support was also associated with more efficient visual strategies. History traces improved users' ability to reconstruct prior work on retest.

1 Introduction

Humans ably incorporate their outside environment into cognitive processes, an ability long recognized for its relevance to the design of effective software tools [1]. It can be more efficient to pick up available information from the external world, including from a virtual display, than to retrieve it from internal long-term memory stores [2]. While static diagrams can be used in this way, interactive ones can present a greater volume and complexity of information to users and also, through recorded interactions, reveal much about those users' visual thinking processes. The present study investigates the impact of video game expertise and availability of search-history traces on reasoning with and retention of a highly interactive social network visualization application. Specifically, we combined interaction measures (gaze tracking and mouse clicks) with traditional search and recall performance measures to examine people's reasoning process and assess how easily they can reconstruct cognitive activities following a one-week hiatus, given various levels of visual support.

1.1 Background

Interactive Diagrams. Network diagrams with over thirty nodes become visually intractable, yet social networks can contain hundreds or thousands of nodes. Interactive visualization tools extend the utility of network diagrams for more complex data sets. One such approach employs 3D displays of large-network diagrams with interactive techniques to allow for view navigation down to the

G. Stapleton, J. Howse, and J. Lee (Eds.): Diagrams 2008, LNAI 5223, pp. 118–126, 2008.

individual node level [3,4]. An alternative approach involves interactive methods for accessing subsets of a complex 2D network diagram. Munzner et al's [5] Constellation system pioneered the use of highlighting to draw attention to topological rather than spatial relations within a node-link diagram. In Constellation, a mouse-over "hover" query on a node causes all nodes connected to it to become highlighted and the links between them to become more distinct. Because these hover queries are very rapid, the time cost of accessing subsets of the graph is much lower than for 3D approaches.

The Motion-Enhanced Graph (or MEgraph) system [6,7] extends the Constellation idea: selected subgraphs are made even more salient using motion, which can be detected more readily in the periphery of the visual field than static cues like color or shape [8]. Clicking on one node triggers motion highlighting of all nodes and intervening links within a given topological range of the selected node. This allows rapid exploration of graphs so dense that such links cannot be discerned without highlighting. Figure 1 illustrates MEgraph with about 300 nodes (see Sec. 2.1); it can display up to a few thousand.

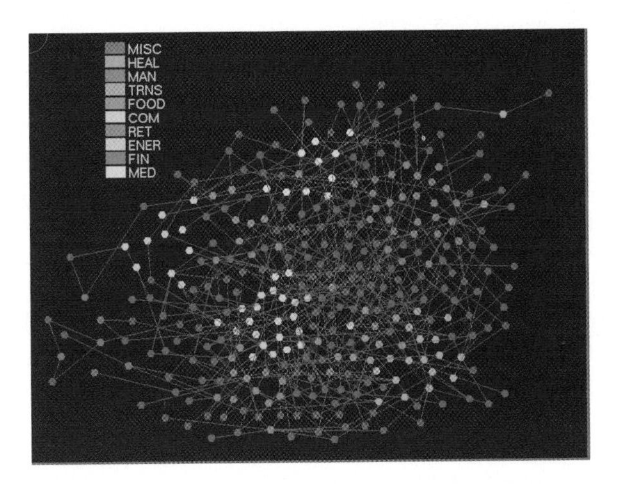

Fig. 1. MEgraph system with 313 company and board-member names

Expertise and Visual Thinking Skills. Prior exposure to interactive displays of a particular kind, namely action video games, can improve basic perceptual skills. Green and Bavelier [9] demonstrated that game players could allocate attentional resources to both the center and the periphery of the visual field more effectively. These results indicate that even low-level visual capabilities can be improved with extended practice.

Usage History. Although the MEgraph approach solves the problem of how to access a medium-sized network diagram of a few hundred nodes, there is a problem with integrating it into a typical work environment. Most knowledge workers switch tasks frequently, leaving partially-completed tasks untouched for hours, days, or weeks. Resuming work requires the analyst to rebuild an entire cognitive system, reassembling materials and reconstructing objectives and any partial solutions developed earlier. The original MEgraph design did little to support such cognitive reconstruction.

One of the benefits that has been proposed for interactive visual displays is mnemonic. When people are shown thousands of pictorial images, even for only a few seconds each, they can later recognize that they have previously seen them at much greater than chance levels [10]. Leaving visual traces (such as a drawing) can boost the memory of people as young as three years old [11]. Also, interactive computer displays of thumbnail icons allow users to rapidly identify specific items among large numbers that have been previously laid out [12].

Wexelblat and Maes [13] make a distinction between active approaches to recording history, in which users must explicitly and deliberately record their actions in some way, and passive approaches, in which activity patterns leave traces on aspects of the data without any explicit user action. For the present work, where users are presented with traces of their own prior activity rather than the activities of others, we have adopted a very light-weight active approach, to keep the recording of history from disrupting the user's train of thought.

2 Method

To investigate visual thinking processes with an interactive diagram, we adapted our MEgraph application to record eye movements and mouse clicks. We posed a set of reasoning tasks about a 313-node social network to users and measured their performance. A week later, we tested their ability to reconstruct task information. Our goals were to better understand the reasoning process, to learn the value of history support in the application, and to identify any advantages conferred by experience with video games.

2.1 Materials and Apparatus

The MEgraph Application. In our MEgraph system, clicking on a node causes a breadth-first search of the network to a specified depth, and this causes all nodes and links within the search range to become highlighted.

Two kinds of motion are employed in this highlighting. All of the nodes and links in the connected subgraph move with a circular motion, with an amplitude of approximately 2 mm—a bit less than the diameter of a node—and frequency of 1 Hz. Also, links that directly connect the selected node with others are made wider and display a radiating sawtooth pattern. At the same time, highlighted links become wider and turn from grey to bright green (see Figure 2). Highlighting times out ten seconds from the original click, unless the user selects another node during that time. If so, the old highlighting pattern is replaced by the new one.

Using a depth of two (as was the case for this experiment), every node one or two links away from a selected node becomes highlighted. Figure 2 shows a section of the display as it appears when the node representing United Technologies has been clicked on, causing the green sawtooth highlighting to appear. One link

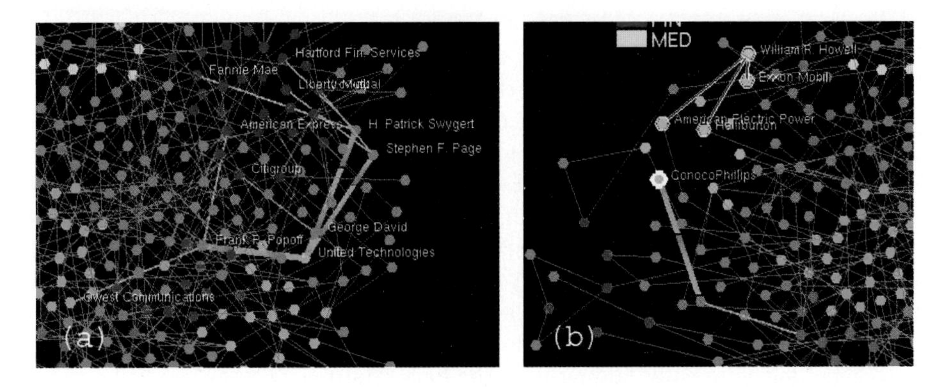

Fig. 2. Screen Fragments. a) Motion highlighting: A mouse click on United Technologies makes all connected links and nodes within a search range of 2 oscillate. b) History: The top of this figure shows a social network the user has already discovered, with one person sitting on the boards of American Electric Power, Halliburton, and Exxon Mobil. The user has just clicked on Conoco Phillips.

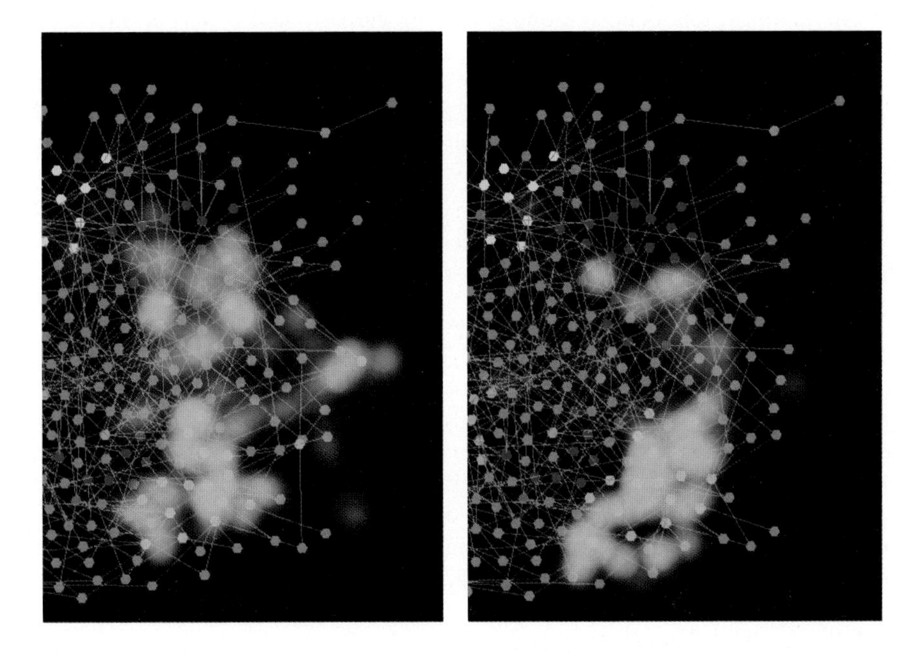

Fig. 3. Accumulated gaze traces for two participants' search for the orange (transportation) node connected to the most red (finance) nodes. The lack of traces over red nodes on the right reflect that user's greater reliance on peripheral vision.

away from United Technologies are three grey nodes showing those members of United Technologies' board who also direct other companies. Two links away are the the nodes representing these connected companies (e.g. Fanny Mae, Liberty

Mutual). Note that Liberty Mutual's label is obscuring a different node's name; users could drag nodes to reposition them if needed.

An iterative design process led to the following "just-enough" history support capability: Every node that is clicked on is added to a history store. Its label persists on the screen after the highlighting times out, as does a circle around the node. If this selected node is connected by a path of length one or two to a node already marked as history, then any intervening nodes and their connecting links are added to the history. History links are highlighted using red lines outlined in white as shown in Figure 2. Right-clicking on a history node removes it from the history store. In history mode, labels are only shown for the selected node, not connected nodes. To reveal the label of a connected node, the user clicks on that node, automatically adding it to the history store.

The Social Network Graph. The network diagram used in the present study was drawn from a list of board members of top US companies [14], many of which are household names, sorted by industry sector. The largest connected component from this network comprised 122 firms and 191 board members serving at least two of the firms. The familiar company names and sectors provided suitable material for social reasoning tasks. Company nodes were grouped and color-coded by sector, with grey nodes indicating board members.

Five simple diagram-reasoning tasks were developed, each having a unique solution, such as the person sitting on the most boards in the energy sector, or the transportation company connected to the most financial firms through shared directors (see Figure 3).

For this study, the MEgraph system was integrated with the software for an Eye-Trac 6000 from Applied Science Laboratories, allowing recording and playback of all eye movements and mouse actions, both sampled at 60Hz.

2.2 Participants and Procedure

Thirty-one current or recently-graduated university students participated in the experiment for US$12 and in some cases course credit. Three participants were unable to complete the experiment, leaving 14 in each condition. Of the 21 who reported their frequency of video-game use, 8 were frequent (twice a week or more) players. Each participant was assigned to either the history or no-history condition, with 3 frequent players in the former and 5 in the latter. Those without history support could only click on nodes to explore the diagram using motion highlighting, while those in the history condition also had the use of search history traces as described in Section 2.1. Test and retest stages were spaced one week apart, and participants went through each stage individually, in the presence of an experimenter.

Test: Each participant received a brief overview of the experiment's goals and the functions of the interface, then searched for an answer to a training question before being calibrated in the eye tracker and performing the five tasks. Participants were asked to return a week later to use the MEgraph system again.

Retest: The surprise recall protocol required participants to report as much information as they could about the previous week's tasks, in phases.

- **Phase 1:** Free recall without visual cues: Participants were asked about the display and its symbols, the mouse interaction, and about the information specific to the tasks they had performed.
- **Phase 2:** Participants were shown the display but not allowed to interact with it.
- **Phase 3:** Participants were allowed to interact with the display.
- **Phase 4:** History-group participants could view the recorded final markings from each of their own five task response screens.
- **Survey:** Participants reported how often they play video games.

2.3 Analysis

For test-stage tasks, answers were recorded and assigned a point if correct and zero if incorrect. For recall scoring, .25 points were awarded for each of: the target's name, sector, linked sector, and type (company or board member name).

For recorded interaction data, fixations were defined as any spans over 50ms where the eye-tracking gaze coordinates remained within 20 min of arc of each other on the screen. Recordings with high amounts of jitter apparent in playback also had high numbers of fixations from 50–100ms long. Recordings were discarded as too jittery if more than 3% of the overall task duration was taken up with these abbreviated fixations.

3 Results

3.1 Test Stage - Week 1

Task Performance. Participants were overall quite successful in finding the unique correct answers to the tasks with an identification rate of 83% (sd=.37). Time spent on each task decreased over the course of the experiment, with participants spending roughly two minutes on each of the first two tasks and just over one minute each on the final three (M1=114s, M2=130s, M3=66s, M4=66s, M5=64s). On a task-by-task basis, greater time spent did not lead to differing accuracy in selecting the right answer ($F(1,117)=.052$, $p > .75$), and presence or absence of history information did not show a significant effect on task duration ($F(1,117)=2.062$, $p > .15$).

Gaze Strategies. Two tasks required participants to find a company node in a target sector with the maximum number of connections to another sector. Since expert video-game players have been shown to make better use of visual attention resources both centrally and peripherally [9], they would be expected to move their eyes less in completing these two tasks. This hypothesis was assessed using a ratio of eye gaze spent in the target sector versus gaze directly on the linked sector. A between-subjects factorial ANOVA showed significantly higher ratios, shown in Table 1, both for high video game use ($p < .05$) and for history condition ($p < .01$). Figure 3 presents user gaze data for for a typical- (1.66) and a high- (4.60) ratio participant.

Table 1. Gaze ratios

History Traces	*High Gaming*	*Low Gaming*	*Total*
Yes	$2.85_{n:6}$	$2.10_{n:13}$	$2.09_{n:25}$
No	$1.65_{n:10}$	$1.22_{n:11}$	$1.48_{n:26}$
Total	$2.10_{n:16}$	$1.70_{n:24}$	$1.78_{n:51}$

3.2 Retest Stage - Week 2

All participants remembered the interaction elements of the display one week later, and none required additional instruction with the MEgraph interface.

On average, people remembered one single company (M=1.04, sd=1.27) among the dozens they had clicked on the week before, with General Motors, MetLife, and Pfizer remembered most often. Companies fared better than board members, though; only two individuals—one participant and one pilot tester—recalled any board member information. The participant with plans to join the US Navy remembered that one board member was an Admiral (Adm. Joseph W. Prueher). The pilot tester, who later joined the experiment team, recalled the board member who shares a name with an established multimedia interface researcher (Michael Johnston).

Looking only at those recall phases (1-3) common to all participants, mean recall scores (out of 100%) per task increased with each phase, though none broke ten percent: M1= 4.1%, M2=6.6%, M3=9.4%. These differences were significant ($F(1,539)=8.034$, $p < .01$) and did not reflect differences in experimental condition at test ($F(1,539)=.646$, $p > .40$). Video game use was not associated with different recall levels ($p > .80$).

However, as anticipated, users recalled much more information with history traces available: by the fourth phase, cumulative recall scores rose to 66.1% on average for history users, compared to 22.2% for the others (significant by F test, $F(1,136)=25.765$, $p < .001$). This considerable effect of history traces goes against strong claims about the mnemonic power of visual displays alone.

4 Conclusion

We had expected that the act of interacting with the diagram would boost long-term memory for task information and that participants would display considerable recall even without the display available to prompt them. However, the low scores from Phase 1 recall (free recall, no display present) showed this not to be the case.

We had further expected that the presence of the graphic display and being allowed to interact with it (Phases 2 and 3) would trigger much greater recall, due to the supposed mnemonic value of graphics and interaction. In fact, only one participant was able to identify individual answer nodes upon seeing the MEgraph layout (Phase 2, display without interaction).

User interaction on retest showed that subjects remembered which sectors of the diagram they had used, as 90% of clicks were on task-relevant nodes, suggesting implicit memory for the large-scale graph layout and/or color scheme. The use of identical nodes and homogeneous small-scale layout in MEgraph may have prevented greater visual memory for specific nodes.

Participants with access to their own search history unsurprisingly did much better, with recall scores increasing to 66%, yet they frequently failed to recall key task details. Nonetheless, this result does suggest that a minimal light-weight history mechanism is an essential component for a tool such as MEgraph if it is to be used in situations where the work flow will often be interrupted.

Recall that the basic interaction strategy with MEgraph is thus: click on a node, see what it is connected to, act accordingly. Our eye-tracking results revealed that while all participants were able to use their peripheral vision—not moving their gaze—to discard some selected nodes with few connections, frequent video-game players used this strategy to a greater extent than non-players. Most of the time, after clicking on a candidate node, frequent players were able to reject it from the pool of potential target nodes without fixating the nodes that were connected to it. This result is fully consistent with the finding of Green and Bavelier [9] that video game players can search the periphery of their visual field more effectively than non-players.

The history condition also resulted in more efficient looking strategies. A possible explanation for this comes from the fact that with the non-history condition clicking on a node caused labels to appear in the periphery whereas in the history condition only the nodes were highlighted (see Figure 2). The moving labels produced considerable overall peripheral motion, and this—combined with any attentional effect of the text labels themselves—may have attracted eye movement. Additionally, the fact that this difference between conditions operated independently of the effect of gaming expertise confirms the importance of specific design choices for real-world interfaces.

Finally, the disparity between the weak recall of information based on available visuospatial cues and the strong effect of passive and very lightweight history traces serves to highlight both the reconstructive nature of memory itself and the importance of history traces in settings where recall of information (and not mere recognition of images) is required.

Acknowledgements

We would like to thank Matthew Plumlee for his work integrating the eye tracker, and Christopher Hookway for his help with collecting data.

This study was supported and monitored by the the Disruptive Technology Office (DTO) and the National Geospatial-Intelligence Agency (NGA) under Contract Number HM1582-05-C-0023.

The views, opinions, and findings contained in this report are those of the author(s) and should not be construed as an official Department of Defense position, policy, or decision, unless so designated by other official documentation.

References

1. Hollan, J., Hutchins, E., Kirsh, D.: Distributed cognition: toward a new foundation for human-computer interaction research. ACM Trans. Comput.-Hum. Interact. 7(2), 174–196 (2000)
2. Plumlee, M.D., Ware, C.: Zooming versus multiple window interfaces: Cognitive costs of visual comparisons. ACM Trans. Comput.-Hum. Interact. 13(2), 179–209 (2006)
3. Fairchild, K.M., Poltrock, S.E., Furnas, G.W.: SemNet: time-dimensional graphic representations of large knowledge bases, pp. 190–206 (1999)
4. Parker, G., Franck, G., Ware, C.: Visualization of large nested graphs in 3D: Navigation and interaction. J. Vis. Lang. Comput. 9(3), 299–317 (1998)
5. Munzner, T., Guimbretiere, F., Robertson, G.: Constellation: A visualization tool for linguistic queries from MindNet. In: IEEE Symposium on Information Visualization, pp. 132–135. IEEE Computer Society Press, Los Alamitos (1999)
6. Ware, C., Bobrow, R.J.: Supporting visual queries on medium sized node-link diagrams. Information Visualization 4(1), 49–58 (2005)
7. Ware, C., Bobrow, R.J.: Motion to support rapid interactive queries on node-link diagrams. ACM Transactions on Applied Perception 1, 1–15 (2004)
8. Bartram, L., Ware, C.: Filtering and brushing with motion. Information Visualization 1(1), 66–79 (2002)
9. Green, C.S., Bavelier, D.: Effect of action video games on the spatial resolution of visuospatial attention. Journal of Experimental Psychology: Human Perception and Performance 32(6), 1465–1478 (2006)
10. Standing, L.: Learning 10,000 pictures. Quarterly Journal of Experimental Psychology 25, 207–222 (1973)
11. Gross, J., Hayne, H.: Young children's recognition and description of their own and others' drawings. Developmental Science 2(4), 476–489 (1999)
12. Robertson, G., Card, S.K., Mackinlay, J.D.: The cognitive coprocessor architecture for interactive user interfaces. In: UIST 1989: Proceedings of the 2nd annual ACM SIGGRAPH symposium on user interface software and technology, pp. 10–18. ACM Press, New York (1989)
13. Wexelblat, A., Maes, P.: Footprints: history-rich tools for information foraging. In: CHI 1999: Proceedings of the SIGCHI conference on human factors in computing systems, pp. 270–277. ACM Press, New York (1999)
14. On, J., Balkin, A.: The They Rule database (2004), http://www.theyrule.net/

Strategy Roadmaps: New Forms, New Practices

Alan F. Blackwell[1], Rob Phaal[2], Martin Eppler[3], and Nathan Crilly[4]

[1] University of Cambridge Computer Laboratory, Cambridge CB3 0FD UK
[2] Institute for Manufacturing, Cambridge
[3] University of Lugano
[4] Engineering Design Centre, Cambridge
{afb21,rp108,nc266}@cam.ac.uk, martin.eppler@lu.unisi.ch

Abstract. Strategy roadmaps are a class of abstract visual representation that is unusually diverse in diagrammatic style. A large corpus of roadmap visualisations was collected, and analysed using a collaborative consultation method involving expert participants. The results indicate organizing principles for the broader study of diagram use within strategic management. They also provide an opportunity for some basic design advice that can be given to professional strategy practitioners or managers who need to create diagrams of this kind. Finally, this novel context of diagram use has required the development of new research methods that may be of value in other Diagrams research.

1 Introduction

In this paper we describe and analyse a class of diagrams that is defined, not by specific visual characteristics and syntax, but by the particular role that it plays within social and organisational contexts. Strategy roadmaps, as described below, include a rather diverse range of diagrammatic representations, all of which can be used to organise and communicate information related to plans for the future. Whereas much research in diagrammatic reasoning starts from an agreed visual convention for a new (or old) type of diagram, the field of strategy roadmaps is different. Indeed, strategy roadmaps are especially interesting because the conventions that support their usage are still in the process of being discovered and negotiated.

The very phrase "strategy roadmap" offers those who might create them (e.g. company directors, management consultants, public policy experts) an apparently intuitive impression of why its construction might be undertaken, how a suitable visual representation might be formed, and what its uses might be. However, there is no simple agreement among strategy practitioners on any of these issues, and they have consequently been lured into the invention of a wide range of different diagrammatic structures.

In the absence of established and consensual best practice, those creating strategy roadmaps have collectively generated a large body of diagrams that are highly heterogeneous in form whilst being highly homogeneous in function. This combination of diversity, continuity and quantity provides us with two distinct yet complementary research opportunities. Firstly, with respect to academic diagram theory, strategy roadmaps provide us with an unusually large pool of visualizations that require new research methods, as described later, that are likely to be of more general utility in Diagrams research. Secondly, with respect to professional practice, strategy roadmappers form a community that

G. Stapleton, J. Howse, and J. Lee (Eds.): Diagrams 2008, LNAI 5223, pp. 127–140, 2008.

develop representations but operate largely without the support of relevant diagrammatic knowledge. This offers potential for diagrams researchers to apply their research in a beneficial way. To address both these opportunities we designed and conducted a research study that both provides new methods and opportunities for diagrams research, and also practical guidance for the roadmapping community.

In this paper we first offer a brief history of the development of strategy roadmapping. We then describe a research corpus that captures the diversity of diagrams developed for roadmapping. We propose a research method that investigates this diversity by taking into account both the communicative intentions of the diagram creators and the effective design options available to them. Finally, we present results from a research project applying that methodology, including both analytic findings, and an experimental application of the findings.

2 Roadmapping – Visualizing Strategy

The 'technology roadmapping' approach was originally developed by Motorola in the 1970s to support improved alignment between technology and product development (Willyard & McClees, 1987). A key feature of this process was the synthesis of the main elements of the strategic plan into a simple high-level visual representation (see Fig. 1). Bob Galvin, who was CEO of Motorola during the period when roadmapping was established, provides the following definition (Galvin, 1998): "A 'roadmap' is an extended look at the future of a chosen field of inquiry composed of the collective knowledge and imagination of the brightest drivers of change in that field".

Following on from Motorola, the roadmapping approach has been adopted widely by many organisations around the world, at company, sector and national levels (for example, Cosner *et al.,* 2007; de Laat & McKibben, 2003). The underlying concepts are very flexible, and roadmapping methods have been adapted to suit many different goals, supporting innovation, strategy and policy development and deployment (Phaal *et al.,* 2004). Subsequent widespread adoption of the method can be readily demonstrated through a search of the Internet. A survey of public-domain roadmap documents by the

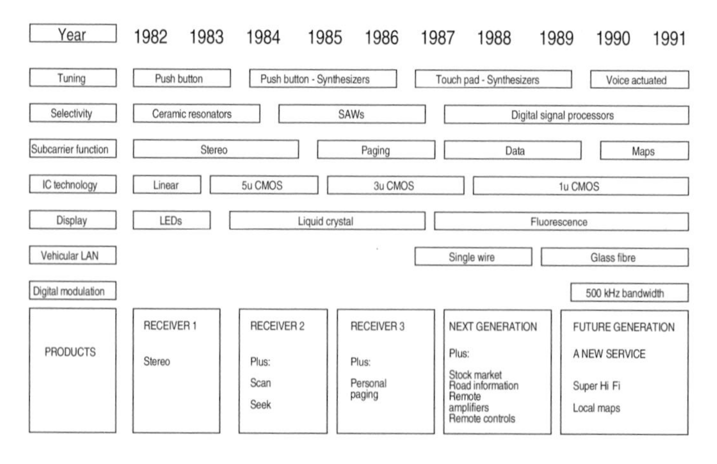

Fig. 1. Motorola roadmap (Willyard & McClees, 1987)

second author (Phaal, 2006) identified more than 900 examples from a wide range of sectors, including energy, transport, aerospace, electronics, ICT, manufacturing, construction, healthcare, defence, materials and science.

The types of issues that stimulate organisations to deploy roadmapping methods tend to share some characteristics that are inherently challenging for managers and policy makers to deal with: 1) the high level of *complexity* of the systems being considered; 2) the high level of *uncertainty* related to the forecasts and assumptions that have to be made; and 3) the high level of *ambiguity* associated with the many different stakeholder views involved in roadmap development.

The most frequently cited benefit of the roadmapping approach is that of communication, enabled primarily by the visual roadmap formats used. The process of roadmap development brings together the various key stakeholders and perspectives needed to develop understanding of complex systems and issues, building consensus about the way forward. Once a roadmap has been developed it can be more widely disseminated, acting as a reference point for ongoing dialogue and action. Design principles that support these kinds of activity have been described, in a more generic sense, by Bresciani *et al.* (2008a, 2008b). Nevertheless, although visual representation of strategy lies at the heart of strategy roadmapping, there has been little research directed towards understanding the visual function and structure of roadmaps.

3 The Research Corpus

In order to address this research question, the second author engaged in more detailed analysis of the 900 documents identified in his survey. Of these, approximately 450

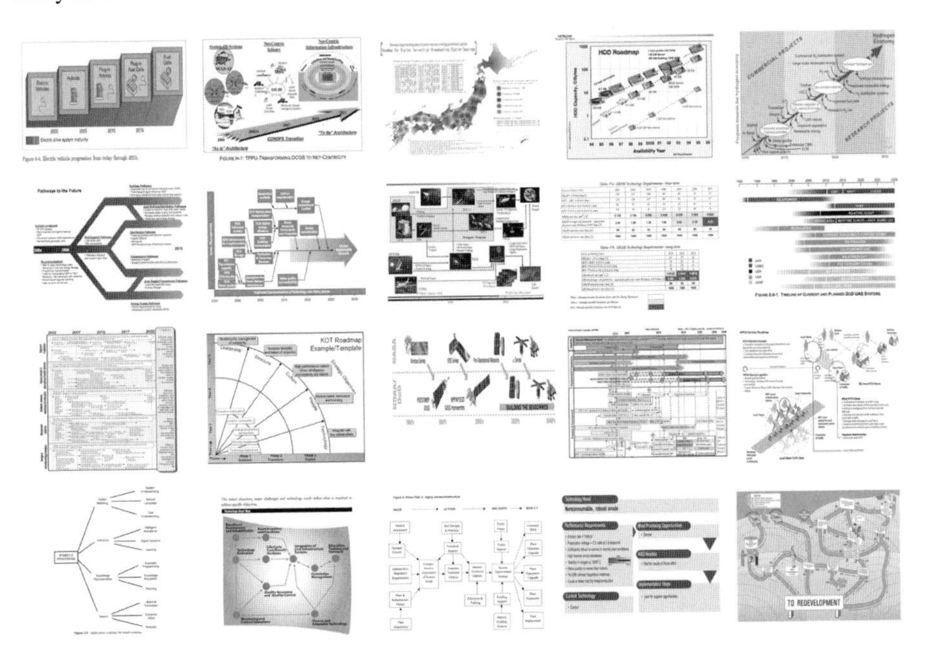

Fig. 2. Selection of visual roadmap representations, illustrating the variety of formats used

were constructed around a central visual representation (many documents described as "roadmaps" simply describe strategy in a text format). These 450 figures were extracted into a corpus including: an A5 scaled version of each figure, the original caption for the figure, and a reference to the source document. Initial examination of this rich corpus revealed that many different formats had been developed for representing strategy, as illustrated in Fig. 2. This diversity demonstrated the research opportunity arising from the strategy roadmap, as a diagrammatic application in which there is not yet an agreed set of visual conventions. It also highlighted a professional need to understand the visual dimensions of roadmapping better – what are the principles for 'good' roadmap design? The motivation and challenge for the main part of our research was to develop a method of characterizing the corpus in ways that would be relevant to the practitioner community from which it had been collected.

4 Method

Participants

The primary objective of the study was to define a typology of roadmap features with a view to characterising best-practice. However, roadmaps take many different forms and there is no authoritative definition with regard to which of these forms is more or less typical, or which of them should be regarded as "proper" members of the set. We therefore decided to proceed with our research on the basis that the set should be defined by consensus of utility to practitioners, through facilitated group meetings with expert informants. Unfortunately, the body of 'expert' practitioners for an evolving diagram class is not easy to define. Many diagrams in our sample might have been drawn by a corporate manager for whom this was the first and last roadmap they ever drew. It may be the case that common practices are emerging only through borrowing of features and slow evolution.

Our approach was therefore to treat aspirations of leadership within this community as being correlated with reflective professional practice. There are a small number of international consultants, academics and policy leaders who see roadmapping as being a central component of their professional skills. We invited a sample of these international leaders in strategic roadmapping, together with several information design specialists, to a one-day workshop held in conjunction with the July 2007 International Summit on Visualisation (Burkhard et. al. 2007). All 12 participants were sufficiently interested in the topic to travel to Zurich at their own expense, and we considered this to be evidence in itself that our informants were motivated to act as representatives of their nascent professional community. For our purposes, this sample, which might otherwise be criticized as biased or self-selected, is desirable because of the fact that it represents (self-identified) professional leadership.

Procedure

At the start of the day, the 12 participants entered a meeting room in which hard copies of the full research corpus (a total of 450 A5 colour copies) had been distributed over all surfaces of the room. Participants were initially formed into three groups of four, each of which selected a sample of 20 roadmaps from the corpus. Each group selected their sample with the objective to illustrate one of three themes designed to address likely practitioner concerns: 1) Purpose of the roadmap, 2) Good visual structure and 3) Design pitfalls. Groups were instructed to make their sample as varied as possible, within the constraint of illustrating the allocated theme. A duplicate set for each sample was then extracted from a second collection of the roadmaps, enabling participants to work in six (changing and balanced) pairs to undertake a series of card sorting activities (Rugg & McGeorge, 2005). This approach was inspired by the classic card sort study by Lohse et. al. (1994), which produced a broad classification of visual representations. The objective of card sorting techniques is organise a set of objects in terms of relative similarity, based on classification decisions made by a number of participants. In our workshop, each informant spent about 30 minutes sorting each set of 20 roadmaps into categories of their own choosing, while their partner recorded those choices.

The 12 independent card sorts for each of the three thematic samples were then statistically combined using hierarchical cluster analysis. On the basis of the cluster analysis, dendrograms were constructed for each theme, with each leaf node of the dendrogram corresponding to one of the roadmap illustrations from the original sample. We then re-drew the dendrograms at large scale, so that the original A5 illustrations could be fixed to a wall-size depiction of the cluster analysis (as shown later in Figures 4 and 5). These large-scale illustrations were prepared during a lunch break, keynote address and social activity in the middle of the day.

The large-scale dendrograms were presented to our expert informants as a "manufactured consensus". We acknowledged that, just as the corpus itself was highly diverse, the developing status of the roadmapping community meant that expert opinion would be diverse rather than uniform. We therefore asked the informants to accept this "consensus" as being a basis for further collaboration, rather than attempting to modify it directly. Instead, we asked groups to work together to find the best possible *labels* for the branch points in the dendrogram, taking the clusters themselves as given. These negotiated labels form a key finding of the facilitated process. Finally, the resulting dendrograms were tested by using the emerging structure to support the rapid development of 'meta roadmaps' (roadmaps of roadmapping research). This was based on a discussion of purpose and good visual structure, taking into account the identified design pitfalls. The most preferred meta roadmap was then presented to other conference attendees (Burkhard et. al. 2007).

Pilot Study

The workshop described above was both methodologically and logistically challenging. The complex and detailed procedure necessitated that it be preceded by a pilot study. In late 2006 a pilot workshop was conducted with 16 graduate students undergoing professional training in engineering design and manufacturing management. The pilot workshop was conducted on the same basis as the main study but differed in two main aspects. Firstly, in the pilot workshop, there were four themes rather than three because the theme that eventually became "design pitfalls" was initially divided into those pitfalls relating to visual structure and those that relate to conceptual purpose. After this caused confusion in the pilot study, the two pitfall types were later merged into a single theme. (For statistical reasons, this reduction in themes also enabled a reduction in the number of participants.) Secondly, the pilot study was divided into two sessions, with the cluster analysis and dendrogram production completed during a two-week break between sessions. This allowed us to explore alternative software packages and representation types, as well as defining the length of the break that would be needed to conduct the analysis during the main one-day workshop.

5 Results

Three kinds of result were achieved from the method described. The first was three hierarchical *classifications* that emerged from statistical cluster analysis of the sorting decisions. The second was the descriptive *labels* for each part of the classifications that were developed by the group through a facilitated discussion process. Our final objective was to present the resulting labeled classifications as decision trees or guides that drew on the three themes, thereby contributing to the design of roadmap diagrams that will a) serve an explicit purpose, b) employ proven graphic practices, and c) avoid common design pitfalls.

Statistical analysis used the Prodax data analysis and visualization tool (www.prodax.ch). The first step was to construct similarity matrices for each theme, in which the 20 roadmaps were arranged on horizontal and vertical axes. Each time a participant sorted two roadmaps in a single group, the number in the corresponding cell of the similarity matrix is increased by one. The similarity matrices of all participants can then be aggregated through cluster analysis. Normalised multi-dimensional scaling (NMDS) was used to transform the similarity ratings of the participants into a Euclidean distance model (Borg & Groenen, 2005: 411) that shows relative similarity of the 20 roadmaps in that theme as relative distance within a 2D plane (roadmaps that are more similar are closer together), as in Figure 3. The reduction of the multi-dimensional data space into this 2D plane results in a degree of "stress", where similar items appear further apart than they should do. The number at the bottom right of the plot indicates the degree of stress for this theme.

We also conducted an outlier analysis, through transformation of the similarity matrices, in order to check whether particular participants made sorting decisions that differed dramatically from all other participants. Figure 4 shows participant 33

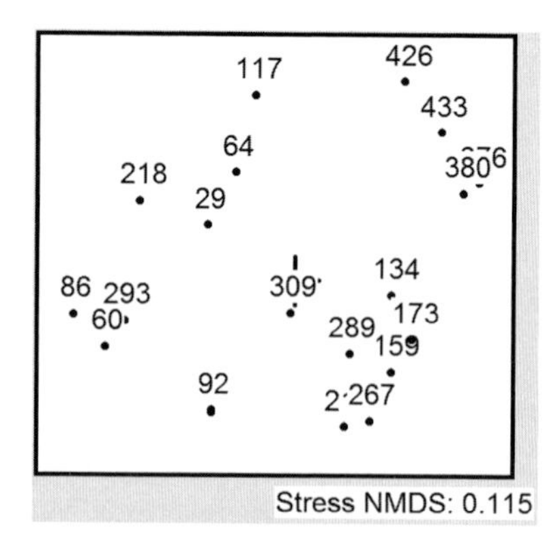

Fig. 3. Example result from multi-dimensional scaling analysis, showing roadmap similarity based on card sorts made within the *purpose* theme

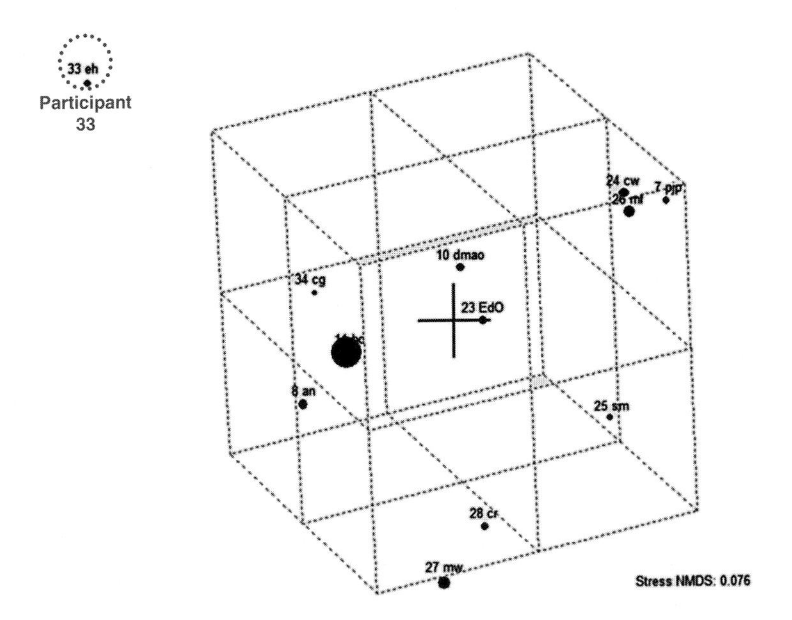

Fig. 4. Example of participant outlier analysis for the *purpose* theme

as an outlier for the *purpose* theme. The results often corresponded to our own observations, with informants whose data appeared as an outlier also disagreeing during other workshop discussion (for example, that participant also appeared as an outlier in another theme). This is likely to be related to different professional or cultural perspectives of those informants, but we chose not to investigate the views of specific participants any further. Instead, we removed outlier sorting decisions when constructing the group consensus.

The average similarity judgments obtained using NMDS were then compared using hierarchical cluster analysis, producing a dendrogram for the purpose theme as shown in Figure 5a). This process starts with the two most similar items – in this case items 376 (partially covered) and 380 in Figure 3. The similarity measure indicates that these two items were grouped into the same class most frequently. These two are linked as the first "cluster" of two items at the left of Figure 5a). The analysis continues to group the most closely related items in order, combining these groups of two into higher level clusters, until the upper levels of the hierarchy show the most generic distinction between different types of roadmap.

Purpose

In the purpose theme, this highest level classification is between roadmaps made for specific scheduling of goal achievement (*with* an explicit time indication) and roadmaps designed for general orientation and visioning (*without* explicit time indications). Figure 5b) shows the large scale version of the dendrogram that was created during the workshop break, onto which the original A5 roadmaps were attached. The labels created by participants (written on Post-it notes) have been reproduced at the left, and the numerical identifiers corresponding to the statistical analysis are shown at the right.

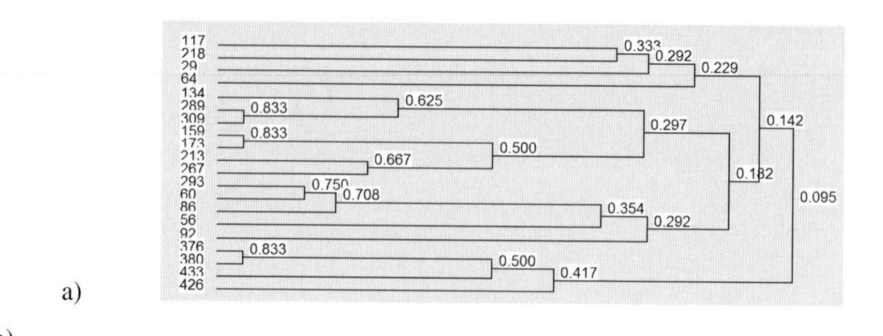

a)

b)

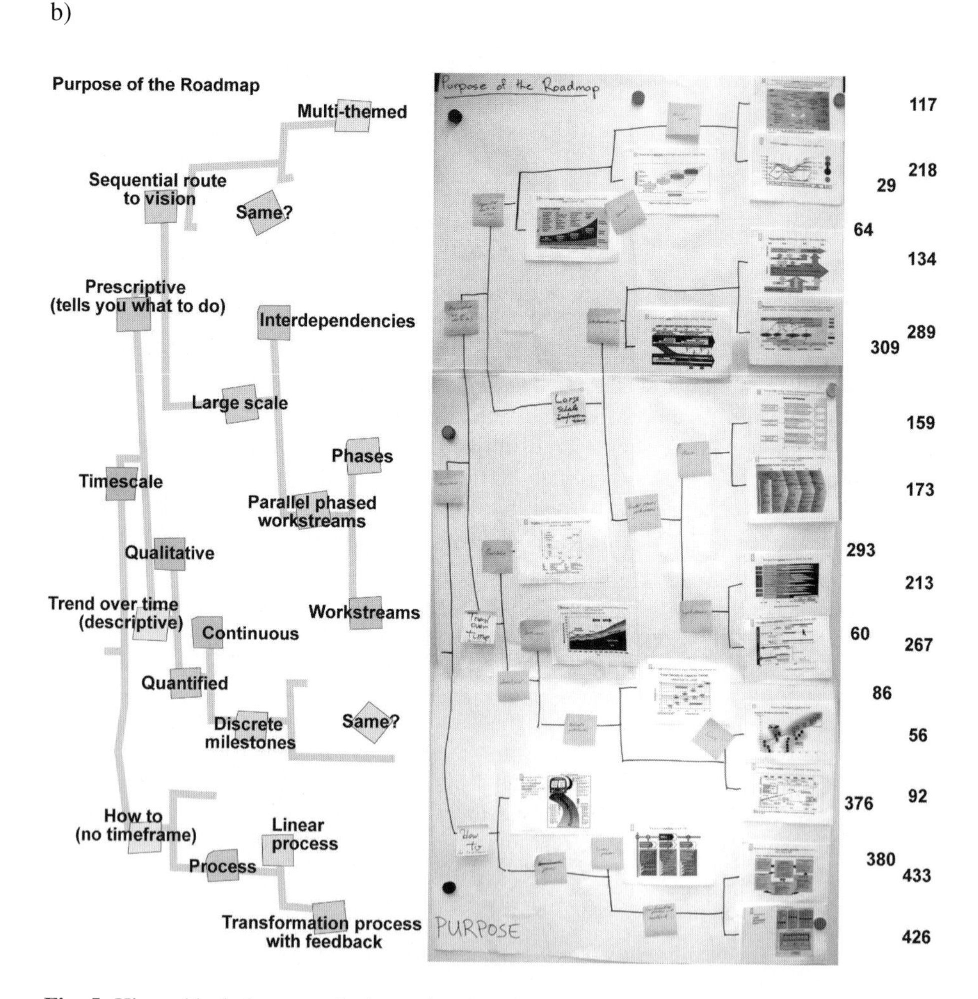

Fig. 5. Hierarchical cluster analysis results a) and resulting category labels b) of the dendrogram for the purpose-oriented classification of roadmap formats. Part b) includes a photograph of the large-scale chart created at the workshop. Note that the chart was reversed with respect to the cluster analysis, in order to emphasise categories. For clarity, labels have been reproduced in a larger font at the left of the photograph, and cluster analysis codes at the right.

a)

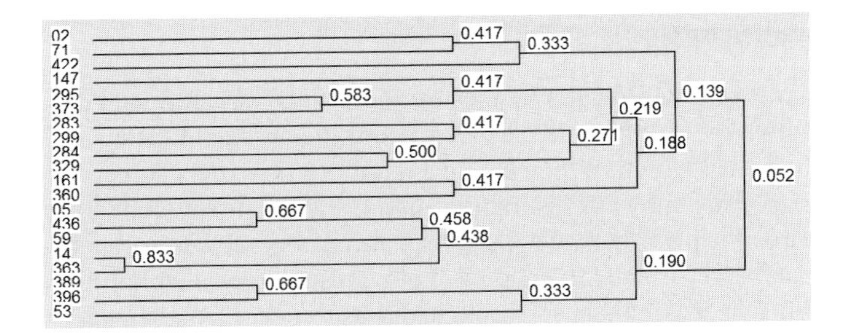

b)

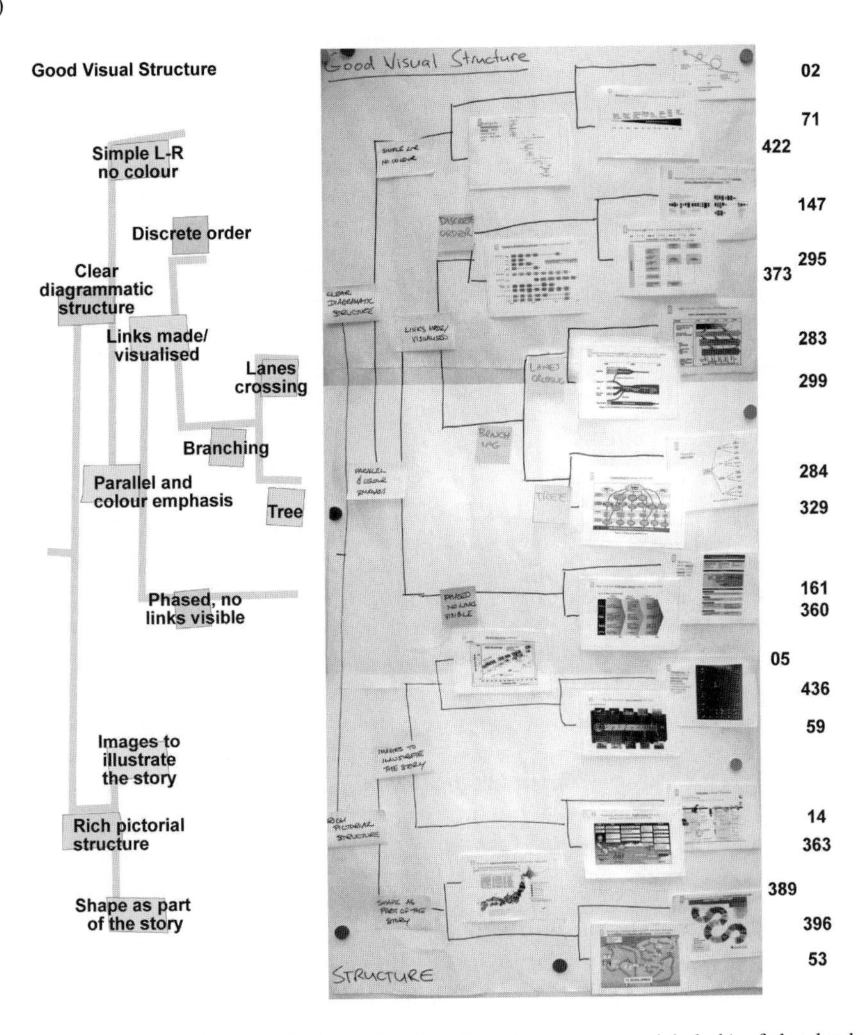

Fig. 6. Hierarchical cluster analysis results a) and resulting category labels b) of the dendrogram for the good visual structure classification of roadmap formats. As before, part b) shows the workshop chart, with labels repeated at the left and cluster analysis codes at the right.

Good Visual Structure

This process was repeated for the themes of good visual structure and design pitfalls. Figure 6a) shows the resulting dendrogram from hierarchical cluster analysis of the sample for the visual structure theme, and Figure 6b) the labels agreed for the higher level categories. This shows a distinction on the highest level among *diagrammatic* and *pictorial* roadmap formats. The roadmap formats perceived to be similar by the most participants are formats 363 and 14, both consisting of a table-like organization of information that was enhanced through the use of icons and imagery. Roadmap number 422, a Gantt chart-like format, achieved the least agreement regarding its grouping of all formats in this selection. This was reflected by the fact that, when labelling the resulting dendrogram, participants did not agree on a suitable label for the branch that contained this figure by itself.

Design Pitfalls

Regarding bad design practices the highest level distinction related to roadmaps that were either *cluttered/overloaded* (too much content) or failed to convey a *consistent* message because they lacked an organizing overall logic or *scheme*. The roadmap formats judged to be similar regarding their design mistakes by the most participants were two formats (337 and 339) that both contained more than 40 unstructured boxes with many connections and two formats (324 and 350) that did not contain any layers but more than ten related sections. We have not included the full set of classified design pitfalls in this paper due to space constraints, but the specific examples mentioned are illustrated in figure 7.

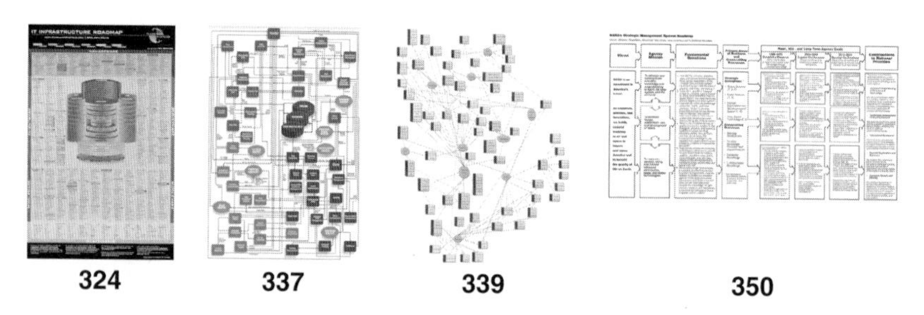

Fig. 7. Examples illustrating common categories of *design pitfalls*

Summary

Overall, we found that dendrograms created using hierarchical cluster analysis provided a valuable tool to streamline group discussion on strategic purpose, and on the types of visual approach that could be used in order to achieve those purposes. Statistical cluster analysis created a manufactured consensus that could readily be interpreted by labelling top level categories. Close correlations at the bottom level of this hierarchy provided an opportunity for participants to recognise similarities among different members of an agreed common class. At intermediate levels, where atypical examples from the corpus were not strongly related to others in the sample, or where

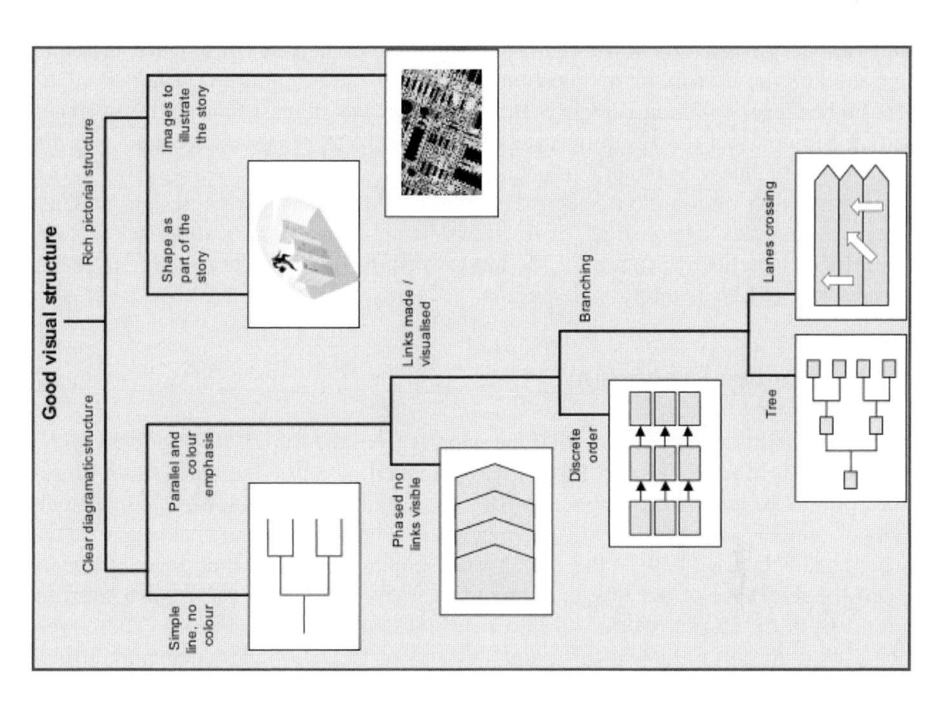

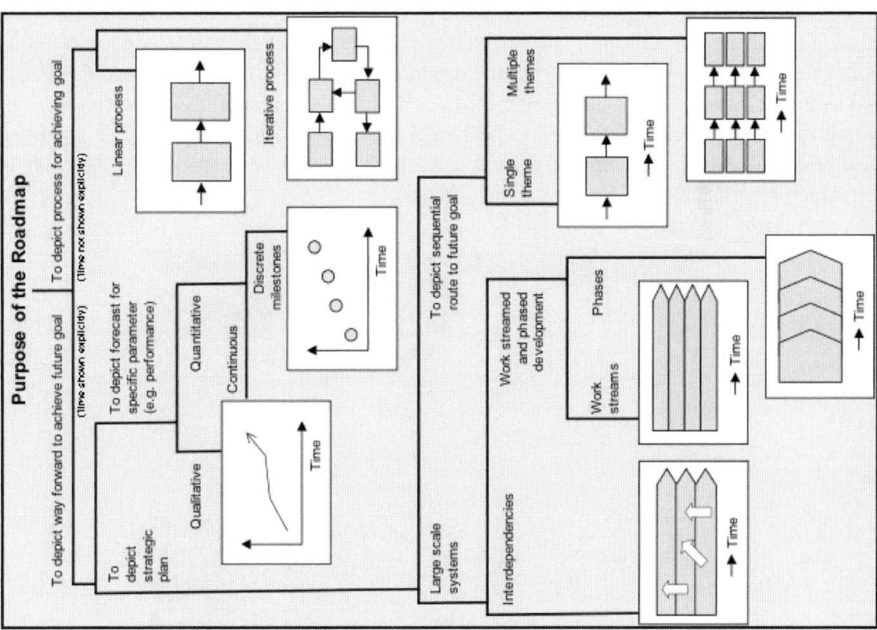

Fig. 8. Decision trees, based on Zurich workshop dendrograms, for use in a practitioner workshop. These illustrate decision trees for roadmap *purpose* and *visual structure*.

there was less consensus with regard to similarity, participants responded either by not labelling the branch, or by questioning whether these might be further examples of a higher level common category. In either case, we suggest that ability of expert participants to label the resulting categories is likely to be well correlated with the utility of those categories to practitioners.

Classification of common design flaws may be of less value for design guidance (specific examples demonstrate likely flaws better than category descriptions), but still provide benefits by prioritising the kinds of warning that might be given to practitioners who wish to avoid possible pitfalls.

6 Practitioner Evaluation

An initial opportunity to test whether the outputs from the Zurich workshop are likely to be relevant to strategy practitioners was provided by an invitation to run a further workshop at an industrial symposium (the 13th Cambridge Technology Management Symposium, in September 2007). Rather than repeating the card sorting exercise (which had been used both in the Zurich workshop, and in the earlier pilot), we chose to test the relevance of our results to practitioners by developing the dendrograms resulting from the Zurich workshop into pictorial decision trees (Fig. 8). These were then used as design frameworks for small groups of workshop participants to develop a roadmap for a strategic management topic of their own choice (Fig. 9).

This practitioner event demonstrated the potential utility of the decision-tree roadmap design concept, although further work is needed to develop the decision tree and associated guidance relating to organizational function. Groups were able to design and construct a roadmap quickly (45 minutes). However, it was observed that groups containing participants with prior experience of roadmapping tended to use roadmap structures influenced by their previous experience (these were somewhat similar to the structure originally developed by Motorola).

Fig. 9. Roadmap design activity at practitioner workshop, after choosing goal and structure from use of decision trees

7 Discussion

Roadmaps are a significant and widespread business tool. We wished to establish some consensus about their desirable visual features and potential applications, at a level that would provide a principled basis for design guidance. In much research into the properties of diagrams, it is hard to establish direct connections between application needs such as these, and available empirical research methods. However, we were successful in creating a summary description of a large and complex representative corpus, with structural categories created by, and recognizable to, domain experts.

In seeking to define normative standards for any type of diagram, there is an inherent trade-off between the desire to obtain objective, balanced guidance and the desire to exploit the personal, subjective judgements of experts who represent the views of the main stakeholders. Our workshops sought to resolve this conflict by applying an (objective) cluster analysis to the (subjective) judgements of relevant experts. As a result, the output from the workshop is a rigorously derived, inter-subjectively agreed, normative typology for roadmap diagrams.

This research approach offered a number of distinct benefits relating to the research corpus, the workshop participants and the selected methodology. Firstly, the unique corpus of material upon which the study was conducted is comprehensive, tangible and accessible. Secondly, the workshop participants represent experts from the various domains most relevant to roadmapping (including managers, visualisation scientists and information designers). Thirdly, the necessarily subjective activity of defining best-practice was tempered by employing the rational technique of cluster analysis to construct a consensus among the various experts. In combination, this approach allows a variety of perspectives to be brought to bear on a large sample of real-world material whilst retaining some objectivity and coherence.

Within the constraints of a workshop session, certain methodological limitations are unavoidable, and consideration must be given to any bias that these limitations impose. Firstly, there is the possibility that under time constraints, the participants involved in the selection and sorting of roadmaps (for 'goodness' or 'quality') will exhibit undue preference for overly simplistic or graphically seductive representations. Secondly, that by viewing these representations out of context, the roadmap sample must be judged with respect to the *inferred* objectives of the roadmap producers (including their intended communicative goal and the anticipated target audience). However by assembling a team of participants – expert in both the production and interpretation of roadmaps – efforts were made to mitigate these potential limitations.

We believe that this study has characterized the domain of strategy roadmaps sufficiently well to allow future investigation of specific applications and visual forms. One contribution from our research is therefore the creation of an initial exploratory structure of the domain. We have also demonstrated that even initial characterization provides a basis for some practical guidance to practitioners, by the simple strategy of turning a research categorization into a decision tree for initial stages of design. Finally, by addressing a context of diagram use that is quite distinctive, yet rather different in nature to more typical educational and technical uses of diagrams, we have found it necessary to develop a qualitative empirical methodology that is well suited to engaging with expert practitioner groups. We expect that this methodology will be of value in future studies of emerging diagram use in specific professional contexts.

Acknowledgements

We thank the workshop participants: students of the EDC and IfM for their assistance in the first workshop, organizers and attendees at the Technology Management Symposium, and expert informants at the Zurich workshop: Brock Craft, Elke den Ouden, Mike Ferril, Clive Goodchild, Elizabeth Harvey, Ralph Lengler, Steve Mann, Andreas Neus, Dominic Oughton, Paul Palmer, Clive Richards, Masayoshi Wanatabe, Colin Winfield. Alan Blackwell's contribution to this project was sponsored by Boeing Corporation.

References

Bresciani, S., Blackwell, A.F., Eppler, M.: A Collaborative Dimensions Framework: Understanding the mediating role of conceptual visualizations in collaborative knowledge work. In: Proc. 41st Hawaii International Conference on System Sciences (HICCS 2008), pp. 180–189 (2008a)

Bresciani, S., Blackwell, A.F., Eppler, M.: Choosing visualisations for collaborative work and meetings: A guide to usability dimensions. Darwin College Research Reports DCRR-007 (2008b), http://www.dar.cam.ac.uk/dcrr/

Borg, I., Groenen, P.J.F.: Modern Multidimensional Scaling, 2nd edn. Springer, New York (2005)

Burkhard, R.A., Andrienko, G., Andrienko, N., Dykes, J., Koutamanis, A., Kienreich, W., Phaal, R., Blackwell, A., Eppler, M., Huang, J., Meagher, M., Grün, A., Lang, S., Perrin, D., Weber, W., Vande Moere, A., Herr, B., Börner, K., Fekete, J.-D., Brodbeck, D.: Visualization Summit 2007: Ten research goals for 2010. Information Visualization 6, 169–188 (2007)

Cosner, R.R., Hynds, E.J., Fusfeld, A.R., Loweth, C.V., Sbouten, C., Albright.: Integrating roadmapping into technical planning, Research Technology Management, November–December, pp. 31–48 (2007)

de Laat, B., McKibben, S.: The effectiveness of technology roadmapping – building a strategic vision, Dutch Ministry of Economic Affairs (2003), http://www.ez.nl

Lohse, G., Biolsi, K., Walker, N., Rueter, H.: A classification of visual representations. Communications of the ACM 37(12), 36–49 (1994)

Phaal, R., Farrukh, C.J.P., Probert, D.R.: Customising roadmapping. Research Technology Management 47(2), 26–37 (2004)

Phaal, R.: Technology and other (mostly sector-level) published roadmaps (2006), http://www.ifm.eng.cam.ac.uk/ctm/trm/documents/published_roadmaps.pdf

Rugg, G., Mc George, P.: The sorting techniques: a tutorial paper on card sorts, picture sorts and item sorts. Expert Systems 22(8), 94–107 (2005)

Willyard, C.H., McClees, C.W.: Motorola's technology roadmap process. Research Management, 13–19 (September–October 1987)

VAST Improvements to Diagrammatic Scheduling Using Representational Epistemic Interface Design

David Ranson and Peter C-H Cheng

Department of Informatics, University of Sussex, Falmer, BN1 9RN, UK
d.j.ranson@sussex.ac.uk, P.C.H.Cheng@sux.ac.uk

Abstract. REpresentational EPistemic Interface Design (REEP-ID) advocates exploiting the abstract structure of a target domain as the foundation for building cohesive diagrammatic representations. Previous research explored the application of this approach to the display and optimisation of solutions to complex, data rich, real world problems with promising results. This paper demonstrates the application of these principles to generate interactive visualisations for solving complex combinatorial optimisation problems, in this case the University Exam Timetabling Problem (ETP). Using the ETP as an example the principles of REEP-ID are applied, illustrating the design process and advantages of this methodology. This led to the implementation of the VAST (Visual Analysis and Scheduling for Timetables) application, enabling individuals to solve complete instances of the ETP using interactive visualisations. Rather than using automated heuristics or algorithms, VAST relies entirely on the user's problem solving abilities, applying their knowledge and perceptiveness to the interactive visualisations maintained by the computer. Results from an evaluation of VAST support the use of the REEP-ID methodology and the case for further research. In the closing discussion these findings are summarised together with implications for future designers.

Keywords: Representational Epistemic Design, Exam Timetabling, Visualisation, Diagrammatic Reasoning.

1 Introduction

Educational Timetabling is a vibrant research field and also a very practical problem faced yearly by almost all educational institutions. Most research on this topic rightly concentrates on the incremental improvement of automated systems and the heuristics used to solve timetabling problems. Typically institutions rely on automated approaches to generating exam timetables; however the solutions generated by such automated systems do not always match real world requirements of a timetable and existing interfaces can make it difficult for practitioners to make minor intuitive changes to a solution without the risk of damaging the timetable in unforeseen ways.

Here, an alternative approach is presented that applies a REpresentational EPistemic approach to Interface Design (REEP-ID). The REEP approach advocates analysing the underlying conceptual structure of a problem domain and finding graphical schemas to directly encode that structure[1]. This approach has been used to design

G. Stapleton, J. Howse, and J. Lee (Eds.): Diagrams 2008, LNAI 5223, pp. 141–155, 2008.

novel representations for science and mathematics learning (e.g., [2]) and to create novel interactive visualisations for event scheduling, personnel rostering and production planning and scheduling (e.g., [1, 3, 4]). This paper is a contribution to the wide research programme on the REEP approach and it is in particular exploring how REEP-ID can be used to create novel representations that enable solutions to instances of the ETP to be found from scratch by humans without resorting to major computer automation of the problem. It builds upon the STARK-Exam [3] systems that successfully support the problem of manually improving existing examination solutions. By combining the problem solving and perceptual abilities afforded to humans with well-designed interactive visualisations maintained by a computer, the aim is to demonstrate certain advantages over purely automated approaches and existing scheduling interfaces. It will be shown that the application of a REEP-ID approach provides one framework that can be applied to generate interactive visualisations capable of supporting these goals.

The following section introduces the ETP in more detail. After this the development of the Visual Analysis and Scheduling for Timetabling (VAST) system is presented in Section 4, together with a discussion of how the REEP-ID methodology was applied. Section 5 summarises some of the results from an evaluation of the VAST system and the paper closes with a brief discussion of the implications of this approach.

2 Exam Timetabling

Many timetabling problems have similar conceptual structures, instances of educational timetabling problems such as the ETP usually differ between institutions making theoretical models and algorithms harder to generalise. As a special case of educational timetabling the ETP is interesting to focus on for several reasons, it is a practical problem that has been the focus of much research for at least four decades. Improvements made to exam timetabling system interfaces could have benefits to practitioners, researchers and students sitting the exams. When scaled up to real life situations the size and algorithmic complexity of the problems make finding valid solutions very difficult. The conceptual complexities inherent in the problem, such as those arising from the sheer number of entities in a typical ETP and the relationships between them, make it difficult for individuals to conceptualise complete instances of the ETP with current widely used interfaces.

2.1 Definition

Fundamentally timetabling is the assignment of events to time periods subject to constraints. In many cases the events also need to be assigned physical locations where they will take place. In the timetable model adopted here [5, 6] a complete ETP representation consists of some abstract timetabling *Problem*, e.g. the University Exam Timetabling Problem, a concrete *Instance* of a problem which defines the available resources, events, time periods, unique constraints and finally a *Solution* or Solutions to an Instance that stores the actual assignment of events. In the case of the ETP there are many concepts, with associated attributes, that need to be represented: Time Periods, Rooms, Students, Exams, the set of Constraints that determine what assignments are valid, and the assignment of solution variables to exam events are all examples of these.

There are also logical relationships between entities in the ETP, for example sets of exams that have the same duration or *Clusters* of exams that do not share any common students. Grouping exams into such Clusters has the benefit that all Exams within a Cluster can be scheduled into a single time period, assuming there is sufficient room capacity. Exams that do share common students are termed *conflicting* exams.

2.2 Constraints

Within the ETP constraints can be either *hard* or *soft*. To be feasible a solution cannot violate any hard constraints, whilst it is undesirable to violate soft constraints they can exist in a feasible solution. It is the goal of exam timetabling to eliminate all hard constraint violations and as many soft violations as possible. Table 1 summarises the most important common constraints that are typically found in the ETP [7]. The *Conflict* constraint, together with *Room Capacity* and *Time Period Duration* constraints are present in all instances of the ETP; we also include *Completeness* as a hard constraint indicating that *all* exams need to be assigned for a timetable to be feasible. The precise details of the other constraints will differ between institutions and this is not an exhaustive list.

Table 1. Common constraints found in most instances of the ETP

Constraint	Description	Hard/Soft?
Completeness	All exams must be timetabled	Hard
Room Capacity	An exam must be timetabled in a location with sufficient capacity for all the assigned students.	Hard
Time Period Duration	An exam must be timetabled in a time period of sufficient duration	Hard
Conflict	A student cannot sit more than one exam simultaneously.	Hard
Consecutive	A student should not sit two or more exams in immediate succession. Depending on the institution this may or may not include overnight gaps.	Typically soft
Order Precedence	An exam should be scheduled before, after or at the same time as another exam.	Depends on instance
Mixed Durations	Exams in the same room should be of the same duration to minimise disruption	Typically soft

To allow an automated solver to evaluate the quality of a solution each constraint violation is typically applied some weight which can then be aggregated to calculate an associated penalty cost for that solution. In this situation an optimal solution with no violations would have a cost of zero whereas a solution with many hard constraint violations would have a very high penalty cost. When using the VAST visualisations this cost function is less important to the user than the more meaningful visual representation of the constraint violations encoded in the interface and possibly combined with other knowledge of an ideal solution not taken into account by the formal evaluation function, e.g. political influences.

2.3 Existing Interfaces

Existing interfaces, including the most widely used commercial applications[8], for inspecting and modifying Exam Timetables are typically based around a traditional timetable layout with days spread across the horizontal dimension and time periods within each day across the vertical dimension. This approach is undesirable from a representational point of view as the concept of time is mapped to two different spatial dimensions of the timetable layout.

This traditional layout, although good for quickly searching for an exam using a day and time as an index, makes some judgments useful to solving these problems very difficult to make, for instance: How much free room capacity remains at any time?, What are the effects on the global timetable of moving a single exam? It is hoped that by taking a representational epistemic approach many of these important concepts, together with the features of the problem that support effectively solving instances of it, can be encoded into a more cohesive representation.

There are also other drawbacks to many of the existing interfaces currently used. Information important to conceptualising the problem, such as exam location, violation of constraints or exam durations are often only represented through text labels or lists. To gain a complete global view information often has to be accessed through individual dialog boxes or by switching between different perspectives, with multiple representations within the interfaces.

3 Representational Epistemic Interface Design

The REEP-ID approach[1] to interface design provides four main steps in the design of a novel interface:

1. Identify entities in the target problem domain and the basic relationships between them.
2. Find the conceptual dimensions, sources of conceptual complexity and levels of granularity present in the problem.
3. Identify the primary conceptual dimensions and try to find a cohesive visual/spatial mapping of these.
4. Design concrete visual representations for the remaining entities and relationships that respect the mapping of the primary conceptual dimensions.

The start point for the design of VAST was our previous STARK-Exam systems [3]. The overall goal was to design a representation that effectively captures the conceptual features and relationships of the ETP without altering or adding to the existing complexities. The first steps in the new design involved making explicit the different levels of granularity of entities and concepts that were implicit in design of STARK-Exam systems. These features of the ETP, based on the entities and relationships identified in the existing ETP model [6] introduced in Section 2.1, contribute to the conceptual complexity of the domain and are summarised in Table 2. At the operational level users will interact with the representation, introducing individual differences and understandings that affect the operational complexity. The functional level contains the features that make the problem so difficult to solve, for example the dynamic interactions between constraints and entities. The molecular level shows some conceptual structures that can emerge from the relationships and groupings of features found at the atomic level, these structures are inherent to the abstract problem so should also be available in an effective closely-mapped representation. Finally the atomic level shows the entities that the representation is attempting to model to solve the ETP, complexity at this level can be found in the magnitudes of the dimensions and the sheer number of entities or functional relationships that can exist in a typical ETP.

Table 2. Different levels of granularity of Conceptual Complexity identified in the ETP

Level	Concepts
Atomic	Students, Exams, Rooms, Times
Molecular	Clusters, Departments
Functional	Conflicts, Capacity, Availability Order, Duration, Completeness
Operational	Heuristics, User Interaction

Once the features of the target domain had been identified, together with the level of granularity at which they contribute to the conceptual complexity of the problem, a cohesive mapping of the primary conceptual dimensions had to be found to provide the foundations to a *global interpretive framework*[1]. The primary conceptual dimensions for the ETP solution were identified as *time periods* and *room capacity* and the mapping is the same used within the STARK-Exam system, and is presented in Figure 1. Other mappings such as the traditional timetable layout were briefly considered but rejected for the reasons previously discussed. If suitable representations for crucial features or concepts in the problem could not be found that fit the visual spatial mapping of the primary conceptual dimensions, then this would be an indication that these mappings might need to be rethought. The following sections describe the particular graphical schemes that have been used to encode each of the levels and concepts described in Table 2.

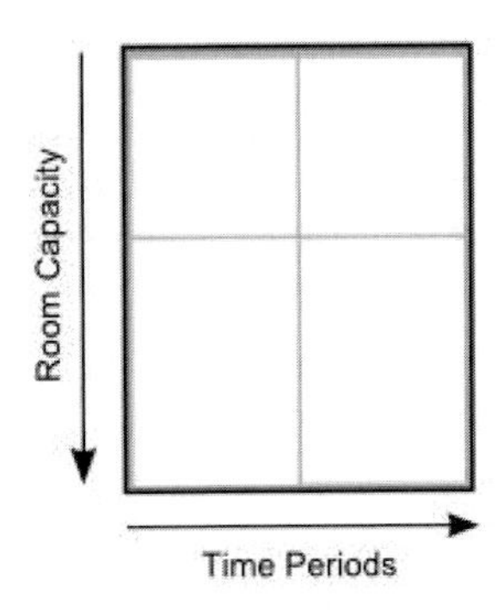

Fig. 1. Mapping the Primary Conceptual dimensions *Time Periods* and *Room Capacity* to a visual spatial representation. Room capacity is represented in the vertical dimension and time periods are shown across the horizontal dimension in chronological order. Breaks in the background indicate the start of a new time period or the start of a new room.

Within the interactive visualisations being developed every graphical entity representing some concept from the problem domain is termed a *glyph*. Each type of glyph should be clearly distinguishable from others and have a clear topological and geometric form. To help maintain a cohesive unambiguous design throughout the visualisations the geometric structure of a glyph cannot change and each glyph always represents the same information, for example once an exam has been encoded as a particular glyph it cannot then be represented by any other glyph or representation within the same visualisation. Interaction with the visualisation is determined by the interactive capabilities of each glyph.

Whilst this visual structure of a glyph cannot change, a glyph can be *decorated* to augment it with further information. Decorations can be used to show different perspectives of information, or perhaps making extra information available in response to some user interaction. Colour is typically used as a decoration on a glyph and as such the colours can change to indicate concepts such as group membership (cluster, department etc), to highlighting a glyph to show it has been selected by the user or if a user's attention needs to be brought to a particular glyph.

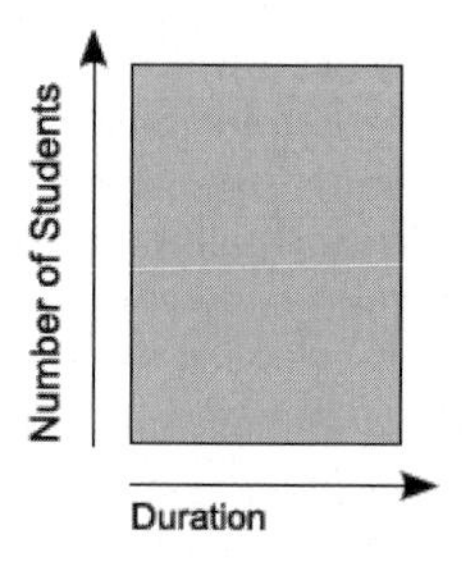

Fig. 2. An Exam Glyph, a concrete graphical representation of an Exam exploiting the spatial mapping of the primary conceptual dimensions. The width is always proportional to the duration of the Exam whilst its height is proportional to the number of students sitting that Exam.

The first glyph we explicitly introduce is the Exam Glyph, shown in Figure 2, this is a visual representation of the Exam event that exploits the mapping of the primary conceptual dimensions to encode the number of students sitting the exam (height along the vertical room period dimension) and the duration of the exam (as width along the horizontal time period dimension). The time period and room location of an assigned exam will be encoded as the glyphs position in the solution view explained in Section 4.2.

In Table 1, and again when examining the sources of functional conceptual complexity, we identified some of the common ETP constraints that need to be encoded in the representation. In this section we show how some of these constraints are already represented by the following the syntactical rules provided by the spatial, geometric and topological mapping of the primary dimensions and exam glyphs, whilst others require new glyphs to be added to the visualisation. For each of the constraints a number of representations were considered such as different forms of lines between violating exams, Shading violating exams, or icons decorating violating exams.

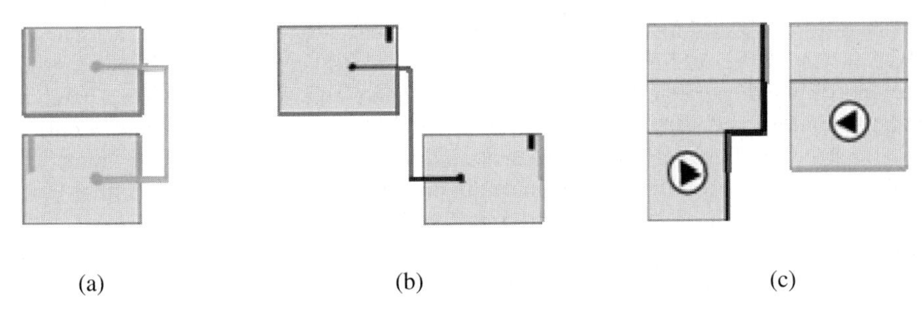

Fig. 3. Representation of constraint violations in VAST, dashed lines indicate time periods. (a) Shows a conflict violation between two exams with a thick line (bright orange in VAST) and size bars on the left to indicate the proportion of students in the conflict. (b) Illustrates a consecutive violation between two exams in successive time periods with a thin black line and size bars on the right. Finally (c) shows both a mixed duration violation, highlighted by the jagged right hand edge in the time period on the left (emphasised here by the thick line), and an order precedence violation between two exams indicated by arrowheads pointing in the direction the exams should be moved to resolve this violation. These violation representations can always be perceived even when applied, in any combination, to the same exams or otherwise overlap.

The room capacity and time duration constraints are encoded by the consistent mapping of time and room capacity in the primary dimensions and the exam glyphs. An exam glyph will not physically fit in a room time period combination in the timetable layout shown in Figure 1 unless that location has sufficient duration or room capacity for that exam. Also the mixed duration constraint is represented by the alignment of the right hand edges of exam glyphs, if exam glyphs in a timetable location have a jagged right hand edge, as illustrated in Figure 3(c), a violation of this mixed duration constraint is present.

It was found that lines were useful to link violations that were spatially close, for example conflict and consecutives violations, shown in Figures 3 (a) and (b), only occur in the same and successive time periods respectively. Drawing lines between points that are not always spatially close, for example between exams with order precedence constraints, adds to visual clutter and it is not always easy to quickly determine which the violating exams are. Care also needed to be taken to ensure that lines do not obscure other features of the visualisation, in VAST all lines follow defined paths between exams and different violations have different thickness so they can always be perceived.

Decorations applied to conflicting exam glyphs are designed to show the magnitude of conflict and consecutive violations, information necessary to strategically resolve such violations. In this scenario a *size bar* is shown which indicates the proportion of students in an exam that contribute to the violation, this allows judgements to be made as to which are the most conflicting exams in a violation. Also order precedence violations are represented by an arrowhead icon; the initial intention had been to represent these with arrow connectors however, as discussed, we wanted to avoid the use of lines except in spatial localities. To resolve this issue we used the arrowhead decoration on violating Exam glyphs to indicate the temporal direction the exams should be moved in order to resolve the conflict, this is also illustrated in Figure 3(c).

Shading of exams to encode constraint violations was also avoided as coloured shading was earlier defined as a decoration that could be changed to indicate group membership and user selection. Shading, in particular an easily distinguished hatching pattern, is however used to highlight all exams that *conflict* with the user's current exam selection. This helps resolve constraints intelligently as users have immediate feedback highlighting where an exam can be moved without introducing further violations. An extension of this, dynamic time period restriction, also shades out entire time periods where scheduling an Exam would cause further violations to be introduced and is illustrated later in Figure 6.

4 VAST

4.1 Instance Visualisation

The next two sections show how the glyphs that have been described are combined in these two interactive visualisations to completely represent the ETP. The VAST system is made up of two main components, corresponding to the *Instance* and *Solution* parts of a timetable model introduced in Section 2.1. A user can switch between the Instance and Solution components at any time by changing between standard tabbed pages and a dialog could be opened to show the glyphs that were currently selected in the instance view if necessary, it might be beneficial to instead show both visualisations on two monitors simultaneously. The purpose of the instance visualisation is to represent exams before they are assigned to the timetable, in this early stage of the problem solving process the concept of time periods is of little help in structuring the ETP entities as no assignments exist.

Rather than presenting this information as an arbitrary list the exams are instead grouped, by columns, into *clusters*. Exams can be selected for scheduling either by clicking individual exams, lassoing any group of exams or by selecting entire clusters by clicking a handle at the top of each column. The intention of cluster selection is to provide a simplification of the conceptual problem to the user. These entire columns can be assigned to single time periods in the solution without introducing any conflict constraints. Given a good enough clustering of the exams and enough space in the timetable a feasible solution could be constructed by assigning each of these columns to different time periods, eliminating the otherwise tedious necessity to consider every individual exam.

The instance view also supports other groupings of the exams by attributes such as department membership or duration. Columns can also be sorted by the number of exams they conflict with, the number of students in the exams or the duration of exams. These groupings and orderings, based on conceptual features of the abstract problem, allow users to make intelligent choices about which exams to select to implement different problem solving strategies.

The final use of the instance view is to represent the completeness constraint, identified in Table 1, that every exam must be assigned to the timetable. As can be seen in Figures 4 and 5 once an exam has been scheduled it is moved from the Instance visualisation to the Solution visualisation however its outline remains in the instance view. This allows users to quickly perceive how many, and precisely which, exams are waiting to be timetabled.

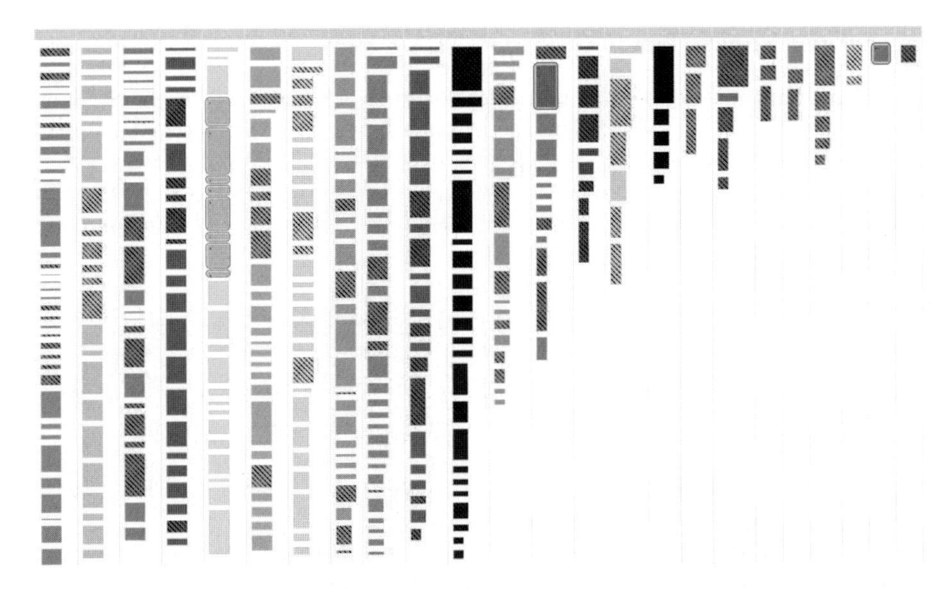

Fig. 4. The VAST Instance visualisation showing unscheduled exams grouped into columns such that exams in the same column do not conflict (clusters). The columns can be changed to group the exams by other attributes such as department membership or their duration. Exams that have been selected by the user are surrounded by a coloured opaque lozenge and exams that share common students (conflict) with the current selection are hatched (See Figure 6).

Fig. 5. The Instance visualisation once all exams have been assigned to timetable locations. Within each outlined exam is a text label reference to the Room and Time Period that the exam has been assigned to so it can easily be located within the Solution visualisation. An exam will reappear here if it is unassigned from the timetable solution.

4.2 Solution Visualisation

The second and most important component of the VAST interface is the Solution Visualisation shown in Figure 6. Time periods are shown across the horizontal dimension grouped into days with gaps representing overnight gaps; rooms and their capacities are represented by the vertical dimension. Most of a user's time will be spent in this visualisation as it supports the initial assignment, perception of violations and subsequent optimisation of ETP solutions.

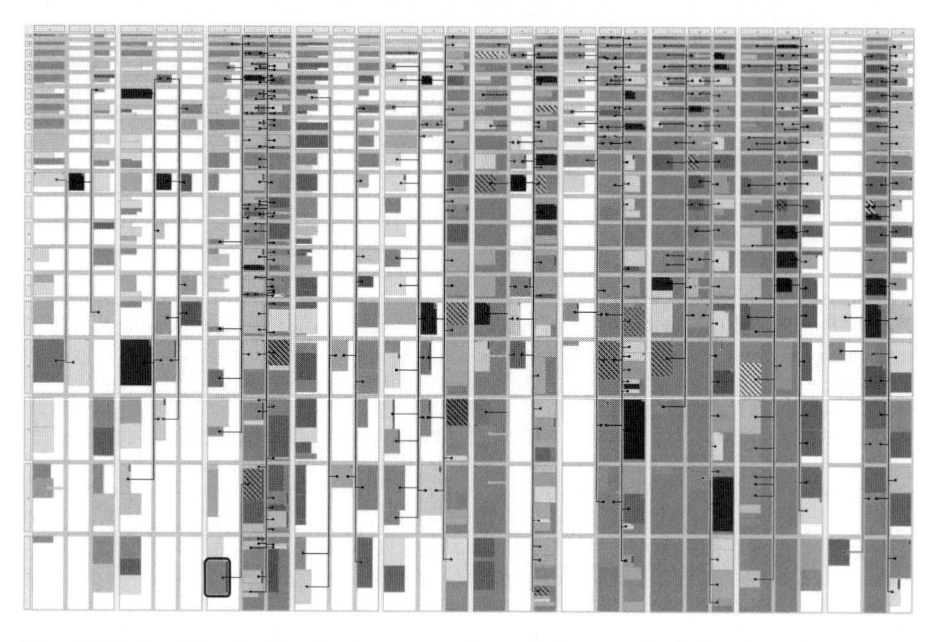

Fig. 6. Solution Visualisation showing a complete solution to an ETP instance. Both conflict and consecutive violations are present in this solution as indicated by the lines between exams (in VAST these appear respectively as orange and black lines). As before the currently selected Exam, in the lower left quartile of the image, is shown surrounded by an opaque lozenge and conflicting exams are hatched out. The grey time periods are the "restricted" time periods and moving the current selection to these locations, or next to these locations, would introduce new conflict or consecutive violations respectively. To resolve the highlighted exam violation in the seventh time period this exam would be moved to any available non-shaded (white) time period that is at least one time period away from a shaded time period.

Within the solution view selected Exams and groups of Exams can be assigned by clicking in the appropriate time period; for this initial assignment exams are automatically located in the smallest room they fit in so that groups can be quickly assigned without worrying about individual room allocation. Exams are moved by clicking first an exam and then its destination room/time period location in the timetable. After selecting an exam to move all conflicting exams are hatched, and time periods containing conflicting exams are shaded, this has the effect of highlighting the time periods that selected Exams can be assigned to without creating new violations. To resolve a conflict violation the Exam causing the problem would be selected and then moved to any non-shaded (white) time period. This mechanism enables conflict violations to be quickly resolved without the need to perform extensive searches for better locations in the timetable. Exams can also be unscheduled and entire time periods of exams can be swapped by clicking on handles at the top of each column. Empirical evidence suggests that, for benchmark problems, after one hour with VAST individuals are typically capable of finding solutions with around twice as many conflict violations as the best known solutions[9].

As this visualisation, shown in Figure 6, is a representation based on the underlying problem domain it supports different ways for users to resolve violations without forcing them to adopt any specific strategy. Whilst the interface immediately suggests non-violating locations for exams to be moved to, users can employ any different strategy they come up with, for example maybe initially moving small exams to create more space to later move large exams into in order to remove violations.

5 Evaluation

An evaluation of the VAST system was undertaken to determine how successful the interface was, in particular how effective the representations were in assisting users to conceptualise and solve ETP instances. Although in a real-world scenario it is likely that interactive visualisations would be used in conjunction with an automated solving system it is interesting to examine the possibility of individuals completely solving these problems without resorting to automation. The problem used for the evaluation was based on the University of Nottingham 1994 exam timetabling dataset (usually known as Nott-94 and available from ftp://ftp.cs.nott.ac.uk/ttp/Data/Nott94-1/) and contained 798 exams, 7896 students, 26 time periods, 18 rooms and a total room capacity of 1810. The data was slightly modified so that the number of students sitting any exam did not exceed the maximum room capacity.

An on-line competition was created allowing participants to compete for a £70 prize by generating and submitting the best solution to an ETP instance using the VAST system. The one month competition allowed participants to spend as much time as they wished solving the problem, being able to log in and out of the system as required, resuming from their previously generated solution. Every action performed by participants was logged together with the generated solutions in a central database. As a further incentive to perform well a web page was maintained which displayed real-time scores for all of the competitors, calculated from the cost function of their current best solution. The following sections briefly summarise the results of this evaluation which are presented in full elsewhere[9].

5.1 Overall Performance

Table 3 summarises the overall performance of the top sixteen participants in the VAST evaluation. In total six participants generated feasible solutions with no hard constraint violations that could have been used without further modification. Two of these six participants, spending around ten hours, generated solutions that were completely optimal in respect to the cost function used in the competition. The solutions

Table 3. Summary of how many participants solved the problem, how long it took them and to what extent they solved the problem

Description	Number of Participants	Mean Active Time
All Exams were assigned to the timetable with no conflict violations	6	6h 51m
Most exams were assigned and the solution may contain some conflicts. Some modification required.	5	2h 38m
Over an hour was spent solving the problem but it still contains many violations and significant work is required.	5	1h 29m

generated by the remaining participants all had some conflict violations, or had failed to schedule all the exams, however many of the participants had generated good initial solutions that could have been refined to remove these hard constraint violations in a short period of time.

5.2 Strategies

It is important to show that the representation supports the use of different individual strategies rather than a biased perspective of the information forcing participants to adopt a particular strategy. When examining the strategies used we considered frequency of group assignments and reassignments compared to the manipulation of individual exams. The number of exams in group manipulations was examined together with the relative frequency of selection, assignment and refinement operations. The ordering of these operations and the impact of group manipulation on the required refinement operations was also looked at. It was found that although there were differences in all these strategic components between participants no single strategy dominated or appeared significantly more effective than any other. The participants who made the most use of the group exam selections and assignments appear to have benefited from being able to construct initial solutions relatively quickly, but the overall amount of time spent refining solutions appears to be the most critical feature of the more successful approaches. Figure 7 shows one strategy, termed *ongoing refinement*, was used to solve the problem. This figure illustrates how the participant constructed a solution by selecting and assigning individual clusters of exams, spending time to optimise each assignment before proceeding to the next cluster.

Previous analysis of REEP interfaces (e.g., [3]) has highlighted the use of recursion in human strategies and this was also discovered in this evaluation. This supports the claim that, rather than random trial and error strategies, in most situations logical relationships exist between successive user actions. One example of this is forward shifting where assigning an exam to a time period creates a conflict violation and resolving this violation in turn creates another conflict violation and the cycle repeats.

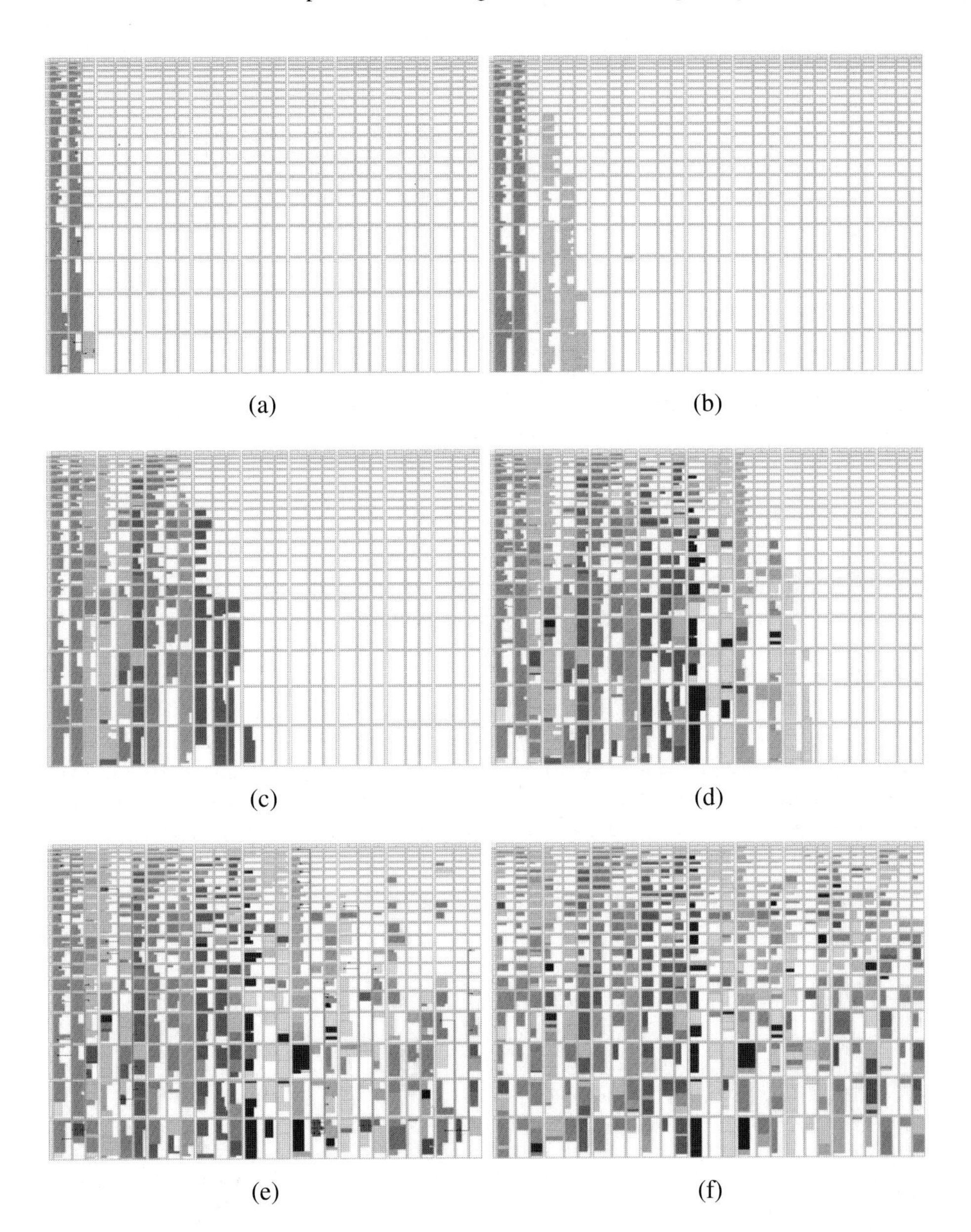

Fig. 7. Screenshots showing the progression of one participant toward solving the ETP instance. In this chronological series (a–e) a constructive approach can be seen where the clustering of exams from the Instance Visualisation has been used to initially assign Exams to time periods from left to right over a period of around 50 minutes. Because of this assignment of clusters to time periods, and careful ongoing refinement, relatively few constraint violations appear in these solutions. Although some consecutive violations can be observed in (e) they have all been resolved in (f).

6 Implications

In this paper we presented the application of the REEP-ID methodology to the generation of novel interfaces, interactive visualisations, to solving the Exam Timetabling Problem. The principles of REEP-ID were outlined together with the design of the VAST interface. In the previous section some of the results of an evaluation were presented indicating how effective VAST is in terms of generating complete solutions to the ETP. Using the representational epistemic approach provides a methodology to design interfaces that effectively support the cognitive problem solving and perceptive abilities of humans. Integration of these visualisations within existing automated systems would greatly increase the levels of available interaction providing benefits such as the ability to perceive exactly what and where the major constraint violations are, and ways of interactively resolving these violations (or other modifications) without introducing new unforeseen problems.

The use of *glyphs* with this approach has generated a set of visual components which could be recombined and extended with new decorations to represent other similar timetabling problems with similar conceptual dimensions. Similarly institutions with unique constraints can introduce new decorations on these glyphs to represent their unique violations without breaking the unambiguous global cohesive framework of the visualisation or having to introduce entirely new representations or perspectives.

This research has provided further evidence to support the use of REEP design approaches to interface design for problem solving in these kinds of conceptually complex data rich domains. Whilst it is not envisaged that the VAST approach will ever replace automated solving systems these interactive visualisations do provide clear benefits to practitioners and suggest one possibility for improving interfaces and levels of interactivity and humanisation within this domain.

References

1. Cheng, P.C.-H., Barone, R.: Representing complex problems: A representational epistemic approach. In: Jonassen, D.H. (ed.) Learning to solve complex scientific problems, pp. 97–130. Lawrence Erlbaum Associates, Mahwah (2007)
2. Cheng, P.C.-H.: Electrifying diagrams for learning: Principles for effective representational systems. Cognitive Science 26(6), 685–736 (2002)
3. Cheng, P.C.-H., et al.: Opening the Information Bottleneck in Complex Scheduling Problems with a Novel Representation: STARK Diagrams. In: Diagrammatic representations and inference: Second International Conference, pp. 264–278. Springer, Berlin (2002)
4. Cheng, P.C.-H., Barone, R.: Representing rosters: Conceptual Integration counteracts visual complexity. In: Diagrammatic representation and inference: Third international conference, Diagrams 2004. Springer, Berlin (2004)
5. Kingston, J.H.: Modelling Timetabling Problems with STTL. In: Selected papers from the Third International Conference on Practice and Theory of Automated Timetabling III, pp. 309–321. Springer, Heidelberg (2001)
6. Ranson, D., Ahmadi, S.: An Extensible Modelling Framework for Timetabling Problems. In: Burke, E.K., Rudova, H. (eds.) Practice and Theory of Automated Timetabling VI, pp. 383–393. Springer, Heidelberg (2007)

7. Burke, E., et al.: Examination Timetabling in British Universities - A Survey. In: Practice and Theory of Automated Timetabling. Springer, Napier University, Edinburgh (1995)
8. McCollum, B., et al.: A Review of Existing Interfaces of Automated Examination and Lecture Scheduling Systems(Abstract). In: Practice and Theory of Automated Timetabling, KaHo St.-Lieven, Gent, Department of Industrial Engineering, Belgium (2002)
9. Ranson, D.: Interactive Visualisations to Improve Exam Timetabling Systems. In: Department of Informatics, University of Sussex (2008)

Enhancing State-Space Tree Diagrams for Collaborative Problem Solving

Steven L. Tanimoto

University of Washington, Seattle WA 98195, USA
tanimoto@cs.washington.edu
http://www.cs.washington.edu/homes/tanimoto/

Abstract. State-space search methods in problem solving have often been illustrated using tree diagrams. We explore a set of issues related to coordination in collaborative problem solving and design, and we present a variety of interactive features for state-space search trees intended to facilitate such activity. Issues include how to show provenance of decisions, how to combine work and views produced separately, and how to represent work performed by computer agents. Some of the features have been implemented in a kit "TStar" and a design tool "PRIME Designer."

1 Introduction

1.1 Motivation

Problem solving and design processes tend to confront more and more complex challenges each year. A solution to a single problem often requires expertise from several domains. For example, Boston's "Big Dig" required expertise from civil engineers, geologists, urban planners, and politicians, among many others.

A group at the University of Washington is studying the use of the classical theory of problem solving as a methodology in support of collaborative design. One part of the approach involves the use of interactive computer displays to anchor the shared experience of a team. This paper describes research in progress on a set of issues and possible features to address them. The issues and features pertain to the use of interactive tree diagrams that show portions of a design space or solution space, and that support a team in managing their work.

1.2 State-Space Search Trees

Problem solving has been cast in the formal framework of "state space search" by early researchers in artificial intelligence, such as Newell and Simon (see Simon, 1969). A solver starts with an initial state, representing an absence of commitments to the particulars of a solution, and step by step, tries adding particular elements or modifying the current state, in an attempt to reach a "goal state". The goal state, or perhaps the sequence of steps taken to reach it, constitute a solution to the problem (Nilsson, 1971). The states visited in such a search can usually be plotted as nodes of a graph. Each move from one state to

G. Stapleton, J. Howse, and J. Lee (Eds.): Diagrams 2008, LNAI 5223, pp. 156–164, 2008.

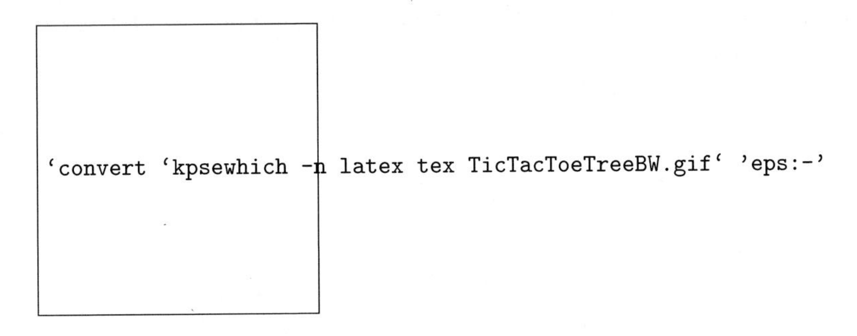

Fig. 1. A familiar kind of state-space search tree. This diagram corresponds to a partial search in the middle of a Tic-Tac-Toe game.

another, in turn, corresponds to an edge of the graph. If we decide that each state is described by the sequence of operators applied, starting at the initial state, then the graph has the structure of a tree, and it can be drawn as a tree. Such diagrams have been used to illustrate textbooks of artificial intelligence, often in conjunction with classical puzzles such as the Eight puzzle (Nilsson, 1971) or the game of Tic-Tac-Toe as shown in Fig. 1. The use of state-space search trees in interfaces was explored in an earlier paper (Tanimoto and Levialdi, 2006), and their use in the design of image-processing procedures was discussed in the context of face-image analysis (Cinque et al, 2007).

1.3 Issues

This paper is concerned with a set of issues that arise when one chooses to use state-space search tree diagrams as interactive representations of the progress of a design or problem-solving team. Some of these issues are the following.

Provenance. Each node represents a partial design or solution. Who contributed what to this design? From what other designs was it derived? Supports for querying and viewing provenance are needed.

Negotiating Shared Views. A designer's personal view of a design tree reflects her priorities and preferences. In a shared view, priorities and preferences may collide, leading to the need for compromises. Algorithms can offer certain compromises automatically, but shared views that are most satisfactory to a group may require give and take. Facilities to manage such negotiation are needed.

Visualization of Opportunity. The members of a team may understand their own components of a design or solution, but it's often difficult to understand the potential interactions of the pieces. In some cases, separately designed pieces can be combined, and the diagram can indicate the compatibility of a piece for a particular role. Some opportunities result from the applicability of operators to newly computed states. Others result from the availability of resources in relation to defined tasks. Helping users see and evaluate opportunities is an important goal for collaborative design and problem-solving tools.

Work by Agents. A collaborative team traditionally consists of people. However, computer agents may be present and available to help out. Agents generally work very differently from people, and special affordances are required to define and display tasks for agents and the results of those tasks.

1.4 Previous Work

While much research has been done separately on problem solving, tree diagram layout, operations on trees, and interacting with trees, we mention in this section some work that bridges these topics.

Flexible Tree Layout. In order to display a large tree so that various parts of the tree can be distinguished, or so that a minimum or limited area is used, many methods have been proposed. A survey of tree display methods was presented by Hanrahan (2001). Popular methods include "hyperbolic trees" (Lamping et al, 1995), subdivision diagrams, and cluster diagrams. Usually, the goal is to display a large tree in such a way that individual nodes get their fair share of screen real estate.

Folder Tree Affordances. A family of interactive techniques for trees has been developed in the context of operating-system file hierarchies (Kobsa, 2004). The best-known example of a file system browser is the Microsoft Windows Explorer (not Internet Explorer). Interior nodes of the tree shown in Explorer correspond to folders in the file system, and the leaves correspond to individual files. The user can click on folder icons (situated at the nodes) to open and close the folders. When a folder that contains files is closed, it has a hotspot labeled with a plus (+) sign. Clicking on the plus sign opens the folder to reveal its contained files and subfolders. A more general set of affordances for hiding and revealing folders is described in a US patent (see Chu and Haynes, 2007). Here it is possible to have two nodes in a relationship where A is an ancestor of B, where A is closed, but B is visible and open. In such a view, all of A's descendants, except B and its descendants (and any other nodes explicitly opened) are hidden.

Interest-Sensitive Layout. There have been several developments that have combined flexible layout with interactivity. A notable example is "Degree of Interest Trees" (Card and Nation, 2002). In such a tree, the user can indicate interest or disinterest in particular nodes of the tree by clicking on them, and the layout of the tree immediately adapts to the new distribution of interest, using a continuous animation to move from the current layout to the new layout.

2 Application to Collaborative Design

2.1 Rationale

We use the classical theory of problem solving for two reasons: (1) provide a common language for the problem solving process to design teams whose members represent different disciplines, and (2) help humans interact, via the computer interface, not only with each other, but with computational agents that perform design or problem-solving services. To put our use of this theory into context, we now explain the general category of atomic actions that we call "design acts."

2.2 Design Acts

We assume that the entire process of design can be broken down into small steps that we call *design acts.* There are several types of design acts, including communication acts, design steps, evaluation acts, and administrative acts. Communication acts include writing and reading messages associated with the project. Design steps are primarily applications of operators to existing nodes to produce new nodes in a search tree. in order to apply an operator, may also be considered a design step. An evaluation act is either a judgment or an application of an evaluation measure to a node or set of nodes. A judgment is a human-created quantitative or qualitative estimation of the value of a node on some scale with regard to some particular characteristic. An evaluation measure is a mathematical function that can be applied to a node to return a value (usually numeric). Such a value may correspond to the degree to which a partial design meets a particular criterion. For example, one evaluation measure for designs of houses is the number of rooms in the house. Another is the square footage of area in the designed house so far. A judgment might correspond to an aesthetic evaluation – how pleasing is this floor layout to the eye of designer Frank? Administrative acts include actions such as the definition of tasks and subgoals for the design team, commitments of effort to tasks, and adjustments to views of the progress made so far.

2.3 Roles of the Diagram

By using tree diagrams, it is possible to provide computer assistance, at some level, for each type of design act mentioned above. Many communication acts relate to parts of a tree. Messages can relate to parts of a tree in two ways: (1) by naming a labeled node or by describe the path to it from the root, and (2) by being embedded in the node as an annotation. Messages can also refer to nodes via descriptions or expressions in a node-specification language. Such a language is a kind of query language with features for identifying nodes not only via their properties and annotations, but via their "kinship" relations.

Design steps are supported via interactive tree-building functionality. The selection of nodes and application of operators is done with clicks and menus. (An example of a design tree built with a tool called PRIME Designer is shown in Fig. 2. The tool supports a four-person team – musical composer, architect, image-puzzle designer, and game-logic programmer – in creating a multimedia game. It has been used in focus-group trials.)

In addition to helping with specific design acts, the diagrams provide an easily browsable record of the history of the design process. They also provide contextual information for particular tasks. Thus a state-space search tree provides an organizing framework to a team for pursuit of solutions.

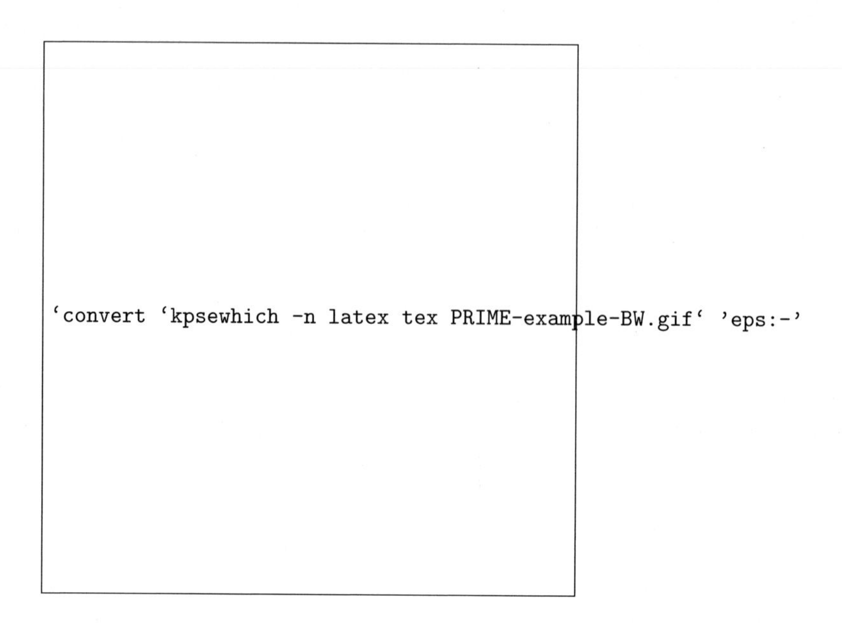

Fig. 2. Design tree showing multiple paths developed by separate members of a team. Also shown are “minified” nodes, an ellipsis under the first minified node (represented with the dotted line), and an enlarged node. The four quadrants in each node correspond to different roles on the design team. This view is an “all-roles” view, and most nodes are empty in three out of the four roles; the “None/0” label means that no components have been added in the role, and none is selected.

3 Layout, Visibility, and View Control

State-space search trees have the potential to grow very large. In order to make large trees manageable on an interactive display, the following techniques are commonly used: scrolling, overall scaling, and opening/closing (revealing/hiding) of subtrees. We have developed variations on each of these techniques that can provide additional user control in the case of search trees.

3.1 Scaling

Overall scaling, equivalent to zooming, is an essential viewing technique for large trees. A challenge for designers of visualizations is providing controls so that users can perform *differential scaling*, so that different parts of the tree can have different scale factors. A form of differential scaling can be found in dynamic tree layout methods such as hyperbolic trees (Lamping et al, 1995) and “degree-of-interest” trees (Card and Nation, 2002). These methods are convenient for giving prominence to regions of a tree, a region being a subtree or the nodes in a graph neighborhood of a given node.

Another approach to differential scaling allows the user to associate, with any node or nodes s/he deems special, a particular scale factor for that node. That factor is then used as a relative factor, and zooming of the overall view maintains the differential. Our TStar software supports this technique.

3.2 Hidden Nodes

While scrolling and scaling are often considered to be geometric operations, they are also means of information hiding, since scrolling is a way of showing some and hiding other information in a diagram, and scaling makes some details visible or invisible and shows more or less of the area of the entire diagram.

Now we describe two additional forms of information hiding, one of which we call "hidden nodes" and the other "ellipsis." They are closely related. They differ in three respects: how affected portions of the diagram are rendered, what kinds of sets of nodes can be hidden, and how the user interacts with them. We make these distinctions in order to provide a rich set of view controls to users, and to suggest additional possibilities for dealing with the complex but important issue of working with state-space tree diagrams.

We define a hidden node to be an invisible, but fully computed node object, that is neither rendered on the screen nor has any effect on the rendering of the rest of the tree. From the standpoint of screen appearance, a hidden node may just as well be a node that was never created or that has been deleted. If it has any children, they, too, must be hidden. In our TStar system, the user can hide a subtree in order to create a simpler view, or unhide a subtree to (re)include it in the diagram. Any overhead that would be associated with deleting or recreating corresponding states is avoided in the hiding and unhiding operations.

In order for the user to know that a hidden node exists in a tree, a small indication is available, on demand, in the form of a highlight on nodes having hidden children. Hiding the root is not permitted in TStar.

3.3 Ellipsis

A node *in ellipsis* is a tree node that is not displayed but is represented by an ellipsis indication. One or more nodes, in an ancestor chain, can be represented by a single ellipsis indication. An *ellipsis* is a collection of all the nodes in ellipsis that are represented by the same ellipsis indication.

An ellipsis therefore represents a set of nodes in a tree whose existence is important enough to indicate, but whose details and whose number are sufficiently unimportant that they are hidden. For example, the nodes in an ellipsis may have an important descendant, displayed in detail, but not have any details of their own worth showing. The technique of ellipsis gives us a means to show nodes that have been explored but which we show in a very abbreviated form. Ellipses not only hide node details and reduce clutter (see, e.g., Taylor et al, 2006) but they save space by reducing the amount of space allocated to the nodes in ellipsis.

While we have just defined an ellipsis as a static set, in practice the user may work with dynamic ellipses. A basic dynamic ellipsis is a changing set of nodes in ellipsis, and a user-controlled dynamic ellipsis is a basic dynamic ellipsis with a boolean variable "currently-in-ellipsis" that can be set true or false by the user to adjust the visibility of the nodes without losing the grouping.

3.4 View Merging

When two members of team merge their trees via basic merging, very few conflicts can arise. The trees share only the root, and root can neither be put into ellipsis nor hidden, at least in our example system. Conflicts are another story when the trees are merged according to node equivalence based on operator sequences. Although a node N_a in tree A is considered path-equivalent to a node N_b in tree B if they are both derived from their (assumed equivalent) roots via the same operator sequence, they have have lots of attributes that are by no means equivalent. They may share different authorship, different annotations, different memberships in ellipses, they may be hidden or not, and they may have different scale factors associated with them.

Attributes of nodes such as authorship, creating timestamps, and descriptive annotations can usually be "unioned" – as entries on lists, the lists can be concatenated to preserve everything. View properties such as node scale factors can also be unioned (kept along for dynamically changing the view), but an actual scale factor for the current view must also be determined. Three different methods are suggested, to be selected by the user at the time view merging is invoked, for different purposes. These methods are, briefly, "min," "max," and "diff." The min method is useful when each view represents work done to reduce scale factors on nodes judged to be unimportant, but each tree represents an incomplete job. The combined view will represent the totality of this work by the two team members. The max method is useful in the complementary situation, where each member has spent some effort to identify particularly important parts of the tree and has put in higher-than-normal scale factors for those nodes. The diff method is a means to highlight those parts of a combined tree in which the ratings of two team members (as expressed by node-specific scale factors) differ notably. Such a method takes the absolute value of the difference of the scale factors and maps it monotonically into a standard range of scale factors (0.1 to 2.0).

3.5 Representing the Work of Agents

Agents can explore large numbers of states, but most states tend to be relatively uninteresting. This calls for display methods that succinctly summarize searches of large subtrees. Agents also blur the boundary between explored states and unexplored states. If an agent has explored the state, has it really been explored? It depends on *how* the agent explores the state as well as the extent to which the user understands what the agent has "seen." One degree of exploration has been reached when a set of states is realized – the states have been computed. But that may be of absolutely no consequence to a user if there is no "end product" of the realization – no additional nodes displayed on the screen, no reports of solutions found or promising nodes reached.

This suggests a need for new ways of rendering subtrees to show different degrees or forms of exploration and results of exploration. If evaluation functions are computed on nodes, color coding according to value is a natural approach. Subtrees currently being analyzed may be shown with flashing or other animation to indicate the locus of activity.

4 Collaboration Operations on Trees

Several topological operations on trees are important in collaborative design as we have structured it: tree merging, path extraction, path merging, path simplification, and path reorganization to enable merging. These operations permit the separate work of individuals and subteams to be combined. Our PRIME Designer system currently the supports the first three of these five operations. Tree merging takes two trees and produces a new one whose root represents both input roots, and having copies of all the other nodes. Path extraction is the separation of a path in a tree as a separate tree. Path merging is the concatenation of two paths to form one long one.

The path merging operation is the most important one in combining the work of subteams, and it is the most problematical in terms of coordination. But it offers interesting challenges for visualization of opportunities for combinations. Two paths A and B cannot be merged unless they are compatible: every operator application used to create a move in B must still be legal when the last state of A is used as the first state of B. The display of compatibility works as follows: after a path A has been selected, all nodes not on path A that begin paths that could be merged with A are highlighted, and the highlighting color gets brighter as the length the potential path is longer. In our future work, we intend to experiment with such compatibility displays, in combination with automatic evaluation of the merit of the potential merged paths.

Acknowledgments

Thanks to Brian Johnson, Richard Karpen, Stefano Levialdi, Tyler Robison, and Linda Shapiro for comments on parts of the software described here. Partial support under NSF Grant 0613550 is gratefully acknowledged.

References

1. Card, S., Nation, D.: Degree-of-interest trees: A component of an attention-reactive user interface. In: Proc. AVI, Trento, Italy. ACM Press, New York (2002)
2. Chu, H., Haynes, T. R.: Methods, Systems and Computer Program Products for Controlling Tree Diagram Graphical User Interfaces and/or For Partially Collapsing Tree Diagrams. US Patent 20070198930
3. Cinque, L., Sellers Canizares, S., Tanimoto, S.L.: Application of a transparent interface methodology to image processing. J. Vis. Lang. Comput. 18(5), 504–512 (2007)

4. Hanrahan, P.: To Draw A Tree. In: Presented at the IEEE Symposium on Information Visualization (2001), http://graphics.stanford.edu/~hanrahan/talks/todrawatree/
5. Kobsa, A.: User experiments with tree visualization systems. In: Proceedings of the IEEE Symposium on Information Visualization 2004, pp. 9–16 (2004)
6. Lamping, J., Rao, R., Pirolli, P.: A focus+context technique based on hyperbolic geometry for visualizing large hierarchies. In: Proc. ACM Conf. Human Factors in Computing Systems, pp. 401–408 (1995)
7. Tanimoto, S.L., Levialdi, S.: A transparent interface to state-space search programs. In: Kraemer, E., Burnett, M.M., Diehl, S. (eds.) Proc. of the ACM 2006 Symposium on Software Visualization (SOFTVIS 2006), Brighton, UK, September 4-5, 2006, pp. 151–152 (2006)
8. Nilsson, N.: Problem-Solving Methods in Artificial Intelligence. McGraw-Hill, New York (1971)
9. Simon, H.: The Sciences of the Artificial. MIT Press, Cambridge (1969)
10. Taylor, J., Fish, A., Howse, J., John, C.: Exploring the Notion of Clutter in Euler Diagrams. In: Barker-Plummer, D., Cox, R., Swoboda, N. (eds.) Diagrams 2006. LNCS (LNAI), vol. 4045, pp. 267–282. Springer, Heidelberg (2006)

Visual Programming with Interaction Nets

Abubakar Hassan, Ian Mackie, and Jorge Sousa Pinto

[1] Department of Informatics, University of Sussex, Falmer, Brighton BN1 9QJ, UK
[2] LIX, École Polytechnique, 91128 Palaiseau Cedex, France
[3] Departamento de Informática/CCTC, Universidade do Minho, Braga, Portugal

Abstract. Programming directly with diagrams offers potential advantages such as visual intuitions, identification of errors (debugging), and insight into the dynamics of the algorithm. The purpose of this paper is to put forward one particular graphical formalism, interaction nets, as a candidate for visual programming which has not only all the desired properties that one would expect, but also has other benefits as a language, for instance sharing computation.

1 Introduction

Interaction nets were introduced in 1990 [8]. The theory and practice of interaction nets have been developed over the last years to provide tools to program and reason about interaction net programs. For instance, a calculus [3], notions of operational equivalence [4], encodings giving implementations of rewriting systems such as term rewriting systems [2] and the lambda calculus [5,9]. However the visual programing aspect of this language has been neglected. The purpose of this paper is to take a fresh look at this formalism from a visual programming perspective and demonstrate that it is a suitable language to develop programs for the following reasons:

- It is a diagrammatic programming language where both programs and data are given the same status: they are both given by diagrams in the same formalism.
- Computation is rule based: the programmer explains the algorithm in terms of rules which are applied by the runtime system. From this perspective, it could be classified as a declarative language.
- All the computation is expressed by the rules: there are no external mechanisms performing parts of the computation, and consequently the diagrams give a full description.
- The rewrite rules transform the diagrams, and this gives a trace of the computation directly at each step. An application to algorithm animation would be a by-product of the approach.

Interaction nets are not new. They have been applied very successfully to representing sharing in computation. In the present paper this is not a feature that we particularly want to develop, but nevertheless it is a convenient plus

G. Stapleton, J. Howse, and J. Lee (Eds.): Diagrams 2008, LNAI 5223, pp. 165–171, 2008.

point that we will come back to later. The purpose of this paper is to show, through examples, that interaction nets are not only a good example of a visual programming language, but they have features that could make them a viable programming paradigm when appropriate tools are developed.

Work closest to ours is in the area of Visual Functional Programming which has addressed different aspects of visual programming. The Pivotal project [6] offers a visual notation (and Haskell programming environment) for data-structures, not programs. Visual Haskell [12] more or less stands at the opposite side of the spectrum of possibilities: this is a dataflow-style visual notation for Haskell programs, which allows programmers to *define* their programs visually (with the assistance of a tool) and then have them translated automatically to Haskell code. Kelso's VFP system [7] is a complete environment that allows to define functional programs visually and then reduce them step by step. Finally, VisualLambda [1] is a formalism based on graph-rewriting: programs are defined as graphs whose reduction mimics the execution of a functional program. As far as we know none of these systems is widely used.

In the next section we introduce the formalism. We then give two examples of the use of this language. We conclude the paper with a discussion about other features of interaction nets, including parallelism, sharing, and perspectives for use as programming language in the larger scale (hierarchical nets).

2 Interaction Nets

We begin by defining the graphical rewriting system, which is a generalisation of interaction nets found in the literature (see [8] for instance). We have a set Σ of *symbols*, which are names of the nodes in our diagrams. Each symbol has an arity ar that determines the number of *auxiliary ports* that the node has. If $ar(\alpha) = n$ for $\alpha \in \Sigma$, then α has $n + 1$ *ports:* n auxiliary ports and a distinguished one called the *principal port.*

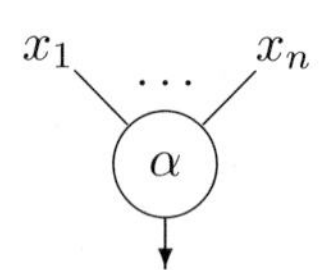

Nodes are drawn variably as circles, triangles or squares, and they optionally have an attribute, which is a value of base type: integers and booleans. We write the attribute in brackets after the name: e.g. $c(2)$ is a node called c which holds the value 2. A *net* built on Σ is an undirected graph with nodes at the vertices. The edges of the net connect nodes together at the ports such that there is only one edge at every port. A port which is not connected is called a *free port.*

Two nodes $(\alpha, \beta) \in \Sigma \times \Sigma$ connected via their principal ports form an *active pair*, which is the interaction nets analogue of a reducible expression (redex). A rule $((\alpha, \beta) \Longrightarrow N)$ replaces the pair (α, β) by the net N. All the free ports are preserved during reduction, and there is at most one rule for each pair of agents. The following diagram illustrates the idea, where N is any net built from Σ.

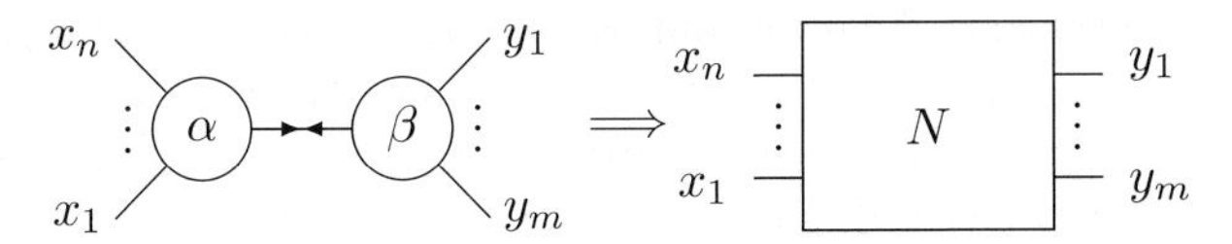

If either (or both) of the nodes are holding a value, then we can use these values to give different right-hand sides of the rule by labelling the arrow with a condition. The condition can be built out of the usual boolean operators ($<$, $>$, $=$, $!=$, etc.). However, the conditions must be all disjoint (there cannot be two rules which can be applied). Each alternative must of course give a net satisfying the property given above for the case without attributes. The most powerful property that this system has is that it is one-step confluent: the order of rewriting is not important, and all sequences of rewrites are of the same length and equal (in fact they are permutations). This has practical aspects: the diagrammatic transformations can be applied in any order, or even in parallel, to give the correct answer.

We next explain how to represent a simple data structure using interaction nets. In the next section we give two examples of algorithms over these data. We can represent a memory location, containing an integer i, simply as a cell holding the value. We can represent a list of cells with the addition of a nil node.

m(i)   nil

In the diagram above, the m node has one principal port that will be used to interact with it, and one auxiliary port to connect to the remaining elements of the list. The nil node just has one principal port, and no auxiliary ports. To simplify the diagrams, we often just write the contents of the node and omit the name when no confusion will arise. For example, here is a list of 4 elements:

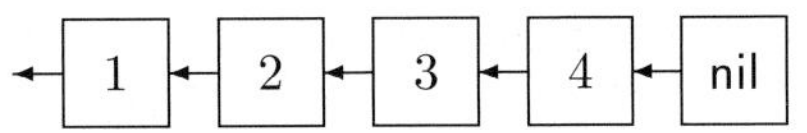

We remark that this diagrammatic representation of the dynamic list data structure is not the same as what one would usually draw, as the arrows are not pointers, but they represent the principal ports which dictates how we can interact with the data structure.

3 Examples: Sorting

Suppose that we wanted to insert an element into a sorted list. In Java we might write the code given below

```
static List insert(int x, List l) {
  if (isEmpty(l) || x <= l.head) return cons(x, l);
  else {
    l.tail = insert(x, l.tail);
    return l;
  }
}
```

When teaching dynamic data structures, we might begin by drawing a diagram for this algorithm, which would consist of the list before the insertion, then a number of modifications of the diagram, yielding the final diagram. To derive the code above from such a diagram is not an automatic process, and is of course subject to error. In interaction nets we would simply write the program directly as the rules needed to transform the data structure:

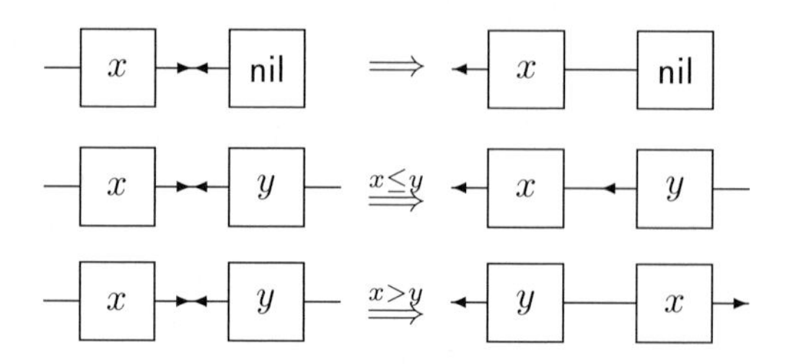

These three rules explain all the computation. We leave the reduction of the following net as an exercise for the interested reader, which will rewrite to the list of 4 elements given above using three applications of the above rules. The implementation of insertion sort is a straightforward extension to this example.

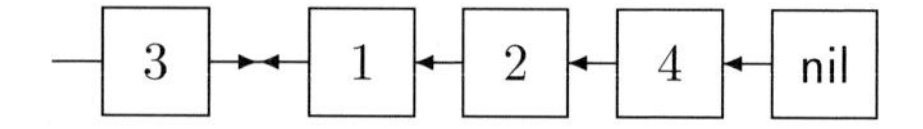

If we compare the interaction rules with the Java program we find that we can follow them step-by-step. However, the three diagrams contain all the implementation details—there are no notions of procedures (methods), conditions, recursion, etc. as syntax, as they are all absorbed into the rules. The relationship between Java programs and interaction nets is not the purpose of this paper; in general we will find that the Java version will contain very different control information than the interaction nets version. However, it is worth remarking that the interaction net program can be seen as directly manipulating the internal data structure, which is one of the main features of this approach.

Our next example to demonstrate how easily one can program with interaction nets also brings out the relationship between the diagrammatic representation of the algorithm and the intuition of the algorithm. We give an implementation of the Quicksort algorithm, using again the linked list structure. We need some auxiliary rules to concatenate two lists, and also to partition a list into two lists. We begin with concatenation, which can be expressed in Java as follows:

```
static List concat(List l1, List l2) {
  if (isEmpty(l1)) return l2;
  else {
    List l = l1;
    while (isCons(l.tail)) l = l.tail;
    l.tail = l2;
    return l1;
  }
}
```

The algorithm performed in both interaction nets and the Java version is essentially the same: traverse the first list and connect its last element to the start of the second list. This is given by the first two rules below. We also give the three rules for the partition of a list, which is the core of the Quicksort algorithm, splitting a list into two lists.

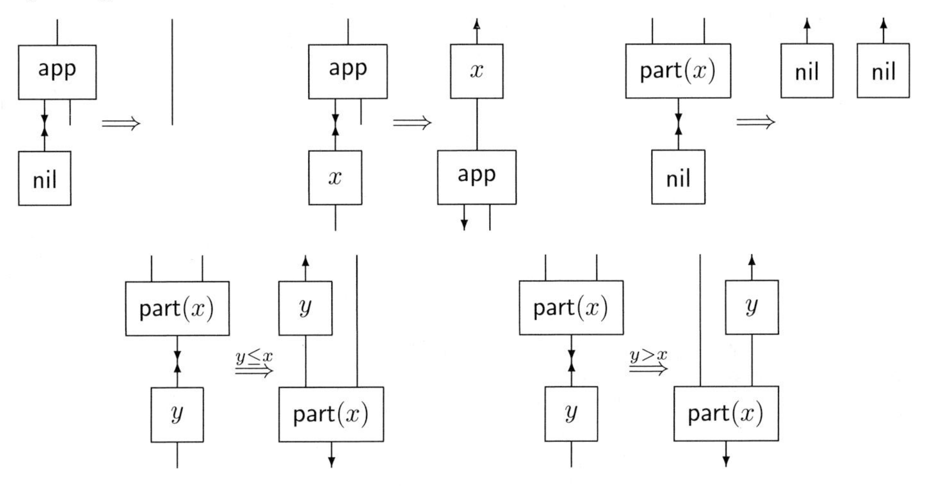

We can now put together the main algorithm for Quicksort which is given by the following two rules. QS is the node to represent Quicksort, and we use app and part from above.

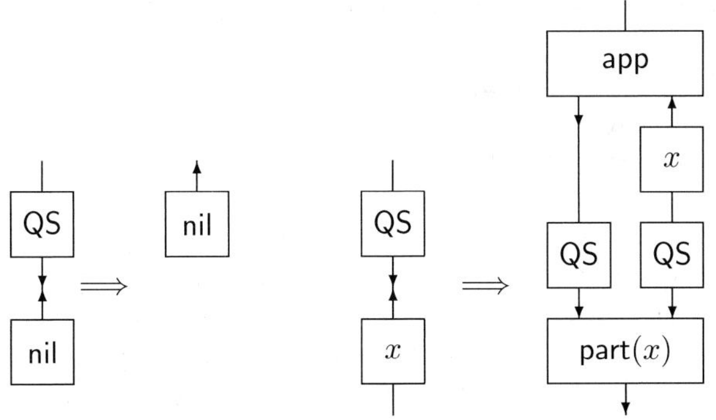

It is worth pointing out some salient features of this implementation of the Quicksort algorithm:

1. The graphical representation of the problem is directly cast into the graphical language. The algorithm given can be understood as programming directly with the internal data structures, rather than some syntax describing it.
2. All rewrite steps correspond to steps in the computation: there is no need to introduce additional data structures and operations that are not part of the problem.
3. Because of point 1 above, if we single-step the computation we get an animation of the algorithm directly from the rewriting system.

4 Discussion

When learning data structures in programming languages, specifically dynamic structures with pointers, we find that the diagram used to explain the problem and the code are very different. In particular, the diagrams do not always show the temporal constraints, and therefore converting the diagrammatic intuitions into programs can be quite difficult. With interaction nets we draw the diagrams once, and this gives the program directly. Because of the confluence result, it does not matter which order we apply the rules, thus eliminating the temporal constraints. It is the programmer that draws the diagrams, and the diagrams explain all of the computation (nothing is replaced by code, like while loops, etc.). The diagrams are then implemented directly (and moreover, they are data structures that are well adapted to implementation).

There are a number of textual programming languages for interaction nets in the literature (see for instance [8,3,10]), and one of the future developments needs to be tools to offer a visual representation of these. Being able to convert the textual representation to diagrams would be a useful tool for debugging, but this does not offer the ability to directly manipulate nets and rules in a uniform way. Some developments are currently underway to address these issues, and preliminary experiments show that this direction is an exciting way to write programs like the ones described in this paper. Nevertheless, this is a topic of current research very much in its infancy.

What we have presented leads to a very direct form of visual programming. We can imagine tools to display animations of traces of executions which will follow the intuitions of the programmer (each step is one rule). Because interaction nets explain all the elements of the computation it is also a useful debugging tool as well as an educational tool for data structures.

Interaction nets have been used for representing sharing in computation. This is obviously still valid in the graphical approach. Having a language which is at the forefront or research into optimal computations is clearly an advantage. As we have hinted above, they are also amenable to parallel computations: the one step confluence property of rewriting together with the fact that all rules are local means that we can apply all redexes at the same time [11]. Interaction nets may therefore have potential to represent parallel algorithms visually.

The next developments that are needed are tools to offer direct manipulation of interaction nets to investigate the potential as a programming language and study usability issues. This relies on the development of editing tools and techniques to write modular programs, hierarchical structures, etc. The development of such an environment will be the subject of a future paper.

5 Conclusion

Interaction nets are a graphical (visual) programming language used successfully in other branches of computer science. Our aim in this paper was to demonstrate by example that they are well adapted to visual programming, with the main goal to introduce this formalism to the *diagrams* community through examples, and hint as some of the possible potential of this formalism that is currently under development. We have achieved this by extending the interaction net formalism, and given some new example programs to exhibit these extensions.

Acknowledgements. This work was partially supported by CRUP, Acção Integrada Luso-Britânica N.B-40/08, and the British Council Treaty of Windsor Programme.

References

1. Dami, L., Vallet, D.: Higher-order functional composition in visual form. Technical report (1996)
2. Fernández, M., Mackie, I.: Interaction nets and term rewriting systems. Theoretical Computer Science 190(1), 3–39 (1998)
3. Fernández, M., Mackie, I.: A calculus for interaction nets. In: Nadathur, G. (ed.) PPDP 1999. LNCS, vol. 1702, pp. 170–187. Springer, Heidelberg (1999)
4. Fernández, M., Mackie, I.: Operational equivalence for interaction nets. Theoretical Computer Science 297(1–3), 157–181 (2003)
5. Gonthier, G., Abadi, M., Lévy, J.-J.: The geometry of optimal lambda reduction. In: Proceedings of the 19th ACM Symposium on Principles of Programming Languages (POPL 1992), January 1992, pp. 15–26. ACM Press, New York (1992)
6. Hanna, K.: Interactive Visual Functional Programming. In: Jones, S.P. (ed.) Proc. Intnl. Conf. on Functional Programming, October 2002, pp. 100–112. ACM Press, New York (2002)
7. Kelso, J.: A Visual Programming Environment for Functional Languages. Ph.D thesis, Murdoch University (2002)
8. Lafont, Y.: Interaction nets. In: Proceedings of the 17th ACM Symposium on Principles of Programming Languages (POPL 1990), January 1990, pp. 95–108. ACM Press, New York (1990)
9. Mackie, I.: Efficient λ-evaluation with interaction nets. In: van Oostrom, V. (ed.) RTA 2004. LNCS, vol. 3091, pp. 155–169. Springer, Heidelberg (2004)
10. Mackie, I.: Towards a programming language for interaction nets. Electronic Notes in Theoretical Computer Science 127(5), 133–151 (2005)
11. Pinto, J.S.: Parallel Implementation with Linear Logic. Ph.D thesis, École Polytechnique (February 2001)
12. Reekie, H.J.: Realtime Signal Processing – Dataflow, Visual, and Functional Programming. Ph.D thesis, University of Technology at Sydney (1995)

Spider Diagrams of Order and a Hierarchy of Star-Free Regular Languages

Aidan Delaney[1], John Taylor[1], and Simon Thompson[2]

[1] Visual Modelling Group, University of Brighton
[2] Computing Laboratory, University of Kent

Abstract. The spider diagram logic forms a fragment of the constraint diagram logic and was designed to be primarily used as a diagrammatic software specification tool. Our interest is in using the logical basis of spider diagrams and the existing known equivalences between certain logics, formal language theory classes and some automata to inform the development of diagrammatic logics. Such developments could have many advantages, one of which would be aiding software engineers who are familiar with formal languages and automata to more intuitively understand diagrammatic logics. In this paper we consider relationships between spider diagrams of order (an extension of spider diagrams) and the star-free subset of regular languages. We extend the concept of the language of a spider diagram to encompass languages over arbitrary alphabets. Furthermore, the product of spider diagrams is introduced. This operator is the diagrammatic analogue of language concatenation. We establish that star-free languages are definable by spider diagrams of order equipped with the product operator and, based on this relationship, spider diagrams of order are as expressive as first order monadic logic of order.

1 Introduction

The study of regular languages, finite automata and associated algebraic formalisms is one of the oldest branches of computer science. Regular languages, which are defined by Type-3 grammars [3], are the least expressive class of phrase structured grammars of the well-known Chomsky-Schützenberger hierarchy. Work by Büchi [2], amongst others, provides a logical characterisation of regular languages. By contrast diagrammatic logics are relatively new. Their formal consideration can arguably be dated to the work of Barwise and Etchemendy [1], Shin [16], and Hammer [9] which in turn builds on the work of Euler [7] and Venn [20]. Spider diagrams [8] are a more recently defined and form a fragment of constraint diagrams [13]. Our interest is in the relationship between an extension of spider diagrams, called spider diagrams of order, and regular languages. This paper builds on our previous work [4,5] and provides a proof that star-free regular languages are definable using spider diagrams of order, when augmented with a product operator. Star-free languages may be described by regular expressions without the use of the Kleene star, a fact from which the name of the language class derives [14]. For example, the language a^* over the alphabet

G. Stapleton, J. Howse, and J. Lee (Eds.): Diagrams 2008, LNAI 5223, pp. 172–187, 2008.

$\Sigma = \{a, b\}$ is star free as it may be written as the star-free expression $\overline{\overline{\emptyset} b \overline{\emptyset}}$ i.e. the complement of the set of all words containing a 'b'. The expression $\overline{\emptyset}$ is the complement of the empty set of words and may be read as the set of all words over Σ, denoted Σ^*. The language $(aa)^*$ over the same alphabet is not star-free [15].

Of most interest to us is the Straubing-Thérin hierarchy (STH), which is one of three infinite hierarchies used to characterise the class of star-free languages. The other two are the dot-depth hierarchy and the group hierarchy. All three hierarchies are recursively constructed from a base case at their respective level 0. Level $\frac{1}{2}$ of each hierarchy is the polynomial closure of level 0, an operation which is explained in section 5. Each of the fractional levels $\frac{1}{2}, 1+\frac{1}{2}, 2+\frac{1}{2}, \ldots$ are similarly formed from the proceeding integer level. Level 1 of each hierarchy is the finite boolean closure of level $\frac{1}{2}$ under the operations intersection, union and complement. In general, integer numbered levels $1, 2, 3, \ldots$ are the finite boolean closure of the half level beneath them [14].

The study of the relationship between spider diagrams of order and regular languages provides a novel view of both subjects. We have previously shown that fragments of spider diagrams (without order) describe classes of regular languages that are incomparable with well-known hierarchies such as the Straubing-Thérin or dot-depth hierarchies [5]. We show, in this paper, that the logic of spider diagrams of order describe classes of languages which correspond to well-known subsets of star-free languages. Conversely, regular languages have helped to inform the development of spider diagrams. Our introduction of the product operator, motivated by [19], allows us to directly compare fragments of spider diagrams of order with fragments of star-free regular languages. By furthering the study of the relationship between diagrammatic logics and formal language theory we hope to "import" well-known results. In particular we would like to formulate a diagrammatic analogue of the Myhill-Nerode theorem [11] in order to characterise minimal unique normal forms for spider diagrams of order.

This paper presents an overview of the syntax and semantics of spider diagrams of order in section 2. In section 3 we define the language of a spider diagram of order. This is largely similar to work in [5] but generalises to cover languages over arbitrary alphabets. The product of spider diagrams is introduced in section 4. The central result of this paper, that all star-free regular languages are definable using spider diagrams of order augmented with the newly introduced product operator, is presented in section 5.

2 Syntax and Semantics of Spider Diagrams of Order

This section provides an overview of the syntax and semantics of spider diagrams of order, originally presented in [5] which in turn extends [12].

The diagrams labelled d_1 and d_2 in figure 1 are *unitary spider diagrams of order*. Such diagrams, like the Euler diagrams they are based on, are wholly contained within a rectangular box. Each unitary spider diagram of order consists of *contours* and *spiders*. Contours are simple closed curves. The spider diagram d_1 contains two labelled contours, P and Q. The diagram also contains three

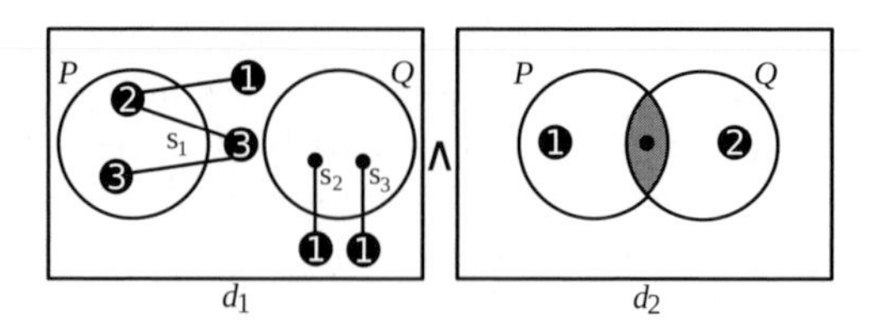

Fig. 1. A spider diagram of order

minimal regions, called *zones*. There is one zone inside the contour P, another inside the contour Q and the other zone is outside both contours P and Q. The unitary spider diagram d_2 contains four zones. One zone is outside both contours P and Q, another is inside the contour P but outside the contour Q, yet another is inside the contour Q but inside the contour P. The final zone is inside both contours and, in this example, is a shaded zone. Each zone in a unitary diagram d can be can be described by a two-way partition of d's contour labels.

In d_1, the zone inside P contains two vertices of one *spider*; spiders are trees whose vertices, called *feet*, are placed in zones. In general, any given spider may contain both *ordered feet* (those of the form ❶, ❷, ❸, ...) and *unordered feet* (those of the form •). In d_1, there is a single four footed spider labelled s_1 and two bi-footed spiders labelled s_2 and s_3. Spider diagrams can also contain *shading*, as in d_2.

In figure 1 d_1 and d_2 are joined by a logical connective, $\wedge$, forming a *compound* spider diagram of order. It depicts the conjunction of statements made by its unitary components d_1 and d_2. An $\vee$ symbol between boxes signifies disjunction between statements whereas a horizontal bar above a diagram denotes negation. We now proceed to define the formal apparatus needed to specify spider diagrams of order.

Definition 1. *We define $\mathcal{C}$ to be a finite set of all contour labels. A* ***zone*** *is defined to be a pair, (in, out), of finite disjoint subsets of $\mathcal{C}$. The set "in" contains the labels of the contours that the zone is inside whereas "out" contains the labels of the contours that the zone is outside. The set of all zones is denoted $\mathcal{Z}$. A* ***region*** *is a set of zones.*

Definition 2. *A* ***spider foot*** *is an element of the set $(\mathbb{Z}^+ \cup \{\bullet\}) \times \mathcal{Z}$ and the set of all feet is denoted $\mathcal{F}$. A* ***spider****, s, is a set of feet together with a number, that is $s \in \mathbb{Z}^+ \times (\mathbb{P}\mathcal{F} - \{\emptyset\})$, and the set of all spiders is denoted $\mathcal{S}$. The* ***habitat*** *of a spider $s = (n, p)$ is the region $habitat(s) = \{z : (k, z) \in p\}$. A spider foot $(k, z) \in \mathcal{F}$ where $k \in \mathbb{Z}^+$ has* ***rank*** *k. The set p is the* ***foot set*** *of spider $s = (n, p)$.*

Spiders are numbered because unitary diagrams can contain many spiders with the same foot set; essentially, we view a unitary diagram as containing a bag of spiders, $S(d)$. However, for consistency with previous work, a unitary spider diagram of order is defined as containing a set of spider identifiers, $SI(d)$.

Definition 3. *A* ***unitary spider diagram of order*** *is a quadruple* $d = \langle C, Z, ShZ, SI \rangle$ *where*

$C = C(d) \subseteq \mathcal{C}$ *is a set of contour labels,*
$Z = Z(d) \subseteq \{(a, C(d) - a) : a \subseteq C(d)\}$ *is a set of zones,*
$ShZ = ShZ(d) \subseteq Z(d)$ *is a set of shaded zones,*
$SI = SI(d) \subsetneq \mathcal{S}$ *is a finite set of* ***spider identifiers*** *such that for all* $(n_1, p_1), (n_2, p_2) \in SI(d)$,

$$(p_1 = p_2 \implies n_1 = n_2) \wedge habitat(n_1, p_1) \subseteq Z(d).$$

Given $(n, p) \in SI(d)$ *there are* n *spiders in* d *with foot set* p*. The set of* ***spiders*** *in unitary diagram* d *is defined to be*

$$S(d) = \{(i, p) : \exists (n, p) \in SI(d)\, 1 \leq i \leq n\}.$$

The symbol $\perp$ *is also a unitary spider diagram. We define*

$$C(\perp) = Z(\perp) = ShZ(\perp) = SI(\perp) = \emptyset.$$

If d_1 *and* d_2 *are spider diagrams of order then* $(d_1 \wedge d_2), (d_1 \vee d_2)$ *and* $\neg d_1$ *are* ***compound spider diagrams of order****.*

It is useful to select feet of spiders when defining the semantics.

Definition 4. *Given a diagram* d *a* ***foot select function*** *is* $FootSelect \colon S(d) \rightarrow \mathcal{F}$ *such that, for all* $(n, p) \in S(d)$, $FootSelect(s) \in p$.

It is also useful to identify which zones could be present in a unitary diagram, given the label set, but are not present; semantically, *missing* zones provide information.

Definition 5. *Given a unitary diagram,* d*, a zone* (in, out) *is said to be* ***missing*** *if it is in the set* $\{(in, C(d) - in) : in \subseteq C(d)\} - Z(d)$ *with the set of such zones denoted* $MZ(d)$*. If* d *has no missing zones then* d *is in* ***Venn form*** *[12].*

The semantics of unitary spider diagrams of order are model based. In essence, contours represent sets and spiders represent the existence of elements. A model for a diagram is an assignment of sets to contour labels that ensures various conditions hold; these conditions are encapsulated by the *semantics predicate* defined below.

Definition 6. *An* ***interpretation*** *is a triple* $(U, \Psi, <)$ *where* U *is a universal set and* $\Psi \colon \mathcal{C} \rightarrow \mathbb{P}U$ *is a function that assigns a subset of* U *to each contour label and* $<$ *is a strict order on* U*. The function* Ψ *can be extended to interpret zones and sets of regions as follows:*

1. *each zone,* $(a, b) \in \mathcal{Z}$*, represents the set* $\bigcap_{l \in a} \Psi(l) \cap \bigcap_{l \in b} \overline{\Psi(l)}$ *and*
2. *each region,* $r \in \mathbb{P}\mathcal{Z}$*, represents the set which is the union of the sets represented by* r*'s constituent zones.*

For brevity, we will continue to write $\Psi\colon \mathcal{C} \to \mathbb{P}U$ but assume that the domain of Ψ includes the zones and regions. Given an interpretation we wish to know whether it is a model for a diagram; in other words, when the information provided by the interpretation agrees with the intended meaning of the diagram. Informally, an interpretation is a *model* for unitary diagram d ($\neq\perp$) whenever

1. all of the zones which are missing represent the empty set,
2. all of the regions represent sets whose cardinality is at least the number of spiders placed entirely within that region and
3. all of the entirely shaded regions represent sets whose cardinality is at most the number of spiders with a foot in that region.
4. the elements represented by the spiders obey the ordering imposed on them by the rank of the spiders' feet.

We now make the notion of a model precise.

Definition 7. *Let $I = (U, \Psi, <)$ be an interpretation and let d ($\neq\perp$) be a unitary spider diagram of order. Then I is a* **model** *for d if and only if the following conditions hold.*

1. ***The missing zones condition*** *All of the missing zones represent the empty set $\bigcup_{z \in MZ(d)} \Psi(z) = \emptyset$.*
2. ***The function extension condition*** *There exists an extension of Ψ to spiders, $\Psi : \mathcal{C} \cup S(d) \to \mathbb{P}U$ which ensures the following further conditions hold.*
   (a) ***The habitats condition*** *All spiders represent elements (strictly, singleton sets) in the sets represented by their habitats:*
   $$\forall s \in S(d)\, \Psi(s) \subseteq \Psi(habitat(s)) \wedge |\Psi(s)| = 1.$$
   (b) ***The distinct spiders condition*** *Distinct spiders denote distinct elements:*
   $$\forall s_1, s_2 \in S(d) \Psi(s_1) = \Psi(s_2) \implies s_1 = s_2.$$
   (c) ***The shading condition*** *Shaded regions represent sets containing elements denoted by spiders:*
   $$\Psi(ShZ(d)) \subseteq \bigcup_{s \in S(d)} \Psi(s).$$
   (d) ***The order condition*** *The ordering information provided by the spiders agrees with that provided by the strict order relation: there exists a foot selection function, $FootSelect\colon S(d) \to \mathcal{F}$, for d such that*
      - *for all $s \in S(d)$, $FootSelect(s) = (n, z)$ implies $\Psi(s) \subseteq \Psi(z)$*
      - *for all $s_1, s_2 \in S(d)$ with $FootSelect(s_1) = (n_1, z_1)$ and $FootSelect(s_2) = (n_2, z_2)$, if $x < y$ where $\Psi(s_1) = \{x\}$ and $\Psi(s_2) = \{y\}$ then either $n_1 = n_2$ or $n_1 = \bullet$ or $n_2 = \bullet$ or $n_1 < n_2$.*

If $\Psi : \mathcal{C} \cup S(d) \rightarrow \mathbb{P}U$ ensures that the above conditions are satisfied then Ψ is a ***valid*** *extension to spiders for d. A foot selection function, $FootSelect : S(d) \rightarrow \mathcal{F}$, that ensures the above conditions are satisfied is also called* ***valid****. If $d = \perp$ then no interpretation is a model for d.*

For compound diagrams, the definition of a model extends in the obvious inductive way.

3 The Language of a Spider Diagram of Order

A language is a set of words over a finite alphabet, typically denoted Σ. In order to discuss the language of a spider diagram of order we associate an alphabet with the contours that may appear in a diagram. First, a function is a fixed which maps elements of $\mathcal{C}$ to sets of letters from the finite alphabet Σ; we require $|\Sigma| \leq 2^{|\mathcal{C}|}$. The use of this function allows us to consider the language of a diagram over an arbitrary alphabet. Previous work [5] considered a much more restricted set of alphabets. In the examples in figure 2 we assume $\mathcal{C} = \{P, Q\}$ and we assign the alphabet $\Sigma = \{a, b, c, d\}$ via a function called *lettermap* in the manner depicted in figure 2(a) i.e. $lettermap(P) = \{b, c\}$ and $lettermap(Q) = \{c, d\}$. It is important to note that the lower-case letters in figure 2(a) are not syntactic elements of spider diagrams of order. The depicted *lettermap* assignment satisfies the following definition:

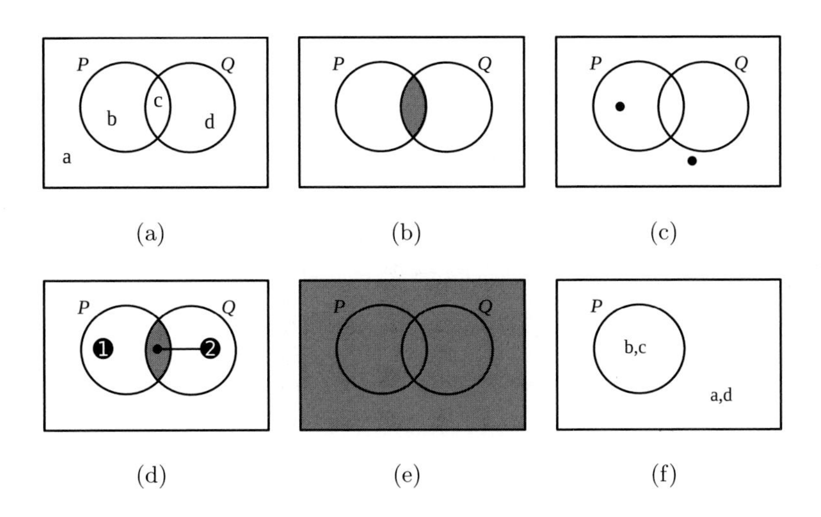

Fig. 2. Example unitary spider diagrams of order

Definition 8. *The function lettermap : $\mathcal{C} \cup \mathcal{Z} \rightarrow \mathbb{P}\Sigma$ is a fixed assignment of sets of letters to contour labels and zones which satisfies the following two conditions:*

1. *Each zone, (in, out), maps to the set of letters inside the set of included contours but outside the set of excluded contours*

$$lettermap(in, out) = \bigcap_{c \in in} lettermap(c) \cap \bigcap_{c \in out} (\Sigma - lettermap(c)).$$

2. *Any zone, (in, out), for which $in \cup out = \mathcal{C}$ is assigned at most one letter.*

$$\text{if } in \cup out = \mathcal{C} \text{ then } |lettermap(in, out)| \leq 1.$$

This ensures that the spider diagram logic is capable of distinguishing each letter.

Let a be a letter in Σ. Then $zone(a) = (inLabels(a), outLabels(a))$ where we define

- $inLabels(a) = \{c \in \mathcal{C} : a \in lettermap(c)\}$ *and*
- $outLabels(a) = \{c \in \mathcal{C} : a \notin lettermap(c)\}$.

We now have a relationship between the spider diagram logic over the labels in $\mathcal{C}$ and the alphabet Σ. Therefore we may now discuss the language of the diagram. Figure 2(a) places no restrictions on the words of the language. This is because it contains no missing zones, no shading and no spiders. As such, the language of this diagram is the set of all words over the alphabet namely Σ^*. Figure 2(b) prevents words from containing a 'c' character. The words "abd", "aaaa" and "dbab" are in the language of the diagram, unlike the words "c" and "abcd". Words in the language of the diagram are said to *correspond* to that diagram and the set of all such words is the *corresponding language*. Figure 2(c) asserts that words must contain a 'b' character and an 'a' character because of the presence of spiders. Words in the language of this diagram have a minimum length of two characters. The words "ab", "ba" and "aaabcd" are elements of the language. The word "acdd" is not, and neither is the word "cdbcd".

In figure 2(d) there are several restrictions placed on words in the language of the diagram. The first restriction imposed by the left-most spider is that the word must contain at least one 'b' character. This is because the spider asserts that the set represented by the zone $(\{P\}, \{Q\})$ is non-empty, thus any words corresponding to the diagram must contain at least one character specified by the previously given *lettermap* function. Further conditions are imposed by the right-most two-footed spider and the shaded zone. The right-most spider provides disjunctive information. In language terms it states that words corresponding to the diagram must contain either a 'c' or a 'd'. The shading indicates an upper bound on the number of 'c' letters in words of the language of the diagram: there can be at most one 'c'. The final constraint is that a letter 'b' must occur before a letter 'd' if 'd' is a letter chosen based on the right-most spider. This is due to the explicit ordering prescribed by the spider feet. Thus the language of the diagram in figure 2(d) may be described by the regular expression

$$(a|b|d)^*b(a|b|d)^*d(a|b|d)^* \cup (a|b|d)^*b(a|b|d)^*c(a|b|d)^* \cup (a|b|d)^*c(a|b|d)^*b(a|b|d)^*.$$

The words of this language contain one ‘b’, at most one ‘c’ and if there is no ‘c’ then ‘b’ must occur before a ‘d’. The diagram in figure 2(e) intuitively places an upper-bound on the number of occurrences of each letter appearing in words of the corresponding language, the corresponding language is the set containing the empty word, $\{\lambda\}$.

Finally, consider the diagram in figure 2(f). The labels in this diagram are drawn from the set $\mathcal{C}$, although its label set contains only P. The models that satisfy the diagram in figure 2(a) also satisfy the diagram in figure 2(f). Similarly, the language that corresponds to the diagram in figure 2(a) also corresponds to the diagram in figure 2(f) due to the depicted assignment of letters to zones i.e. $lettermap(P) = \{b, c\}$ and $lettermap(Q) = \{c, d\}$. In general, semantically equivalent diagrams have the same language under any *lettermap* function.

In order to define how an interpretation models a word it is useful to treat a word as an array.

Definition 9. *Let w be a word of some language. $Array(w)$ is a set of pairs (a, i) where each i is a position in word w and a is the letter at position i.*

For example, consider the word $w = ab$ then $Array(w) = \{(a, 1), (b, 2)\}$.

The interpretation $I = (U, \Psi, <)$ is a model for word $w = ab$ given $lettermap(P) = \{a\}$ and $lettermap(Q) = \{b\}$ where

$$U = \{1, 2\} \qquad \Psi(P) = \{1\}, \Psi(Q) = \{2\} \qquad < = \{(1, 2)\}.$$

An interpretation I models w because a bijection exists between U and the letters in w, namely $\{1 \mapsto a, 2 \mapsto b\}$, which respects the order relation $<$ and the order of ‘a’ and ‘b’ in w. Furthermore, the bijection and Ψ are in agreement on the assignment of letter sets to the interpretation of contours and zones. The following is a definition of a model of a word.

Definition 10. *An interpretation $I = (U, \Psi, <)$ is a* **model** *for a word w if there exists a bijection, f, between $Array(w)$ and U such that*

1. *the order of the letters in w is respected by $<$*

$$\forall (a_i, j), (a_k, l) \in Array(w) : f(a_i, j) < f(a_k, l) \Rightarrow j < l, \text{ and}$$

2. *the element in U chosen for the letter a_i is in a set represented by the zone that partitions $\mathcal{C}$ and contains a_i*

$$\forall (a_i, j) \in Array(w) : f(a_i, j) \in \Psi(zone(a_i)).$$

We may now define the language of a spider diagram of order.

Definition 11. *The* **language of a diagram**, *D, denoted $\mathcal{L}(D)$, is defined to be the set of words that are modelled by at least one interpretation that models D.*

Lemma 1. *The following properties hold for spider diagrams:*

– *The language of the disjunction of two diagrams is the union of languages of the components: $\mathcal{L}(D_1 \vee D_2) = \mathcal{L}(D_1) \cup \mathcal{L}(D_2)$.*

- *Similarly, the language of the conjunction of two diagrams is the intersection of the components:* $\mathcal{L}(D_1 \wedge D_2) = \mathcal{L}(D_1) \cap \mathcal{L}(D_2)$.
- *The language of the negation of a diagram is the set complement of the language of the diagram with respect to the universe* Σ^**, formally:* $\mathcal{L}(\neg D_1) = \Sigma^* - \mathcal{L}(D_1) = \overline{\mathcal{L}(D_1)}$.

4 The Product of Spider Diagrams

The languages corresponding to the diagrams in figures 3(a) and 3(b) are $\{$"*aba*"$\}$ and b^* (star-free $\overline{\overline{\emptyset}a\overline{\emptyset}}$) over $\Sigma = \{a, b\}$ respectively. The language $aba(b)^*$ is the concatenation of the word "aba" and words from the language b^*. It too is star-free and may be written as $aba\overline{\overline{\emptyset}a\overline{\emptyset}}$. Words in this language include any word which begins with the subword "*aba*" followed zero or more '*b*' characters eg: "aba", "abab" and "ababb". The language $abab^*$ is be defined by a spider diagram of order which is the conjunction of three unitary diagrams defining the languages $b^*ab^*ab^*$, $\overline{((a|b)^*a(a|b)^*b(a|b)^*b(a|b)^*a(a|b)^*)}$ and $\overline{((a|b)^*b(a|b)^*a(a|b)^*a(a|b)^*)}$, see figure 3(c). We show how diagrams define languages, such as these three, below. Furthermore, we strongly conjecture that there are star-free regular languages, for example $aba(a|b)^*$, that do not correspond to any spider diagram of order.

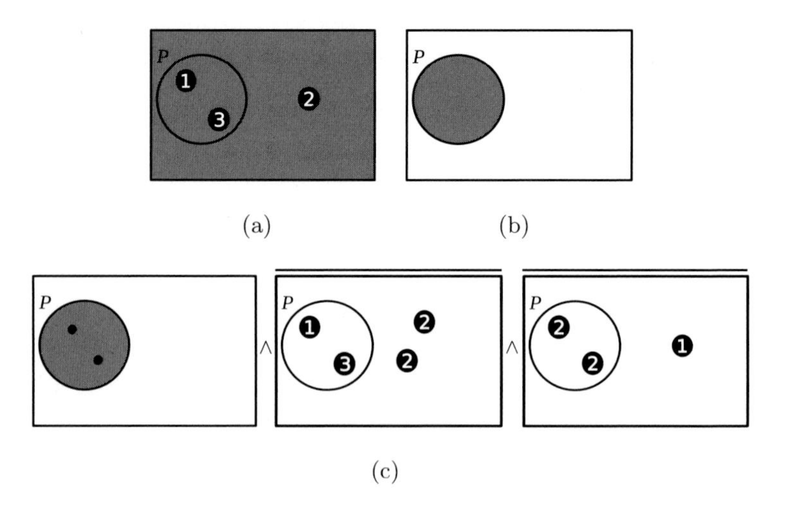

Fig. 3. Two unitary spider diagrams of order and a more complex example

To remove this expressiveness limitation we introduce the *product* of spider diagrams of order denoted $\lhd$ as depicted in figure 4. We first extend the syntax of spider diagrams of order to include $\lhd$ as a boolean operation.

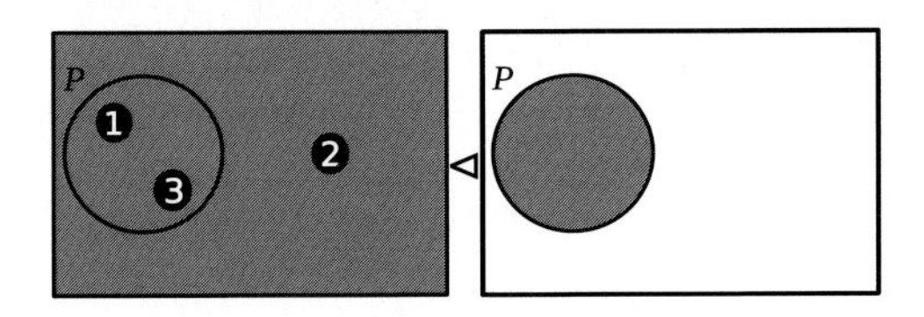

Fig. 4. A spider diagram of order with a product operator and language $abab^*$

Definition 12. *A diagram is a spider diagram of order augmented with a product operator if it*

- *is a spider diagram of order,*
- *is of the form $(D_1 \lhd D_2)$ where D_1 and D_2 are spider diagrams of order augmented with a product operator.*

In order to extend the semantics of spider diagrams of order we recall the following definition from [6].

Definition 13. *The* ***ordered sum*** *of two interpretations $m_1 = (U_1, \Psi_1, <_1)$ and $m_2 = (U_2, \Psi_2, <_2)$, denoted $m_1 \lhd m_2$, where U_1 and U_2 are disjoint is the interpretation $m = (U, \Psi, <)$ such that*

$U = U_1 \cup U_2$,
$\Psi(c) = \Psi_1(c) \cup \Psi_2(c)$ *for all* $c \in \mathcal{C}$,
$< = <_1 \cup <_2 \cup \{(a, b) : a \in U_1, b \in U_2\}$.

The definition for the model of the product of diagrams follows from the ordered sum of interpretations.

Definition 14. *Let m be an interpretation. Then m is a* ***model*** *for $D_1 \lhd D_2$ if there exist interpretations m_1 and m_2, such that $m = m_1 \lhd m_2$, and m_1 and m_2 are models for D_1 and D_2 respectively.*

Finally, we extend the properties of the language of a spider diagram (lemma 1) to include the following clause:

- The language of the product of two diagrams is the concatenation of its components: $\mathcal{L}(D_1 \lhd D_2) = \mathcal{L}(D_1) \cdot \mathcal{L}(D_2)$.

The key insight underlying this is that the elements of a model of the compound diagram will be the disjoint union of elements for models of the constituent diagrams, with all the elements of the first being less than all the elements of the second under the ordering relation of the compound model. So, the language denoted by the compound diagram will be the concatenation of all words corresponding to the individual diagrams. We now see that the language corresponding to the diagram in figure 4 is $abab^*$. We now have a complete picture of the syntax and semantics of spider diagrams of order augmented with a product operator. Furthermore we have seen that the product of diagrams is a diagrammatic analogue of language concatenation.

5 Comparing Classes of Regular Languages and Spider Diagrams

In this section we prove that all star-free regular languages are definable using spider diagrams of order. We prove this by induction on the levels of the Straubing-Thérin hierarchy (STH). Level 0 is our base case, which contains the languages $\{\}$ and Σ^*. We assume that any language at level n of the STH is definable using spider diagrams of order augmented with a product operator. As the STH has whole numbered levels $0, 1, 2, \ldots$ and half levels $0 + \frac{1}{2}, 1 + \frac{1}{2}, 2 + \frac{1}{2}$ our proof shows that languages at level $n + \frac{1}{2}$ and $n + 1$ are definable using spider diagrams of order augmented with product. We also pay particular attention to the structure of diagrams which define languages at level $\frac{1}{2}$ and level 1 as these sets of languages are the well-known shuffle-ideal [10] and piecewise-testable [18] classes respectively.

Lemma 2. *The languages at level* 0 *of the Straubing-Thérin hierarchy are definable using spider diagrams of order.*

Proof. Let L be a language at level 0 of the STH. Then, by definition, L is either $\{\}$ or Σ^*. Considering the diagram d_1 in figure 5 it can be shown that $lettermap(\{\}, \{\}) = \Sigma$ so the language of the diagram d_1 is Σ^*, which is one element of level 0. Given any *lettermap* function $\mathcal{L}(\perp) = \{\}$ where $\perp$ is a unitary spider diagram of order, which is the other element of level 0.

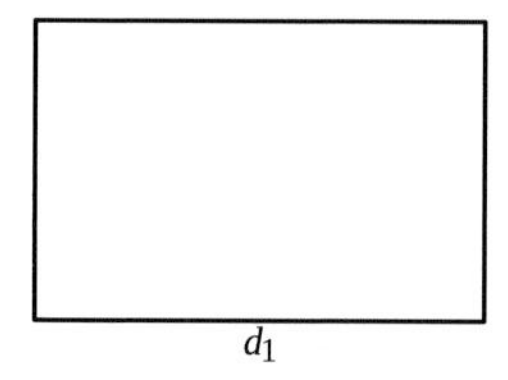

Fig. 5. Unitary spider diagram of order to which Σ^* corresponds

We denote the diagram d_1 in figure 5, to which the language Σ^* corresponds, by the symbol. □

The following lemma shows that the diagram in figure 6 may be constructed such that only the word "bd" corresponds to it. Given the *lettermap* function as depicted in figure 2(a) the zones of the diagram map to singleton sets or the empty set. In order to construct the diagram we examine the first letter of the word and place a spider with single foot of rank 1 in the zone that *lettermap* states it is contained in. We repeat this exercise for the second letter however, this time, placing a spider with single foot of rank 2.

Lemma 3. *Any language containing a single word is definable by a unitary spider diagram of order.*

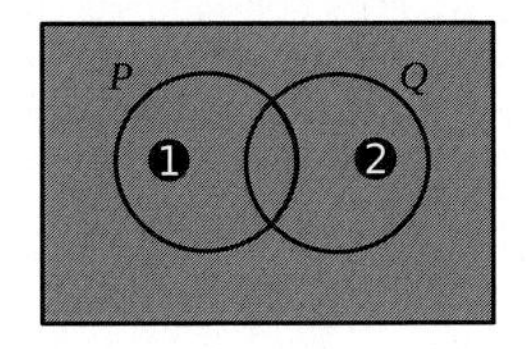

Fig. 6. The language $\{bd\}$ defined by a spider diagram of order

Proof. Let L be a set of words over alphabet Σ containing exactly one word w of length n. Furthermore, let d be a unitary spider diagram of order in Venn form where all contours from $\mathcal{C}$ are present and all of the zones in d are shaded. The diagram contains all contours and has no missing zones. Therefore, as a consequence of definition 8, *lettermap* must assign each zone to a singleton set containing a letter or the empty set. For each (a, i) in $Array(w)$ of the word w in L we place a single-footed spider with a foot of rank i in the zone $zone(\{a\})$ of d.

We denote a fully shaded diagram, as constructed in the above lemma, with corresponding single word language $\{w\}$ by the symbol $\blacksquare_w$.

Corollary 1. *Any arbitrary finite set of words of finite length is definable by a spider diagram of order.*

Proof. Let L be an arbitrary finite set of finite length words $\{w_1, w_2, \ldots, w_n\}$ over an alphabet Σ. By lemma 3 we may create a diagram $\blacksquare_{w_i}$ for each $w_i \in L$. The finite disjunction of these diagrams $\blacksquare_{w_1} \vee \blacksquare_{w_2} \vee \ldots \vee \blacksquare_{w_n}$ is the diagram to which L corresponds.

The fractional levels of the STH are the polynomial closure of languages at the integer level beneath them.

Definition 15. *A language is in the* **polynomial closure** *of a set of languages S if it is a finite union of languages of the form*

$$L_0 a_1 L_1 \ldots L_{n-1} a_n L_n$$

where each $L_i \in S$ and $a_j \in \Sigma$.

We now consider the form of diagrams that define the class of shuffle-ideal languages.

Lemma 4. *The languages at level $\frac{1}{2}$ of the Straubing-Thérin hierarchy (the class of shuffle-ideal languages) are definable by spider diagrams of order augmented with a product operator.*

Proof. Let L be a language at level $\frac{1}{2}$ of the STH. The set of languages at level $\frac{1}{2}$ is the polynomial closure of languages at level 0 [14]. The polynomial closure of level 0 results in finite unions of languages of the form $\Sigma^* a_1 \Sigma^* \ldots \Sigma^* a_n \Sigma^*$ [10].

By lemma 3 we may construct a diagram $\blacksquare_{a_i}$ to which the language $\{a_i\}$ corresponds. Furthermore, by lemma 2 we may construct a diagram $\square$ to which the language Σ^* corresponds. Therefore, L corresponds to finite disjunctions of diagrams of the form

$$\square \lhd \blacksquare_{a_1} \lhd \square \lhd \blacksquare_{a_2} \ldots \square \lhd \blacksquare_{a_n} \lhd \square.$$

Lemma 4 may be developed into a more succinct characterisation.

Theorem 1. *The set of shuffle-ideal languages are definable by spider diagrams of order where each diagram is unshaded and the only binary connectives used are* $\vee$ *and* $\lhd$.

Proof. Let D be a diagram from lemma 4 to which a shuffle-ideal language corresponds. Then D is a finite disjunction of diagrams of the form

$$\square \lhd \blacksquare_{a_1} \lhd \square \lhd \blacksquare_{a_2} \ldots \square \lhd \blacksquare_{a_n} \lhd \square.$$

The diagram D' is semantically equivalent to D, where D' is a finite disjunction of diagrams of the form

$$\square \lhd \blacksquare_{a_1} \lhd \square \lhd \square \lhd \blacksquare_{a_2} \lhd \square \ldots \lhd \square \lhd \blacksquare_{a_{n-1}} \lhd \square \lhd \square \lhd \blacksquare_{a_n} \lhd \square.$$

In D' each diagram $\blacksquare_{a_i}$ is proceeded by a diagram $\square$ and is followed by $\square$. The diagram $\blacksquare_{a_{i+1}}$ is proceeded by a diagram $\square$ which is not the diagram that follows $\blacksquare_{a_i}$. In language terms we are stating that the language

$$\Sigma^* a_1 \Sigma^* a_2 \Sigma^* \ldots \Sigma^* a_n \Sigma^*$$

is equivalent to the language

$$\Sigma^* a_1 \Sigma^* \Sigma^* a_2 \Sigma^* \ldots \Sigma^* a_{n-1} \Sigma^* \Sigma^* a_n \Sigma^*.$$

Each $\square \lhd \blacksquare_{a_i} \lhd \square$ may be collapsed into a single diagram $\square_{a_i}$ where $\square_{a_i}$ is $\blacksquare_{a_i}$ in Venn form with the shading removed from all zones. The language $\Sigma^* a_i \Sigma^*$ corresponds to the diagram $\square_{a_i}$. Therefore D is equivalent to finite disjunctions of diagrams of the form

$$\square_{a_1} \lhd \square_{a_2} \lhd \ldots \lhd \square_{a_n}$$

where each $\square_{a_i}$ is unitary and contains no missing or shaded zones.

Lemma 5. *The languages at level* 1 *of the Straubing-Thérin hierarchy are definable by spider diagrams of order augmented with a product operator.*

Proof. Languages at level 1 are formed by taking boolean combinations of languages at level $\frac{1}{2}$. The boolean operators $\cup$ and $^-$ over languages at level $\frac{1}{2}$ are semantically equivalent to $\vee$ and $\neg$ over the diagrams to which the languages at level $\frac{1}{2}$ correspond.

Given the example of a star-free language ab^* and the example of the regular, but not star-free, language $(aa)^*$ over alphabet $\Sigma = \{a, b\}$ presented in the Introduction, a reader may see how a spider diagram may be constructed such that its corresponding language is ab^*. The language ab^* may be written as the star-free expression $a\overline{\overline{\emptyset}a\overline{\emptyset}}$ or the diagram $\blacksquare_a \lhd (\neg(\square \lhd \blacksquare_a \lhd \square))$ (by theorem 1 a more succinct diagram may be derived). It may also be seen that it is impossible to construct a spider diagram of order to express $(aa)^*$ as it is not a star-free language. We now provide a lower bound on the class of languages definable by spider-diagrams of order

Theorem 2. *All star-free languages are definable by spider diagrams of order augmented with a product operator.*

Proof. From [19] we know that the Straubing-Thérin hierarchy, in the limit, contains all and only the star-free languages. We, therefore, prove the theorem inductively where lemma 2 is the base case. It is assumed that languages at level n of the Straubing-Thérin hierarchy to correspond to spider diagrams of order. We prove that languages at level $n + \frac{1}{2}$ and languages at level $n + 1$ correspond to spider diagrams of order.

Let L be a language, over Σ, at level $n + \frac{1}{2}$. Then L is an element of the polynomial closure of level n i.e. finite unions of languages of the form

$$L_0 a_1 L_1 a_2 \ldots a_{n-1} L_{n-1} a_n L_n$$

where each $a_i \in \Sigma$ and the languages L_j are languages at level n. We may take the diagram D_j to which language L_j corresponds and create a diagram $\blacksquare_{a_i}$ (by lemma 3) to which the language $\{a_i\}$ corresponds. Then diagrams to which languages at level $n + \frac{1}{2}$ correspond are finite disjunctions of diagrams of the form

$$D_0 \lhd \blacksquare_{a_1} \lhd D_1 \lhd \blacksquare_{a_2} \lhd \ldots \lhd \blacksquare_{a_{n-1}} \lhd D_{n-1} \lhd \blacksquare_{a_n} \lhd D_n.$$

Let L now be a language at level $n + 1$ of the STH. Languages at level $n + 1$ are the finite boolean closure (over union and complement) of languages at level $n + \frac{1}{2}$. Therefore L corresponds to a diagram in the finite boolean closure (over $\vee$ and $\neg$) of diagrams to which languages at level $n + \frac{1}{2}$ correspond. Hence all star-free languages are definable by spider diagrams of order augmented with a product operator.

From [17] we know that spider diagrams without order are expressively equivalent to monadic first order with equality, denoted $MFoL[=]$. Our extension to spider diagrams, following Thomas' work [19], adds an order relation to the semantic models. Thus, an alternative statement of the theorem 2 may be made in terms of first order logic.

Theorem 3. *Spider diagrams of order augmented with a product operator are at least as expressive as logical sentences in monadic first order logic equipped with an order relation $MFoL[<]$.*

Proof. We recall from [19] that star-free regular languages are definable using $MFoL[<]$ and the result that languages are definable using $MFoL[<]$ if and only if they are star-free. We have shown that all star-free languages correspond to spider diagrams of order augmented with product. Therefore, spider diagrams of order are at least as expressive as $MFoL[<]$.

Given the established relationship between spider diagrams of order, star-free regular languages and monadic first order logic over structures containing an order relation we may now establish the expressive power of spider diagrams of order.

Theorem 4. *Spider diagrams of order augmented with a product operator are expressively equivalent to monadic first order logic of order.*

6 Conclusion

In this paper we have introduced a product operation on spider diagrams of order. This operation is a spider diagram analogue of language concatenation. We have further shown that spider diagrams of order, when augmented with this product operation, are as expressive as monadic first order logic of order. This increase in expressiveness was suggested by previous work in [5]. Our intention now is to further the results from [4] and develop an algorithm to construct a minimal finite state automaton given a spider diagram of order. Such an algorithm will answer questions concerning the succinctness of description afforded by spider diagrams of order when compared to finite automata.

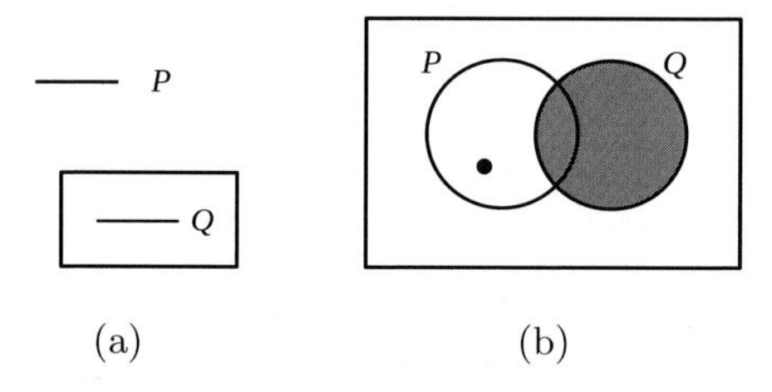

(a) (b)

Fig. 7. Both the existential graph and spider diagram have language $(a|b)^*b(a|b)^*$

We see no reason why our approach to establishing expressiveness may not be applied to other model-based diagrammatic reasoning systems. For example in figure 7(a), consider the sheet of assertion in Pierce's Beta system to be the set of all words Σ^*. Cuts and assertions on the sheet restrict the language corresponding to the diagram to be a subset of Σ^*. In the case of the Beta system *lettermap* would assign sets of letters to predicate symbols.

Acknowledgements

We would like to thank Gem Stapleton and the anonymous reviewers for their useful feedback.

References

1. Barwise, J., Etchemendy, J.: Hyperproof. CSLI Press (1994)
2. Büchi, J.: Weak second order arithmetic and finite automata. Z. Math. Logik und Grundl. Math. 6, 66–92 (1960)
3. Chomsky, N.: Three models for the description of language. IRE Transactions on Information Theory, 113–124 (1956)
4. Delaney, A., Stapleton, G.: On the descriptional complexity of a diagrammatic notation. In: Visual Languages and Computing (September 2007)
5. Delaney, A., Stapleton, G.: Spider diagrams of order. In: International Workshop on Visual Languages and Logic (September 2007)
6. Ebbinghaus, H.-D., Flum, J.: Finite Model Theory, 2nd edn. Springer, Heidelberg (1991)
7. Euler, L.: Lettres a une princesse dallemagne sur divers sujets de physique et de philosophie. Letters 2, 102–108 (1775); Berne, Socit Typographique
8. Gil, J., Howse, J., Kent, S.: Formalising spider diagrams. In: Proceedings of IEEE Symposium on Visual Languages, Tokyo, September 1999, pp. 130–137. IEEE Computer Society Press, Los Alamitos (1999)
9. Hammer, E.: Logic and Visual Information. CSLI Publications (1995)
10. Héam, P.-C.: On shuffle ideals. Theoretical Informatics and Applications 36, 359–384 (2002)
11. Hopcroft, J.E., Ullman, J.D.: Introduction to Automata Theory, Languages and Computation. Addison-Wesley Publishing, Reading (1979)
12. Howse, J., Stapleton, G., Taylor, J.: Spider diagrams. LMS Journal of Computation and Mathematics 8, 145–194 (2005)
13. Kent, S.: Constraint diagrams: Visualizing invariants in object oriented modelling. In: Proceedings of OOPSLA 1997, October 1997, pp. 327–341. ACM Press, New York (1997)
14. Pin, J.-E.: Syntactic semigroups, pp. 679–746. Springer-Verlag, New York (1997)
15. Pin, J.-E., Straubing, H., Thérien, D.: Some results on the generalized star-height problem. Information and Computation 101, 219–250 (1992)
16. Shin, S.-J.: The Logical Status of Diagrams. Cambridge University Press, Cambridge (1994)
17. Stapleton, G., Thompson, S., Howse, J., Taylor, J.: The expressiveness of spider diagrams. Journal of Logic and Computation 14(6), 857–880 (2004)
18. Stern, J.: Characterizations of some classes of regular events. Theoretical Computer Science 35, 17–42 (1985)
19. Thomas, W.: Classifying regular events in symbolic logic. Journal of Computer and System Sciences 25, 360–376 (1982)
20. Venn, J.: On the diagrammatic and mechanical representation of propositions and reasonings. Phil. Mag. (1880)

Diagrammatic Reasoning System with Euler Circles: Theory and Experiment Design

Koji Mineshima, Mitsuhiro Okada, Yuri Sato, and Ryo Takemura

Department of Philosophy, Keio University,
2-15-45 Mita, Minato-ku, Tokyo 108-8345, Japan
{minesima,mitsu,sato,takemura}@abelard.flet.keio.ac.jp

Abstract. In this paper we are concerned with logical and cognitive aspects of reasoning with Euler circles. We give a proof-theoretical analysis of diagrammatic reasoning with Euler circles involving unification and deletion rules. Diagrammatic *syllogistic* reasoning is characterized as a particular class of the general diagrammatic proofs. Given this proof-theoretical analysis, we present some conjectures on cognitive aspects of reasoning with Euler diagrams. Then we propose a design of experiment for a cognitive psychological study.

1 Introduction

We study diagrammatic reasoning with Euler circles composed of unification and deletion inferences. A primitive unification inference step is specified as a unification of two Euler diagrams.[1] We define the notion of diagrammatic *proof* (*d-proof*, in short), which is considered as a (possibly long) chain of unification and deletion steps. A major difficulty in the Euler-style diagrammatic proofs consists in the fact that the complexity of the diagrams increases during the processes of diagrammatic proof constructions; especially, unification often multiplies *disjunctive ambiguity* (i.e., the ambiguity described by duplicating a point, say x, at many different positions in a unified diagram with linking them, or, instead, by multiplying diagram-pages; cf. Peirce [12]). By contrast, when the Euler-style diagrammatic reasoning is restricted to the *syllogistic* inferences, essentially no disjunctive ambiguity appears during any diagrammatic proof construction process. We shall present a diagrammatic proof system which has essentially no disjunctive ambiguity, and which includes the syllogistic proofs as special cases.

In Section 2, we consider a diagrammatic representation system for Euler circles in which disjunctive ambiguity is not allowed. (One may call disjunction-free diagrams "one-page diagrams", and disjunctive diagrams "multiple-page

[1] Some Euler-style diagrammatic reasoning systems (e.g. Hammer [9]) do not have unification rule, hence cannot deal with syllogistic reasoning. But we do not consider such a simple case in this paper.

G. Stapleton, J. Howse, and J. Lee (Eds.): Diagrams 2008, LNAI 5223, pp. 188–205, 2008.

diagrams.") We give a definition of an Euler diagrammatic syntax and a set-theoretical semantics for it.

In Section 3, we provide a diagrammatic inference system consisting of unification and deletion rules, where any conclusion of a (possibly long) d-proof is always representable on a one-page diagram. The *syllogistic* d-proof system is characterized as a specific subsystem of our d-proof system (where unification and deletion appear alternately without repeating in a proof). Compared with linguistic syllogistic reasoning, the diagrammatic reasoning with Euler diagrams in our system has some distinctive features: linguistic syllogistic reasoning involves explicit operations with logical negation and with the "subject-predicate" distinction, whereas the reasoning with Euler diagrams in our system does not.

In Section 4, we outline an experimental design to test whether these differences are also shown by the performance level of human reasoning. Our pilot experiment suggests that the actual diagrammatic syllogistic reasoning processes have some characteristics predicted by our inference system.

2 Diagrammatic Representation System EUL for Euler Circles and Its (Set-Theoretical) Semantics

In this section, we introduce a graphical representation system EUL for Euler diagrams in which disjunctive ambiguity is not allowed. In Section 2.1, we clarify the main features of EUL by comparing it with other systems. Then, we define a syntax of EUL (Section 2.2) and its formal semantics (Section 2.3).

2.1 A Classification of Diagrammatic Representation Systems

Since Euler's invention of graphical representation of syllogisms (Euler [7]), and its major modification and extension by Venn [24] and Peirce [12], many diagrammatic representation systems have been developed. (For a recent survey, see Stapleton [17].) We give a classification of these systems, focusing on how to represent logical negation, existence and disjunction graphically.

(1) **On negation:**
 a. Systems with a syntactic device to express logical negation. (e.g. "shading" in Venn diagrams)
 b. Systems which make use of the topological relations between circles to represent negative information (Euler-style systems).

(2) **On existence:**
 a. Systems with syntactic devices for indicating the existence of objects. (e.g. "x" introduced by Peirce [12])
 b. Systems which assume that every region in a diagram represents a non-empty set. (One of Euler's original systems; cf. Hammer and Shin [10]).
 c. Systems in which existential claims cannot be made (e.g., Hammer [9]).

(3) **On disjunction:**
 a. Systems which can represent disjunctive information (e.g., by using a linking between duplicated existential points as in Peirce [12]).
 b. Systems which cannot represent disjunctive information.

Among various possibilities, we introduce a system with the following features: regarding negation, we adopt (1b) and focus on the Euler-style diagrammatic reasoning without explicit negation; regarding existence, we adopt (2a) and use "named points" to indicate the existence of a particular element; regarding disjunction, we adopt (3b) and focus on disjunction-free, one-page diagrams.

We call our system the EUL representation system for Euler diagrams. In the next subsection, we give a precise definition of the syntax of EUL.

2.2 Diagrammatic Syntax of EUL

Let us start by defining the diagrams of EUL.

Definition 1 (EUL-diagrams). An EUL-*diagram* is a 2-dimensional ($\mathbb{R}^2$) plane with a finite number (at least two) of *named simple closed curves*[2] (denoted by $A, B, C, \dots$) and *named points* (denoted by $x, y, z, \dots$), where each named simple closed curve or named point has a unique and distinct name.

EUL-diagrams are denoted by $\mathcal{D}, \mathcal{E}, \mathcal{D}_1, \mathcal{D}_2, \dots$.

In what follows, we sometimes call a named simple closed curve a *named circle.* Moreover, named circles and named points are collectively called *objects.* We use a rectangle to represent a plane for a EUL-diagram.

The binary relations $A \sqsubset B, A \vdash\!\dashv B, A \bowtie B, x \sqsubset A, x \vdash\!\dashv A$, and $x \vdash\!\dashv y$ mean, respectively, "the interior[3] of A is inside of the interior of B," "the interior of A is outside of the interior of B," "there is at least one crossing point between A and B," "x is inside of the interior of A," "x is outside of the interior of A," and "x is outside of y (i.e. x is not located at the point of y)".

Proposition 1. *Given an* EUL-*diagram* $\mathcal{D}$,

1. *for any distinct named simple closed curves A and B, exactly one of $A \sqsubset B, B \sqsubset A, A \vdash\!\dashv B$, and $A \bowtie B$ holds;*
2. *for any named point x and any named simple closed curve A, exactly one of $x \sqsubset A$ and $x \vdash\!\dashv A$ holds;*
3. *for any distinct named points x and y, $x \vdash\!\dashv y$ holds.*

For example, consider the EUL-diagram $\mathcal{D}_1$ below, composed of A, B, C, and x.

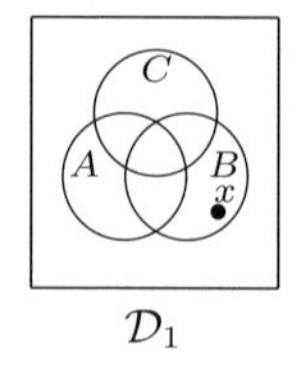

$\mathcal{D}_1$

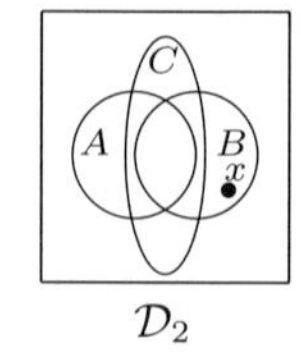

$\mathcal{D}_2$

Fig. 1.

[2] See [3,18] for a formal definition of simple closed curves on $\mathbb{R}^2$.

[3] Here, the interior of a named circle A means the region strictly inside of A. Cf. [3].

The relations $A \bowtie B, A \bowtie C, B \bowtie C, x \vdash A, x \sqsubset B$, and $x \vdash C$ hold on $\mathcal{D}_1$. The same relations also hold for $\mathcal{D}_2$.

An EUL-diagram which has only two objects is called a *minimal diagram* (or an *atomic diagram*).

Given an EUL-diagram $\mathcal{D}$ and two objects, say s and t, on $\mathcal{D}$, a diagram obtained from $\mathcal{D}$ by deleting all objects other than s and t is called a *component minimal diagram* of $\mathcal{D}$. Given $\mathcal{D}$ and objects s and t, the component minimal diagram thus obtained is determined uniquely up to isomorphism. Given an EUL-diagram $\mathcal{D}$, then, the set of component minimal diagrams is determined (up to isomorphism). The set of component minimal (atomic) diagrams of $\mathcal{D}$ is called the *decomposition set* of $\mathcal{D}$. As a direct corollary of Proposition 1, for any minimal (atomic) diagram, say $\mathcal{D}$, composed of two objects s and t, exactly one of $s \sqsubset t, t \sqsubset s, s \vdash t, s \bowtie t$ holds. For example, the set $\{\mathcal{E}_1, \mathcal{E}_2, \mathcal{E}_3, \mathcal{E}_4, \mathcal{E}_5, \mathcal{E}_6\}$ is, up to isomorphisms, the decomposition set of $\mathcal{D}_1$ in Fig. 1 above.

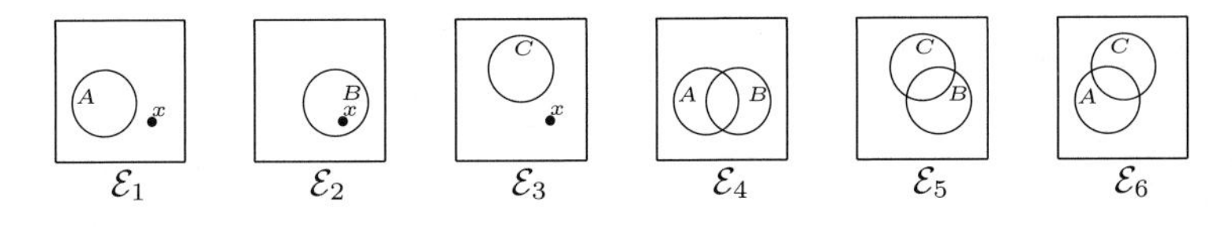

Fig. 2. The decomposition set of $\mathcal{D}_1$

2.3 Set-Theoretical Semantics of EUL

Diagrams of EUL can be used as an auxiliary device to supplement linguistic representations such as formulas of predicate logic. A named circle plays the role of a predicate, and a named point plays the role of a constant symbol. In order to capture such roles of diagrams in a precise way, we give a formal semantics for EUL. Here, we adopt the standard set-theoretical semantics.[4]

Definition 2 (Model). A model $\mathcal{M}$ of EUL is a pair (U, I), where U is a non-empty set (the domain of $\mathcal{M}$), and I is an interpretation function such that

- $I(x) \in U$ for any named point x;
- $I(A) \subseteq U$ for any named simple closed curve A.

Definition 3 (Truth-conditions)

(I) For any minimal (atomic) diagram $\mathcal{D}$ and for any model $\mathcal{M}$,

1. when $x \vdash y$ holds on $\mathcal{D}$, $\mathcal{M} \models \mathcal{D}$ if and only if $I(x) \neq I(y)$;
2. when $x \sqsubset A$ holds on $\mathcal{D}$, $\mathcal{M} \models \mathcal{D}$ if and only if $I(x) \in I(A)$;
3. when $x \vdash A$ holds on $\mathcal{D}$, $\mathcal{M} \models \mathcal{D}$ if and only if $I(x) \notin I(A)$;
4. when $A \sqsubset B$ holds on $\mathcal{D}$, $\mathcal{M} \models \mathcal{D}$ if and only if $I(A) \subseteq I(B)$;

[4] For similar set-theoretical approaches to the semantics of Euler diagrams, see Hammer [9], Hammer and Shin [10], and Swoboda and Allwein [23]. Compared to them, our semantics is distinctive in that diagrams are interpreted as a set of binary relations, and thus that not every region in a diagram has a meaning.

5. when $A \dashv\vdash B$ holds on $\mathcal{D}$, $\mathcal{M} \models \mathcal{D}$ if and only if $I(A) \cap I(B) = \emptyset$;
6. when $A \bowtie B$ holds on $\mathcal{D}$, $\mathcal{M} \models \mathcal{D}$.[5]

(II) For any EUL-diagram $\mathcal{D}$, where $\{\mathcal{D}_1, \mathcal{D}_2, \ldots, \mathcal{D}_n\}$ is the decomposition set of $\mathcal{D}$, $\mathcal{M} \models \mathcal{D}$ if and only if $\mathcal{M} \models \mathcal{D}_1$ and $\mathcal{M} \models \mathcal{D}_2 \ldots \mathcal{M} \models \mathcal{D}_n$.

Note that well-definedness of Definition 3 (II) of the truth-conditions follows from Proposition 1, which assures that the decomposition set is uniquely determined for a given diagram $\mathcal{D}$.

Definition 4 (Validity). Euler diagram $\mathcal{E}$ is a *semantically valid consequence* of $\mathcal{D}_1, \ldots, \mathcal{D}_n$ (written as $\mathcal{D}_1, \ldots, \mathcal{D}_n \models \mathcal{E}$) when for any model $\mathcal{M}$ such that $\mathcal{M} \models \mathcal{D}_1, \ldots, \mathcal{M} \models \mathcal{D}_n$ hold, $\mathcal{M} \models \mathcal{E}$ holds.

For example, consider the diagram $\mathcal{D}_1$ in Fig. 1 in Section 2.2. The decomposition set of $\mathcal{D}_1$ is $\{\mathcal{E}_1, \mathcal{E}_2, \mathcal{E}_3, \mathcal{E}_4, \mathcal{E}_5, \mathcal{E}_6\}$ shown in Fig. 2. Here, $x \dashv\vdash A$ holds on $\mathcal{E}_1$, $x \sqsubset B$ on $\mathcal{E}_2$, $x \dashv\vdash A$ on $\mathcal{E}_3$, $A \bowtie B$ on $\mathcal{E}_4$, $B \bowtie C$ on $\mathcal{E}_5$, and $A \bowtie C$ on $\mathcal{E}_6$. Thus, given a model $M = (U, I)$, the diagram $\mathcal{D}_1$ is true if and only if $I(x) \notin I(A), I(x) \in I(B)$, and $I(x) \notin I(C)$. Note that $\mathcal{E}_4$, $\mathcal{E}_5$, and $\mathcal{E}_6$ do not contribute to the truth-condition of $\mathcal{D}_1$; these express tautological truth in our semantics.

Since the truth-conditions for diagrams are defined in terms of the binary relations $\sqsubset, \dashv\vdash$, and $\bowtie$ on diagrams, any two diagrams, say $\mathcal{D}$ and $\mathcal{E}$, are identified when exactly the same relations hold on both $\mathcal{D}$ and $\mathcal{E}$. Let us consider the diagrams $\mathcal{D}_1$, $\mathcal{D}_2$, and $\mathcal{D}_3$ in Fig. 3 below.

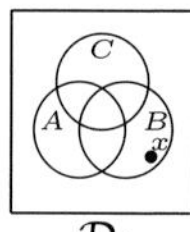
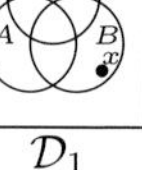

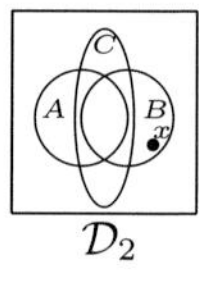

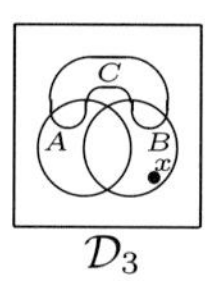

Fig. 3. Identification of EUL-diagrams

$\mathcal{D}_2$ and $\mathcal{D}_3$ are semantically identified with $\mathcal{D}_1$, since exactly the same relations as $\mathcal{D}_1$ hold on $\mathcal{D}_2$ and $\mathcal{D}_3$.

From a syntactic viewpoint, such an identification of diagrams can be explained in terms of "continuous transformation" of named circles, which does not change any of the binary relations $\sqsubset, \dashv\vdash$, and $\bowtie$. The named circle C in $\mathcal{D}_1$ can be continuously transformed, without changing the binary relations with A and with B, in such a way that C covers the intersection region of A and B as it does in $\mathcal{D}_2$. Similarly, C in $\mathcal{D}_1$ can be continuously transformed, without changing the binary relations with A and with B, in such a way that C is disjoint from the intersection region of A and B as it is in $\mathcal{D}_3$.

Note that the intersection region of A and B has no name in our EUL-diagram representation system, which gives the effect of the identical meaning of $\mathcal{D}_1, \mathcal{D}_2$

[5] Informally speaking, $A \bowtie B$ may be understood as $I(A) \cap I(B) = \emptyset \vee I(A) \cap I(B) \neq \emptyset$, which is true as stated in 6 here.

and $\mathcal{D}_3$ (in Fig. 3) on our semantics. In what follows, we assume such an identification by continuous transformation when we use particular diagrams.

3 A Diagrammatic Inference System for Generalized Syllogistic Reasoning

In this section, based on the graphical representation system EUL, we introduce a diagrammatic inference system for EUL, called Generalized Diagrammatic Syllogistic inference system GDS. In Section 3.1, we give an informal description of the inference rules of GDS (Unification and Deletion). In Section 3.2, we give the definition of GDS. We also show that GDS is sound with respect to the formal semantics of EUL given in Section 2.3. Finally, in Section 3.3, we show that the diagrammatic inferences for Aristotelian categorical syllogisms are characterized as a specific subclass of the diagrammatic proofs of GDS.

3.1 An Introduction to GDS Inference System

Let us consider the following question. Given the diagrams $\mathcal{D}_1, \mathcal{D}_2, \mathcal{D}_4, \mathcal{D}_6, \mathcal{D}_7$, how can the diagrammatic information on A, C and x be obtained? Fig. 4 below represents one way of solving the question.

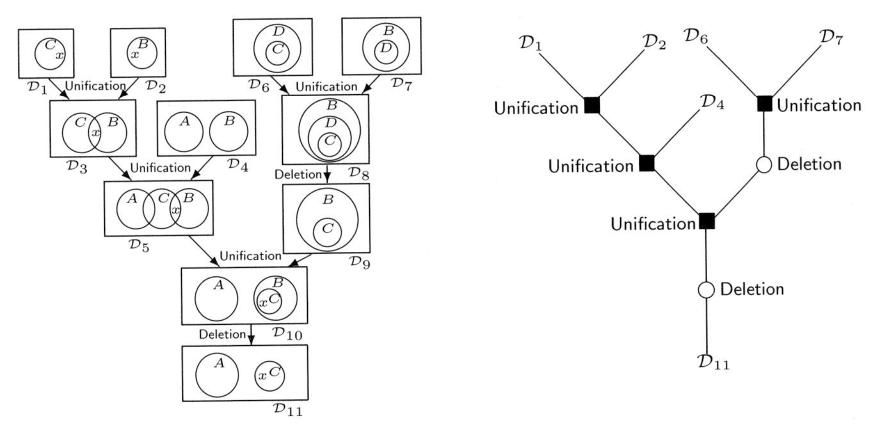

Fig. 4. A diagrammatic proof (d-proof) in GDS

Fig. 5. An underlying tree structure of a d-proof in GDS

Here, two diagrams $\mathcal{D}_1$ and $\mathcal{D}_2$ are unified to obtain $\mathcal{D}_3$, where x in $\mathcal{D}_1$ and x in $\mathcal{D}_2$ are identified, and B is added to $\mathcal{D}_1$ so that x is inside of B and that B overlaps with C without any implication of a relationship between C and B. Then, $\mathcal{D}_3$ is combined with another diagram $\mathcal{D}_4$ to obtain $\mathcal{D}_5$. The diagram $\mathcal{D}_9$ is obtained from $\mathcal{D}_6$ and $\mathcal{D}_7$ by unifying them and by deleting some unnecessary information. The diagrams $\mathcal{D}_5$ and $\mathcal{D}_9$ share two circles C and B: $C \bowtie B$ holds on $\mathcal{D}_5$ and $C \sqsubset B$ holds on $\mathcal{D}_9$. Since the semantic information of $C \sqsubset B$ on $\mathcal{D}_9$ is more accurate than that of $C \bowtie B$ on $\mathcal{D}_5$ according to our semantics for EUL (recall that $C \bowtie B$ means "true" in our semantics), one keeps the diagrammatic relation $C \sqsubset B$ in the unifying process of $\mathcal{D}_5$ and $\mathcal{D}_9$. Finally, by deleting the unnecessary B, one obtains diagram $\mathcal{D}_{11}$ on A, C and x. The above process to obtain $\mathcal{D}_{11}$ from $\mathcal{D}_1, \mathcal{D}_2, \mathcal{D}_4, \mathcal{D}_6$, and $\mathcal{D}_7$ is composed of unifying steps and deleting steps, whose structure could be shown by Fig. 5.

In general, when two diagrammatic representations are unified, a disjunctive device, such as linking-points and/or multiple pages is required. When unifying steps are repeatedly performed, the number of disjunctive linkings/pages is augmented significantly, which would often cause complexity beyond the actual performance of human's handling diagrams. In order to avoid any augmentation of the disjunctive complexity during unifying steps, some constraint should be imposed on unification, which we call the constraint for *determinacy*. For example, it is not permitted to unify two diagrams $\mathcal{D}_1$ and $\mathcal{D}_2$ when, as is shown in Fig. 6, they share only one circle A, and $B \sqsubset A$ holds on $\mathcal{D}_1$ and $x \sqsubset A$ holds on $\mathcal{D}_2$. Note that there are two possible ways of locating B and x: one in which $x \sqsubset B$ holds and one in which $x \vdash\dashv B$ holds. (Cf. Stenning's discussion of "case identifiability" in his [19].)

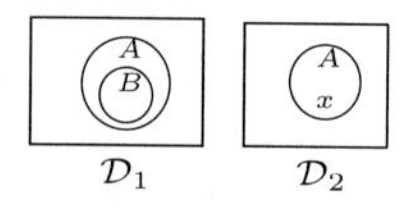

Fig. 6. Indeterminacy

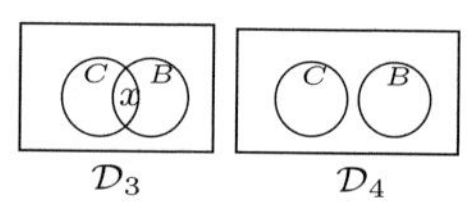

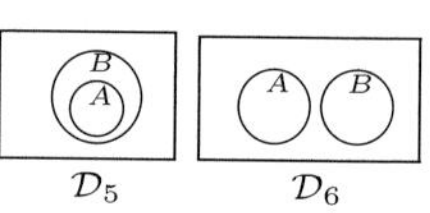

Fig. 7. Graphical inconsistency

In order to avoid complexity due to conflicting graphical informations represented in one diagram, there is another constraint to be imposed on unification, which we call the constraint for *graphical consistency*. For example, it is not permitted to unify two diagram $\mathcal{D}_3$ and $\mathcal{D}_4$ when, as is shown in Fig. 7, they share two circles C and B, and $x \sqsubset C$ and $x \sqsubset B$ holds on $\mathcal{D}_3$ and $C \vdash\dashv B$ holds on $\mathcal{D}_4$. Note that these relations $x \sqsubset C, x \sqsubset B$, and $C \vdash\dashv B$ are incompatible. The diagrams $\mathcal{D}_5$ and $\mathcal{D}_6$ in Fig. 7 are also not permitted to be unified in our system, and hence the graphical inconsistency does not always imply logical inconsistency.

There is some difficulty in defining unification rules in such a general way that it allows us to combine *any* two diagrams. Consider:

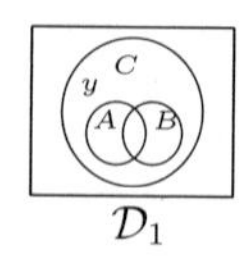

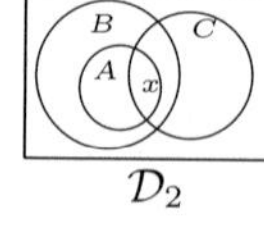

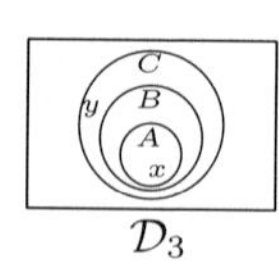

Here, it is expected that $\mathcal{D}_1$ and $\mathcal{D}_2$ should be unifiable into $\mathcal{D}_3$. One strategy to do this is to decompose $\mathcal{D}_1$ into the decomposition set and to add each member of this set to $\mathcal{D}_2$ one by one. However, it is impossible to do so because it violates the constraint of indeterminacy. In order to characterize diagrammatic syllogistic reasoning, by contrast, it is enough to consider unification of an *atomic diagram* and a (possibly complex) diagram. Thus, we define unification of two diagrams by stipulating that one of the premises is an atomic diagram.

3.2 Generalized Diagrammatic Syllogistic Inference System GDS

In this subsection, we introduce *Unification* and *Deletion* of GDS. In the following definition, in order to indicate occurrence of some objects in a context on a diagram, we write the indicated objects explicitly and the context by "dots" as in $\mathcal{D}_1$ below.[6] For example, when we need to indicate only A and x on $\mathcal{D}_1$ of the left-hand side, we could write $\mathcal{D}_1$ in the manner shown in the right-hand side.

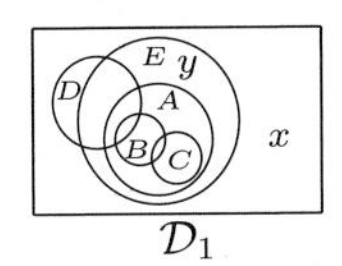

$\mathcal{D}_1$

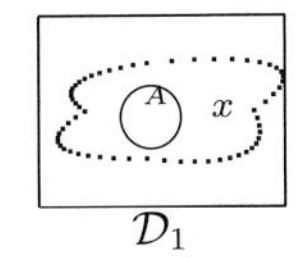

$\mathcal{D}_1$

Using this notation, we describe general patterns of each rule and present some examples of its application. In the following definition, we assume the identification of EUL-diagrams by continuous transformation explained in Section 2.3.

Definition 5 (Inference rules of GDS). *Unification* and *Deletion* of GDS are defined as follows:

Unification: For simplicity, we consider unification of an EUL-diagram only with an atomic diagram in this paper. Atomic diagrams are denoted by $\alpha, \beta, \ldots$ for readability. The unified diagram of $\mathcal{D}$ with α is denoted as $\mathcal{D} + \alpha$.

When the relation which holds on α also holds on $\mathcal{D}$, $\mathcal{D} + \alpha$ is $\mathcal{D}$ itself. The other unification rules with premises $\mathcal{D}$ and α to obtain $\mathcal{D} + \alpha$ are listed as U1–U10 below. In each rule, the first sentence attached to the two premise diagrams expresses the constraint which should be satisfied before performing the unification, and the sentence attached to the conclusion diagram $\mathcal{D}+\alpha$ expresses the actual operation to be performed at the unification step. We distinguish the following two cases: (I) $\mathcal{D}$ and α share one object; (II) $\mathcal{D}$ and α share two circles.

(I) $\mathcal{D}$ and α share one object:

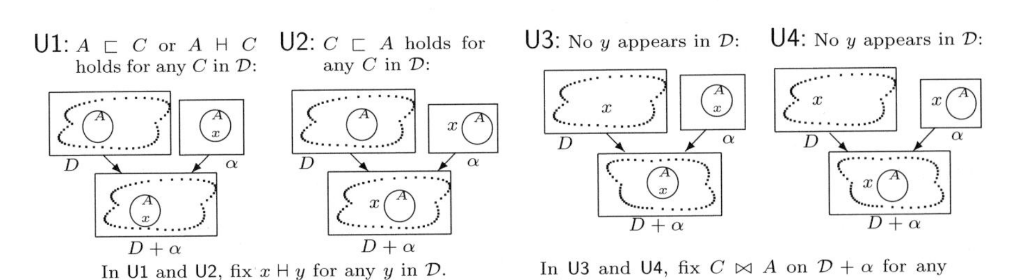

In U1 and U2, fix $x \vdash\dashv y$ for any y in $\mathcal{D}$.

In U3 and U4, fix $C \bowtie A$ on $\mathcal{D} + \alpha$ for any circle C in $\mathcal{D}$.

[6] Note that the dots notation is used only for abbreviation of a given diagram. For a formal treatment of such "backgrounds" in a diagram, see Meyer [11].

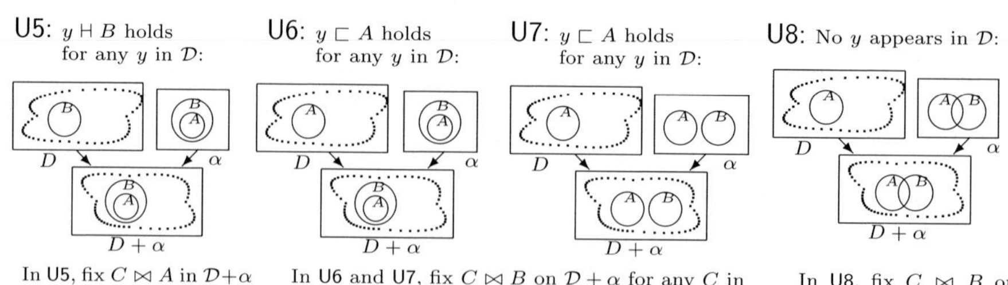

In U5, fix $C \bowtie A$ in $\mathcal{D}+\alpha$ for any C on $\mathcal{D}$ such that $C \sqsubset B$ or $C \bowtie B$ holds.

In U6 and U7, fix $C \bowtie B$ on $\mathcal{D}+\alpha$ for any C in $\mathcal{D}$ such that $A \sqsubset C$ or $A \vdash\dashv C$ or $A \bowtie C$ holds.

In U8, fix $C \bowtie B$ on $\mathcal{D}+\alpha$ for any C in $\mathcal{D}$.

For example, U1, U4, U5 and U7 rules are applied as follows:

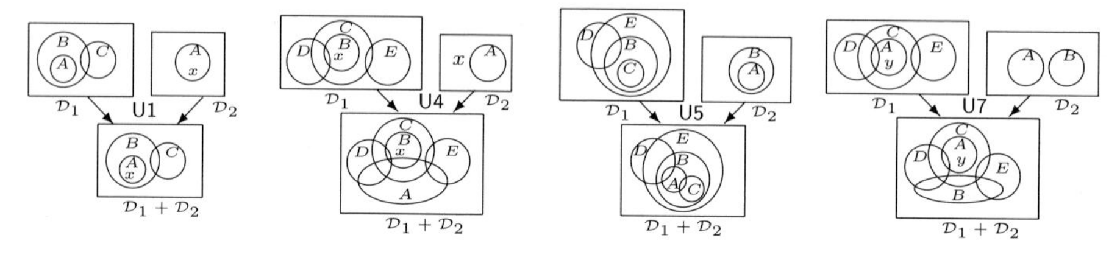

(II) $\mathcal{D}$ and α share two circles:

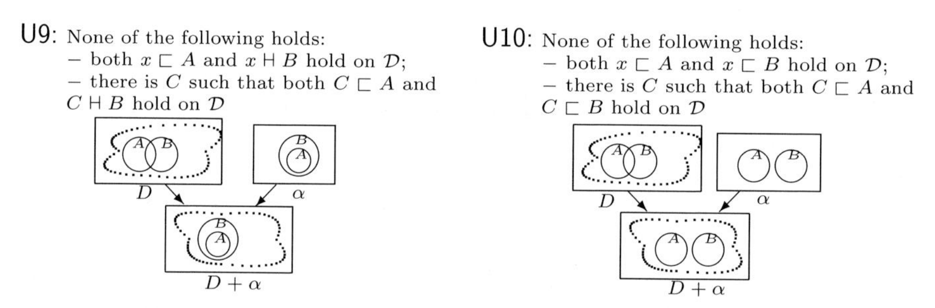

For example, U9 and U10 rules are applied as follows:

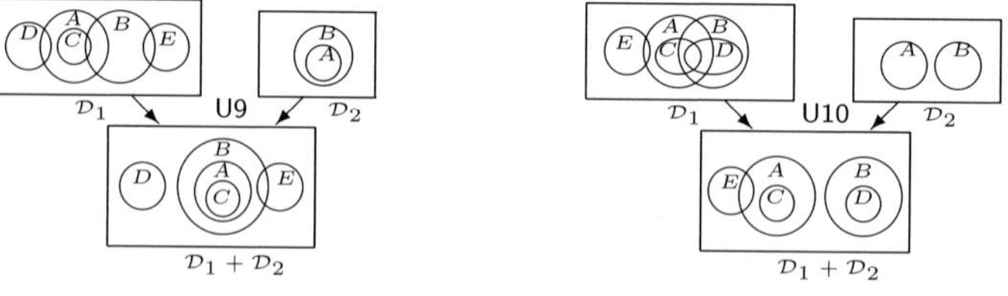

Deletion: There are two deletion rules: the circle-deletion rule and the point-deletion rule. Let t be an object, i.e., a named circle or a named point. For any EUL-diagram $\mathcal{D}$ and for any object t in $\mathcal{D}$, applying deletion rule results in $\mathcal{D} - t$ under the constraint that $\mathcal{D} - t$ has at least two objects.

We give an inductive definition of diagrammatic proofs (d-proofs) of GDS.

Definition 6 (Diagrammatic proofs of GDS). A diagrammatic proof (d-proof, for short) π of GDS is defined inductively as follows:

1. A diagram $\mathcal{D}$ is a d-proof from the premise $\mathcal{D}$ to the conclusion $\mathcal{D}$.
2. Let π_1 be a d-proof from $\mathcal{D}_1, \ldots, \mathcal{D}_n$ to $\mathcal{F}$ and π_2 be a d-proof from $\mathcal{E}_1, \ldots, \mathcal{E}_m$ to $\mathcal{E}$, respectively. If $\mathcal{D}$ is obtained by an application of Unification of $\mathcal{F}$ and $\mathcal{E}$, then (i) is a d-proof π from $\mathcal{D}_1, \ldots, \mathcal{D}_n, \mathcal{E}_1, \ldots, \mathcal{E}_m$ to $\mathcal{D}$ in GDS.
3. Let π_1 be a d-proof from $\mathcal{D}_1, \ldots, \mathcal{D}_n$ to $\mathcal{E}$. If $\mathcal{D}$ is obtained by an application of Deletion to $\mathcal{E}$, then (ii) is a d-proof π from $\mathcal{D}_1, \ldots, \mathcal{D}_n$ to $\mathcal{D}$ in GDS.

Here $\overset{\pi}{\mathcal{D}}$ means a d-proof π with $\mathcal{D}$ as the conclusion. The *length* of a d-proof is defined as the number of the applications of inference rules.

Fig. 4 of Section 3.1 is an example of d-proof from $\mathcal{D}_1, \mathcal{D}_2, \mathcal{D}_4, \mathcal{D}_6, \mathcal{D}_7$ to $\mathcal{D}_{11}$.

The above inference rules of GDS are well-defined.

Proposition 2 (Well-definedness of inference rules). *For any* EUL-*diagrams $\mathcal{D}_1, \ldots, \mathcal{D}_n$, if there is a d-proof from $\mathcal{D}_1, \ldots, \mathcal{D}_n$ to $\mathcal{E}$, then $\mathcal{E}$ is an* EUL-*diagram.*

Proof. The proposition is obvious for the deletion rules. The proposition for the unification rules can be easily proved using the operation indicated at the conclusion of each unification rule. ■

The soundness theorem of GDS holds with respect to the formal semantics given in Section 2.3. It is shown by induction on the length of a given d-proof.

Theorem 1 (Soundness of GDS). *For any* EUL-*diagrams $\mathcal{D}_1, \ldots, \mathcal{D}_n, \mathcal{E}$, if there is a d-proof from $\mathcal{D}_1, \ldots, \mathcal{D}_n$ to $\mathcal{E}$ in* GDS*, then $\mathcal{E}$ is a semantically valid consequence of $\mathcal{D}_1, \ldots, \mathcal{D}_n$.*

Proof. By induction on the length of a given d-proof from $\mathcal{D}_1, \ldots, \mathcal{D}_n$ to $\mathcal{E}$ using the constraints of each rule. It is sufficient to show that each rule of U1–U10, D1 and D2 is sound in the sense that if each premise of the rule is true in a model, then the conclusion is also true in the model. We describe the EUL-relations which holds on the conclusion diagram of each rule. Then the soundness of each rule follows from the interpretation of premises by set-theoretical calculations. We treat only U1-rule. The other cases can be proved in a similar way.

By applying U1-rule to $\mathcal{D}$ and an atomic diagram α in which $x \sqsubset A$ holds, since x is added to $\mathcal{D}$, the EUL-relations between x and each circle C in $\mathcal{D}$ are augmented to those of $\mathcal{D}$. Note that, under the constraint of U1, $A \sqsubset C$ or $A \vdash\dashv C$ holds for any C in $\mathcal{D}$. Thus, the binary relations on $\mathcal{D} + (x \sqsubset A)$ is those of $\mathcal{D}$, $x \sqsubset A$, $x \sqsubset C$ for any C such that $A \sqsubset C$ holds on $\mathcal{D}$, and $x \vdash\dashv C$ for any C such that $A \vdash\dashv C$ holds on $\mathcal{D}$. Then, in any model $\mathcal{M}$ for $\mathcal{D}$ and for α, we have $I(x) \in I(C)$ for any C such that $A \sqsubset C$ holds on $\mathcal{D}$, and we have $I(x) \notin I(C)$ for any C such that $A \vdash\dashv C$ holds on $\mathcal{D}$. ■

The converse of soundness, i.e., completeness of GDS does not hold as shown by $\mathcal{D}_1$ and $\mathcal{D}_2$ in the following example:

There is no d-proof from $\mathcal{D}_1$ to $\mathcal{D}_2$ in GDS, even though $\mathcal{D}_2$ is a valid consequence of $\mathcal{D}_1$ with respect to our formal semantics. Note that in the case of $\mathcal{D}_3$ and $\mathcal{D}_4$, we can obtain a d-proof from $\mathcal{D}_3$ to $\mathcal{D}_4$ by deleting B from $\mathcal{D}_3$ and by applying U8-rule with an atomic diagram in which $x \vdash\!\dashv B$ holds.

Although GDS is not complete, it is expressive enough to derive all valid patterns of syllogisms as will be shown in the next subsection.

3.3 Aristotelian Categorical Syllogisms

In this subsection, we show that the diagrammatic inferences for Aristotelian categorical syllogisms are characterized as a specific diagrammatic proofs of GDS.

We first introduce, in Fig. 8, the correspondence between the statements of Aristotelian categorical syllogisms and a class of EUL-diagrams under the identification by continuous transformation explained in Section 2.3.

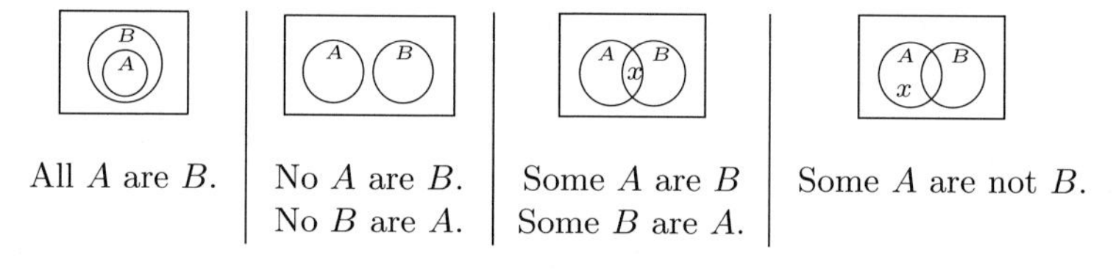

Fig. 8. Syllogistic diagrams

We call the diagrams of the forms given in Fig. 8 *syllogistic diagrams*. For any statement S of Aristotelian categorical syllogism, we denote as $S^{\mathcal{D}}$ the corresponding syllogistic diagram given in Fig. 8.

Next, we define a particular class of d-proofs of GDS called *syllogistic normal d-proofs* as follows:

Definition 7 (Syllogistic normal d-proofs). For any syllogistic diagrams $\mathcal{D}_1, \ldots, \mathcal{D}_n, \mathcal{D}$, a d-proof π from $\mathcal{D}_1, \ldots, \mathcal{D}_n$ to $\mathcal{D}$ of GDS is in *syllogistic normal form* if a unification rule and a deletion rule appear alternately in π.

Fig. 9 illustrates a syllogistic normal d-proof, where each pair of a unification rule and a deletion rule application corresponds to a valid pattern of syllogisms. For example, the sub-proof from $\mathcal{D}_1$ and $\mathcal{D}_2$ to $\mathcal{D}_4$ is a diagrammatic representation of syllogism of the form: *Some C are B. No A are B.* Therefore *Some C are not A.* (This valid pattern is sometimes symbolized as EI2O. See [1,15] for the notation.) Indeed, each syllogistic normal d-proof can be considered as a chain of valid patterns of Aristotelian categorical syllogisms. Note that, compared with the underlying proof tree of GDS in Fig. 5 in Section 3.2, the tree of the syllogistic

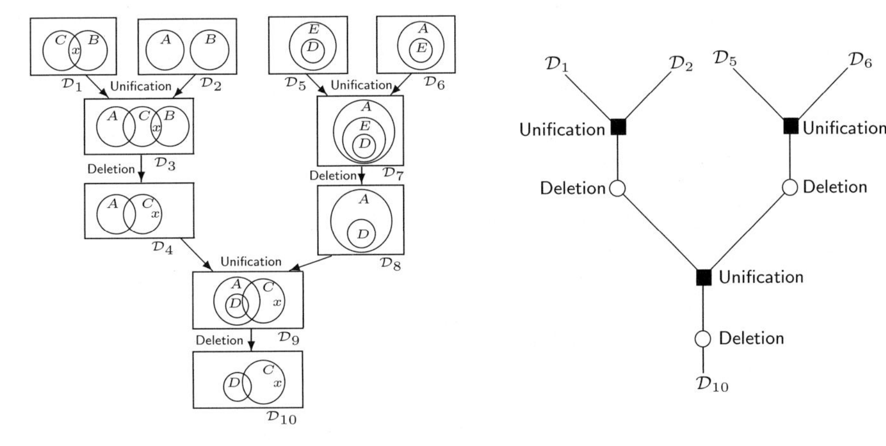

Fig. 9. A syllogistic normal d-proof

Fig. 10. An underlying tree structure of a syllogistic normal d-proof

normal d-proof has a canonical form, where a unification node, denoted by ■, and a deletion node, denoted by ○, appear one after the other.

In order to characterize the syllogistic normal d-proofs, we introduce a subsystem DS of GDS:

Definition 8 (DS)**.** A diagrammatic syllogistic inference system DS is a subsystem of GDS, where:

1. Unification is restricted to U1–U7, and their premises are restricted to syllogistic diagrams sharing no common named point;
2. Deletion is applicable only when its conclusion is a syllogistic diagram.

Essentially, DS is a subsystem of GDS where only syllogistic diagrams are considered. It is shown that DS corresponds to the Aristotelian categorical syllogisms. Let S be a statement of syllogism and $S^{\mathcal{D}}$ be the corresponding syllogistic diagram in Fig. 8. We have the following correspondence:

Proposition 3 (Syllogisms and DS**).** *Let $S_1, \ldots, S_n, S$ be statements of Aristotelian categorical syllogism. Then S is a valid conclusion of Aristotelian categorical syllogisms from the premises $S_1, \ldots, S_n$ if and only if, there is a d-proof of $S^{\mathcal{D}}$ in* DS *from the premises $S_1^{\mathcal{D}}, \ldots, S_n^{\mathcal{D}}$.*

As is indicated in Fig. 8, the correspondence between diagrams in EUL and linguistic forms of syllogistic categorical statements is not one-to-one. This is due to the fact that linguistic statements have the subject-predicate distinction, whereas diagrammatic representations in EUL do not. In general, compared to linguistic reasoning, diagrammatic reasoning with Euler circles in EUL has the following distinctive features:

$C1$ Diagrammatic reasoning with Euler circles in EUL does not have explicit operations with negation.

$C2$ Diagrammatic reasoning with Euler circles in EUL does not have explicit operations with "subject-predicate" distinction.

4 Experiment Design for Syllogistic Reasoning with Euler Diagrams

In order to see whether the theoretical consideration carried out in the last section can be shown at the performance level of people's reasoning, we propose a design of experiment for a cognitive psychological study. We consider the question of how the differences in *figures* of syllogistic reasoning affect subjects' performance of syllogistic deductive reasoning[7] and reasoning with Euler diagrams. The questions of whether and how the use of Euler diagrams aids subjects to solve syllogistic reasoning tasks do not have a clear answer in psychological contexts (for a negative answer, see Cavillo et al. [4]; Rizzo and Palmonari [14]). In this situation, it is useful to first give a logical reconstruction of reasoning with Euler diagrams, and then make empirical hypotheses about people's performance in reasoning tasks, as is done in a series of pioneering researches of Stenning's group (For a recent study, see Stenning and van Lambalgen [21]). Following Stenning's approach, we make some predictions on the role of diagrams in syllogism solving tasks, which are based on our analysis of reasoning with Euler diagrams in the previous sections.

4.1 Predictions from Theoretical Considerations

There are significant differences between diagrammatic reasoning in GDS and linguistic syllogistic reasoning. In linguistic reasoning, syllogisms are classified according to their *figures*, i.e., a four-way classification in terms of the arrangement of terms in the premises. For example, EI1O syllogism (*No B are A*, *Some C are B*; therefore *Some C aren't A*) and EI2O syllogism (*No A are B*, *Some C are B*; therefore *Some C aren't A*) have different figures. It is well known that difference in figures give rise to errors in syllogistic reasoning (e.g. Dickstein [6]). Furthermore, syllogisms like EI2O require inferences involving negation,[8] and it is known that such inferences cause some difficulties in linguistic reasoning tasks (cf. Braine and O'Brien [2]; Rips [13]). By contrast, in diagrammatic reasoning in GDS, the linguistic reasoning processes for EI1O and EI2O correspond to the same diagrammatic proof in GDS. (See the diagrammatic syllogistic proof in Fig. 9 in Section 3.3, where the step from the premises $\mathcal{D}_1$ and $\mathcal{D}_2$ to the conclusion $\mathcal{D}_4$ corresponds to both EI1O and EI2O syllogisms. Thus, given $C1$ and $C2$ in Section 3.3, we give the following conjectures:

[7] In our research, categorical syllogisms are employed as a benchmark to characterize cognitive processes in deductive reasoning. A large number of empirical researches concerning deductive reasoning have dealt with categorical syllogisms (for a review, see Evans et al. [8]).

[8] In Sato et al. [15] we give an analysis of the role of negation in syllogistic reasoning, and show that syllogisms like EI2O have Natural Deduction proofs involving $\bot$-rule in Minimal Logic, i.e., a subsystem of Gentzen-type natural deduction system.

No B are A. Some C are B. 1. Some C are A. 2. Some C are not A. 3. No C are A. 4. None of these are valid.	No A are B. Some C are B. 1. Some C are A. 2. Some C are not A. 3. No C are A. 4. None of these are valid.

Fig. 11. Examples of the standard test sheets of syllogisms (EI1O/EI2O)

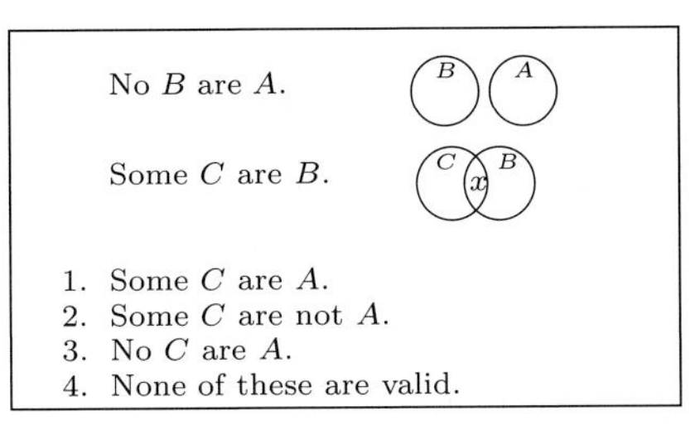

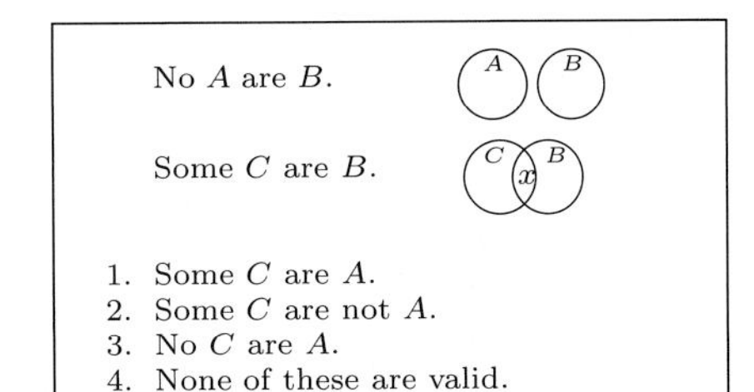

Fig. 12. Examples of the test sheets of syllogisms with Euler diagrams (EI1O/EI2O)

$C1'$ In diagrammatic reasoning, there will be no difference in reasoning accuracy between the syllogisms with explicit operations on negation and those without any.

$C2'$ In diagrammatic reasoning, there will be no *figure*-specific difference.

Regarding $C2'$, Stenning [19] also discusses differences between linguistic and diagrammatic representations with respect to the subject-predicate distinction.

4.2 A Design of Experiment: The BAROCO Test

In order to provide empirical evidence for the two predictions made above, we devise an experiment, called BAROCO test, whose design is similar to the one in [4]. Fig. 11 shows examples of the standard test sheets, which are presented to one group of participants. Fig. 12 shows examples of the test sheets with Euler diagram representations, which are presented to the other group of participants. On the left is EI1O, and on the right is EI2O in both Fig. 11 and Fig. 12. We hypothesize that the former group will show a low accuracy rate with respect to the problem of EI2O on the right, whereas in the case of the latter group, there will be little or no significant difference between the two problems.

In addition to the standard test sheets of the syllogistic reasoning, we prepare a sheet with a list of Euler diagrams corresponding to each quantifier *uniquely*. As Stenning and Oberlander [20] also point out, a single Euler diagram often corresponds to various different first order logical propositions (in fact, Rizzo and Palmonari [14] used these representations); this situation would often cause essential difficulties when one wants to make a direct comparison in an experiment between the diagrammatic and linguistic reasoning. Therefore, our experiment uses the EUL representation system for Euler diagrams. Additionally, we conduct a pretest to check whether a participant can correctly understand EUL representation system to a certain degree. This seems to be necessary because Euler diagrams do not deliver the meaning of a sentence so transparently as "coffee cups" in the Tower of Hanoi puzzle used by Zhang and Norman [25], which is a classical study showing that the physical properties of experiment tasks affect how the problems look to the participants as well as the level of difficulty.

Then, we compare cases in which participants only have a standard test sheet with the cases in which participants have a diagram sheet in addition to the standard test sheets. In order to show that effects which appear in linguistic reasoning do not appear in diagrammatic reasoning, even when it is the group of participants who have diagrams, the participants should be given the premise not only in diagrammatic but also in linguistic representation and asked to produce a linguistic conclusion in the last result. Although [4] have used an evaluation task, which is suitable for examining belief effect, our participants' responses are measured by choosing from a list of possible conclusions as follows: each item has four response options, such as *All-*, *No-*, *Some-* and *NoValid* in AE-, EA-types; *No-*, *Some-*, *Some-not-*, *NoValid* in the other types. Since, as is well known, some Aristotelian interpretations take it that *Some-not-* is derivable from *No-*, we exclude *No-* from the response options of AE- and EA-type to avoid situations in which there are two possible correct answers. In accordance with this treatment, the other types do not have option *All-*, thus having four options, *No-*, *Some-*, *Some-not-*, and *NoValid*, since *All-* was not chosen in the literature (e.g. Chater and Oakford [5]).

4.3 Preliminary Results of Our Experiment

Based on the above design we conducted a preliminary experiment outlined as below:[9]

Methods, Participants: 19 first- and second-year undergraduates in an elementary logic class, who gave a consent to their cooperation in the experiment, took part in the experiment. The Diagrammatic reasoning group consists of ten people, among whom five scored more than 8 out of 10 points on the pretest. The Linguistic reasoning group originally consisted of nine, from whom we excluded one person who left the last four questions unanswered.

Materials: All the participants were given either Aristotelian categorical syllogisms with Euler diagrams (such as shown in Fig. 12) or ones without the diagrams (such as shown in Fig. 11). In total 21 syllogisms were given, of which 14 are valid and 7 are invalid. The test was an 18-minute power test, and the items were presented in a random order. The pretest for the Diagrammatic reasoning group consists of 8 diagrammatic representations. The participants were asked to choose, from a list of four possibilities, a sentence corresponding to the representations given.

[9] For more detail, see http://abelard.flet.keio.ac.jp/person/sato/index.html

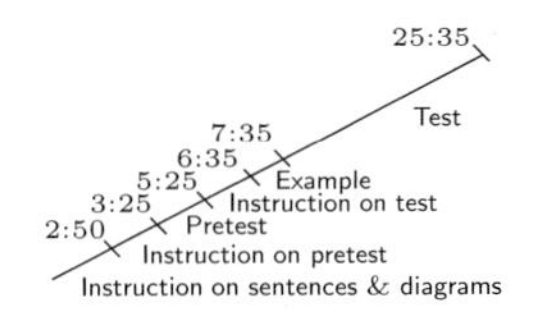

Fig. 13. Time schedule for the Diagrammatic reasoning group

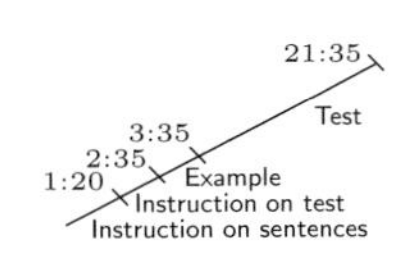

Fig. 14. Time schedule for the Linguistic reasoning group

Procedures: The Diagrammatic reasoning group answered the syllogism test after the pretest, following the time-schedule shown in Fig. 13. The Linguistic group, as shown in Fig. 14, took the test without taking the pretest.

Results and Discussion: Table 2 shows the accuracy rates for the syllogistic problems in each group. In the valid syllogisms, the use of diagrams seems to have improved solving performance. Table 3 compares the accuracy rates in the two groups for the syllogisms which require operations with negation with the syllogisms which do not. Only in diagrammatic reasoning of EIO syllogisms, there is nearly no difference between the two, although this tendency did not appear in all valid syllogisms. Table 3 seems to give partial evidence for $C1'$.

Table 2. Accuracy for Valid and Invalid

	Linguistic	Diagrammatic
Valid	0.73	0.74
Invalid	0.42	0.74
Total	0.63	0.74

Table 3. Accuracy for Negation and No-neg

All patterns | EIO type

	Linguistic	Diagrammatic
Negation	0.70 | 0.50	0.76 | 0.50
No-negation	0.75 | 0.68	0.72 | 0.60
Difference	0.05 | 0.18	0.04 | 0.10

The standard deviations for EIO-type, which has all four figures, in the linguistic group was 0.34 and the ones in diagrammatic group was 0.45. The standard deviation for four figures in diagrammatic reasoning was higher than those in linguistic reasoning. Therefore, Table 3 does not seem to give evidence for $C2'$. Moreover, this experiment is a preliminary one involving only 19 participants. Thus, we have yet to obtain sufficient empirical evidence for $C1'$ and $C2'$.

5 Conclusion

In this paper, we have studied logical and cognitive aspects of reasoning with Euler diagrams. As a logical study, we gave a proof-theoretical analysis of diagrammatic reasoning using Euler diagrams, based on EUL representation system and its inference system GDS. The analysis of the underlying structure of diagrammatic proofs, i.e., proof trees, provides a characterization of diagrammatic *syllogistic* reasoning as a particular class of the diagrammatic proofs in GDS. We showed that our inference system is sound with respect to the formal (set-theoretical) semantics. Compared with linguistic syllogistic reasoning, diagrammatic reasoning with Euler diagrams in our system has a distinctive feature, namely, it does not involve explicit operation with logical negation or with the "subject-predicate" distinction. In order to see whether these differences are also shown by the performance level of human reasoning, we outlined a design of experiment for a cognitive psychological study.

References

1. Ando, J., Shikishima, C., Hiraishi, K., Sugimoto, Y., Takemura, R., Okada, M.: At the crossroads of logic, psychology, and behavioral genetics. In: Okada, M., et al. (eds.) Reasoning and Cognition, pp. 19–36. Keio University Press (2006)
2. Braine, M., O'Brien, D.: Mental logic. Lawrence Erlbaum Associates, Mahwah (1998)
3. Blackett, D.: Elementary Topology. Academic Press, London (1983)
4. Calvillo, P.D., DeLeeuw, K., Revlin, R.: Deduction with Euler Circles. In: Baker-Plummer, D., et al. (eds.) Diagrams 2006. LNCS (LNAI), vol. 4045, pp. 199–203. Springer, Heidelberg (2006)
5. Chater, N., Oakford, M.: The probability heuristics model of syllogistic reasoning. Cognitive Psychology 38, 191–258 (1999)
6. Dickstein, L.S.: The effect of figure on syllogistic reasoning. Memory and Cognition 6, 76–83 (1978)
7. Euler, L.: Lettres à une Princesse d'Allemagne sur Divers Sujets de Physique et de Philosophie. De l'Académie des Sciences, Saint-Pétersbourg (1768)
8. Evans, J., Newstead, S.E., Byrne, R.: Human Reasoning. Lawrence Erlbaum Associates, Mahwah (1993)
9. Hammer, E.: Logic and Visual Information. CSLI Publications (1995)
10. Hammer, E., Shin, S.: Euler's visual logic. History and Philosophy of Logic 19, 1–29 (1998)
11. Meyer, B.: Diagrammatic evaluation of visual mathematical notations. In: Anderson, M., Meyer, B., Olivier, P. (eds.) Diagrammatic Representation and Reasoning, pp. 261–277. Springer, Heidelberg (2001)
12. Peirce, C.S.: Collected Papers IV. Harvard University Press (1897/1933)
13. Rips, L.J.: The Psychology of Proof. MIT Press, Cambridge (1994)
14. Rizzo, A., Palmonari, M.: The mediating role of artifacts in deductive reasoning. In: Poster in the 27th Annual Conference of the Cognitive Science Society (2005)
15. Sato, Y., Takemura, R., Mineshima, K., Shikishima, C., Sugimoto, Y., Ando, J., Okada, M.: Some remarks on deductive syllogistic reasoning studies. In: Okada, M., Takemura, R., Ando, J. (eds.) Reports on Interdisciplinary Logic Inference Studies, pp. 3–32. Keio University Press (2008)
16. Shin, S.-J.: The Logical Status of Diagrams. Cambridge University Press, Cambridge (1994)
17. Stapleton, G.: A survey of reasoning systems based on Euler diagrams. In: Euler Diagrams 2004. ENTCS, vol. 134, pp. 127–151. Elsevier, Amsterdam (2005)
18. Stapleton, G., Rodgers, P., Howse, J., Taylor, J.: Properties of Euler diagrams. In: Stapleton, G., Rodgers, P., Howse, J. (eds.) Proc. Layout of Software Engineering Diagrams. ECEASST, vol. 7, pp. 2–16 (2007)
19. Stenning, K.: Seeing Reason. Oxford University Press, Oxford (2002)
20. Stenning, K., Oberlander, J.: A cognitive theory of graphical and linguistic reasoning. Cognitive Science 19, 97–140 (1995)

21. Stenning, K., van Lambalgen, M.: Human Reasoning and Cognitive Science. MIT Press, Cambridge (2007)
22. Swoboda, N., Allwein, G.: Using DAG transformations to verify Euler/Venn homogeneous and Euler/Venn FOL heterogeneous rules of inference. Journal on Software and System Modeling 3(2), 136–149 (2004)
23. Swoboda, N., Allwein, G.: Heterogeneous reasoning with Euler/Venn diagrams containing named constants and FOL. ENTCS, vol. 134, pp. 153–187. Elsevier, Amsterdam (2005)
24. Venn, J.: Symbolic Logic. Macmillan, Basingstoke (1881)
25. Zhang, J., Norman, D.A.: Representations in distributed cognitive tasks. Cognitive Science 18(1), 87–122 (1994)

A Normal Form for Euler Diagrams with Shading

Andrew Fish*, Chris John, and John Taylor

Visual Modelling Group, University of Brighton, Brighton, UK
{Andrew.Fish,John.Taylor}@brighton.ac.uk

Abstract. In logic, there are various normal forms for formulae; for example, disjunctive and conjunctive normal form for formulae of propositional logic or prenex normal form for formulae of predicate logic. There are algorithms for 'reducing' a given formula to a semantically equivalent formula in normal form. Normal forms are used in a variety of contexts including proofs of completeness, automated theorem proving, logic programming etc. In this paper, we develop a normal form for unitary Euler diagrams with shading. We give an algorithm for reducing a given Euler diagram to a semantically equivalent diagram in normal form and hence a decision procedure for determining whether two Euler diagrams are semantically equivalent. Potential applications of the normal form include clutter reduction and automated theorem proving in systems based on Euler diagrams.

1 Introduction

In logic, there are various normal forms for formulae. In propositional logic, for example, a formula is in *disjunctive normal form* if is is a disjunction of a conjunction of literals or *conjunctive normal form* if it is a conjunction of a disjunction of literals. The formula $(A \wedge \neg B) \vee (\neg A \wedge B)$ is in disjunctive normal form and $(A \vee B) \wedge (\neg A \vee \neg B)$ is a logically equivalent formula in conjunctive normal form. In predicate logic, a formula is in *prenex normal form* if all the quantifiers 'come first'; for example, $\forall x \exists y \forall z\, (P(x) \vee (Q(y,z) \rightarrow R(z))$ is in prenex normal form. Also, there are algorithms that take an 'input' formula and produce a logically equivalent formula in the required normal form.

Normal forms have a wide variety of uses in logical systems. For example, simplification of propositional logic formulae, with applications in circuit design, using Karnaugh maps or the Quine-McCluskey algorithm assume an input formula in disjunctive normal form. In predicate logic, Gödel's proof of the completeness of first order logic supposes that all formulae are expressed in prenex normal form. Automated theorem provers make extensive use of normal forms in the manipulation, or rewriting, of formulae.

Euler diagrams are the basis for a number of different notations such as higraphs [1], spider diagrams [2], Euler/Venn diagrams [3] and constraint diagrams [4]. Euler

* This research was part-funded by EPSRC grant EP/EO11160: Visualisation with Euler Diagrams.

G. Stapleton, J. Howse, and J. Lee (Eds.): Diagrams 2008, LNAI 5223, pp. 206–221, 2008.

diagrams and Venn diagrams (see [5] for a comprehensive survey of Venn diagrams) have been applied in a variety of areas including file-information systems [6], library systems [7] and system specification [8]. With Euler diagrams augmented with shading (to denote empty sets), there are different ways of representing certain relationships between sets. Making choices between different representations may involve factors such as user preferences, drawability criteria [9] and 'clutter' metrics [10].

Compound diagrams are built from unitary ('single') diagrams using disjunction, conjunction and (sometimes) negation. For example, $d_1 \vee (d_2 \wedge \neg d_3)$ is a compound diagram (where each d_i is a unitary or compound diagram). Two normal forms for compound diagrams, called 'literal conjunctive/disjunctive normal form', were introduced in [11]. These normal forms are akin to conjunctive and disjunctive normal forms in propositional logic and were used in [11] to obtain a completeness proof for a system of compound Euler diagrams with shading.

In this paper, we introduce a normal form for unitary Euler diagrams with shading. Thus our normal form concerns the 'internal' structure of unitary diagrams rather than the 'global' structure of compound diagrams (how the compound diagram is 'built' from unitary diagrams). We also provide an algorithm that produces, for any diagram, a semantically equivalent diagram in normal form. The normal form is minimally cluttered of all semantically equivalent diagrams (using any of the three clutter metrics introduced in [10]).

The structure of this paper is as follows. We introduce the syntax and semantics of Euler diagrams with shading in section 2. Section 3 introduces various concepts and syntactic manipulations needed to define the normal form. In section 4 we give a syntactic characterisation the semantic property of redundancy of contours and we define the normal form. Finally, conclusions and directions of future work are discussed in section 5.

2 Euler Diagrams with Shading

2.1 Syntax

This section provides a standard description of Euler diagrams with shading; similar descriptions of related systems can be found in [2,10,12,13] for example. Euler diagrams are built using contours to represent sets. The topological relationships of 'containment' and 'separation' between contours represent the relations of subset and disjointness between sets respectively. We augment Euler diagrams with shading, where shaded regions represent the empty set.

Example 1. Figure 1 shows two Euler diagrams (with shading). The diagram d_1 expresses the following properties of the sets represented in the diagram.

$$A \cap C = \varnothing, \quad E \subseteq A \cap \overline{C} \cap \overline{F}, \quad F = \varnothing.$$

Although the contour B appears in the diagram, d_1 makes no semantic assertion about the corresponding set B. In this sense the contour B is 'redundant'.

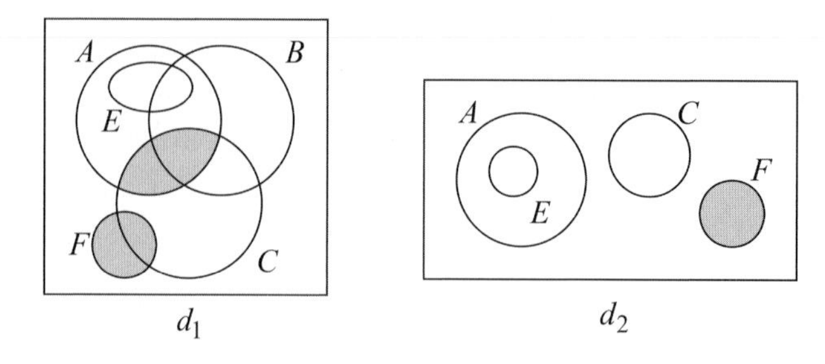

Fig. 1. Two Euler diagrams representing the same information

The diagram d_2 represents the same information as d_1 but, intuitively, it is visually simpler. In fact, d_2 is in normal form as defined in section 4.

Informally, an *Euler diagram with shading* in the plane $\mathbb{R}^2$ comprises a collection of simple closed curves called *contours*. We draw the contours of our Euler diagrams inside a bounding rectangle which is not part of the diagram but serves to indicate 'where the diagram ends'. The contours divide the part of the plane inside the bounding box into connected *regions*. A *zone* is a region that lies inside some of the contours and outside the remaining contours of the diagram. Thus a zone can be described as a pair (in, out) where *in* and *out* are sets of contours that partition the set of contours of d. The diagram d_2 in figure 1 can be specified as follows.

- The contours are: A, C, E and F.
- The five zones are: $(\{A\},\{C,E,F\})$, $(\{A,E\},\{C,F\})$, $(\{C\},\{A,E,F\})$, $(\{F\},\{A,C,E\})$ and $(\varnothing,\{A,C,E,F\})$.
- There is one shaded zone: $(\{F\},\{A,C,E\})$.

We call this specification of d_2 in figure 1 an *abstract diagram.* The benefits of distinguishing clearly between *concrete diagrams* – that is, diagrams that are drawn on paper or realised on some other medium such as a computer screen – and their abstract descriptions is well-documented; see [12], for example.

It will be convenient for the contour labels of our diagrams to be drawn from a fixed, countably infinite set $\mathcal{L}$. Given $\mathcal{L}$, we can define an *abstract zone* to be an ordered pair (a, b) where a and b are finite subsets of $\mathcal{L}$ such that $a \cap b = \varnothing$. Abstract zones exist independently of any diagram. We may think of the zones of a diagram d as being drawn from the set of abstract zones (a, b) where $a \cup b$ is the set of contour labels in the diagram.

Definition 1. *The set $\mathcal{L}$ is a countably infinite set of* **contour labels** *from which all contour labels of diagrams will be drawn.*

An **abstract zone on $\mathcal{L}$** *is an ordered pair $z = (in(z), out(z))$ where $in(z)$ and $out(z)$ are disjoint, finite subsets of $\mathcal{L}$ called the set of contour labels that* **contain** *z and* **exclude** *z respectively. The set of abstract zones is $\mathcal{Z} = \{(a,b) \in \mathbb{F}(\mathcal{L}) \times \mathbb{F}(\mathcal{L}) : a \cap b = \varnothing\}$ where $\mathbb{F}(\mathcal{L})$ denotes the set of all finite subsets of $\mathcal{L}$.*

An **abstract region on $\mathcal{L}$** *is a set of abstract zones. The set of regions on $\mathcal{L}$ is $\mathcal{R} = \mathbb{P}(\mathcal{Z})$, the power set of $\mathcal{Z}$.*

Definition 2. *An* ***abstract Euler diagram with shading*** *d (with labels in $\mathcal{L}$) is an ordered pair $\langle Z, Z^\bullet \rangle$ whose components are defined as follows.*

1. *$Z = Z(d) \subseteq \mathcal{Z}$ is a finite set of* ***zones*** *such that, for some finite set of contour labels $L \in \mathbb{F}(\mathcal{L})$, the following three conditions are satisfied.*
   (a) For all $z \in Z$, $in(z) \cup out(z) = L$; the set $L = L(d)$ is called the ***contour labels*** *in d;*
   (b) For all $\ell \in L$, there is a zone $z \in Z$ such that $\ell \in in(z)$;
   (c) $(\varnothing, L(d)) \in Z(d)$.
2. *$Z^\bullet = Z^\bullet(d) \subseteq Z(d)$ is the set of* ***shaded zones.***

Henceforth, we will use 'Euler diagram' to mean 'Euler diagram with shading'. We now introduce some terminology that will be used in the rest of the paper. Let $d = \langle Z, Z^\bullet \rangle$ be an Euler diagram.

- We denote the set of **unshaded zones** of d by $Z^\circ(d) = Z(d) - Z^\bullet(d)$.
- A **region** is a non-empty set of zones; the set of regions in d is $R = R(d) = \mathbb{P}(Z) - \{\varnothing\}$. We also define $R^\circ(d) = \mathbb{P}(Z^\circ) - \{\varnothing\}$ and $R^\bullet(d) = \mathbb{P}(Z^\bullet) - \{\varnothing\}$ to be the sets of **unshaded regions** and **shaded regions** in d respectively.
- Given any region $r \in R(d)$, we define its **unshaded subregion** to be $r^\circ = \{z \in r : z \in Z^\circ(d)\} = r \cap Z^\circ(d)$ and its **shaded subregion** to be $r^\bullet = \{z \in r : z \in Z^\bullet(d)\} = r \cap Z^\bullet(d)$.
- If a diagram d contains all the possible zones – that is, if $Z(d) = \{(x, L-x) : x \subseteq L\}$ – then d is a **Venn diagram.**
- The abstract zones in the set $Z^m(d) = \{(x, L-x) : x \subseteq L\} - Z(d)$ are said to be **missing from *d*** and $Z^m(d)$ is called the **missing zone set** of d.

2.2 Semantics

The zones and regions in an Euler diagram represent sets; missing and shaded zones represent the empty set. The following definitions, modified from those given in [2], make this precise. We first interpret the abstract labels, zones and regions as subsets of some universal set U and then define a 'semantics predicate' that captures the conditions that missing and shaded zones represent the empty set. Augmenting Euler diagrams with shading increases expressiveness (for example, no unshaded Euler diagram can express '$A = \varnothing$') and overcomes some drawability problems (for example, we can express '$A = B$' without using concurrent contours). Also, some notations based on Euler diagrams, such as spider or constraint diagrams, use shading to represent upper bounds on set cardinalities rather than just denoting the empty set.

Definition 3. *An* ***interpretation*** *is a pair, (U, Ψ) where U is a set, called the* ***universal set,*** *and $\Psi\colon \mathcal{L} \cup \mathcal{Z} \cup \mathcal{R} \to \mathbb{P}(U)$ is a function that interprets contour labels, zones and regions as subsets of U such that the images of the zones and regions are completely determined by the images of the contour labels as follows:*

1. *for each zone,* $(x, y) \in \mathcal{Z}$, $\Psi(x, y) = \bigcap_{\ell \in x} \Psi(\ell) \cap \bigcap_{\ell \in y} \overline{\Psi(\ell)}$ *where* $\overline{\Psi(\ell)} = U - \Psi(\ell)$.
   When a *is empty we define* $\bigcap_{\ell \in a} \Psi(\ell) = U$ *and* $\bigcup_{\ell \in a} \overline{\Psi(\ell)} = U$.
2. *for each region,* $r \in \mathcal{R}$, $\Psi(r) = \bigcup_{z \in r} \Psi(z)$.

Definition 4. *Let* d *be an Euler diagram and let* $I = (U, \Psi)$ *be an interpretation. We define the* **semantics predicate** *of* d*, denoted* $P_d(I)$*, to be the conjunction of the following conditions.*

1. ***Shading Condition***
   Each shaded zone represents the empty set: $\bigwedge_{z \in Z^{\bullet}(d)} \Psi(z) = \emptyset$.
2. ***Missing Zones Condition.***
   Each missing zone represents the empty set: $\bigwedge_{z \in Z^{m}(d)} \Psi(z) = \emptyset$.

We say I *is a* **model** *for* d*, denoted* $I \models d$*, if and only if* $P_d(I)$ *is true.*

Definition 5. *Let* d_1 *and* d_2 *be Euler diagrams. Then* d_1 ***semantically entails*** d_2 *(and* d_2 *is a* ***semantic consequence*** *of* d_1*), denoted* $d_1 \models d_2$*, if every interpretation that is a model for* d_1 *is also a model for* d_2*. If* $d_1 \models d_2$ *and* $d_2 \models d_1$ *then we say that* d_1 *and* d_2 *are* ***semantically equivalent****, denoted* $d_1 \equiv_{\models} d_2$*. The set of Euler diagrams semantically equivalent to* d_1 *is the* ***semantic equivalence class*** *of* d_1*, denoted* $\langle d_1 \rangle_{\models}$*.*

3 Manipulating Diagrams

In this section, we describe ways of manipulating diagrams by adding, moving or removing syntactic elements and we consider the semantic consequences of these manipulations. The manipulations we introduce here will be used in the next section to obtain the normal form. We are primarily interested in Euler diagrams for information representation and reasoning, so we require our diagram manipulations to respect the semantics in the sense that if d_2 is obtained from d_1 by a syntactic manipulation then $d_1 \models d_2$. Thus, the manipulations described in section 3.1 are examples of 'reasoning rules'; see [2] for a complete set of reasoning rules.

We define four operations: add and remove a contour label and add and remove a shaded zone. These four operations are sufficient to be able to 'navigate' the semantic equivalence class of a diagram $\langle d \rangle_{\models}$; any diagram semantically equivalent to d may be obtained from d by applying a sequence of these four syntactic operations.

3.1 Adding and Removing Syntactic Elements

We begin by defining a syntactic operation that removes a label from a diagram. For $\ell \in \mathcal{L}$, we first define a *remove label* function on the set of abstract zones by

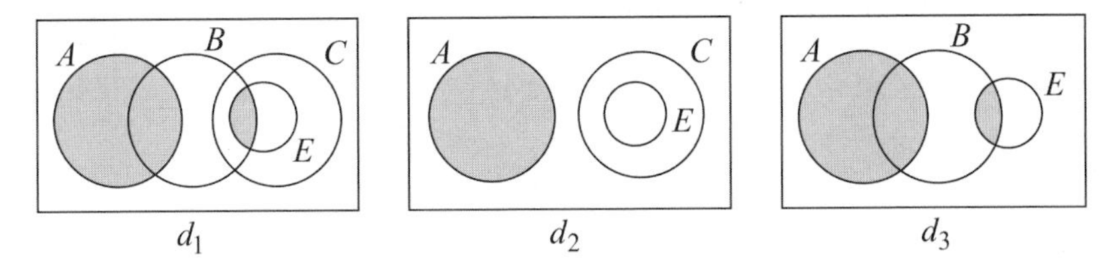

Fig. 2. Removing a contour: losing information

$r_\ell : \mathcal{Z} \to \mathcal{Z}$, $r_\ell(a,b) = (a - \{\ell\}, b - \{\ell\})$. The function extends naturally to the set of abstract regions $r_\ell : \mathcal{R} \to \mathcal{R}$ by defining $r_\ell(r) = \{r_\ell(z) : z \in r\}$ for $r \in \mathcal{R}$.

Example 2. In figure 2, removing the contour labelled B from d_1 results in the diagram d_2. Note that d_2 is not semantically equivalent to d_1; for example, d_1 expresses that $\overline{A} \cap B \cap C \cap E = \varnothing$ but d_2 makes no statement about B. Just erasing the contour B from d_1 would produce a diagram in which the zone inside contour E is partially shaded. In order to obtain a well-defined diagram the partial shading is removed and this zone is unshaded in d_2.

The zone set of d_2 is $r_B(Z(d_1))$. In this case the mapping $r_B : Z(d_1) \to Z(d_2)$ is two-to-one. For example, $r_B(\{B,C,E\},\{A\}) = r_B(\{C,E\},\{A,B\}) = (\{C,E\},\{A\})$. Only one of this pair of zones in d_1 is shaded so the corresponding zone in d_2 is left unshaded. Note that, if $(x,y) \in Z(d_2)$ then the 'corresponding zones' in d_1 are

$$r_B^{-1}(x,y) \cap Z(d_1) = \{(x \cup \{B\}, y), (x, y \cup \{B\})\} \cap Z(d_1) \subseteq R(d_1).$$

Since d_1 has missing zones, the set $r_\ell^{-1}(x,y) \cap Z(d_1) \subseteq R(d_1)$ may contain only a single zone. For example, removing the contour labelled C from d_1 produces d_3; the zone $(\{A,B\},\{E\})$ in d_3, has inverse image $r_C^{-1}(\{A,B\},\{E\}) \cap Z(d_1) = \{(\{A,B\},\{C,E\})\}$ because the zone $(\{A,B,C\},\{E\})$ is missing from d_1.

In both d_2 (removing B) and d_3 (removing C), the shaded zones are those for which the zone or zones in $r_\ell^{-1}(x,y) \cap Z(d_1)$ (for $\ell = B$ or C) are shaded.

Definition 6. *Let $d = \langle Z(d), Z^\bullet(d) \rangle$ be an Euler diagram and let $\ell \in \mathcal{L}$. The Euler diagram **d with ℓ removed**, denoted $r_\ell(d) = d - \ell$, is d' where*

1. *$L(d') = L(d) - \{\ell\}$;*
2. *$Z(d') = r_\ell(Z(d))$ and*
3. *$Z^\bullet(d') = \{(x,y) \in Z(d') : r_\ell^{-1}(x,y) \cap Z(d) \subseteq Z^\bullet(d)\}$.*

It is straightforward to generalise definition 6 to removing a set of labels $\mathscr{L} \subset \mathcal{L}$. It can be shown that removing a pair of labels is equivalent to removing first one label and then the other (in either order). It then follows by a simple inductive argument that removing the labels $\mathscr{L} \subseteq L(d)$ is equivalent to removing the labels in $\mathscr{L}$ one at a time (in any order).

We may add a contour to a diagram. In order to respect the semantics, the contour must have a label that is not already present in the diagram and it needs to be added in such a way that it splits each zone into two new zones. Adding the contour labelled E to the diagram d_1 in figure 3 produces d_2.

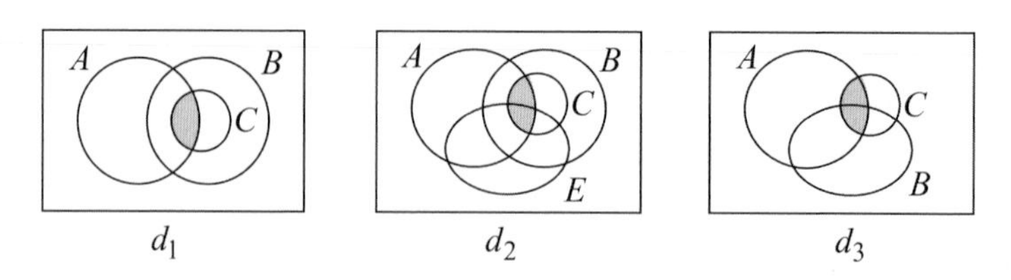

Fig. 3. Adding a contour

Definition 7. *Let d be an Euler diagram and let $\ell \in \mathcal{L}$ be a contour label not in $L(d)$. The Euler diagram* **d with ℓ added**, *denoted $d + \ell$, is d' where*

1. $L(d') = L(d) \cup \{\ell\}$;
2. $Z(d') = \{(x \cup \{\ell\}, y) : (x, y) \in Z(d)\} \cup \{(x, y \cup \{\ell\}) : (x, y) \in Z(d)\}$; *and*
3. $Z^\bullet(d') = \{(x \cup \{\ell\}, y) : (x, y) \in Z^\bullet(d)\} \cup \{(x, y \cup \{\ell\}) : (x, y) \in Z^\bullet(d)\}$.

Note that the operations 'add ℓ' and 'remove ℓ' do not in general commute. Although $(d+\ell)-\ell$ always gives the original diagram d, $(d-\ell)+\ell$ does not. For example, in figure 3, $(d_1 - B) + B$ produces the diagram d_3 which is different (syntactically and semantically) from d_1.

In our system, semantically both shaded and missing zones denote the empty set. This allows for a variety of representations of set theoretic relationships. Systems based on Euler diagrams with shading may allow users a choice of representation. Hence it is desirable, at the syntactic level, to be able to move between different representations of the empty set; that is, to introduce a missing zone into a diagram as a shaded zone and to delete a shaded zone from a diagram. For example, referring back to figure 1, the diagram d_2 can be obtained from d_1 by first removing B and then removing the two shaded zones in $d_1 - B$.

Definition 8. *Let d be an Euler diagram and let $z \in Z^m(d)$ be a missing zone of d. The Euler diagram* **d with z added**, *denoted $d + z$, is d' where*

1. $L(d') = L(d)$;
2. $Z(d') = Z(d) \cup \{z\}$; *and*
3. $Z^\bullet(d') = Z^\bullet(d) \cup \{z\}$.

Removing a shaded zone is complicated by the fact that it may result in the removal of one or more labels from the diagram. For example, in figure 2, removing the shaded zone $(\{A\}, \{C, E\})$ from d_2 has the effect of removing A entirely from the diagram. This is because the removed zone is the only zone in the diagram that contains the contour label A. The resulting diagram is not semantically equivalent to d_2 since it does not assert $\Psi(A) = \varnothing$.

Definition 9. *Let d be an Euler diagram and let $z \in Z^\bullet(d)$ be a shaded zone of d. Let $\mathscr{L} \subseteq in(z)$ be the set of labels for which the zone z is the only zone in d that has any labels of $\mathscr{L}$ in its containing set. The Euler diagram* **d with z removed**, *denoted $d - z$, is d' where*

1. $L(d') = L(d) - \mathscr{L}$;
2. $Z(d') = \{(x, y - \mathscr{L}) : (x, y) \in Z(d) - \{z\}\}$; *and*
3. $Z^\bullet(d') = \{(x, y - \mathscr{L}) : (x, y) \in Z^\bullet(d) - \{z\}\}$.

It is worth noting that, if $\mathscr{L} = \varnothing$ (so that no labels are removed from d), then $Z(d') = Z(d) - \{z\}$ and $Z^\bullet(d') = Z^\bullet(d) - \{z\}$. Also, if $\mathscr{L} \neq \varnothing$ then $d - z$ is the same diagram as that obtained by removing the label set $\mathscr{L}$, $d - \mathscr{L}$.

Recall that we require our diagram manipulations to be valid in the sense that if we manipulate d_1 to obtain d_2 then d_2 is a semantic consequence of d_1, $d_1 \vDash d_2$. The following theorem summarises the situation. Each of the four manipulations described above is valid although it is not always the case that a semantically equivalent diagram results.

Theorem 1. *Let d be an Euler diagram.*

1. *If $\ell \in L(d)$ is a contour label in d then $d \vDash d - \ell$.*
2. *If $\ell \in \mathcal{L} - L(d)$ is a contour label not in d then $d \equiv_\vDash d + \ell$.*
3. *If $z \in Z^m(d)$ is a zone missing from d then $d \equiv_\vDash d + z$.*
4. *If $z \in Z^\bullet(d)$ is a shaded zone then $d \vDash d - z$.*

In general, $d - \ell \nvDash d$ as Example 2 shows. The conditions under which $d - \ell$ is semantically equivalent to d are considered in section 4.1 below, summarised in Theorem 3. The conditions under which removing a shaded zone results in a semantically equivalent diagram are easier to describe: $d \equiv_\vDash d - z$ if and only if $\mathscr{L} = \varnothing$ in definition 9.

3.2 Nomads

If every zone 'inside' a contour is shaded, the diagram asserts that the set assigned to the label of the contour is empty.

Example 3. In figure 4, each of the diagrams d_1, d_2, d_3 and d_4 asserts $A \cap C = \varnothing$ and $E = \varnothing$. Essentially, the only difference between the diagrams is the placing of the entirely shaded contour labelled E within the diagram. Provided there is at least one zone inside E and all the zones inside E are shaded, there are many semantically equivalent diagrams to those in figure 4 with different placings of the contour E. In an informal way we can think of E as being 'free to wander around' the diagram. Following [14], we refer to the contour E as a *nomad*.

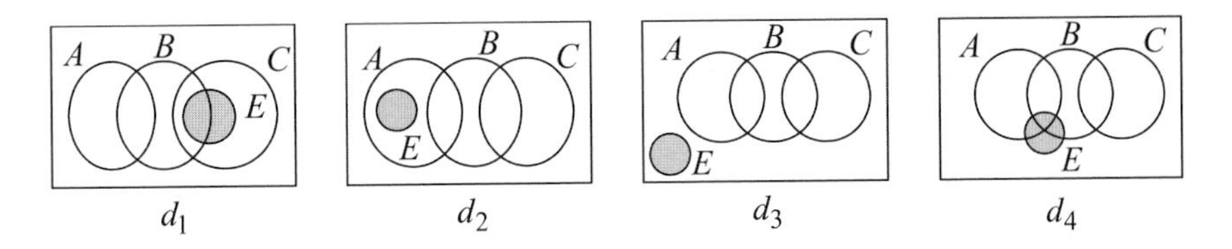

Fig. 4. Four diagrams where E is a 'nomad'

Definition 10. *Let d be an Euler diagram and $\ell \in L(d)$ be a label. We say ℓ is a* ***nomad*** *in d if the set of zones within ℓ is entirely shaded, $\{(x, y) \in Z(d) : \ell \in x\} \subseteq Z^{\bullet}(d)$. We denote the set of nomads in d by $\mathcal{N} = \mathcal{N}(d)$.*

Example 4. Consider the diagram d_1 in figure 5 which has two 'separated' nomads labelled A and B. In d_2, these contours are concurrent. The diagrams d_1 and d_2 are semantically equivalent since both diagrams assert $A = \varnothing$ and $B = \varnothing$.

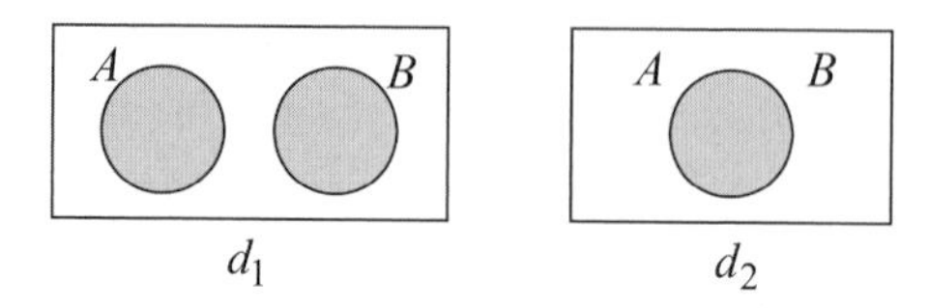

Fig. 5. Moving nomads

Combining the ideas in examples 3 and 4, an Euler diagram containing several nomads is semantically equivalent to a diagram where all of the nomads have been moved to the outside zone and 'overlaid' to form a single shaded zone. This is illustrated in figure 6 where diagram d_1 has three nomads, the contours labelled E, F and G. In the semantically equivalent diagram d_2, these three contours form a single shaded zone situated in the outside zone. We refer to this as *exiling the nomads* and we say that d_2 has *exiled nomads*.

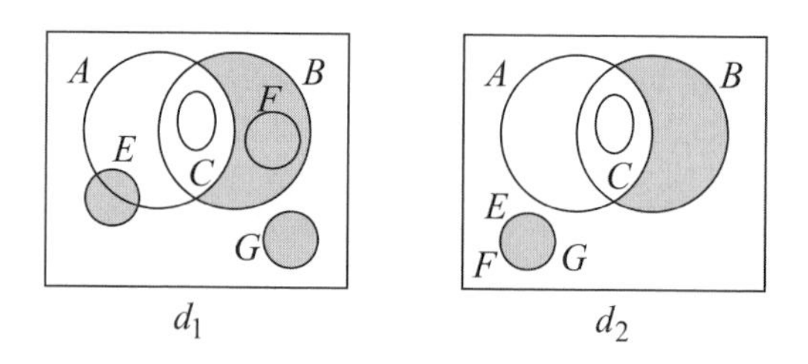

Fig. 6. Moving several nomads

Definition 11. *Let d be an Euler diagram with a non-empty set of nomads $\mathcal{N}(d)$.*

1. *The set of zones $Z(d)$ partitions into the zones inside nomads $Z_{in_{\mathcal{N}}}(d) = \{(x, y) \in Z(d) : x \cap \mathcal{N} \neq \varnothing\}$ and the set of zones outside nomads $Z_{out_{\mathcal{N}}}(d) = \{(x, y) \in Z(d) : x \cap \mathcal{N} = \varnothing\}$. The set of shaded zones partitions similarly.*
2. *The Euler diagram* ***d with exiled nomads*** *is d' where*
   (a) *the unshaded zones of d' are the unshaded zones of d, $Z^{\circ}(d') = Z^{\circ}(d)$;*
   (b) *the shaded zones in d' are the shaded zones outside nomads in d together with an additional shaded zone inside all the nomads, $Z^{\bullet}(d') = Z^{\bullet}_{out_{\mathcal{N}}}(d) \cup \{(\mathcal{N}(d), \varnothing)\}$.*

Theorem 2. *Let d_1 be an Euler diagram containing nomads, $\mathcal{N}(d_1) \neq \varnothing$ and let d_2 be d_1 with exiled nomads. Then d_1 is semantically equivalent to d_2, $d_1 \equiv_{\models} d_2$.*

4 The Normal Form

In this section we develop our normal form for Euler diagrams. We first develop a syntactic characterisation of when a contour label is semantically redundant and may therefore be removed from a diagram. Then we use the manipulations introduced in the previous section to obtain, from a given diagram d, a semantically equivalent diagram d' that has a simple syntactic description and where the properties of d' define our normal form.

4.1 Redundant Contour Labels

Definition 12. *Let d be an Euler diagram. A contour label $\ell \in L(d)$ is* ***redundant*** *in d if removing it results in a semantically equivalent diagram, $d \equiv_{\models} d - \ell$.*

More generally, a set of contour labels $\mathscr{L} \subseteq L(d)$ is ***redundant*** *in d if removing it results in a semantically equivalent diagram, $d \equiv_{\models} d - \mathscr{L}$.*

By Theorem 1, adding a contour label ℓ to a diagram d produces a semantically equivalent diagram $d + \ell$; thus ℓ is redundant in $d + \ell$. For example, in figure 3, the contour labelled E is redundant in $d_2 = d_1 + E$. When adding E to d_1, each zone in d_1 is split in two. Hence, in d_2, every zone that is contained by E has a corresponding zone, which (following [14]) we call its 'E-twin', that is excluded by E. Furthermore, the E-twin of every shaded zone is also shaded. We say that the contour E 'completely splits' both the shaded and unshaded regions of d_2. We now make these ideas more precise for abstract diagrams.

Definition 13. *1. For $\ell \in \mathcal{L}$, we define the* ***move label*** *function on the set of abstract zones by*

$$m_\ell : \mathcal{Z} \to \mathcal{Z},\ m_\ell(x, y) = \begin{cases} (x - \{\ell\}, y \cup \{\ell\}) & \text{if } \ell \in x \\ (x \cup \{\ell\}, y - \{\ell\}) & \text{if } \ell \in y \\ (x, y) & \text{otherwise.} \end{cases}$$

2. A pair of distinct zones $z_1, z_2 \in \mathcal{Z}$ are called ***ℓ-twins*** *if $z_2 = m_\ell(z_1)$ (and $z_1 = m_\ell(z_2)$). Thus a pair of distinct zones are ℓ-twins if they are of the form $(x - \{\ell\}, y \cup \{\ell\})$ and $(x \cup \{\ell\}, y - \{\ell\})$.*

3. A zone $z \in Z(d)$ is an ***ℓ-single*** *in d if its ℓ-twin does not belong to d.*

4. Let $r \in R(d)$ be a region in an Euler diagram d and let $\ell \in L(d)$. We say that ℓ ***completely splits*** *r if, for each zone $z \in r$, its ℓ-twin is also in r, $m_\ell(z) \in r$.*

The presence or otherwise of ℓ-twins in a diagram is important in determining whether ℓ is redundant but not, perhaps, in the most obvious way. Since $d \equiv_{\models} d + \ell$ (where ℓ is not a label in d), it follows that ℓ is redundant in $d + \ell$ and, by

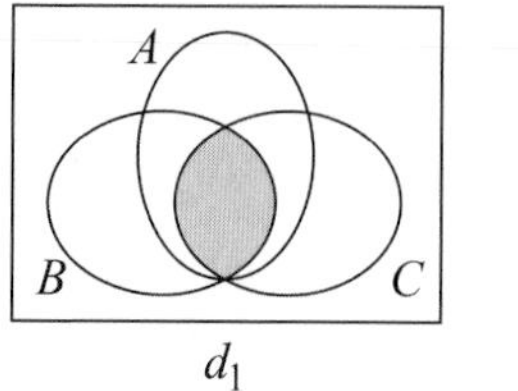

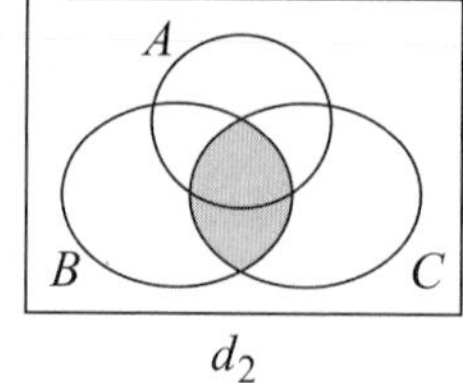

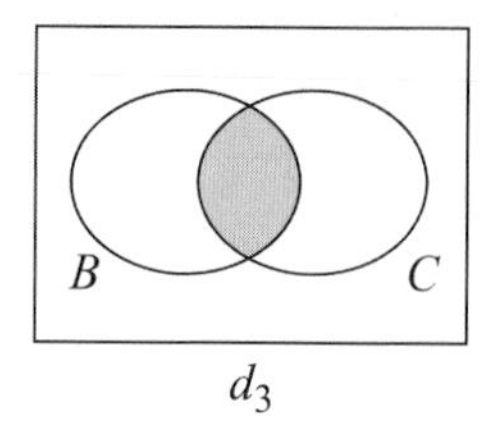

Fig. 7. Missing twin

construction, ℓ completely splits both the shaded and unshaded regions of $d+\ell$. Therefore we might be tempted to conjecture that completely splitting both the shaded and unshaded regions is the syntactic condition for contour redundancy. However, this is not correct.

Consider the four diagrams containing a nomad in figure 4. In each diagram, the contour labelled B is redundant but B does not split the unshaded regions in d_2 or d_3. Similarly, the contour labelled A is redundant in both d_1 and d_2 in figure 7 below; removing A from either of these diagrams gives the semantically equivalent diagram d_3. Although A completely splits the shaded and unshaded regions in d_2, it does not split the shaded region in d_1 because the zone $(\{B, C\}, \{A\})$, which is the A-twin of the shaded zone, is missing in d_1. In fact, we only need consider the unshaded zones to determine whether a contour is redundant.

Definition 14. *Let d be an Euler diagram.*

*A label $\ell \in L(d)$ is a **splitting label**[1] for d if ℓ completely splits $Z^{\circ}(d)$, the unshaded region of d.*

The following are straightforward consequences of definition 14:

- if ℓ is a splitting label for d then, for each shaded zone $z \in Z^{\bullet}(d)$, its ℓ-twin $m_{\ell}(z)$ is either shaded or missing from the diagram;
- if $n \in L(d)$ is a nomad in d then n is a splitting label for d if and only if d is entirely shaded.

As the following theorem shows, the notion of splitting label is the syntactic characterisation we are seeking for a contour label to be redundant. The proof is omitted although our approach in establishing the implication 'if ℓ is not a splitting label then it is not redundant' is illustrated in the example that follows.

Theorem 3. *A contour label ℓ is redundant in an Euler diagram d if and only if ℓ is a splitting label for d.*

An interesting consequence is that an Euler diagram with an odd number of unshaded zones has no redundant contours.

[1] In [14], the term 'weak mirror' is used.

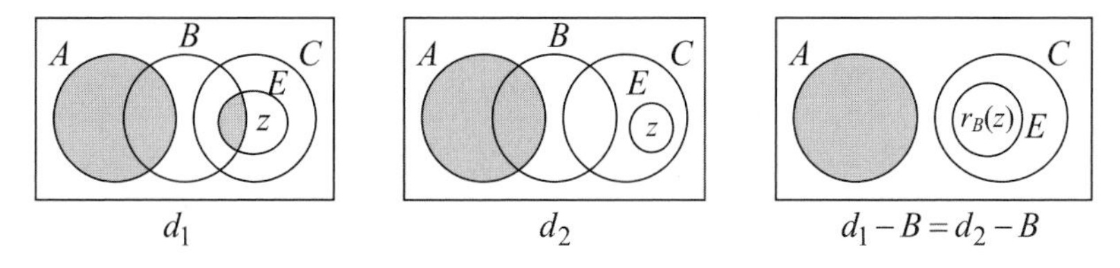

Fig. 8. B does not split the unshaded region

Example 5. Suppose that ℓ is not a splitting label in d. Then it does not split the unshaded region $Z^\circ(d)$ so there is an unshaded zone in $Z^\circ(d)$ such that its ℓ-twin is either shaded or missing. These two possibilities are illustrated in figure 8 for the zone $z = (\{C, E\}, \{A, B\})$. In d_1, the B-twin of z is shaded and in d_2 this B-twin is missing. Note that the annotations 'z' and '$r_B(z)$' are not part of the syntax of the diagrams.

To show that $d_1 - B$ is not semantically equivalent to d_1, we construct a model for $d_1 - B$ for which $\Psi(m_B(z)) \neq \varnothing$. This is not a model for d_1 since the B-twin of z, $m_B(z)$, is shaded in d_1 so $\Psi(m_B(z)) = \varnothing$ in any model for d_1.

We start with a model (U, Ψ) in which $\Psi(r_B(z))$ is non-empty. We take the universal set to be the positive integers $\mathbb{Z}^+$ and define a set assignment $\Psi : \mathcal{L} \to \mathbb{Z}^+$ by $\Psi(A) = \varnothing$, $\Psi(C) = \{1, 2\}$, $\Psi(E) = \{2\}$ and $\Psi(\ell) = \mathbb{Z}^+$ for all $\ell \in \mathcal{L} - L(d_1 - B)$. (To obtain this model, we listed the unshaded zones inside contours, $z_1, z_2, \ldots$ and defined Ψ such that $\Psi(z_i) = \{i\}$ and $\Psi(z^\bullet) = \varnothing$ for any shaded zone $z^\bullet$.)

It is easy to verify that this interpretation $(\mathbb{Z}^+, \Psi)$ is a model for $d_1 - B$ but this is not quite the model we need. Since $B \in \mathcal{L} - \{A, C, E\}$, we defined $\Psi(B) = \mathbb{Z}^+$. We now redefine $\Psi(B)$ as follows (but not changing $\Psi(\ell)$ for $\ell \neq B$):

$$\Psi(B) = \Psi(r_B(z)) = \Psi(C) \cap \Psi(E) \cap \overline{\Psi(A)} = \{1, 2\} \cap \{2\} \cap \mathbb{Z}^+ = \{2\}.$$

This new interpretation $(\mathbb{Z}^+, \Psi)$ is still a model for $d_1 - B$ since neither the shading condition nor the missing zones condition in $d_1 - B$ involve $\Psi(B)$.

However, in d_1, $m_B(z) = (\{B, C, E\}, \{A\})$ so

$$\Psi(m_B(z)) = \Psi(B) \cap \Psi(C) \cap \Psi(E) \cap \overline{\Psi(A)} = \{2\} \cap \{1, 2\} \cap \{2\} \cap \mathbb{Z}^+ = \{2\}.$$

Since $\Psi(m_B(z)) \neq \varnothing$ but $m_B(z)$ is shaded in d_1, this interpretation is not a model for d_1 (or d_2). Therefore $d_1 \not\equiv_\vDash d_1 - B$ so B is not redundant in d_1.

4.2 Defining the Normal Form

In this section, we describe our normal form. Each Euler diagram will be semantically equivalent to a unique diagram in normal form. A diagram d in normal form will have a minimal number of contour labels and a minimal number of zones amongst all of the diagrams in its semantic equivalence class $\langle d \rangle_\vDash$. Although it is 'structurally' simple at the abstract level, the normal form need not

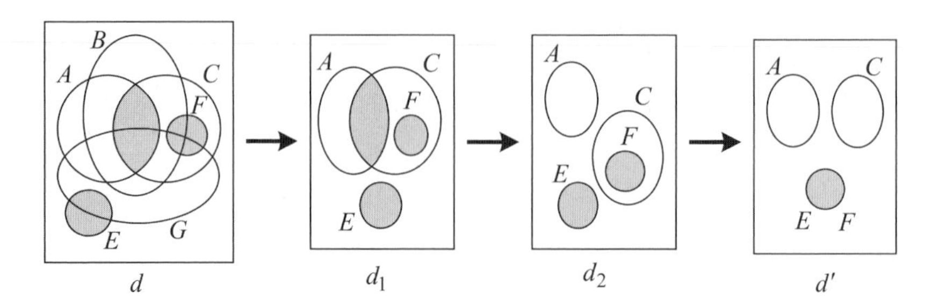

Fig. 9. Obtaining the normal form

have a 'nice' drawing. For example, the diagram d' in figure 9 below is in normal form and the contours E and F are concurrent, a feature that is often considered problematic in concrete diagrams.

Definition 15. *An Euler diagram d is in* **normal form** *if it satisfies the following three properties:*

1. *d had no redundant contour labels;*
2. *d has no shaded zones apart, possibly, from zones within nomads;*
3. *if d has any nomads, these are represented by a single zone 'situated in the outside zone'.*

In fact, there is an algorithm to obtain, from a diagram d, a semantically equivalent diagram d' in normal form. The steps in obtaining d' are described informally as: remove any redundant contour labels; remove any shaded zones that do not lie within a nomad; finally, exile any nomads (see theorem 4).

Figure 9 illustrates the algorithm. The diagram d has two redundant contours labelled B and G. The first step is to remove these to obtain d_1. Next the shaded zone $(\{A, C\}, \{E, F\})$ in d_1 that is not part of a nomad is removed to obtain d_2. Finally, the two nomads E and F are moved to form a single zone 'situated in the outside zone'. This gives the diagram d' which is in normal form.

The following theorem follows from theorems 2 and 3 and the equivalence $d \equiv_{\models} d - z$ when $\mathscr{L} = \varnothing$ (see definition 9).

Theorem 4. *Let d be an Euler diagram and let d' be the Euler diagram in normal form obtained by applying the following three steps to d.*

1. *Remove all redundant contours: replace d with $d_1 = d - \mathscr{L}$ where $\mathscr{L} = \{\ell \in L(d) : \ell \text{ is redundant in } d\}$.*
2. *Remove any shaded zones that do not lie within a nomad: replace d_1 with the diagram d_2 obtained by removing in turn each zone in $Z^{\bullet}_{out\mathcal{N}}(d_1)$.*
3. *Exile all nomads: if $\mathcal{N}(d_2) \neq \varnothing$, replace d_2 by d' which is d_2 with exiled nomads.*

Then d is semantically equivalent to d'.

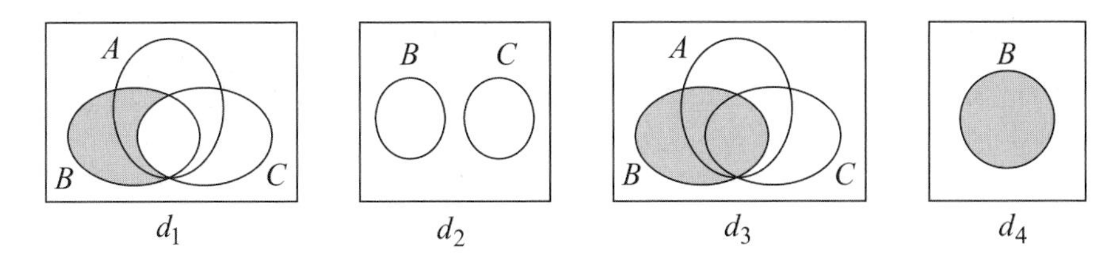

Fig. 10. Reasoning with diagrams

5 Conclusion and Further Work

The multiplicity of representations of set relationships afforded by Euler diagrams with shading means that the semantic equivalence class of a diagram $\langle d \rangle_{\models}$ may contain many diagrams. In this paper we have obtained a syntactic characterisation of when a contour is semantically redundant and we have defined a normal form for unitary Euler diagrams. The normal form provides a 'canonical representative' of each semantic equivalence class and gives a decision procedure for semantic equivalence: $d_1 \equiv_{\models} d_2$ if and only if their normal forms are equal.

The diagrams in $\langle d \rangle_{\models}$ represent the range of ways of displaying the same information of which the normal form is structurally simple but may not be nicely drawable. The Euler diagram generation problem (EDGP) is that of generating a concrete diagram from an abstract description, under some specified set of wellformedness conditions. The EDGP has been solved for some set of wellformedness conditions [9]. The effect of imposing certain wellformedness conditions can lead to there being no concrete diagram that complies with a given abstract diagram; informally, the abstract diagram is not 'nicely drawable'. An alternative approach to the problem could be to consider a variant of the EDGP: instead of generating a diagram with exactly the correct set of zones, generate a nicely drawable diagram which has some extra (shaded) zones. One way to attempt this could be to start with a diagram in our normal form and then add shaded zones until a nicely drawable diagram is obtained.

Normal forms in symbolic logic are an important tool in automated theorem proving and we envisage the same will be the case for diagrammatic systems. One can imagine an automated reasoning system for Euler diagrams that, in some reasoning step, combines diagrams d_1 and d_2 in figure 10 (taken in conjunction) to produce the 'conclusion' diagram d_3. Although there are no redundant contours in either d_1 or d_2, both A and C are redundant in d_3 which has a simple normal form in d_4. Furthermore, designers of a diagrammatic reasoning system (whether or not based on Euler diagrams) are faced with choices of possible rules, often with no systematic way of choosing between competing rule sets. Identification of semantic equivalence classes may guide designers to identify rules that enable 'traversal' of equivalence classes and those that move between classes.

Euler diagrams provide the basis for a number of diagrammatic systems that add new syntactic elements. For example, projections provide a means to give 'local' information within Euler diagram based systems [15,14]. For Euler diagrams, including projections does not increase expressiveness but does provide much greater variety of representation so that the size of semantic equivalence classes are increased. The notion of normal form can be extended to Euler diagrams with projections by combining the normal form for the underlying diagram without projections with 'local' normal forms for the projections within (underlying) zones. Further work with projections will also relate the extended normal form to nesting of diagrams [13] and measuring clutter [10,14].

The ideas presented in this paper could also have wider applicability to visual languages that allow redundancy and multiple representations of information. Having a 'canonical' representation within such systems can provide a useful tool both for understanding different representations of the same information as well as distinguishing between representations with different information content.

References

1. Harel, D.: On visual formalisms. In: Glasgow, J., Narayan, N.H., Chandrasekaran, B. (eds.) Diagrammatic Reasoning, pp. 235–271. MIT Press, Cambridge (1998)
2. Howse, J., Stapleton, G., Taylor, J.: Spider diagrams. LMS Journal of Computation and Mathematics 8, 145–194 (2005)
3. Swoboda, N., Allwein, G.: Using DAG transformations to verify Euler/Venn homogeneous and Euler/Venn FOL heterogeneous rules of inference. Journal on Software and System Modeling 3(2), 136–149 (2004)
4. Kent, S.: Constraint diagrams: Visualizing invariants in object oriented modelling. In: Proceedings of OOPSLA 1997, October 1997, pp. 327–341. ACM Press, New York (1997)
5. Ruskey, F.: A survey of Venn diagrams. Electronic Journal of Combinatorics (1997, updated 2001, 2005), www.combinatorics.org/Surveys/ds5/VennEJC.html
6. DeChiara, R., Erra, U., Scarano, V.: VennFS: A Venn diagram file manager. In: Proceedings of Information Visualisation, pp. 120–126. IEEE Computer Society, Los Alamitos (2003)
7. Thièvre, J., Viaud, M., Verroust-Blondet, A.: Using euler diagrams in traditional library environments. In: Euler Diagrams 2004. ENTCS, vol. 134, pp. 189–202 (2005)
8. Howse, J., Schuman, S.: Precise visual modelling. Journal of Software and Systems Modeling 4, 310–325 (2005)
9. Flower, J., Fish, A., Howse, J.: Euler diagram generation. Journal of Visual Languages and Computing (to appear, 2008), http://dx.doi.org/10.1016/j.jvlc.2008.01.004
10. John, C., Fish, A., Howse, J., Taylor, J.: Exploring the notion of clutter in Euler diagrams. In: Barker-Plummer, D., Cox, R., Swoboda, N. (eds.) Diagrams 2006. LNCS (LNAI), vol. 4045, pp. 267–282. Springer, Heidelberg (2006)
11. Stapleton, G., Masthoff, J.: Incorporating negation into visual logics: A case study using Euler diagrams. In: Visual Languages and Computing 2007, Knowledge Systems Institute, pp. 187–194 (2007)
12. Howse, J., Molina, F., Shin, S.J., Taylor, J.: Type-syntax and token-syntax in diagrammatic systems. In: Proceedings of 2nd International Conference on Formal Ontology in Information Systems, Maine, USA, pp. 174–185. ACM Press, New York (2001)

13. Flower, J., Howse, J., Taylor, J.: Nesting in Euler diagrams: syntax, semantics and construction. Software and Systems Modelling 3, 55–67 (2004)
14. John, C.: Measuring and Reducing Clutter in Spider Diagrams with Projections. Ph.D thesis, University of Brighton (2006)
15. Gil, J., Howse, J., Kent, S., Taylor, J.: Projections in Venn-Euler diagrams. In: Proc. IEEE Symposium on Visual Languages, September 2000, pp. 119–126. IEEE Computer Society Press, Los Alamitos (2000)

Ensuring Generality in Euclid's Diagrammatic Arguments

John Mumma

Carnegie Mellon University

Abstract. This paper presents and compares **FG** and **Eu**, two recent formalizations of Euclid's diagrammatic arguments in the *Elements*. The analysis of **FG**, developed by the mathematician Nathaniel Miller, and that of **Eu**, developed by the author, both exploit the fact that Euclid's diagrammatic inferences depend only on the topology of the diagram. In both systems, the symbols playing the role of Euclid's diagrams are discrete objects individuated in proofs by their topology. The key difference between **FG** and **Eu** lies in the way that a derivation is ensured to have the generality of Euclid's results. Carrying out one of Euclid's constructions on an individual diagram can produce topological relations which are not shared by all diagrams so constructed. **FG** meets this difficulty by an enumeration of cases with every construction step. **Eu**, on the other hand, specifies a procedure for interpreting a constructed diagram in terms of the way it was constructed. After describing both approaches, the paper discusses the theoretical significance of their differences. There is in **Eu** a context dependence to diagram use, which enables one to bypass the (sometimes very long) case analyses required by **FG**.

Each of the arguments in Euclid's *Elements*, as given in [1], comes equipped with a geometric diagram. The role of the diagram in the text is not merely to illustrate the geometric configuration being discussed. It also furnishes a basis for inference. For some of Euclid's steps, the logical form of the preceeding sentences is not enough to ground the step. One must consult the diagram to understand what justifies it. Consequently, Euclid is standardly taken to have failed in his efforts to provide exact, fully explicit mathematical proofs. Inspection of geometric diagrams is thought to be too vague and open-ended a process to play any part in rigorous mathematical reasoning.

This assumption has recently been disproved. **FG**, developed by Nathaniel Miller and presented in [3], and **Eu**, developed by the author in [4], are formal systems of proof which possess a symbol type for geometric diagrams. Working within each system, one can reconstruct Euclid's proofs in an exact and fully explicit manner, *with* diagrams. In this paper I compare the two systems as accounts of Euclid's diagrammatic reasoning. In the first part, I discuss the feature of Euclid's diagram use which makes both formalizations possible. In the second, I explain how both systems codify this use in their rules. In **FG**,

G. Stapleton, J. Howse, and J. Lee (Eds.): Diagrams 2008, LNAI 5223, pp. 222–235, 2008.

the content of single diagram is context independent, and the result of a geometric construction is a disjunctive array of such diagrams. Alternatively, in **Eu**, the content of a diagram depends systematically on the way it is constructed in a proof. The need for the disjunctive arrays of **FG**, which can become very large, is thus avoided.

1 What a Diagram Can Do for Euclid

A close reading of the *Elements* reveals that the significance of a diagram in a proof is neither vague nor open-ended for Euclid. The first to discover this is Ken Manders, who laid out his insights on ancient geometric proof in [2].[1]

To explain the division of labor between text and diagram in ancient geometry, Manders distinguishes between the *exact* and *co-exact* properties of geometric diagrams. Any one of Euclid's diagrams contains a collection of spatially related magnitudes—e.g. lengths, angles, areas. For any two magnitudes of the same type, one will be greater than another, or they will be equal. These relations comprise the *exact* properties of the diagram. How these magnitudes relate topologically to one another—i.e. the regions they define, the containment relations between these regions—comprise the diagram's *co-exact* properties. Diagrams of a single triangle, for instance, vary with respect to their exact properties. That is, the lengths of the sides, the size of the angles, the area enclosed, vary. Yet with respect to their co-exact properties the diagrams are all the same. Each consists of three bounded linear regions, which together define an area.

The key observation is that Euclid's diagrams contribute to proofs *only* through their co-exact properties. Euclid never infers an exact property from a diagram unless it follows directly from a co-exact property. Exact relations between magnitudes which are not exhibited as a containment are either assumed from the outset or are proved via a chain of inferences in the text. It is not difficult to hypothesize why Euclid would have restricted himself in such a way. Any proof, diagrammatic or otherwise, ought to be reproducible. Generating the symbols which comprise it ought to be straightforward and unproblematic. Yet there seems to be room for doubt whether one has succeeded in constructing a diagram according to its exact specifications perfectly. The compass may have slipped slightly, or the ruler may have taken a tiny nudge. In constraining himself to the co-exact properties of diagrams, Euclid is constraining himself to those properties stable under such perturbations.

For an illustration of the interplay between text and diagram, consider proposition 35 of book I. It asserts that any two parallelograms which are bounded by the same parallel lines and share the same base have the same area. Euclid's proof proceeds as follows.

[1] The paper was written in 1995 but published only recently, in *Philosophy of Mathematical Practice*, edited by Paolo Mancosu (Clarendon Press, 2008). Despite the fact the paper existed only as draft for most of its 13 years, it has been influential to those interested in diagrams, geometry and proof. Mancosu describes it an 'underground classic.'

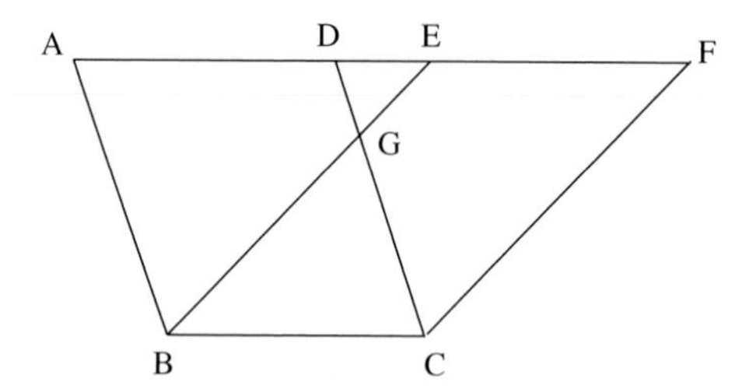

Let ABCD, EBCF be parallelograms on the same base BC and in the same parallels AF, BC.
Since ABCD is parallelogram, AD equals BC (proposition 34). Similarly, EF equals BC.
Thus, AD equals EF (common notion 1).
Equals added to equals are equal, so AE equals DF (Common notion 2).
Again, since ABCD is a parallelogram, AB equals DC (proposition 34) and angle EAB equals angle FDC (proposition 29).
By side angle side congruence, triangle EAB equals triangle FDC (proposition 4).
Subtracting triangle EDG from both, we have that the trapezium ABGD equals the trapezium EGCF (common notion 3).
Adding triangle GBC to both, we have that ABCD equals EBCF (common notion 2).
QED

The proof is independent of the diagram up until the inference that AE equals DF. This step depends on common notion 2, which states that if equals are added to equals, the wholes are equal. The rule is correctly invoked because four conditions are satisfied: $AD = EF$, $DE = DE$, DE is contained in AE, and DE is contained in DF. The first pair of conditions are exact, the second pair co-exact. Accordingly, the first pair of conditions are seen to be satisfied via the text, and the second pair via the diagram. Similar observations apply to the last two inferences. The applicability of the relevant common notion is secured by both the text and the diagram. With just the textual component of the proof to go on, we would have no reason to believe that the necessary containment relations hold. Indeed, we would be completely in the dark as to the nature of containment relations in general.

The standard line is that this situation needs to be rectified with something like a betweenness relation. Manders's opposing thesis is that diagrams function in the *Elements* as reliable symbols because Euclid only invokes their co-exact features. Though we may not be able to trust ourselves to produce and read off the exact properties of diagrams, we can trust ourselves to produce and read off co-exact properties. Thus, Euclid seems to be within his rights to use diagrams to record co-exact information. If Manders's analysis is correct, Euclid's proofs ought to go through with diagrams which are equivalent in a co-exact sense (hereafter *c.e. equivalent*), but differ with respect to their exact properties. This turns out to be the case. The proof of proposition 35, for instance, still works if we substitute either of the following for the given diagram.

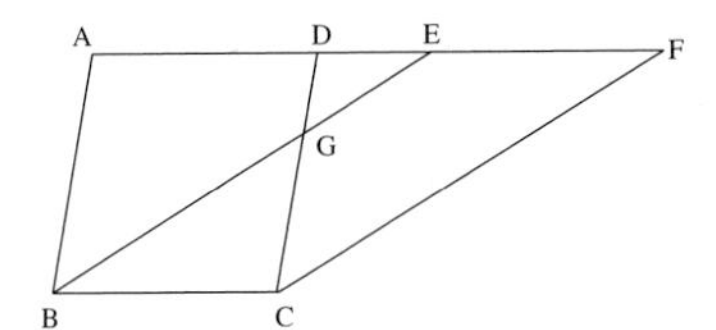

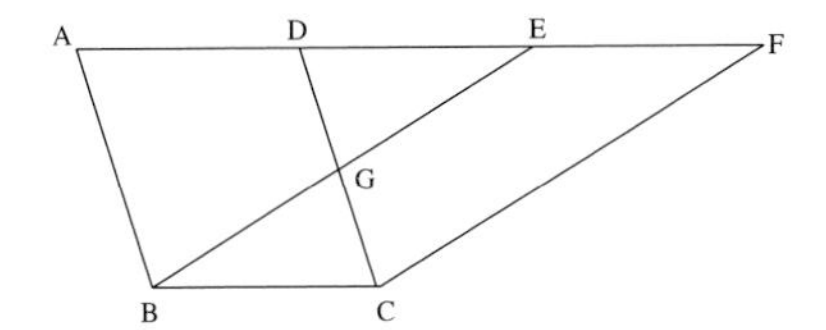

The diagram need not even satisfy the stipulated exact conditions. The diagram

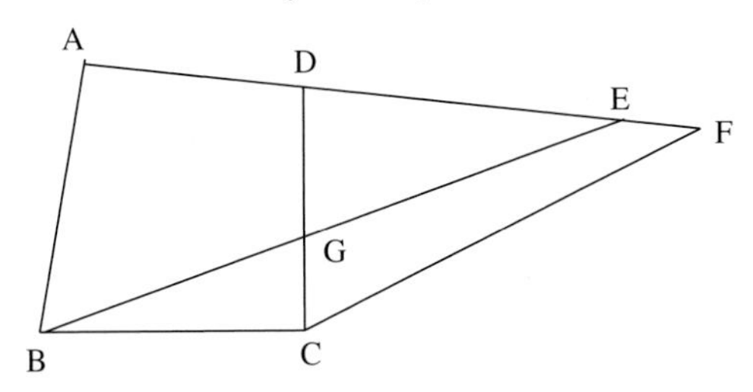

also fulfills the role the proof demands of it. The diagram's burden is to reveal how certain co-exact relationships lead to others. It is not used to show exact relationships. This is the job of the text. The proof must invariably employ a particular diagram, with particular exact relationships. But since the proof only calls on the co-exact relationships of the diagram, it holds of *all* diagrams which are c.e. equivalent to it.

Manders' observations naturally suggest the general approach of both **FG** and **Eu** in formalizing Euclid. As topological objects, Euclid's diagrams are discrete. What identifies them, topologically, is the way their lines and circles partition a bounded region of the plane into a finite set of regions. Thus, the discrete syntactic objects which are to function as diagrams in a formalization ought to be individuated in the same manner. The first challenge, then, is to define such syntactic objects precisely. The second is to formulate suitable rules for how objects so defined are to be used in proofs.

Though **FG** and **Eu** meet the first challenge in different ways, my focus is on their different approaches to the second challenge. With respect to the goal of defining syntactic objects which express the information Euclid relies on his diagrams to express, the diagrams of **FG** and **Eu** are equally sufficient. **Eu** diagrams are arguably closer to the diagrams of the Euclidean tradition, in that **Eu** lines and **Eu** circles are not purely topological.[2] I will not explore, however, whether this difference between **FG** and **Eu** is a significant one. Similarly, I

[2] The only initial constraint on the line and circle segments of **FG** diagrams is that they be one-dimensional. Though the definition of a well-formed **FG** diagram imposes restrictions on how such syntactic elements can relate to one another topologically, it still leaves room for diagrammatic lines and circles which twist and turn in wildly non-linear and non-circular ways. In contrast, **Eu** diagrams possess an underlying array structure, which allows for linearity and convexity to built into the definitions of lines and circles from the start. For the precise formal characterization of diagrams in **FG** and **Eu**, see pp. 21-34 in [3] and pp. 14-40 in [4]. For a discussion which raises doubts about the faithfulness of **FG** 's purely topological diagrams to the Euclidean tradition, see [5].

will pass over the fact that **FG** is a purely diagrammatic proof system, and **Eu** is a heterogeneous one. Both differences are not essential in understanding the central difference between the proof rules of **Eu** and **FG**.

2 Euclid's Constructions in FG and Eu

The diagram of a Euclidean proof rarely displays just the geometric elements stipulated at the beginning of the proof. A proof often has a construction stage dictating how new geometric elements are to be built on top of the given configuration. The demonstration stage then follows, in which inferences from the augmented figure can be made. The building up process is not shown explicitly. All that appears is the end result of the construction on a particular configuration.

As the proof of proposition 35 has no construction stage, it fails to illustrate this common feature of Euclid's proofs. The diagram of the proof contains just those elements which instantiate the proposition's general co-exact conditions. We are thus justified in grounding the result on the co-exact features of the diagram, given that we only apply the result to configurations which are c.e. equivalent to the diagram.

The soundness of Euclid's co-exact inferences is much less obvious when the proof's diagram contains augmented elements. The construction is always performed on a particular diagram. Though the diagram is representative of a range of configurations—i.e. all configurations c.e. equivalent to it—it cannot avoid having particular exact properties. And these exact properties can influence how the co-exact relations within the final diagram work out. When the same construction is performed on two diagrams which are c.e. equivalent but distinct with respect to their exact features, there is no reason to think that the two resulting diagrams will be c.e. equivalent.

Consider for example the construction of proposition 2 of book I of the *Elements*. The proposition states a construction problem: given a point A and a segment BC, construct from A a segment equal to BC. Euclid advances the following construction as a solution to the problem:

From the point A to the point B let the straight line AB be joined; and on it let the equilateral triangle DAB be constructed.

Let the straight lines AE, BF be produced in a straight line with DA and DB.

With center B and radius BC let the circle GCH be described; and again, with center D and radius DG let the circle GKL be described.

If the construction is performed on the *particular* point A and a *particular* segment BC

A

B C

the result is the diagram

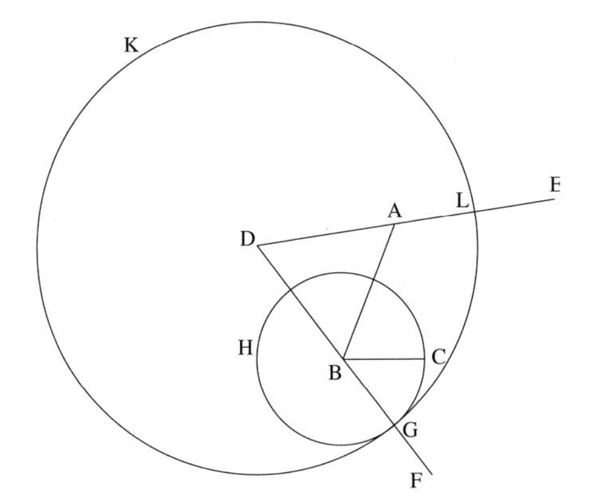

If however the construction is performed on the different particular configuration

A

B C

the result is

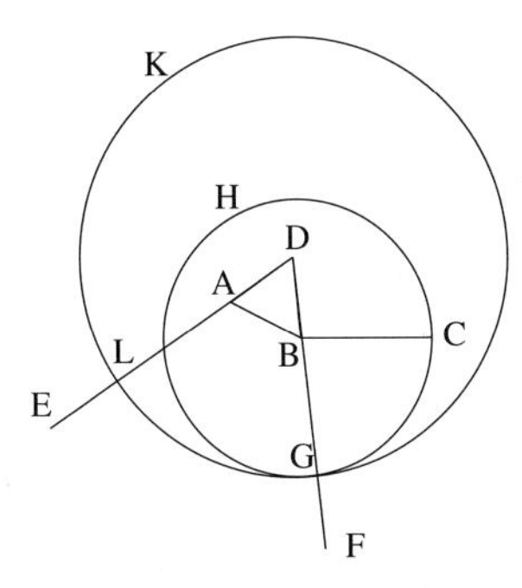

This diagram is distinct from the first diagram, topologically. Euclid nevertheless uses the co-exact features of one such diagram to argue that the construction does indeed solve the stated construction problem. The crucial step in the argument is the inference $AL = BG$. This follows from an application of the equals subtracted from equals rule. And for this to be applicable, A must lie on the segment DL and B must lie on the segment DG. So with these two diagrams we see that with two of the possible exact positions A can have to BC the topology needed for the proof obtains. But *prima facie* we have no mathematical reason to believe that it obtains for *all* the other positions A can have to BC.

And so, the vexing question is: how do we know that the co-exact features that Euclid isolates as general in a constructed diagram are shared by *all* diagrams which could result from the construction? Though Euclid never mistakes a property particular to an individual diagram as general, he does not provide any explicit criteria for how the separation of the general from the particular is to be made. What the formalizations of **FG** and **Eu** must do, if they are to count as formalizations, is furnish such criteria via its rules of proof.

2.1 The Diagram Arrays of FG

The rules of **FG** do this via disjunctive diagram arrays. A Euclidean construction in **FG** is not carried out via a single, representative diagram, but via an array of representative diagrams. In applying a construction step (such as joining two points in a segment, or drawing a circle on a radius) to a diagram D, one must produce the array representing all topological cases which could possibly result from applying the step to a figure represented by D. For instance, if D is the **FG** diagram

then constructing a circle on the segment of D produces in **FG** the array

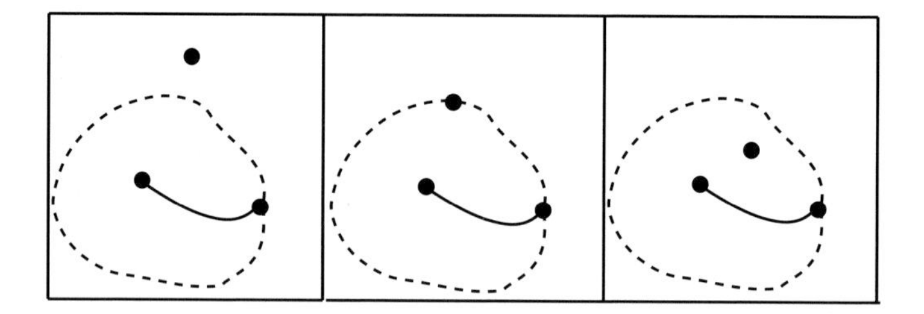

As only topological features can be read off from a diagram, D contains no information about the distance of the left-endpoint of the segment to the point off the segment. And so, it is consistent with D that the latter point sit outside, on, or inside the constructed circle.

Applying a construction step to an array A produces the array of all diagrams obtained by applying the step to each diagram of A. Thus, the array produced after n construction steps contains all topological cases which could possibly result from those n steps. Logically, the array is similar to a propositional statement in disjunctive normal form. It asserts that the geometric figure of the proof has the properties of one of the diagrams in the array. Once all contradictory diagrams (diagrams whose markings equate the part of a whole to the whole) are thrown out, one is then in a position to discern what holds in general. This consists in those properties manifest in all diagrams of the array.

The central technical achievement of **FG** is the specification of a mechanical procedure which given any initial diagram and any geometric construction as input outputs the appropriate array. Though it is clear what all the cases are when a circle is added to a diagram consisting only of a segment and a point, it is not clear if there is a general method for enumerating cases given any construction step and any diagram. The diagram does not have to become that much more complex for the range of possible cases consistent with a construction step to become obscure. Even if some cases can be seen, one usually lacks a guarantee that these constitute *all* cases to be considered.

The purely topological character of his diagrams, however, allows Miller to specify a method which has such a guarantee built in. The reason that the range of cases which come with a construction step is obscure is that it is not immediate how the metric symmetries of lines and circles restrict what is and isn't possible topologically. Yet once we allow line and circles to bend any which way, the obscurity vanishes. The range of cases emerges as the range of all topological possibilities consistent with the conditions Miller stipulates of the diagrams of his system. These conditions comprise the definition of a *nicely well-formed diagram* in **FG**. They ensure that lines and dotted lines form configurations which behave in a rough topological way like Euclidean lines and circles. One condition, for instance, ensures that two circles intersect no more than twice. And so, specifically, the array in an **FG** formalization of an Euclidean proof consists of all nicely well-formed diagrams obtainable from the proof's construction. These can be generated by a straightforward, if tedious, procedure, implemented by Miller in a computer program named **CDEG** (for Computerized Diagrammatic Euclidean Geometry).

2.2 Diagrammatic Inferences in Eu

The diagrammatic proof method of **Eu** is based on the principle that what is general in a diagram depends on how it was constructed.[3] Consider the diagram

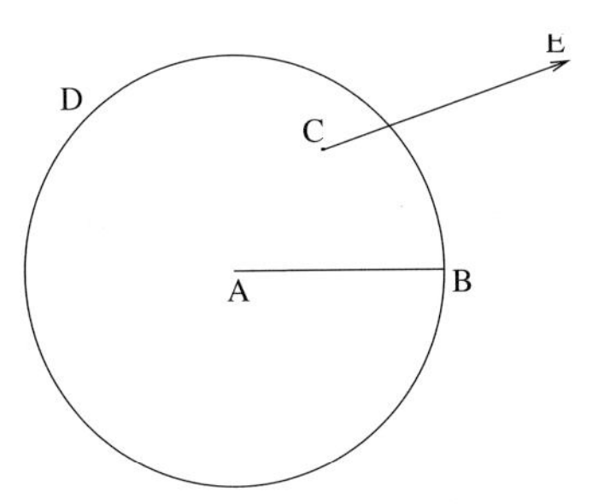

Many distinct constructions could have produced it. For instance, the initial configuration could have been the segment AB, and the construction steps leading to the diagram could have been:

[3] The principle is perhaps close to what Kant is talking about when he speaks of the "the universal conditions of the construction" in the passage quoted above. For a discussion which relates Kant's philosophy of mathematics to Euclid's geometric constructions, see [6].

- draw the circle D with center A and radius AB.
- pick a point C in the circle D, and a point E outside it.
- produce the ray CE from the point C.

Call this construction **C1**. Alternatively, it is possible that the initial configuration consists of the segment AB and the points C and E, while the construction consists of the following two steps:

- draw the circle D with center A and radius AB.
- produce the ray CE from the point C

Call this construction **C2**.

Now, if **C1** is responsible for the diagram, we are justified in taking the position of C within D as a general property of the diagram. The act of picking C in D fixes the point's position with respect to the circle as general. And since we know the position of C relative to D is general, we can pick out the point of intersection of the ray CE with D with confidence. It always exists in general, since a ray originating inside a circle must intersect the circle. In contrast, none of these inferences are justified if **C2** is responsible for the diagram. Nothing is assumed from the outset about the distance of the point C to A. And so, even though C lies within D in this particular diagram, it could possibly lie on D or outside it. Further, as the position of C relative to D is indeterminate, the intersection point of CE and D cannot be assumed to exist in general, even though one exists in this particular diagram.

Viewing proposition 2 in this way, we can satisfy ourselves that Euclid's diagrammatic inferences are sound. Though the position of segment BC with respect to the triangle ADB is indeterminate, what that segment contributes to the proof is the circle H, whose role in turn is to produce an intersection point G with the ray DF. The intersection point always exists no matter the position of BC to the ray DF. We can rotate BC through the possible alternatives, and we will always have a circle H whose center is B. And this is all we need to be assured that the intersection point G exists. The ray DF contains B, since it is the extension of the segment DB, and a ray which contains a point inside a circle *always* intersects the circle.

A similar argument shows that the intersection point L of the ray DE and the circle K always exists. The argument does not establish, however, that A lies *between* D and L. Here a case analysis is forced upon us. We must consider the case where A coincides with L, or the case where L lies between A and D. These latter two possibilities, however, are quickly ruled out, since they imply that $DL = DA$ or that $DL < DA$. This contradicts $DA < DL$, which follows from the equalities $DA = DB$, $DG = DL$ and the inequality $DB < DG$. (The equalities follow from the properties of equilateral triangles and circles. The inequality $DB < DG$ is entailed by the fact that B lies between D and G, which holds because G was stipulated to lie on the extension of DB.) Thus, Euclid's construction in I,2 can always be trusted to produce a configuration

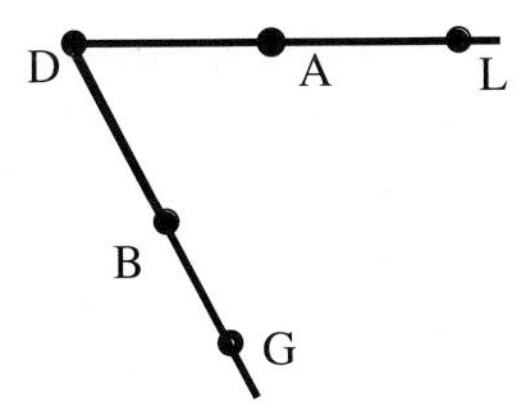

where $DA = DB$ and $DG = DL$. Accordingly, the equals-subtracted-from-equals rule is applicable, and we can infer that $AL = BG$.

Thus runs the proof of proposition 2 in **Eu**. Though the informal version given here is much more compact, each of its moves is matched in the formal version. Generally, proofs of propositions in **Eu** are two tiered, just as they are in the *Elements*. They open with a construction stage, and end with a demonstration stage. The rules which govern the construction stage are relatively lax. One is free to enrich the initial diagram Δ_1 by adding points, joining segments, extending segments and rays, and constructing a circle on a segment. Presented as a sequence of **Eu** diagrams, the construction stage for proposition 2 is as follows.

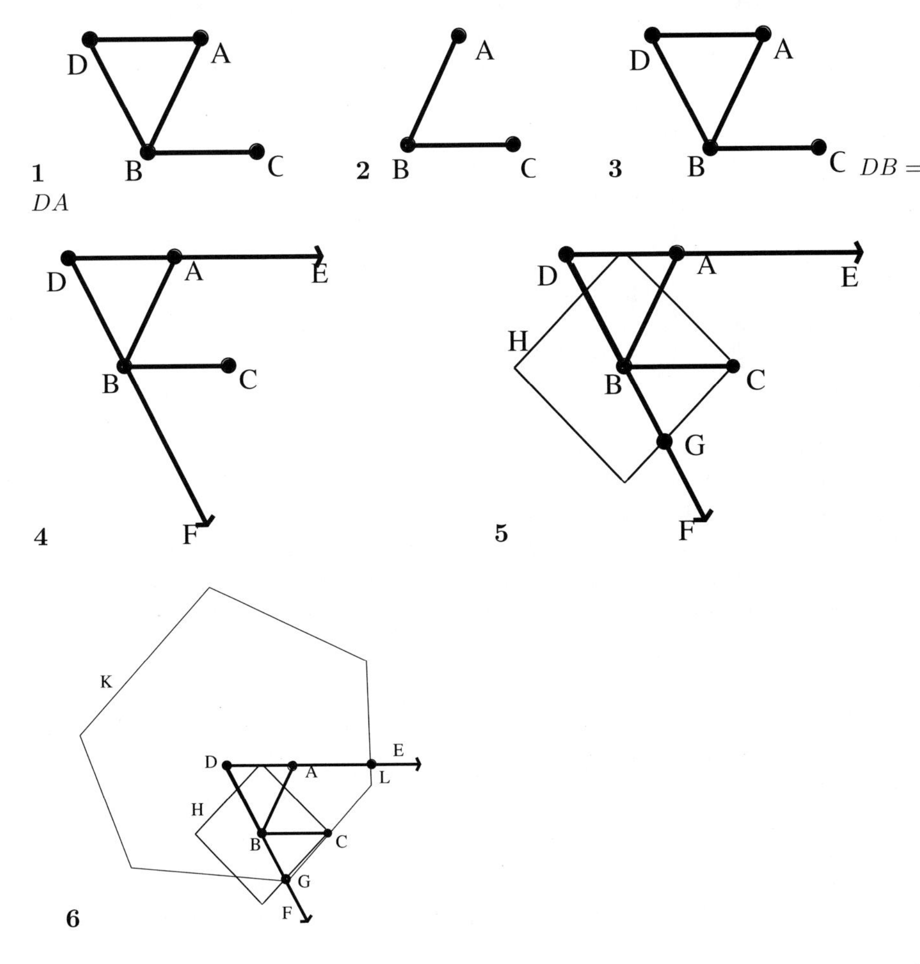

The last step in the construction yields a diagram Σ, which contains all the objects to be reasoned about in the demonstration. But it is not Σ alone, but the whole construction history of Σ, which determines what can be inferred in the demonstration.

The sequence of steps by which Σ was constructed determine a partial ordering $\triangleright$ of its geometric elements. An element x in the diagram *immediately precedes* y if the construction of y utilized x. For instance, if the points A and B are joined in a construction, the points A and B immediately precede the segment AB. Likewise, if a circle H is constructed with radius BC, the segment BC immediately precedes H. The complete partial ordering $\triangleright$ is simply the transitive closure of the *immediately precedes* relation. It serves to record the dependencies among the elements of Σ. For example, for the representation of I,2 in **Eu**, the partial ordering works out to be:

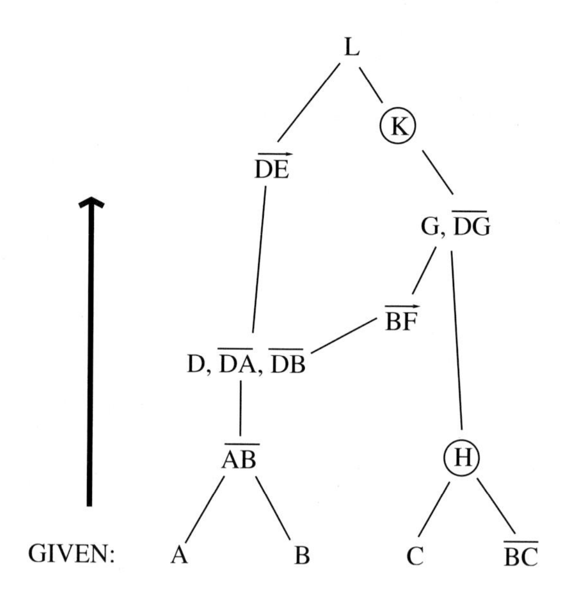

As the elements A, B, C and BC are part of the initial diagram Δ_1, they were not constructed from any elements in Σ, and so nothing precedes them. The rest appear somewhere above these, according to the way they were introduced.

The construction thus produces a tuple

$$\langle \Sigma, M, \triangleright \rangle$$

which is called the *context* of the proof. The term M is the metric assertion which records the exact relationships stipulated from the beginning or introduced during the course of the construction. These three pieces of data serve as input for the demonstration stage. Rules of this stage are of two types: positional and metric.

An application of a positional rule results in a sub-diagram of Σ. Deriving a sub-diagram amounts to confirming the generality of the co-exact relationships exhibited in it. As such, the application of the rules are constrained by $\triangleright$. One can introduce as a premise any sub-diagram of the initial diagram Δ_1—i.e. any sub-diagram consisting of elements which have nothing preceeding them by $\triangleright$. Any other sub-diagram must be derived from these by the positional rules, where the derivations proceed along the branches laid out by $\triangleright$. For instance, in the proof of proposition 2, one derives from the segment BC the sub-diagram

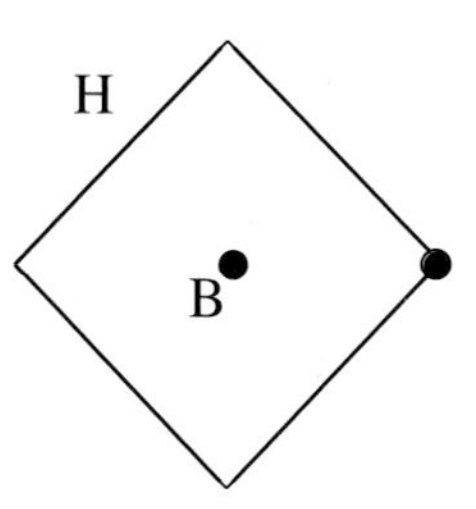

and from the points A and B the sub-diagram

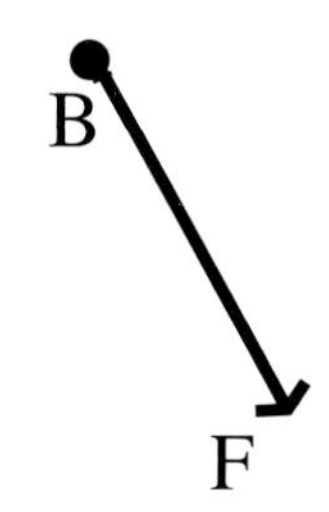

From these two sub-diagrams we can then derive

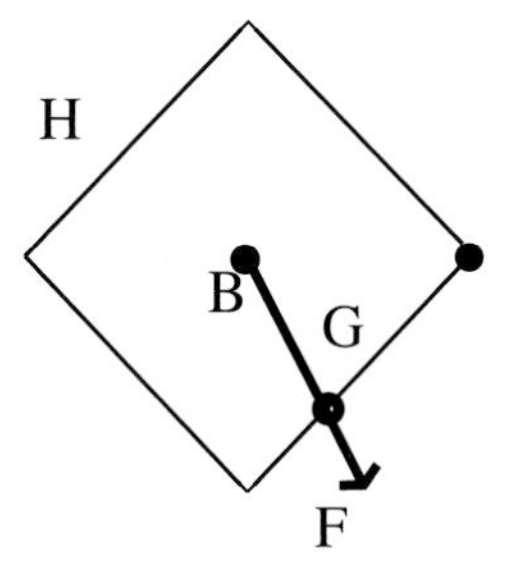

from a rule which encodes the general condition for the intersection of a ray and a circle.

The complete **Eu** formalization of proposition 2 can be viewed on pp. 95-109 of [4]. Both the construction and the demonstration in the formalization contain spatially separated diagrams. They are spatially separated, however, only to make clear the rule being applied at each step. The formalization is intended to model a process where one *single* diagram Σ is being constructed and inspected.

Any given diagram in the **Eu** construction represents a certain point in the construction of Σ. Any given diagram in the demonstration represents the part of Σ being considered at a certain point in the proof. As far as it possible, then, **Eu** formalizes Euclid's diagrammatic reasoning as *agglomerative* (in the sense of the term used by K. Stenning in [7]). This is in contrast to **FG**, which requires a disjunctive array of *different* diagrams every time the topology of a construction is not unique.

Understanding the logical properties of **Eu**, and implementing the system in a computer program, are goals of current and future work. It is straightforward to check the relative consistency of **Eu** to a modern axiomatization of elementary geometry. Claims formulated in **Eu** have a natural interpretation in terms of first order formulas composed of the primitives of such an axiomatization. The soundness of **Eu**'s proof rules is then easily checked in terms of this interpretation. The completeness of **Eu** with respect to a modern formalization is a more difficult question. To address it, a non-diagrammatic system of proof designed to model **Eu**, termed ***E***, has been developed (and will soon be presented in a forthcoming paper by Jeremy Avigad, Ed Dean and myself). The new system can be shown to be complete. That is, it can be shown that a modern axiomatization of elementary geometry is a conservative extension of ***E***. The question of whether **Eu** is complete then reduces to the question of how well it is modelled by ***E***. Work is now being done to implement ***E***. A goal of future research is to implement **Eu** directly—i.e. to develop a program where diagrams are codified just as they are formally defined in **Eu**.

3 Concluding Remarks

From a logical perspective, requiring disjunctive arrays of diagrams as **FG** does is a natural way to ensure generality in a diagrammatic formalization of Euclid. Yet if the goal is to understand what underlies the single diagram proofs of the *Elements*, the approach is far less natural. When a case-heavy **FG** formalization is laid beside Euclid's original version, the original does not appear deficient. Rather, the multitude of cases generated by the rules of **FG** appear excessive. The geometric differences recorded by a case-branching often do not seem material to the issue the proof decides.

This comes out if we compare **FG**'s formalization of proposition 2 with **Eu**'s. In [3] Miller does not explicitly discuss how **FG** handles the proof, but one can work out by hand the cases generated in the **FG** proof, at least for the first few steps of the construction. My efforts yielded 57 cases at the fourth step. Pushing further with the whole construction on three of these yielded 50 more. (At this point my will to continue gave out.) What distinguishes the cases are positional relations which are irrelevant to the inferences Euclid makes later, in that the relations need not be attended to for the soundness of the inferences to be confirmed. As the **Eu** formalization shows, we can focus in on certain relations in a single representative diagram (Σ in the **Eu** formalization) and ignore others in verifying the generality of the result. We do not have to check that the result holds in a long list of cases. That Euclid's proof allows us to do this does not seem to be an accidental feature. It seems, rather, to be a key mathematical insight of the proof. With proposition 2 and others throughout the *Elements*, Euclid seems deliberately to frame his arguments so that it suffices, or almost suffices, to consider a single diagram. The formal account of **Eu** respects this feature of the *Elements* while the formal account of **FG** does not. In **FG**, the one and only way to secure a general result with diagrams is by a brute enumeration of cases.

Eu avoids the need for such enumerations (most of the time) by allowing the content of a diagram to be context dependent. If the same geometric diagram is to communicate the same topological information every time it appears in a proof, then an enumeration of cases seems the only option in formalizing Euclid's diagrammatic constructions. With any faithful formal characterization of Euclid's diagrams there will be instances where what is representative in a diagram depends on how it is constructed. Specifically, there will be diagrams which when understood as a result of one of Euclid's constructions manifest non-representative features, and when understood as the result of another construction manifest nothing non-representative. Since such diagrams can in *some* instances be understood as representing all the topological relations it embodies, it seems ad hoc and overly restrictive to stipulate a partial content of the diagram in *all* instances. And so, if we must stipulate a content for all instances, the only systematic approach which seems available is that of **FG**: have a diagram communicate all topological relations manifest in it, and require case branching with every construction step. What **Eu** shows is that we need not do this. It is possible for the content of a diagram to vary across proofs. The information a diagram holds in a proof can be understood to depend systematically on particular features of the proof. Whether or not it is illuminating to understand diagram use in other mathematical contexts in this way seems a question worth pursuing.

References

1. Heath, T.: The Thirteen Books of Euclid's Elements—translated from the text of Heiberg. Dover Publications, New York (1956)
2. Manders, K.: The Euclidean Diagram. In: Mancosu, P. (ed.) Philosophy of Mathematical Practice, pp. 112–183. Clarendon Press, Oxford (2008)
3. Miller, N.: Euclid and His Twentieth Century Rivals: Diagrams in the Logic of Euclidean Geometry. Center for the Study of Language and Information, Stanford (2007)
4. Mumma, J.: Intuition Formalized: Ancient and Modern Methods of Proof in Elementary Euclidean Geometry. Ph.D Dissertation, Carnegie Mellon University (2006), `www.andrew.cmu.edu/jmumma`
5. Mumma, J.: Review of Euclid and His Twentieth Century Rivals: Diagrams in the Logic of Euclidean Geometry. Philosophica Mathematica 16(2), 256–264 (2008)
6. Shabel, L.: Kant's Philosophy of Mathematics. In: Guyer, P. (ed.) The Cambridge Companion of Kant, 2nd edn. Cambridge University Press, Cambridge (2006)
7. Stenning, K.: Distinctions with Differences: Comparing Criteria for Distinguishing Diagrammatic from Sentential Systems. In: Anderson, M., Cheng, P., Haarslev, V. (eds.) Diagrams 2000. LNCS (LNAI), vol. 1889, pp. 132–148. Springer, Heidelberg (2000)

Depicting Negation in Diagrammatic Logic: Legacy and Prospects

Fabien Schang[1] and Amirouche Moktefi[2]

[1] LHSP Henri Poincaré (UMR 7117), 23 Bd Albert 1er, 54100 Nancy France
schang.fabien@voila.fr
[2] IRIST, 7 rue de l'Université, 67000 Strasbourg France
amirouche.moktefi@gersulp.u-strasbg.fr

Abstract. Here are considered the conditions under which the method of diagrams is liable to include non-classical logics, among which the spatial representation of non-bivalent negation. This will be done with two intended purposes, namely: to review the basic concepts involved in the definition of logical negation; to account for a variety of epistemological obstacles against the introduction of non-classical negations within diagrammatic logic. It will be mainly argued that non-classical logics don't challenge dichotomy as such but merely show that a logical operator may be a negation without operating as a dichotomy.

Keywords: bivalence, complementation, dichotomy, internalization, logical negation.

1 From Dichotomy to Bivalence

In the primary spatial diagrams suggested in the logic of classes ([2], [7]), a class is represented by a closed curve. The individuals belonging to the class are *within* the curve while the individuals not-included in the class are *outside*. Thus, the proposition

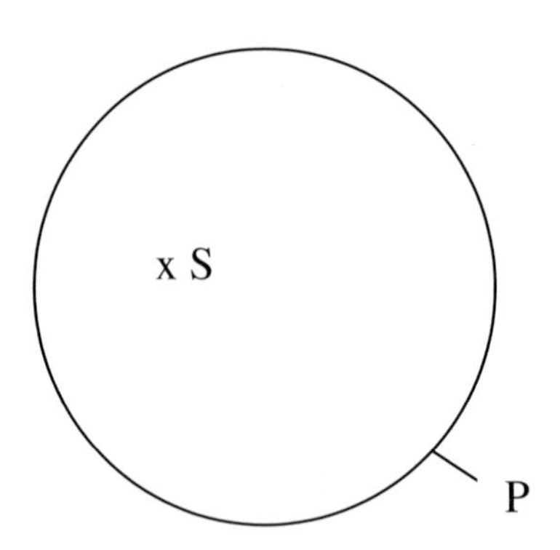

Fig. 1. The set of individuals that are not P (i.e. the non-P's) are not located in a specific space since only the closed curve is taken into account

G. Stapleton, J. Howse, and J. Lee (Eds.): Diagrams 2008, LNAI 5223, pp. 236 – 241, 2008.

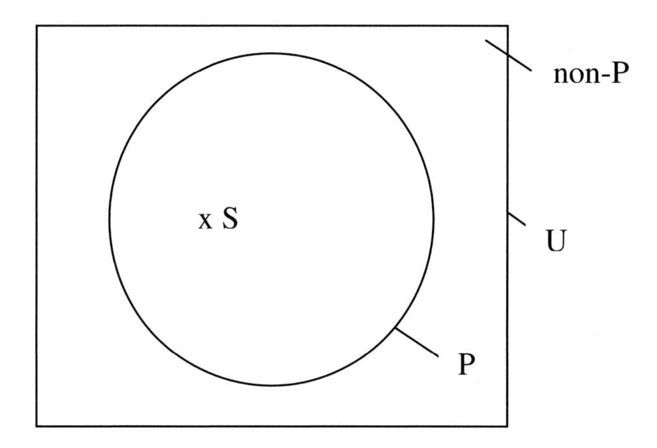

Fig. 2. All the non-P's correspond to the region located outside the closed curve and within the *universe of discourse* U

S is P may be represented by a circle symbolizing the class P in which an element S occurred; i.e. *S is P* is true if and only if the individual value occurs *in* the region purported to figure the class of individuals which satisfy P. A translation of classes into propositions is possible, provided that the class P is considered as a predicate term attached to a subject term S. As to the class of individuals which don't satisfy P, they must occur *outside* the circle of P's. A first obstacle to the representation of logical negation is the lack of any universe of discourse in these diagrams.

We call by *dichotomy* the opposition between the whole individuals located within (the class) P and the whole remaining ones located outside (the class) P. The opposition between P and non-P thus characterizes logical negation as a dichotomy; it is presented in diagrammatic logic as a spatial contrast *in* vs. *out*.

As to the semantic notion of *bivalence*, it extends dichotomy in splitting the universe into two classes of truth-values. If S is located in the region of P's, then the proposition *S is P* is true; if S is not located in the region of P's, then *S is P* is false. Now since S cannot be both in and outside P, the truth of the proposition *S is P* entails the falsehood of its negation *S is not P*, and the falsehood of the proposition *S is P* entails the truth of its negation *S is not P*. Consequently, every S belongs to the universe U: either it is in the region P, or it is in the region non-P. Another way to state these properties of bivalence is to say that a diagram is *complete* and *consistent*.

It seems difficult to challenge bivalence in logical diagrams, assuming that it implies for any element S in U both its occurrence and non-ubiquity. Nevertheless, the logic of diagrams gave way to the introduction of non-classical constants in its modern variants, including non-classical negation. How to embed the latter without questioning both preceding requirements of completeness and consistency?

Our answer is: by internalizing truth-values within the logic of diagrams, and by making a distinction between two readings of negation. Our final claim is that non-classical negation is less revolutionary than it might appear first, insofar as it doesn't really challenge the general property of dichotomy.

2 Non-bivalent Logics

By an *internalization*, we mean the process that consists in introducing a metalanguage notion within the object-language. While truth-values belong to metalanguage in the logic of diagrams, given that they don't occur in the logical space U, some non-classical logics make use of the following process in order to depict non-classical constants, namely: to see the general proposition *p* of the form *S is P* as an element in a class of truth-value (a "semantic class", say); e.g. *[S is P] is true*, with

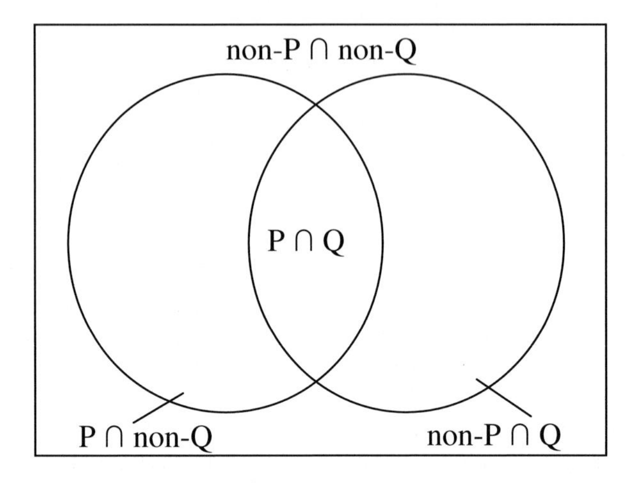

Fig. 3. A diagram is presented as a range of four possible relations between two classes; by analogy with the classes of propositions, the classes of truth-values equally propose four possible relations between the class T of true propositions and the class F of false propositions

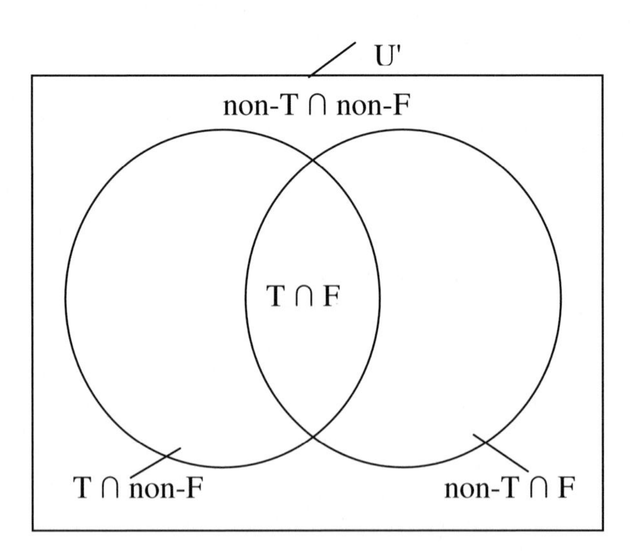

Fig. 4. In a non-bivalent space, a proposition can be neither true nor false or both true and false

both expressions *S is P* and *true* to be substituted for S and P, respectively. We obtain the subsequent illustrations of non-classical logic, where the Venn scheme is internalized in the semantic level of truth-values.

Any proposition can thus be considered as: true and non-false, false and non-true, true and false, non-true and non-false.

In this sense, the space U' of non-classical logics is not complete but *paracomplete*; it is not consistent, either, but *paraconsistent*. A sample of paracomplete three-valued logics are Kleene's intuitionistic logic in [3] or Łukasiewicz's logic of indetermination in [4], in which some propositions are neither true nor false; a case of paraconsistent logic is Priest's dialetheist logic in [6], in which a proposition can be paradoxical and, therefore, both true and false.

However, paraconsistency doesn't mean that a proposition may be both *in* and *outside* one and the same class: if it is both true and false, then it is located within the class TF that intersects T and F. Paracompleteness doesn't mean any more that a proposition may be absent from the universe of discourse, given that propositions that are neither true nor false *are* located in U'. Therefore, the introduction of non-classical diagrams does not entail the revision of dichotomy as such: if a proposition is not true, for instance, then it is located in the non-true (in –T) and, thus, outside the true (i.e. outside T); although it may be both true and false, such a case does not constitute any more an infringement on dichotomy than for an element S to be both in the class P and the class Q. In other words, the change caused by non-classical logics may generate some plausible confusion between the concepts of *bivalence* and *dichotomy*: the former states that any true proposition is non-false and any false proposition is non-true, hence an obvious connection between such a formulation and dichotomy. However, dichotomy merely requires the distinction between an arbitrary class and any other one that is not itself: P and non-P, true and non-true, false and non-false, and so on.

In order to bring out the distinction between mere *difference* and *incompatibility*, it can be said that while the class T of *strictly* true propositions and the class F of *strictly* false propositions are dichotomous in classical logic and even in most of non-classical logics, T's and F's are no more dichotomous in non-classical logics since TF is a new class in U' including both true and false propositions. Consequently, this means that even if non-classical negations violate bivalence, they still follow dichotomy.

3 What Is Logical Negation?

Borrowing a terminology from Terence Parsons in [5], the difference is made here between negation as a *choice operation* and negation as an *exclusion operation*. While exclusion negation generally consists in projecting any proposition from a class (whether semantic or not) into another class, choice negation turns a proposition into another proposition but doesn't necessarily project it from a semantic class to another one. Łukasiewicz's or Priest's logics are an illustration of it, since the (choice) negation of a proposition that is neither strictly true nor strictly false (say, indeterminate: –T–F = I) results in an unchanged truth-value when the negation is said to be *normal*. That is: let v be a valuation function from a set of propositions to a set of truth values; then $v(p) = v(\sim p) = I$, so that p and non-p do have the same truth-value. Such a non-classical negation does not exhibit any more the property of exclusion that characterizes dichotomy. Let us note that, while not every logical negation acts as an exclusion operation, some non-classical negations still do. These are said to be *non-normal*, e.g. Post's cyclic negation: for an ordered series of {1, ..., *i*, ... *n*} truth-values, $v(\sim P) = i+1$ whenever $v(P) = i$, so the truth-values of P and not-P are never the same. How to characterize logical negation intensionally, if not every negation is to be depicted as an exclusion operation?

As Brady put it in [1], we can consider an intensional property of logical negation that is more general than dichotomy and does justice to both classical and non-classical negations, that is: negation as a *mirror-image* concept with an axis of symmetry. If one depicts a range of three truth-values with T and F as opposite sides and I in the middle, then the negation of I results in I just as the mirror-image of any point located on an axis of symmetry results in this point itself.

Thus, not every logical negation is an exclusion property when applied to propositions with a non-classical value; nevertheless, the distinction between classical and non-classical remains as a dichotomy in itself between {T,F} and not-{T,F}.

The overall situation may be summarized as follows:

Firstly, Exclusion negation '–' is an *intensional* property: it projects a proposition –(*p*) outside the semantic class of *p*, but without determining the class in which –(*p*) should be located; and conversely, choice negation '~' is *extensional* since it determines a specific semantic class for ~(*p*) without projecting it necessarily outside the semantic class of *p*. Priest's paraconsistent negation, for example, locates the negation of a strictly true proposition in the strictly false region F, the negation of a strictly false proposition in the strictly true region T, and the negation of a paradoxical proposition (i.e. both true and false) in the paradoxical region TF.

Secondly, assimilation of logical negation to dichotomy is due to the behavior of choice negation within classical, bivalent logic: given that only two strictly separate semantic classes occur in it (i.e. T and F), classical negation behaves exactly like an exclusion negation and the distinction between choice '–' and exclusion '~' is thus made impossible extensionally speaking, contrary to the case with most of non-classical logics and especially the three-valued ones.

Thirdly, being *inconsistent* does not mean the same as being *contradictory* if, by contradictory, we mean the possibility to be both X and non-X whatever the syntactic category of X may be (whether a class of propositions or a class of truth-values). Paraconsistent logic does not include both true and non-true propositions or false and not-false propositions: T–T and F–F cannot occur in U'.

Such a non-classical (choice) negation doesn't infringe on dichotomy at all: it seems to do so only for whoever goes on to think of it in a classical universe of discourse U without including the more complex universe U'. Thinking about TF in U is an impossible thing to do, just as thinking about true contradictions even in U'.

4 Conclusion: Non-classical Negations Are Harmless for Dichotomy as Such

In sum, the introduction of non-classical logics into diagrams entails neither a significant revolution within the logical space, nor the obligation to modify our representation of such a space so as to allow the ubiquity of propositions both in and outside one and the same class. The process of internalization merely helps to make propositions *relatively* true and false and to go beyond the strictly separate relation between T and F within a bivalent framework; now if such a process if refused, then it appears as literally impossible to think of negation from a non-classical perspective, given the spatial relation in-out as a undebatable statement of dichotomy. In a nutshell: either logical negations are dichotomous or not, as choice operations; but they don't infringe on dichotomy in a logical space at all, because such an infringement would assume a substantial change in the very geometrical properties of the logical space. Nothing similar occurs in paraconsistent or paracomplete logics in U', anyway.

One task remains to be accomplished in non-classical diagrammatic logic once the internalization is accepted, namely: assuming that not every non-classical logic is an exclusion operation and, thus, an operation of *complementation*, how does one represent the non-complementary laws of projection between *p* and not-*p* for paraconsistent and paracomplete negations in U', and which of these projections should replace the dichotomous relation in-out that marks classical negation? Such a task will be explored in a forthcoming work.

References

1. Brady, R.T.: On the formalization of the Law of non-Contradiction. In: Priest, G., Beall, J.C., Armour-Garb, B. (eds.) The Law of Non-Contradiction (New Philosophical Essays), pp. 41–48. Clarendon Press, Oxford (2004)
2. Euler, L.: Lettres à une Princesse d'Allemagne, vol. 2. Academie Imperiale des Sciences, Saint Petersburg (1768)
3. Kleene, S.: Introduction to Metamathematics. Van Nostrand, New York (1952)
4. Łukasiewicz, J.: On three-valued logic. Ruch Filozoficzny 5, 170–171 (1920)
5. Parsons, T.: Assertion, Denial, and the Liar Paradox. Journal of Philosophical Logic 13, 137–152 (1984)
6. Priest, G.: The Logic of Paradox. Journal of Philosophical Logic 8, 219–241 (1979)
7. Venn, J.: On the diagrammatic and mechanical representation of propositions and reasonings. Philosophical Magazine 10, 1–18 (1880)

Transforming Descriptions and Diagrams to Sketches in Information System Design

Barbara Tversky[1], James E. Corter[1], Jeffrey V. Nickerson[2], Doris Zahner[2], and Yun Jin Rho[1]

[1] Teachers College, Columbia University
[2] Stevens Institute of Technology
bt2158@columbia.edu, corter@tc.edu, jnickerson@stevens.edu, dzahner@stevens.edu, yjr2101@columbia.edu

Abstract. Sketching is integral to information systems design. Designers need to become fluent in translating verbal descriptions of systems to a variety of kinds of sketches, notably sequential and logical, and to translate among the kinds. Here, we investigated these cognitive skills in design students, asking them to design a system configuration starting from either a sequential diagram or a sequential description. Although the two source descriptions were logically equivalent, the diagram led to designs that corresponded more closely to the source description – that is, designs with fewer omissions of crucial components and links. Text descriptions led to more variable and less accurate designs, most likely because they require more cognitive steps from problem representation to problem solution.

Keywords: Diagrammatic reasoning, sketches, descriptions, problem representation, information systems design, topological diagrams.

1 Introduction

Sketching is critical to design, whether for design of the concrete, products or buildings, or for design of the abstract, information systems or corporate structures. In designing, formative sketches show the components, walls, for example, or computers, and their relations, spatial, functional, or temporal. In emphasizing the components and their relations, sketches abstract out information not needed in early stages of design such as the surface of the walls or the specifics of the servers. One common task of designers, especially when working with clients, is translating a verbal description of the requirements for a building or an information management task into a sketch. On other occasions, designers need to translate one kind of sketch into another, for example, transforming a plan into an elevation in the case of architecture, changing perspective, or transforming a temporal sketch of information transmission into a structural sketch of the components in the case of information design, changing time to space. For design of information systems, the focus here, the design task is made more challenging by the fact that it is the topology of the system

G. Stapleton, J. Howse, and J. Lee (Eds.): Diagrams 2008, LNAI 5223, pp. 242–256, 2008.

that is critical, not the Euclidean properties. The topology reflects the functional organization of the system. This fact causes difficulty for novices, since sketches are inherently Euclidean, and beginning designers bring Euclidean assumptions and habits to the situation.

Designers must master many types of external representations. Often a client will give the requirements to the designer verbally, and the designer will present the initial design as a sketch. More conversation occurs, followed by more sketches. Designers must be facile translators to and from diagrammatic representations. For the design of information systems, there are two general types of requirements, sequential and structural. Sequential requirements are step-by-step constraints, for example, the steps from a customer's request to the shipment of an item or from placing an order with a supplier to the arrival of the goods. Structural requirements fulfill a large set of sequential requirements, insuring that the right information gets to the right sources efficiently, and that the wrong information does not get to the wrong sources.

Because information systems involve links that function near the speed of light, the Euclidean distance between components is usually not important. Instead, the number of connections between system components and the nature of those connections are critical. The components – computers, servers, and the like – and their connections or links represent the functional structure of the system. Systems are represented topologically; the components, called actors, as nodes and the connections or links as edges. The placement of the nodes is of little importance, but the patterns of edge connection are of high importance because they carry the functional structure. Because information systems can have many interconnections, diagrams of them have developed conventions to increase transparency by lowering clutter. One convention in particular, the logical bus, reduces the number of edges that need to be drawn, yielding a connectivity pattern that is often misinterpreted by novices. This convention is widely used to represent Local Area Networks (LANs). Within a LAN all system components are interconnected, however those interconnections are not explicitly shown, but rather indicated by a superordinate line connecting all of them, as shown in Fig. 5c.

These abstractions are not always well understood by the users of technology. Nor are they well understood by novice designers [1]. Experts are expected to be able to understand the deep functional structure of a system, that is, its underlying topology, from many different forms of surface representations, including sequential diagrams, structural diagrams, natural language descriptions, and source code. They are also expected to be adept at going from one surface representation to another.

What are the consequences of translating from one external representation to another? It is well known that the form of an external representation affects interpretation and inference from it; think of doing multiplication with Roman numerals (e. g., [2][3]). The present focus is a common task faced by designers, translating one form of external representation to another, notably text to sketch and diagram to sketch.

Here, students in a course in system design were asked to sketch the configuration of two systems. In one case, the source for the design of the system structure was a diagram of a sequence of operations the system should support. In the other case, the source for the design of the structure of a system was text, a verbal description of the same sequence of operations the system should support. Two questions are of interest: Which source leads to better design sketches, and are there qualitative differences between sketches produced from diagrams and sketches produced from text? Especially for novices, there should be advantages to producing a sketch from a diagram, that is, going from a sequential diagram to a structural design sketch. The source sequential diagram will have already abstracted all the critical components of the system. In addition, because the source sequential diagram shows one possible temporal route through the information system, it also shows at least part of the connectivity, the topological relations. A verbal description of the sequence requires the design student to abstract the components and the links. Novice designers are likely to miss some of them. Because verbal descriptions have fewer explicit components and spatial constraints, they are likely to lead to more variable design sketches.

The source, whether diagram or text, specifies one sequence of operations to be accomplished by the system, but not all of them. Discussions with clients often begin that way; the client outlines the major task the system is expected to accomplish. That sequence does not completely specify the configuration. In order to construct a complete system, the designer must bring in other considerations deriving from general knowledge of information systems. For example, experienced designers might realize that certain groups of components should be grouped as a LAN, with restricted access. There are two ways that student designs could deviate from the minimal constraints of the source: they could omit nodes or edges, or they could add them. Omissions are always errors. Because the source problem does not completely stipulate the design, additions could be errors or they could be creative or wise design considerations.

The source representation is also expected to affect the actual spatial layout of the design sketch, which is not specified by the source. Because systems design depends on connectivity, the locations of components in the Euclidean world are not relevant; consequently, the spatial locations of components in the sketches need not reflect their locations in the world. One default for determining sketch location in the absence of Euclidean constraints is reading order (e. g., [1][4]). Reading order predicts that the layout of components in sketches where the source is text should correspond to the order of mention in the text, and that the layout of components in sketches where the source is a sequence diagram should correspond to the left-right order of the sequence diagram.

A challenge for research on sketches is to develop a coding system, because there is so much variability in spontaneous productions. Coding sketches is especially challenging for topological sketches, where it is desirable to extract the underlying logical structure from sketches that may differ in only superficial ways. Once this logical structure is abstracted, it becomes possible to establish the equivalence of diagrams, and to separately measure the surface and the logical dissimilarity of diagrams. To these ends, we developed a technique for establishing logical equivalences as well as similarities between different systems sketches and diagrams. This involved the construction of a *canonical graph*, the nodes and links that correspond to the minimal solution that incorporates the source constraints. The student solutions can be compared to the canonical graph to assess omissions and additions.

Since topological diagrams are used not only in information systems, but also in such diverse fields as electrical systems, transportation systems, systems biology, and geography (c.f. [6][7][8]), the coding system and insights into the production and understanding of information systems sketches will have broader implications.

2 Methods

The predictions were tested in a Master's level course in the design of information systems. The thirty-six student participants had varying levels of expertise: some were relative newcomers to information systems, and others were working professionals with many years of system design experience. During the course, students solve a series of increasingly complex design problems and bring them in for critique every week [9].

Two problems were given to students in the present experiment. Each student sketched the configuration design for two information systems, using a sequential diagram as the source for one problem and using a text description as the source for the other problem. The problems were chosen to involve the same number of nodes (actors in UML terminology [5]), but to be in different domains. One problem, called "FastStuff," asked students to design a system that delivers purchased products to the customer's door. The other problem, called "HedgeFund," asked students to design a new Internet presence for hedge fund investors (see Figures 1 and 2). A source diagram and a source text were developed for each of the problems. These different

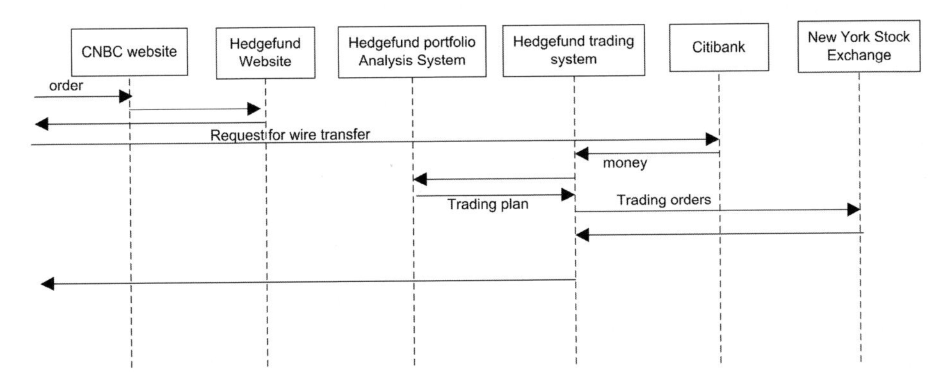

Fig. 1. HedgeFund Diagram, given to half the participants, and HedgeFund text, given to the other half. The diagram is in the Unified Modeling Language (UML) sequence diagram format [5].

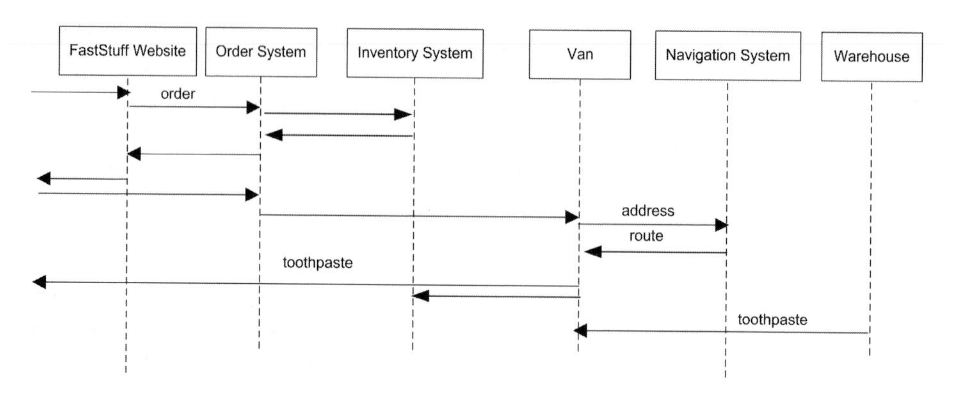

Fig. 2. FastStuff Diagram, given to half the participants, and FastStuff text, given to the other half

problem representations are only approximately equivalent. Complete equivalence would have resulted in either stilted text or unconventional diagrams, because the two types of representation have differing conventions. For example, sequence diagrams have an implied user on the left of the diagram that is never labeled, while text descriptions of design problems usually explicitly mention the user. Textual representations will often provide semantic information on the nature of an interaction (describing the type of message transmitted), but may not say explicitly where the interaction originates. However, this implicit information can usually be derived from the context. In contrast, sequence diagrams show all interactions explicitly, but do not give semantically meaningfully labels to the type of message transmitted. Rather than try to exactly equate the types of information provided by the two problem formats, we chose to use naturalistic problem descriptions that resemble those that students would encounter in a design course or in actual work settings.

Half the students were given the text source for FastStuff and the diagram source for HedgeFund; the other half were given the diagram source for FastStuff and the text source for HedgeFund. Because of the small sample size, all students first used a diagram source and then a text source. If there is any consequential bias on the results from solving a problem presented as a diagram first and a problem presented as text second, the additional experience should favor text as source.

3 Results

3.1 Omissions and Additions

Both the sequence diagram and the text versions of each problem explicitly define a set of actors (nodes in the diagram) and a set of interactions (edges). Thus, adequate performance in each problem involves constructing a labeled graph to represent these actors and interactions. The graph that represents all and only these explicitly defined actors and relationships will be termed the *canonical graph*. The distance between two graphs can be defined as the edits (additions or deletions) that will transform one graph into another [10]. In this paper, cases of leaving out any explicitly mentioned actors or communication links are termed *omissions*, and regarded as errors.

However, these design problems also offer opportunities to be creative, to go beyond the problem information by envisioning additional actors and relationships, additional system capabilities that might be beneficial, and by optimizing system performance in other ways. Thus, additional nodes and edges not explicitly stipulated in the problem statement are not necessarily errors. Rather they may reflect more or less relevant elaborations to the conception of the problem or the solution. However, irrelevant additions may be considered to be a form of error, because they have costs (time, expense, security concerns) and do not bring benefits.

In addition to omissions and additions, the spatial layout of the produced sketch is of interest: does the left-right organization of its components correspond to the order of presentation in the text or the left-right organization of the sequence diagram?

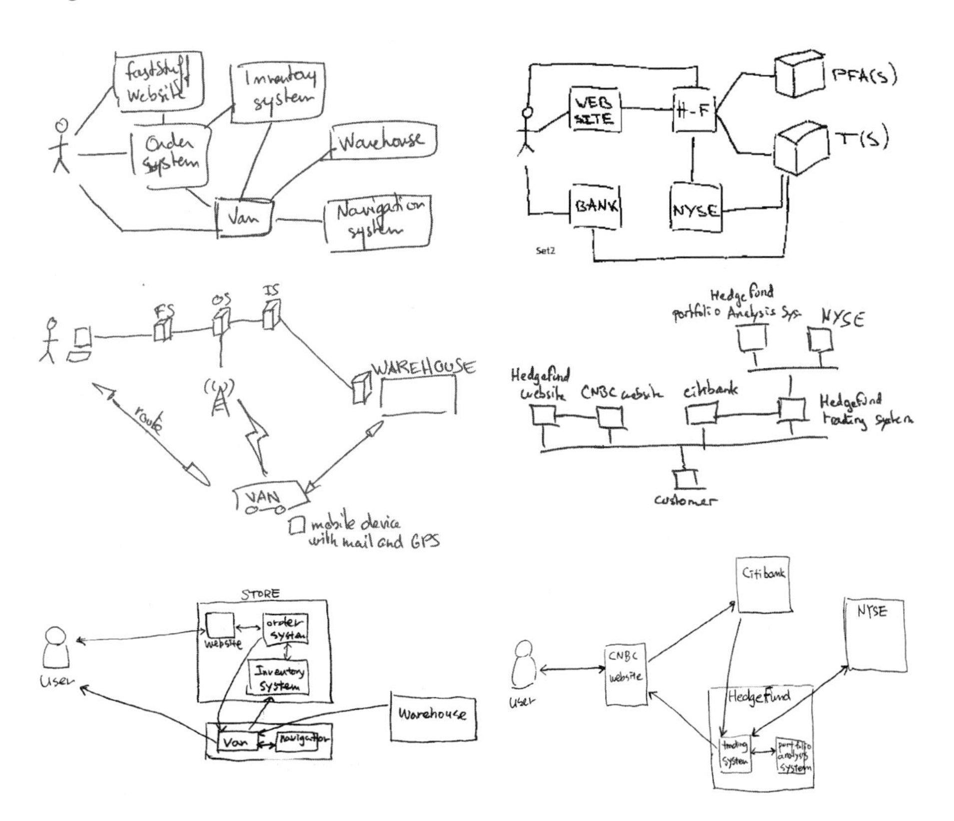

Fig. 3. Example student sketches for the FastStuff (left) and HedgeFund (right) problems

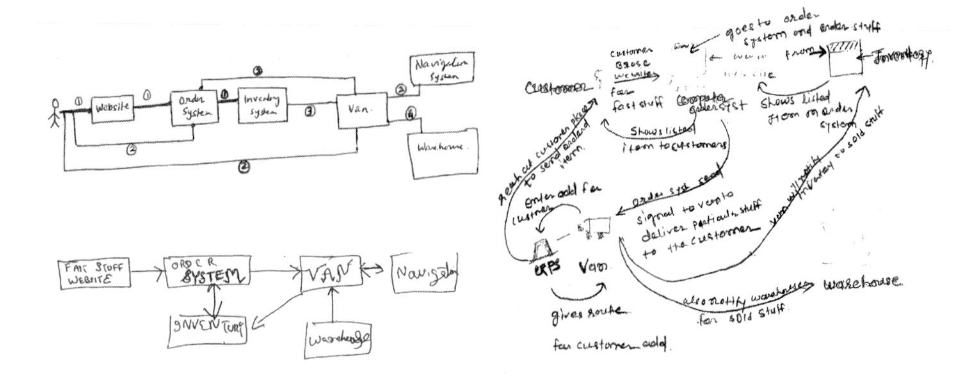

Fig. 4. Sketches generated for the FastStuff problem that are highly similar with respect to Euclidean node positions and other surface characteristics (the two on the left), with a third sketch (right) that is very dissimilar. In contrast, with respect to logical structure the sketch on the right is almost identical to the sketch at top left.

To evaluate produced designs, we first analyzed the topological information provided to students in the form of a sequence diagram or text for both problems. For both problems, the text and diagram versions provide identical information about actors (nodes) and their pattern of connectivity. We call the logical graph that can be drawn for each of the problems, shown in Figures 1 and 2, the *canonical graph* of the problem, because this standardized representation of the problem is analogous to canonical data models used in systems integration (e. g. [11]).

We then developed a scheme for coding the student-produced sketches. Some examples of student-produced sketches appear in Fig. 3. It is obvious that the diagrams vary in their surface details, but without a formal analysis, it is less obvious if they differ in their logical structure. In order to test the predictions, we need a method to compare the student-produced sketches at the logical level. This requires coding the sketches in terms of their graph topology.

As an illustration of the coding issues, we show three more sketches in Fig. 4. In terms of surface structure, the two sketches on the left are very similar to each other. In fact, they are the two closest sketches in the data set as measured by the Euclidean distances between corresponding nodes in the sketches. For example, the position of the node *Van* on one sketch is compared with the position of the node *Van* on the other sketch. The sketches on the left are very dissimilar to the sketch on the right. But by analyzing the logical connections between nodes (the *edges* in the terminology of graph theory), we find that the two sketches on the left are quite dissimilar to each other with respect to logical structure, and the sketch on the top left is very similar to the sketch on the right.

However, analyzing the logical structure of the sketched diagrams involves more than simply coding the student-produced diagrams for their graph topology. Specifically, the use of diagrammatic conventions, such as sketching a logical bus to represent a LAN, means that logical connectivity in the system and graph topology do not have a one-to-one correspondence. Thus, we recoded various diagrammatic

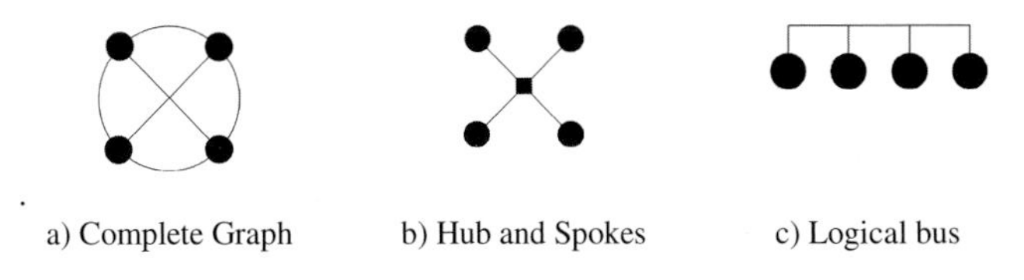

a) Complete Graph b) Hub and Spokes c) Logical bus

Fig. 5. Alternative network representations topologically equivalent at the logical level

devices used to represent networks, such as logical buses, the Internet, and satellite links, in terms of the network connectivity implied by the device and its constituents. This was done in the following way. First, all examples of use of LANs and other types of networks were identified. Two people coded these network types, and discussed any discrepancies until consensus was achieved regarding the presence and type of network connections in each student solution. The next step in the process was automatic – the graph representation of the problem was transformed by software we wrote into a standardized form. All devices directly connected to LANs were assumed to be directly connected to each other, and the resulting connections were represented in the *logical form* of the graph for each student solution.

For example, Fig. 5c shows the standard convention for representing a LAN. The convention means that all nodes can communicate to each other, and so is logically equivalent to the canonical form shown in Fig. 5a. To code this case, we formed a new graph, G', in which the edges implied by the networks were added, and the depicted node (if any) representing the network was deleted. We call G' the logical form of the graph G, or the *logical graph* for short. In this way, we can compare the logical connections in the diagrams produced for a particular problem, regardless of the student's choice of convention.

Several students created sketches whose logical graph exactly matched the canonical graph. Most students, however, either omitted edges or added new edges (sometimes as a result of adding new nodes). We analyzed the differences between the students' logical graphs and the canonical graph.

For the FastStuff problem, five graphs were produced that matched exactly the canonical graph. One of these solutions is depicted in Fig. 3 (top left), the other four

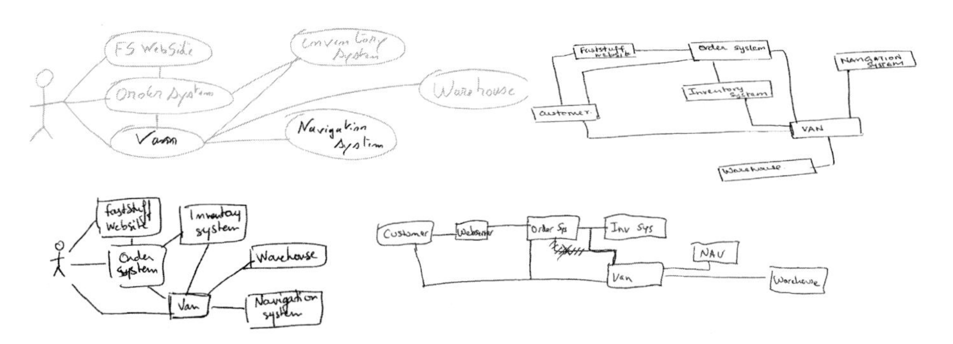

Fig. 6. The four additional graphs that (along with the top left graph shown in Fig. 3) are topologically identical to the canonical graph for the FastStuff question

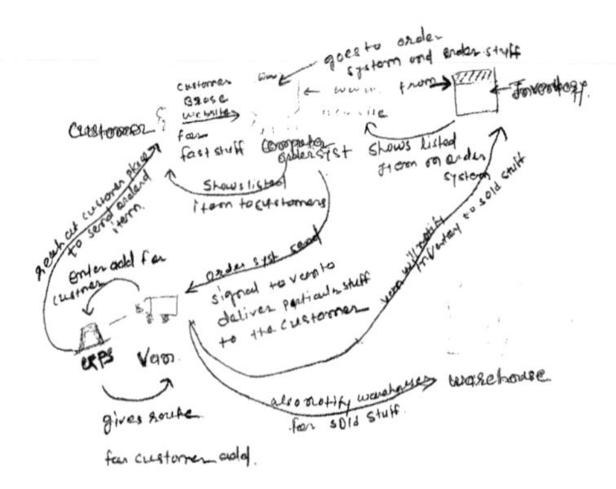

Fig. 7. A graph that is logically almost identical to those in Fig. 6 – an edge has been added between the GPS device and the customer

are shown in Fig. 6. Remarkably, these five matching graphs were all drawn in the Diagram condition. Notice that very different styles are used to represent the same underlying structure. The graphs exhibit different degrees of linearity, with the bottom right graph mapping closely to the positions of the nodes shown in the problem diagram, and the top right graph exhibiting a weaker correlation with the positions of nodes in the problem diagram.

Fig. 7 shows a diagram that is almost identical to those in Fig. 6. An edge has been added from the GPS system to the customer. This addition is a creative idea: the customer could then know exactly where the truck was at all times. The student did not consider a better alternative, that the information could be transmitted through the truck's network back to the website so that customers could track the truck's location without overloading the GPS system.

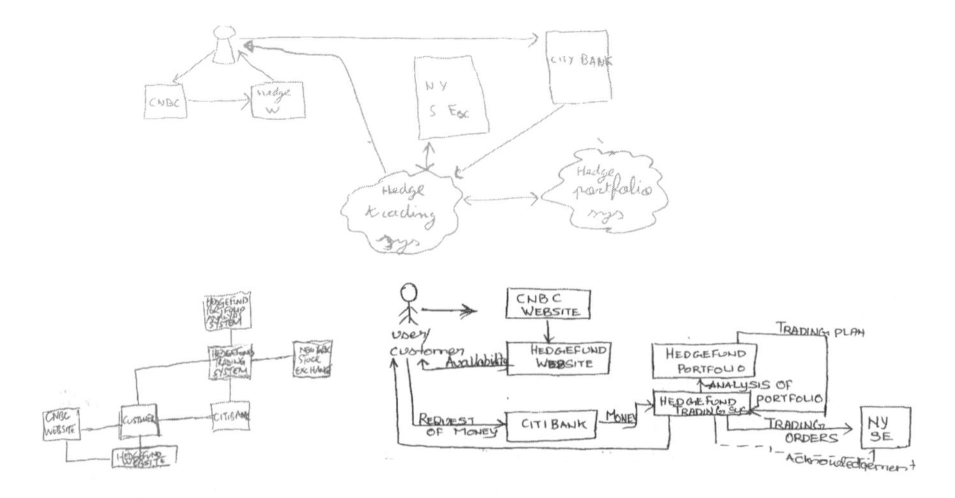

Fig. 8. Sketches that match the canonical graph for the HedgeFund problem

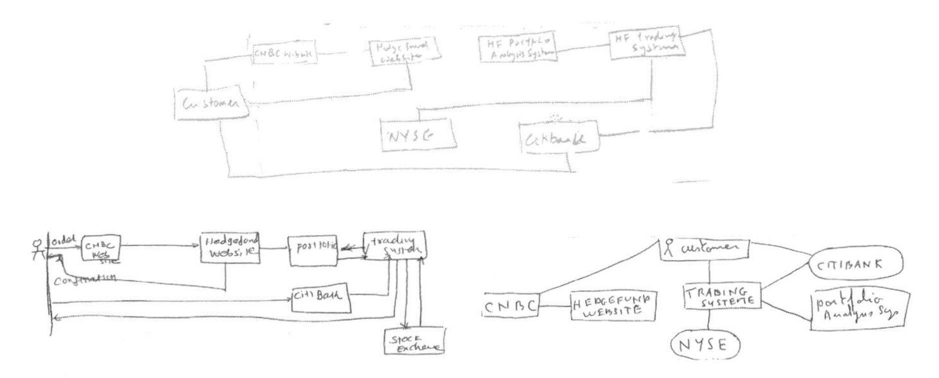

Fig. 9. Graphs that are one edit distant from the HedgeFund canonical graph. The graphs are not identical to each other. There is an extra edge on the diagram on the left, and missing edges in the two other diagrams.

For the HedgeFund problem, students produced three sketches whose logical graphs matched the canonical graph; these are shown in Fig. 8. Again, these all were drawn in the Diagram condition. Thus, for this problem too, no student with text as source created a graph that matched the canonical graph. There were three figures that differed by one edge from the canonical graph; these are shown in Fig. 9.

The sketch coding scheme just described allows comparisons of sketches produced from diagram and text sources for omissions and additions. The number of omitted edges (relative to the canonical graph) was analyzed in a replicated 2 x 2 Latin Square

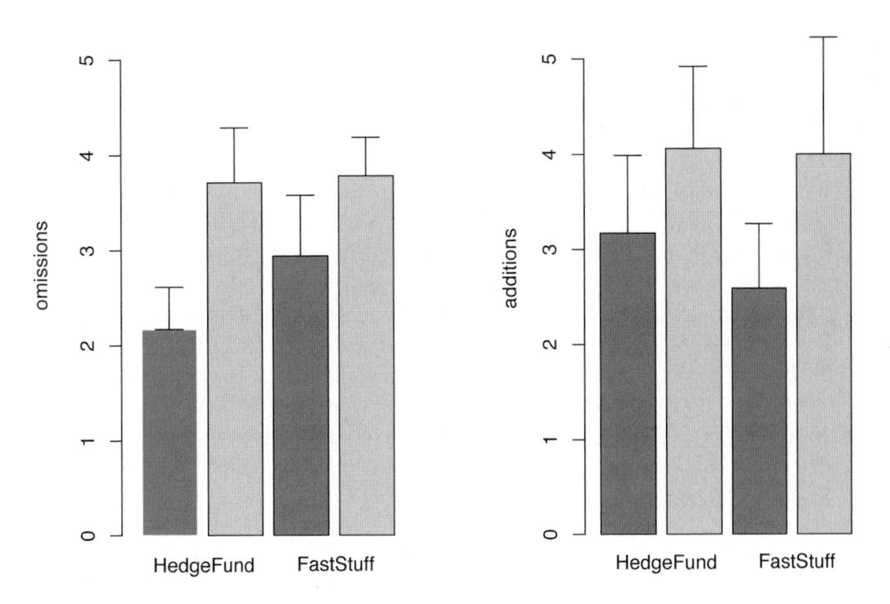

Fig. 10. Number of omissions and additions by Modality and Problem. The dark bars represent sketches drawn from diagrams, the light bars sketches drawn from text. The error bars show standard error.

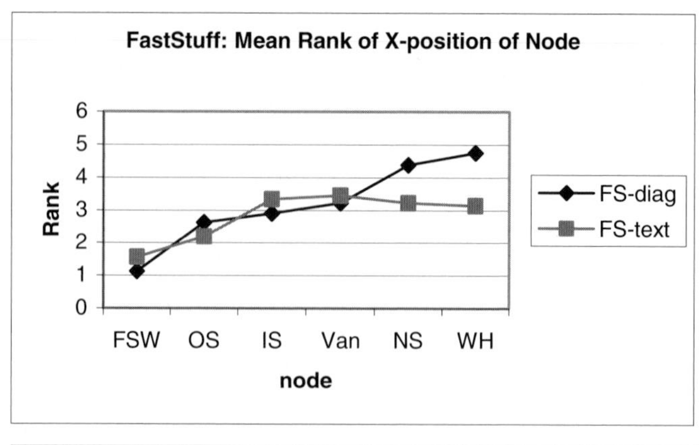

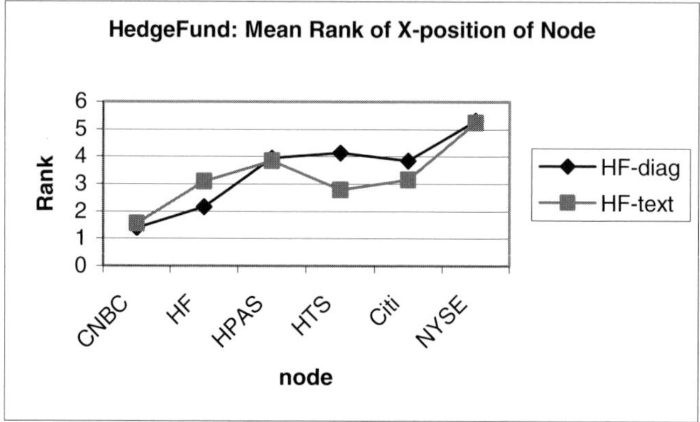

Fig. 11. Rank X-axis positions of the objects of Figures 1 and 2 for the two problem scenarios

design, with Condition as the between-subjects factor. The within-subjects factors were Problem (FastStuff or HedgeFund) and problem Format (diagram or text). Means and standard errors for the four conditions are shown in Fig. 10.

There were fewer sketch errors with a diagram source than a text source. For omissions, this was significant by a repeated-measures ANOVA, $F(1,33) = 14.37$, $p=.001$. The effects of Problem and Condition were not significant, $F(1,33) = 1.82$, $p=.186$ and $F(1,33) = 0.27$, $p=.605$. For additions, the differences due to Source were not significant, $F(1,33) = 2.64$, $p=.114$, nor were the effects of Problem and Condition, $F(1,33) = 2.64$, $p=.656$ and $F(1,33) = 0.05$, $p=.816$.

3.2 Reading Order Bias

Because there were no constraints on the positions of objects in the sketches, the left-right organization of components was expected to correspond to the reading order of the source text or the left-right organization of the source diagram. To test for reading order bias, we first coded the horizontal locations of all nodes in the source diagrams shown in Figures 1 and 2, and coded the locations in the problem texts of the same objects, treating the text as a continuous string. Next, we coded the vertical and horizontal locations of each node in the sketches produced by students. Then we compared the positions of objects in each sketch to the reading order of the source, whether diagram or text (Fig. 11).

As is evident from Fig. 11, there was a reading order bias. The X-axis of the figure shows the rank-order left-right position of each node in the problem description and the Y-axis shows the rank-order horizontal position of each node in the student's sketch for those students who depicted all six actors explicitly mentioned in the problem text. For the FastStuff problem, when the source was a diagram the order of nodes in the students' sketches was exactly the same as the order in the source diagram. By a permutation test [12], this ordering can be shown to deviate significantly from random in the direction of the reading order, p=.001. When the source was text, the correspondence between the order of nodes in the sketch and the order of mention in the source text was nearly as strong, with only one pairwise inversion of node order compared to the reading order. This too differs from random ordering, p=.008.

The pattern was similar for the HedgeFund problem (Fig. 11), showing closer correspondence of source ordering to sketch ordering when the source was a diagram compared to when it was text. For the diagram presentation, the mean rank order of nodes in the sketch differed from the diagram order by only a single inversion, which again represents a significant degree of deviation from random ordering, p=.008. For the text condition, the mean rank order differed from the order of nodes in the text by three pair inversions. This correspondence with the reading order was only marginally greater than chance, p=.068. In both conditions, the deviation of node order in the design sketches differed from the source ordering mainly in the location of "Citibank." The source diagram located it to the right of HedgeFund, HTS (the HedgeFund trading system) and HPAS (HedgeFund's portfolio analysis system), but students' sketches locate it close to HedgeFund and to the left of HPAS. It could be argued that this is a better placement, because the HTS and perhaps the HPAS node need to communicate frequently with the New York Stock Exchange, the rightmost node, while the customer (at the extreme left in the sequence diagram) needs to communicate directly with Citibank.

4 Conclusions and Implications

Students in a class in information systems were presented two comparable design tasks and asked to produce sketches of their solutions. One problem was presented in the form of a diagram of a sequence of operations the system should perform. Another problem was presented as a description of the identical sequence. Thus, students were presented with the same information, but in different modes. We sought to investigate if the mode of the problem source, diagram or text, would affect students' designs. The results showed that the mode of the source for the design problem, diagram or sketch, in fact affected students' design sketches in two major ways. First, a diagram source led to designs that were more accurate in the sense of more closely matching the specified problem structure. That is, students omitted fewer explicitly mentioned actors and connections when designing from a diagram, probably because the source diagram preprocesses that information for the student. There is some indication that students also made fewer additions when the source was a diagram than when the source was text. Additions, however, are not necessarily erroneous; they may be creative and wise elaborations. Next, both the left-right spatial arrangement of actors in the source diagram and the order of mention of actors in the source text affected the left-right order of actors in students' design sketches. However, that correspondence was stronger when the source was a diagram than when the source was text. Together the findings indicate that students' design representations are more variable and more flexible when created from text than when created from diagrams. There are more mental steps to translate text to a design sketch than to translate a diagram to a design sketch. Each mental step provides an opportunity for error but also an opportunity for creativity. Thus, diagrammatic representations of design problems constrain design solutions more than verbal representations of design problems.

Students' design sketches are wonderfully variable, posing problems for data analysis. What's more, what is critical in system design is not the surface connections of the diagrams but the underlying functional connections, which are logical and topological, not Euclidean. In order to analyze and compare students' sketches, we developed a coding system that captured the underlying topology That coding system allowed us to count and characterize omissions and additions. This coding system should prove useful for many other cases of diagrams whose structure is topological, including electrical systems, transportation systems, systems biology, and geography.

Diagrammatic representations of design problems provide a scaffold for designers by selecting the relevant actors and specifying some of the relevant links. This scaffold serves as training wheels for beginning designers, making sure that they stay upright. However, the scaffold, like training wheels, also places limits and constraints. Verbal descriptions of design problems are freer of constraints, and are likely to lead to more flexible and creative performance in the long run. In any case, just as a bicyclist must eventually abandon the training wheels, designers must eventually become expert at translating diagrammatic and verbal representations of problems to good designs. And, in fact, it seems likely that the problem representations with fewer constraints will eventually lead to more flexible and creative designs. Ambiguity allows invention (e. g., [13]), but successful invention requires expertise, skills and knowledge.

Sketches are integral to design, of products, of buildings, and even of abstract information systems. They are used to translate clients' desires into initial designs, they are used by designers to articulate and revise design ideas, and they are used to present the design ideas as they progress for discussions with colleagues and clients. In information design, clients often specify a series of temporal steps; from that, the designer must configure a set of components. This is one of the challenges novice designers face in order to become expert, a challenge examined here. For systems designers, another challenge is appropriate use and interpretation of sketches and diagrams that are manifest in space but that do not always support Euclidean assumptions. The spatial configuration of components and connections, though highly salient in a sketch, does not carry useful information about functional relations. Rather, the information on functional relations is carried by the network connections among the components. In typical information systems, those connections are dense; expressing each and every one would quickly clutter a sketch, making it difficult to follow. To cope, systems designers have developed conventions that summarize a set of connections, notably, a logical bus to represent a LAN. A LAN is drawn as a single line from which a set of components hang like clothes on a clothesline, but its meaning is that all the hanging components are interconnected. Such sketches may bear similarities to familiar route maps, but their interpretation is quite different. In a LAN, to get from the left-most component to the right-most component, information does not have to go through the middle components; it goes directly. Using this and related conventions also causes problems for design students, because the conventions cannot be approached with the Euclidean assumptions people usually bring to bear in interpreting sketches. The difficulty this causes for design students was revealed in a previous study [1]. The difficulty was echoed in the current study: for example, one student had problems sketching a LAN connected to the Internet when working from text, even though the student drew a correct network when working from a diagram. Thus, working from diagrams can provide support for beginning designers, but also can mask conceptual difficulties from instructors.

Designers of information systems must learn to resist Euclidean interpretations of sketches and become fluent in the conventions. They also must become adept at going back and forth between ideas and sketches, and between two kinds of sketches, temporal and logical. Design, of products, buildings, and systems, is a cognitive activity, and successful design entails honing cognitive skills. Some of these skills require learning to benefit from different external representations, to translate among them, and to use each to its advantage.

Acknowledgments. We are grateful for the support of the National Science Foundation under grant IIS-0725223 as well as to NSF REC-0440103 and the Stanford Regional Visualization and Analysis Center.

References

1. Nickerson, J.V., Corter, J.E., Tversky, B., Zahner, D., Rho, Y.: Diagrams as Tools in the Design of Information Systems. In: Third International Conference on Design Computing and Cognition (DCC 2008). LNCS, Springer, Berlin (2008)
2. Zhang, J., Norman, D.A.: A Representational Analysis of Numeration Systems. Cognition 57, 271–295 (1995)
3. Hayes, J.R., Simon, H.A.: Psychological Differences Among Problem Isomorphs. In: Castellan, N.J., Pisoni, D.B., Potts, G.R. (eds.) Cognitive Theory, vol. 2, pp. 21–41. Erlbaum, Hillsdale (1977)
4. Taylor, H.A., Tversky, B.: Descriptions and Depictions of Environments. Memory and Cognition 20, 483–496 (1992)
5. Fowler, M.: UML Distilled: A Brief Guide to the Standard Object Modeling Language. Addison-Wesley, Reading (2004)
6. Egenhofer, M., Franzosa, R.: Point-Set Topological Spatial Relations. International Journal of Geographical Information Systems 5, 161–174 (1991)
7. Egenhofer, M., Mark, D.: Naive Geography. In: Frank, A., Kuhn, W. (eds.) COSIT 1995. LNCS, vol. 988, pp. 1–15. Springer, Heidelberg (1995)
8. Stevens, A., Coupe, P.: Distortions in Judged Spatial Relations. Cognitive Psychology 10, 422–437 (1978)

9. Nickerson, J.V.: Teaching the Integration of Information Systems Technologies. IEEE Transactions on Education 49, 1–7 (2006)
10. Sanfeliu, A., Fu, K.S.: A Distance Measure Between Attributed Relational Graphs for Pattern Recognition. IEEE Trans. Systems, Man, and Cybernetics 13, 353–362 (1983)
11. Bergamaschi, S., Castano, S., Vincini, M.: Semantic Integration of Semistructured and Structured Data Sources. SIGMOD Rec. 28, 54–59 (1999)
12. Rosander, A.C.: The Use of Inversions as a Test of Random Order. Journal of the American Statistical Association 37, 352–358 (1942)
13. Suwa, M., Tversky, B.: What do Architects and Students Perceive in their Design Sketches? A Protocol Analysis. Design Studies 18, 385–403 (1997)

Graphical Revelations: Comparing Students' Translation Errors in Graphics and Logic

Richard Cox[1], Robert Dale[2], John Etchemendy[3], and Dave Barker-Plummer[3]

[1] Department of Informatics, University of Sussex, Falmer, BN1 9QJ, UK
richc@sussex.ac.uk

[2] Center for Language Technology, Macquarie University, Sydney, NSW 2019, Australia
rdale@ics.mq.edu.au

[3] CSLI, Stanford University, Stanford, CA, 94305, USA
{dbp,etch}@csli.stanford.edu

Abstract. We are interested in developing a better understanding of what it is that students find difficult in learning logic. We use both natural language and diagram-based methods for teaching students the formal language of first-order logic. In this paper, we present some initial results that demonstrate that, when we look at how students construct diagrammatic representations of information expressed in natural language (NL) sentences, the error patterns are different from those observed when students translate from NL to first-order logic (FOL). In the NL-to-diagram construction task, errors associated with the interpretation of the expression *not a small dodecahedron* were manifested much more frequently with respect to the object's size than with respect to its shape. In the NL-to-FOL task, however, no such asymmetry was observed. We hypothesize a number of possible factors that might be implicated here: differences between the NL-to-diagram and NL-to-FOL tasks; the reduced expressivity of diagrams compared to language; scoping errors in participants' NL parsing; and the visuospatial properties of the blocks-world domain. In sum, constructing a diagram requires the student to provide an instantiated representation of the meaning of a natural language sentence; this tests their understanding in a way that translation into first-order logic does not, by ensuring that they are not simply carrying out a symbol manipulation exercise.

Keywords: errors, natural language, graphical representations, first-order logic.

1 Introduction

In this paper we present findings concerning the errors that students make when learning first order logic. In our teaching we use both natural language and diagrammatic representations to aid the students in learning this material. Here we consider exercises in which students had to construct a blocks world (a particular kind of diagrammatic representation) in which twelve English sentences are true, in addition to translating each of the sentences into first-order logic. We find that the consideration of the diagrammatic modality and the ability to compare two channels of information flow through deduction (natural language into diagrams and natural language into first-order logic)

G. Stapleton, J. Howse, and J. Lee (Eds.): Diagrams 2008, LNAI 5223, pp. 257–265, 2008.

identifies problems in students' understanding that are not obvious simply from their first-order logic translations.

Our data are derived from student-generated solutions to exercises in *Language, Proof and Logic* (LPL) [4], a courseware package consisting of a textbook together with desktop applications which students use to complete exercises presented in the text.[1] Students may submit answers to 489 of LPL's 748 exercises[2] to The Grade Grinder (GG), a robust automated assessment system that has assessed approximately 1.8 million submissions of work by more than 38,000 individual students over the past eight years. These submissions form an extremely large corpus of high ecological validity which we wish to exploit in order (*inter alia*) to gain insights into cognitive processes during formal reasoning and to extend our research on individual differences in reasoning (e.g. [8]), to improve logic teaching.

Some exercises require the student to translate an English sentence, such as **d** *is a small dodecahedron unless* **a** *is small*, into first-order logic (FOL). Tarski's World (TW)[3] allows the student to enter candidate solutions, and then to manipulate diagrams of worlds containing blocks on a checkerboard to see whether the situations in which the English statement is true also make the proposed translation true. Blocks in TW can have one of three shapes: tetrahedron, cube and dodecahedron, represented by the predicates Tet, Cube and Dodec; and one of three sizes: small, medium and large, represented by the predicates Small, Medium and Large.[4]

If the truth of the student's translation matches the truth of the English statement in the worlds under consideration, then the student has evidence that the translation is a good candidate for the correct answer. However, the student's answer may yet be incorrect, perhaps because they have not considered a relevant situation. The student can only obtain a definitive answer by submitting the proposed solution to GG.

In recent work we analysed students' errors in translating natural language (NL) sentences into first-order logic [2]. In that work, we demonstrated that students had particular difficulties with distinguishing the conditional from the biconditional, were sensitive to source-sentence word-order effects during translation, and were sensitive to factors associated with the naming of constants. In [2], we noted that students had particular problems in translating sentences of the form *P unless Q* into FOL on a sentence-by-sentence basis. In the present paper, we once again find this same form to be a significant source of errors, this time in a 'deeper' reasoning context—one which requires students to engage in a chain of inference steps in order to build a diagrammatic representation.

2 The Focus of This Study

For the exploration described in this paper, we chose to focus on LPL Exercise 7.15, which, like the exercise discussed in [2], addresses conditionals, and involves the translation of sentences from NL to FOL. However, it also requires the student to make

[1] See http://lpl.stanford.edu.
[2] The other exercises require that students submit their answers on paper to their instructors.
[3] TW is one of LPL's three desktop applications.
[4] Blocks also have a position on the checkerboard, leading to predicates such as LeftOf, but these are unused in the work described in this paper.

"...translate the following English sentences (into FOL)...(build) a world in which the 12 English sentences are true. Verify that your translations are true in this world as well. Submit both your sentence file and your world file."

1. *If* **a** *is a tetrahedron then* **b** *is also a tetrahedron.*
2. **c** *is a tetrahedron if* **b** *is.*
3. **a** *and* **c** *are both tetrahedra only if at least one of them is large.*
4. **a** *is a tetrahedron but* **c** *isn't large.*
5. *If* **c** *is small and* **d** *is a dodecahedron, then* **d** *is neither large nor small.*
6. **c** *is medium only if none of* **d**, **e**, *and* **f** *are cubes.*
7. **d** *is a small dodecahedron unless* **a** *is small.*
8. **e** *is large just in case it is a fact that* **d** *is large if and only if* **f** *is.*
9. **d** *and* **e** *are the same size.*
10. **d** *and* **e** *are the same shape.*
11. **f** *is either a cube or a dodec, if it is large.*
12. **c** *is larger than* **e** *only if* **b** *is larger than* **c**.

Fig. 1. The natural language sentences in Exercise 7.15

#	N	% incorrect	Correct FOL translation
1	183	1.7	$\mathsf{Tet(a)} \rightarrow \mathsf{Tet(b)}$
2	755	7.3	$\mathsf{Tet(b)} \rightarrow \mathsf{Tet(c)}$
3	2739	26.5	$(\mathsf{Tet(a)} \land \mathsf{Tet(c)}) \rightarrow (\mathsf{Large(a)} \lor \mathsf{Large(c)})$
4	865	8.4	$(\mathsf{Tet(a)} \land \neg\mathsf{Large(c)}$
5	2093	20.2	$(\mathsf{Small(c)} \land \mathsf{Dodec(d)}) \rightarrow (\neg\mathsf{Large(d)} \land \neg\mathsf{Small(d)})$
6	3762	36.4	$\mathsf{Medium(c)} \rightarrow (\neg\mathsf{Cube(d)} \land \neg\mathsf{Cube(e)} \land \neg\mathsf{Cube(f)})$
7	3258	31.6	$\neg\mathsf{Small(a)} \rightarrow (\mathsf{Small(d)} \land \mathsf{Dodec(d)})$
8	4055	39.2	$\mathsf{Large(e)} \leftrightarrow (\mathsf{Large(d)} \leftrightarrow \mathsf{Large(f)})$
9	224	2.2	$\mathsf{SameSize(d,e)}$
10	236	2.3	$\mathsf{SameShape(d,e)}$
11	1175	11.4	$\mathsf{Large(f)} \rightarrow (\mathsf{Cube(f)} \lor \mathsf{Dodec(f)})$
12	2477	24.0	$\mathsf{Larger(c,e)} \rightarrow \mathsf{Larger(b,c)}$

Fig. 2. FOL translations of the sentences in Figure 1

inferences from the sentences (all of which concern the sizes and shapes of objects **a** through **f**) and then to build a blocks world in which these sentences are true. In order to complete the exercise, students are required to submit both their FOL sentences and the constructed world.

A translation for a sentence (which we refer to here as a **solution**) is considered correct if it is equivalent to a **reference solution** known to GG.[5] Figure 1 shows the sentences in this exercise; example correct FOL translations are presented in Figure 2. A submission for the NL-to-FOL translation task is considered correct if all twelve of

[5] There are infinitely many correct answers for any sentence, so GG employs a theorem prover to determine equivalence.

Fig. 3. Most frequent above-median (upper) & below-median (lower) diagrams

the student's FOL sentences in the submission are equivalent to their corresponding reference sentences.

The twelve sentences uniquely determine the sizes and shapes of the blocks with names **a** through **f**. The submitted world is correct if the blocks have these sizes and shapes, or, equivalently, if all of the reference solutions evaluate to true in the submitted world. The blocks world shown in the upper part of Figure 3 is an example of a (correct) world in which all of the NL sentences in Figure 1 are true.

The corpus of submissions for Exercise 7.15 made by students during the calendar years 2000–2007 contains more than 29,500 submissions, of which 18,609 submissions (61%) were erroneous. Of the erroneous submissions, 7,918 were missing the world file, and 372 were lacking the sentence file. We discarded these from the analysis, leaving 10,319 submissions. These submissions were made by 5,176 different students.

3 Information Flow through Deduction

Unlike the NL-to-FOL translation task, in which the sentences may be translated independently of one another, building the blocks world requires the use of the sentences in concert. We can trace the information flow required to complete the task by looking at the sentences and determining the inferences that can be drawn. This is a heterogeneous deduction of the kind implemented in our Hyperproof program [3].

There is only one place to start in the deduction, namely with Sentence 4, the only sentence that contains unconditional information: **a** *is a tetrahedron but* **c** *isn't large*. It is a conjunction of two facts, one of which tells us that **a** is a tetrahedron. This fact

N	FOL	Error type
350	$(\mathsf{Small(d)} \wedge \mathsf{Dodec(d)}) \rightarrow \neg\mathsf{Small(a)}$	ACREV
235	$\neg\mathsf{Small(a)} \rightarrow \mathsf{Dodec(d)}$	Missing conjunct
219	$\mathsf{Small(a)} \rightarrow (\mathsf{Small(d)} \wedge \mathsf{Dodec(d)})$	Missing negation
195	$\mathsf{Small(a)} \rightarrow \neg(\mathsf{Small(d)} \wedge \mathsf{Dodec(d)})$	Moved negation
189	$(\mathsf{Dodec(d)} \wedge \mathsf{Small(d)}) \rightarrow \neg\mathsf{Small(a)}$	ACREV
137	$\mathsf{Small(a)} \rightarrow \neg(\mathsf{Dodec(d)} \wedge \mathsf{Small(d)})$	Moved negation
117	$(\mathsf{Small(d)} \wedge \mathsf{Dodec(d)}) \rightarrow \mathsf{Small(a)}$	ACREV; missing negation
104	$\mathsf{Small(a)} \rightarrow (\mathsf{Dodec(d)} \wedge \mathsf{Small(d)})$	Missing negation
80	$\mathsf{Small(a)} \rightarrow (\neg\mathsf{Small(d)} \wedge \neg\mathsf{Dodec(d)})$	Moved negation
79	$(\mathsf{Small(d)} \wedge \mathsf{Dodec(d)}) \leftrightarrow \neg\mathsf{Small(a)}$	ACREV; Biconditional

Fig. 4. 10 highest-frequency FOL errors on S7. ACREV = antecedent–consequent reversal.

can be combined with the conditional information in Sentence 1 (*If* **a** *is a tetrahedron then* **b** *is also a tetrahedron*) to infer the shape of **b**, and once the shape of **b** is known, it can be combined with the conditional information in Sentence 2 (**c** *is a tetrahedron if* **b** *is*) to infer that **c** is also a tetrahedron. A more complex inference is now required, one which uses Sentence 3: **a** *and* **c** *are both tetrahedra only if at least one of them is large.* We know that **a** and **c** are indeed tetrahedra, and from Sentence 4 (again) we know that **c** is not large, and so we conclude that **a** is large. These four results—that **a**, **b** and **c** are tetrahedra, and that **a** is large—are obtained in 8,650 (84%) of all of the (erroneous) submitted worlds. In other words, up to this point, most students do not have any problems in carrying out the task.

The next inference, however, is the focus of this paper. In a correct deduction, Sentence 7 (**d** *is a small dodecahedron unless* **a** *is small*) should now be used to infer that **d** is a small dodecahedron (since the only occasion when it might not be is when **a** is small, and we know that it is not). Students find this inference relatively difficult. 2,552 (29%) of the submitted worlds show an error concerning the size and/or shape of **d**.

4 FOL Translations

Figure 2 shows the NL-to-FOL translation error rate for each of the 12 sentences. Six of the twelve sentences account for 85% of all errors (S3, S5, S6, S7, S8, S12). The length of the natural language sentences (number of words per sentence) is significantly positively correlated with percentage error rate ($r = .71, p = .01, n = 12$). Sentences containing conditional or biconditional connectives (*i.e.* Sentences 3, 5-8, 11 and 12) are also associated with high rates of translation error.

Sentence 7—**d** *is a small dodecahedron unless* **a** *is small*—is the third most error-prone FOL translation (Figure 2). It is a relatively simple sentence, and the only one in this exercise that includes the term *unless*. The LPL textbook suggests that *P unless Q* is best translated into FOL as $\neg\mathsf{Q} \rightarrow \mathsf{P}$; by this rule, the corresponding translation into FOL is $\neg\mathsf{Small(a)} \rightarrow (\mathsf{Small(d)} \wedge \mathsf{Dodec(d)})$, although any (propositionally) equivalent formula is accepted by GG.

Sentence 7 results in 372 different forms of FOL translation error, for a total of 3,258 errors. Figure 4 shows the ten most frequently occurring forms (accounting for 52% of the errors) and that, of the 10 highest-frequency errors, four involve antecedent–consequent reversals.

The fact that Sentence 7 presents a problem to students is, perhaps, not surprising. We found in our earlier work that the *unless* form is problematic for students. One contributing feature is that the translation of Sentence 7 (**d** *is a small dodecahedron unless* **a** *is small*) into $\neg\mathsf{Small(a)} \rightarrow (\mathsf{Small(d)} \wedge \mathsf{Dodec(d)})$ does not preserve word order. When word order is not preserved, antecedent–consequent errors are more likely [2]. Six error forms involved misplacement or omission of the negation symbol. Errors involving negation ranked as the fourth most common (accounting for 9.2% of all errors) in a different translation exercise involving conditional sentences [2].

An important feature of almost all of the mistranslations (and the correct translation) is that the sentences are **symmetrical** with respect to size and shape for **d** (the exception being $\neg\mathsf{Small(a)} \rightarrow \mathsf{Dodec(d)}$), which is to say that the two atoms appear in the same configuration: typically as two conjuncts; sometimes with the conjunction negated, and sometimes with each conjunct negated. Only in the FOL error type $\neg\mathsf{Small(a)} \rightarrow \mathsf{Dodec(d)}$ do we see any evidence that the two atoms are being treated differently from one another; but note that this error type represents only a small percentage (4.3%) of the large variety of other forms of FOL error, almost all of which refer to both the shape and size of **d**.[6]

5 Translation Errors and TW Diagrams

The observations above lead us to conjecture that the students (correctly) understand the noun phrase *a small dodecahedron* as concerning two properties of an object both applying the the same way to the same object. However, it turns out that the errors students make in the graphical domain tell a different story: the graphical products of students' reasoning reveal more errors related to block size compared to block shape.

By analyzing the flow of information through deductions across the Exercise 7.15 sentences, the source of the size/shape asymmetry can be tracked down to Sentence 7. Of the submitted worlds that contain an error resulting from a mistaken inference using this sentence, 2,346 (92%) of the errors concern the size of **d**, while only 623 (24%) concern the shape of **d**; 417 (16%) make an error on both properties. Thus the evidence based on the graphical data indicates that the information in Sentence 7 concerning the size of **d** is handled differently than that concerning the shape of **d**.

The properties of the objects in the correct blocks world are shown in the top row of Figure 5. The ten most popular incorrect blocks worlds, accounting for 3,155 (42%) of the 7,489 erroneous worlds, are also presented in Figure 5 (those with count values), with incorrect values shown in bold.

Students' blocks-world diagrams vary in terms of their **diagram accuracy**, which we assess by scoring each diagram according to the number of the twelve correct properties (size and shape for each of six blocks) it possesses (scores can therefore range

[6] Note that the sentence $\mathsf{Small(d)} \rightarrow \neg\mathsf{Small(a)}$ *does* appear in the incorrect translation set with a smaller frequency (19 occurrences).

Count	a		b		c		d		e		f	
	Tet	Large	Tet	Large	Tet	Medium	Dodec	Small	Dodec	Small	Dodec	Large
607	Tet	Large	Tet	Large	Tet	Medium	Dodec	Small	Dodec	Small	**Tet**	**Small**
493	Tet	Large	Tet	Large	Tet	Medium	Dodec	Small	Dodec	Small	Dodec	**Small**
423	Tet	Large	Tet	Large	Tet	Medium	Dodec	Small	Dodec	Small	**Tet**	**Medium**
365	Tet	Large	Tet	Large	Tet	Medium	Dodec	Small	Dodec	Small	**Cube**	Large
351	Tet	Large	Tet	**Small**	Tet	Medium	Dodec	**Large**	Dodec	**Large**	Dodec	Large
256	Tet	Large	Tet	**Small**	Tet	**Small**	Dodec	**Medium**	Dodec	**Medium**	**Cube**	Large
175	Tet	Large	Tet	Large	Tet	Medium	Dodec	Small	Dodec	Small	Dodec	**Medium**
168	Tet	Large	Tet	**Small**	Tet	**Small**	Dodec	Small	Dodec	Small	**Cube**	Large
162	Tet	Large	Tet	Large	Tet	Medium	Dodec	**Large**	Dodec	**Large**	Dodec	Large
155	Tet	Large	Tet	Large	Tet	Medium	Dodec	Small	Dodec	Small	**Tet**	Large

Fig. 5. Ten most popular incorrect blocks worlds

from 0–12). Students whose blocks-world diagrams scored below the 50th percentile tended to make many more errors (an average of .29 errors per student) in translating Sentence 7 than students whose diagrams scored above the median (.18 errors per student). For example, Sentence 7 evaluates as true in the upper diagram in Figure 3 (as do all 12 sentences), but in the lower (below-median-score) diagram it is the only sentence of the twelve to evaluate as false.[7]

The size of **d** must be inferred from Sentence 7. Students infer that **d** is not a small dodecahedron, but they are much more likely to infer that it is not small than infer that it is not a dodecahedron. In other words their inference is asymmetrical, and impacts the size dimension much more than the shape dimension.

6 Discussion

We have explored the translation of NL sentences into two different modalities, first-order logic and graphical. As the preceding discussion demonstrates, this provides a much clearer picture of the nature of scoping errors during NL interpretation than the study of, say, translations from one linguistic modality (natural language) into another one (FOL).

Errors in the translation into graphics turn out to be consistent with a wrongly-scoped reading of Sentence 7 (**d** *is a small dodecahedron unless* **a** *is small*) which leads these students to conclude that **d** is not a small dodecahedron; further, this is incorrectly scoped as akin to **d** *is (not a small) dodecahedron* rather than **d** *is (not a small dodecahedron)*. However, there is no evidence of this misunderstanding in the results of the NL-to-FOL task.

The absence of shape/size-asymmetry in the NL-to-FOL case is possibly due to the cognitively 'shallower' sentence-by-sentence nature of the NL-to-FOL translation task (referred to as **immediate inference** by Newstead [6]). In other words, the scoping error is associated only with sentence comprehension in the context of *inference across sentences*, plus the need to recall from working memory parameters of the blocks (*e.g.* the

[7] The lower diagram in Figure 3 corresponds to the row of Figure 5 with count = 351.

shape of **a**) established from earlier sentences (in particular, Sentence 4) to supplement new information (*i.e.* the size of **a** given in Sentence 7). It appears that the expression *small dodecahedron* is understood as a whole in the FOL case, but increased working memory load causes it to become fractured when building a blocks-world diagram, resulting in *small* being treated differently from *dodecahedron*).

The student's focus on only the size of **d** rather than on the size *and* shape of **d** may be due to size being the only aspect of **a** on the right hand side of the sentence that is explicitly referred to. The shape of **a** is unstated in Sentence 7, and so has to either be extracted by the student from her blocks-world-in-progress, or retrieved from visuospatial working memory. It should also be noted that, in terms of the flow of information through deduction, the shape of **a** is the very first block parameter to be established (from Sentence 4) and is therefore the piece of information that has been longest in memory by the time Sentence 7 becomes the focus of the student's reasoning.

Several factors may interact to produce the effects we report. The first concerns differences between the tasks. The NL-to-FOL task differs from the NL-to-diagram task in that the former requires the translation of one sentence at a time, whereas the latter requires (a) the integration of information from several sentences used in concert and (b) the active construction of a blocks world diagram. The difference between the tasks, together with the observation that graphics are more limited than linguistic representations in terms of their ability to express abstraction [9], may go some way towards explaining the finding. The NL-to-FOL translation process occurs on a sentence-by-sentence basis involving the translation or mapping of an English sentence's subject, object, nouns, verbs, adjectives, adverbs, and so on into antecedent, consequent clauses, atoms, constants, connectives, negation symbols, and so on. To some extent this translation task can be seen as one which only requires the application of NL-to-FOL transformation rules; however, as we have shown in [2], NL features such as word order, types of connective and the labelling of constants can have systematic negative effects upon translation accuracy. Constructing a blocks world, on the other hand, requires information to be deduced from a collection of sentences, a task that imposes a considerably greater load on working memory's phonological, visuospatial and central executive (*e.g.* attention management) components [1]. The graphical task requires the production of a blocks world in which twelve NL sentences, describing the sizes and shapes of objects **a** through **f**, are true. Analysing the flow of information through the deduction process shows that whereas the shape of block **a** is explicitly given (in Sentence 4), other blocks' size and shape parameters entail quite lengthy chains of reasoning. For example, the size of **d** involves inference across five sentences (1, 2, 3, 4 and 7). Determining the shape of **f** seems to be the most taxing: it requires inference across eleven sentences (1–7 and 9–12).[8]

The findings we report here are from a blocks-world construction task that involves *visual* or *visuospatial* block parameters such as *size* (for example, Large(a)) and *shape* (for example, Tet(a)) and which do not involve *spatial* parameters (board positions). Size is a dimension that possesses a natural commensurability and ordinality (smaller–larger) whereas shape doesn't have this property (it is nominal). In the context of the exercise reported here, however, the natural commensurability of size is irrelevant to the

[8] Note that the block **f** is the one on which the most shape errors are observed (Figure 5).

logical reasoning. These semantic factors contribute to the tendency of visual images to be either 'not critical' or 'interfering' in deductive reasoning (whereas spatial representations help), as argued by [5]. This effect may also contribute to the differences in error patterns observed for size *vs* shape on the graphical task. Mental imagery effects such as these are not predicted by rule-based theories of deductive reasoning (e.g. [7]).

At this stage of our research, we can only speculate about which of several competing explanations account for these effects. We aim to more clearly delineate their relative effects in future work. However, an important pedagogical implication is already clear from these preliminary results: the ambiguity of NL and scoping during NL interpretation are topics that should be given more attention in the logic curriculum, and the problems evident here may not be so clear if we only consider NL-to-FOL translation exercises. The insights into students' reasoning that these analyses provide are also useful for improving the Grade Grinder and for enriching the type of feedback that it can provide to students.

Acknowledgements

RC gratefully acknowledges support provided by an ESRC TLRP (UK) & SSRC (US) Visiting Americas Fellowship; RD gratefully acknowledges the support of the Australian Research Council. Albert Liu assisted with the collection of the data. Thanks too to three anonymous reviewers for their constructive suggestions.

References

1. Baddeley, A.: Working memory, thought, and action. Oxford University Press, Oxford (2007)
2. Barker-Plummer, D., Cox, R., Dale, R., Etchemendy, J.: An empirical study of errors in translating natural language into logic. In: Sloutsky, V., Love, B., McRae, K. (eds.) Proceedings of the 30th Annual Cognitive Science Society Conference. Lawrence Erlbaum Associates, Mahwah (2008)
3. Barwise, J., Etchemendy, J.: Hyperproof. CSLI Publications, University of Chicago Press (September 1994)
4. Barwise, J., Etchemendy, J., Allwein, G., Barker-Plummer, D., Liu, A.: Language, Proof and Logic. CSLI Publications, University of Chicago Press (September 1999)
5. Knauff, M., Johnson-Laird, P.N.: Visual imagery can impede reasoning. Memory and Cognition 30, 363–371 (2002)
6. Newstead, S.: Interpretational errors in syllogistic reasoning. Journal of Memory and Language 28, 78–91 (1989)
7. Rips, L.J.: The Psychology of Proof. MIT Press, Cambridge (1994)
8. Stenning, K., Cox, R.: Reconnecting interpretation to reasoning through individual differences. The Quarterly Journal of Experimental Psychology 59, 1454–1483 (2006)
9. Stenning, K., Oberlander, J.: A cognitive theory of graphical and linguistic reasoning: Logic and implementation. Cognitive Science 19, 97–140 (1995)

Learning from Animated Diagrams: How Are Mental Models Built?

Richard Lowe and Jean-Michel Boucheix

[1]Curtin University, Australia
r.k.lowe@curtin.edu.au
[2]University of Burgundy, France
J-M Boucheix@u-Bourgogne.fr

Abstract. Current approaches to the design of educational animations too often appear to be largely founded upon intuition rather than research-based principles. Animated diagrams designed to be behaviourally realistic run the risk of learners overlooking vital high relevance information that has low intrinsic perceptual salience. The information that learners extract from such representations is a poor basis upon which to build high quality dynamic mental models. For animated diagrams to be effective as tools for learning, their design should be based upon explicit and principled modeling of the way learners process such depictions. This paper synthesizes recent research to propose a theoretical framework for learners' perceptual and conceptual processing of animated diagrams. A five-stage model is presented that characterizes the role of different levels of processing in building dynamic mental models of the depicted content.

Keywords: Animated diagrams, theoretical framework, perceptual and cognitive processing, mental model construction, complex dynamic content.

1 Introduction

In this paper we propose a preliminary theoretical model for learner processing of complex animated diagrams. We offer this model because theoretical developments with respect to learning from animation have tended to lag behind the rapid rise of empirical activity in this field. Animations are increasingly included in multimedia learning resources. Although the broader area of research on learning with multimedia is well served with established processing models [e.g. 1], there is no comparable coherent account of how animations *per se* are processed by learners. The origins of current multimedia processing models can be traced back to learning resources that consisted of text accompanied static pictures [e.g. 2,3,4]. These models were subsequently recruited for the purpose of explaining how learners process multimedia resources incorporating animations. However, relatively little allowance has been made for the key difference between static and animated graphics – the presence of temporal change. There is evidence that this dynamic aspect of animations can have a most compelling effect

G. Stapleton, J. Howse, and J. Lee (Eds.): Diagrams 2008, LNAI 5223, pp. 266–281, 2008.

on the way the displayed graphic material is processed because of its powerful perceptual influence [5]. In particular, *dynamic contrast* effects [6] can be an important driver of learner attention and hence play a significant role in whether information of high task relevance is extracted from an animation or not.

Learners are likely to be more susceptible to negative effects from animation's dynamic properties if they are novices with respect to the subject matter depicted and if the display is complex. These negative effects range from those that result from the high level of information processing demands imposed by animations to those that can be attributed to animation's level of expressiveness as a representational medium [5,7]. Considerable domain expertise is required in order to deal effectively with such complexity. Without the specialist background knowledge required for domain-specific top-down guidance of selective attention, novices are heavily dependent on domain general approaches with a strong influence from bottom-up processing [8]. Most of the initial empirical research on learning from animation has involved domain novices working with relatively straightforward content in which both visuospatial and dynamic complexity were comparatively low. Further, the presentations used tended to be single exposures to system controlled animations typically accompanied by written or spoken textual information. Under these circumstances, it is hardly surprising that perceptual issues did not emerge as a major concern in studies of learning from multimedia resources containing animation. Rather, the research in this area tended to focus almost exclusively on higher level cognitive processing. However, strong perceptual effects were found when those without domain expertise were asked to learn from more complex and specialised animated meteorological diagrams in the absence of explanatory text [9]. Although non-meteorologists sometimes invoked everyday knowledge of physical phenomena to ascribe generalised causality to what they observed (and inappropriately attributed events to mechanical interactions), the bulk of their processing appeared to be perceptually dominated. Subsequent research using varied complex animated content, both mechanical and natural (piano mechanism functioning, [10]; Newton's Cradle dynamics, [11]; Kangaroo locomotion, [12]) suggests that these perceptual effects are widely applicable.

A major impetus for the present paper has been a program of empirical research on learning from an animated diagram of a piano mechanism. This subject matter was selected for the research because it is a sophisticated, highly specialised system that relies on a set of complex interrelated dynamic changes in order to function effectively across a wide range of performance conditions. Although the piano animation is abstract in a visuospatial sense, it is behaviourally realistic. That is, the order and nature of the events depicted in the animation correspond with those of the referent in the represented domain. When a piano is played, its functioning involves repeated operation of a three-stage mechanical cycle that is typically executed with many different specific parameters across a musical performance. The extreme range of performance demands that are imposed on the mechanism are met by the precise coordination of interlinked causal chains. Within each operational cycle, the sets of temporal changes comprising these causal chains overlap and rapidly cascade so that the human information processing demands from the piano animation are very high. For this reason, a number of our studies give learners repeated exposure to the animation and allow them to manipulate its playing regime. Comparisons of novice and expert interrogation of the animation suggested that various phases were involved in its processing. For the novices, perceptually-driven processes tended to be dominant with little evidence of progression to a higher level conceptual characterisation of the piano's functioning. In contrast, the experts (piano repairers) exhibited far more sophisticated top-down modulation of attention and processing activity consistent with a high level understanding of the mechanism's varying functional requirements. Although this comparison of experts and novices indicates very different approaches at these extremes, one cannot of course infer a simple linear path in the development of expertise. For this reason, trainee piano technicians at various stages of their studies are being included in our program of investigation. These participants will allow us to probe contributions such as insight and reconceptualisation of domain content to this development process.

Examples from learner processing of the piano mechanism animation will be used to illustrate aspects of the theoretical model that we present in this paper. Figure 1 shows the components of an upright piano mechanism and indicates the three functional stages that comprise the mechanism's operational cycle. Each of the piano's keys operates such a mechanism in order to produce its corresponding musical note. In Stage 1 of an operational cycle (*Strike*), depressing the right-hand end of the piano key causes the whippen to pivot. The whippen's role in the mechanism's operation is crucial because it is responsible for both (i) setting the hammer in motion so that it will strike the string to produce a musical note, and (ii) withdrawing the felt-covered damper from the string just before the hammer's impact so it is free to sound. From the whippen, anticlockwise motion is transferred to the hammer via the jack, and clockwise motion to the damper via the spoon. Stage 2 of the operational cycle (*Rebound*) occurs once the string has been struck and is extended for as long as the piano key is still depressed. During this stage, a subset only of the piano's components move (i.e., the hammer and the jack), while all others remain fixed. Collision between the balance and the backcheck stops the hammer's backward movement once it has rebounded from the string, with the jack 'riding' freely around the hammer butt. Because the damper continues to be held away from the string during this stage, the duration of the note is not artificially limited. However, in Stage 3 (*Reset*), the pianist's release of the key allows the damper to return to the string and end the note as all components of the mechanism are restored to their original positions in readiness for another operational cycle.

The examples used to illustrate aspects of the proposed animation processing model are taken from current studies that use a manipulable replica piano mechanism to probe how information from the piano animation is extracted. This replica is made from a set of pivoted clear Perspex pieces that correspond in appearance, arrangement, and movement possibilities to the operational components shown in the piano animation. Participants in these studies are Teacher

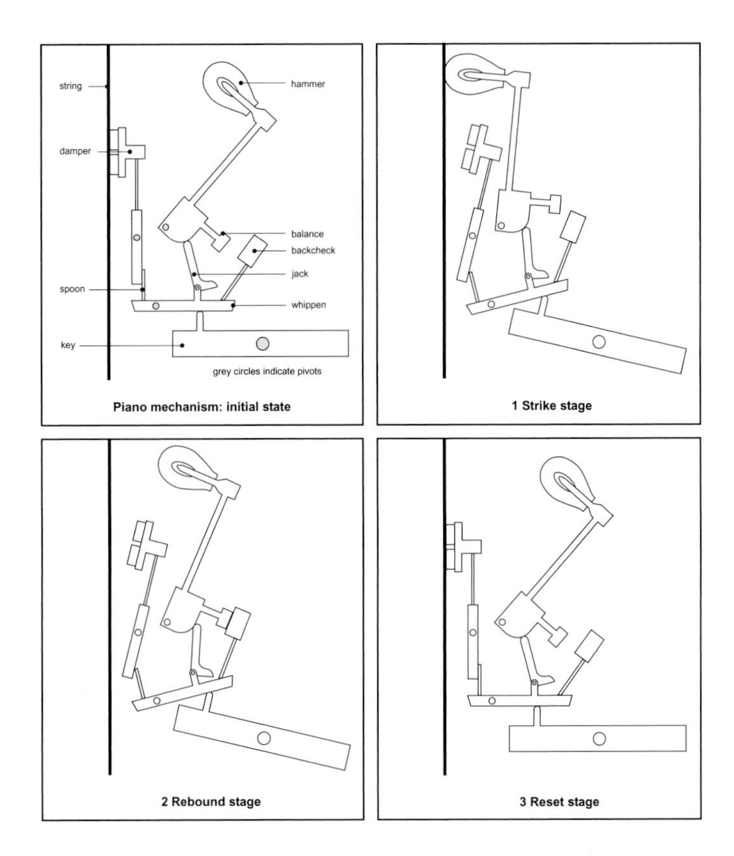

Fig. 1. When the key is pressed down to play a note, its opposite end tips the whippen anticlockwise. As a result, the spoon pushes the tail of the damper causing it to move clear of the string. Pivoting of the other end of the whippen in the opposite direction causes the jack to impel the hammer forwards. It strikes the string just after the damper has been retracted and produces the required musical note (Stage 1). Having struck the string, the hammer rebounds but its backward movement is arrested when the balance reaches the backcheck. Then, for as long as the key remains depressed, the mechanism stays 'frozen' in position (Stage 2). Once the key is released, the mechanism's components return to their starting places and are reset ready to respond to the next key press (Stage 3).

Education students who have little or no knowledge of a piano's internal mechanism. In the present investigation, they watch the piano animation repeatedly with the instruction to demonstrate what they have observed after each viewing by physically manipulating the Perspex piano parts. Their manipulations are videoed from below through a glass-topped table and analysed to determine which parts were moved, the direction and timing of movements, how movements were grouped or related, and changes in these aspects across successive demonstrations.

One of our main research aims is to find empirically-based approaches to the design of animations that facilitate the acquisition of expertise in subject domains requiring comprehension of complex dynamic content. A comprehensive account of how complex animations are processed is an essential foundation for this research. The model presented here is a first step in developing such an account and touches upon all main phases of processing (from initial parsing of the animation through to ultimate development of a high quality dynamic mental model). We assume that information extracted from the external animated display complemented by information available in a viewer's existing knowledge structures allows the referent system to be modeled internally via a set of mental tokens [13]. The quality of the mental model constructed as a result of viewing the animation (in terms of how accurate, comprehensive, and appropriate it is) will determine its utility for performing specific cognitive tasks [14]. If the externally presented material depicts content beyond the familiar and everyday, it is unlikely that non-specialists will be able to generate a high quality mental model without additional support. For this reason, our primary focus in this paper is the earlier perceptually-based phases rather than details of top down aspects for which successful processing is heavily dependent on domain-specific background knowledge. We assume that the learners who must deal with complex animations can often be considered as domain novices so their processing would typically be dominated by bottom-up activity and domain general interpretative frameworks. This is particularly likely where the animation presents the referent subject matter in a behaviourally realistic manner [9], with all its attendant dynamic complexity. Although the following discussion presents the five phases of the model successively, this is not to suggest that processing of a particular animation by a particular learner occurs strictly in this sequence. Further, we do not consider that processing necessarily continues through all five stages. Rather, it may effectively terminate earlier, depending on the specific characteristics of the animation and the individual learner.

2 Phase 1: Localised Perceptual Exploration

A complex animation depicting a process or procedure can be characterised as a hierarchically organized temporal macro structure. As such, it should be amenable to segmentation into component events on the basis of a learner's perceptual and conceptual processing [15]. The first phase of our model focuses upon the initial bottom-up processes that are fundamental to perceptually-based parsing of the animation into these individual events. Segmentation of a continuous flow of dynamic information into its component embedded events appears to be founded in the observer's attempts to anticipate upcoming information. The observer recognizes that a new event has begun when there is a transient increase in the difficulty of making accurate predictions [16].

Identification of the events that comprise an animation ultimately results in a series of inter-event boundaries distributed along its time course. However, it is unlikely that a learner's segmentation efforts would proceed in a strict linear sequence from the animation's beginning to its end. Rather, the order in which the learner actually identifies the animation's component events is likely to depend on their relative perceptual conspicuity [5]. Throughout the animation's time course, the various graphic entities engaged in these events typically vary in relative perceptual salience. Due to the particular combinations of visuospatial and temporal properties they exhibit at any instant, some entities will become more conspicuous while others become less conspicuous [c.f. 17]. This changing *perceptibility profile* [18] should then be a key influence on the processing priority accorded to different events throughout the animated presentation. If the progress of segmentation proceeds according to the perceptibility profile of the display's component entities, processing would be expected to begin with aspects of the animation that are initially most perceptually salient [6]. However, shifts in the locus of segmentation could occur due to perceptibility profile changes along the animation's time course.

Initial perceptual parsing of a complex animation into the individual event units that will be the basis for further processing is likely to be demanding for learners due to competition for limited processing resources. Segmentation of a display that presents multiple moving entities at once requires the learner to track these objects over space and time. The number of objects that can be tracked successfully in this way is very limited and reduces with increasing presentation speed [19, 20]. As a result, the initial parsing of a complex animation is likely to favour perceptual processing that is localised rather than dispersed. We suggest that processing proceeds by way of a succession of small groups of neighbouring graphic entities, with each processing cycle resulting in the generation of an event unit. A processing cycle can be interrupted before all entities within the group have been dealt with if information that is more perceptually salient intrudes via peripheral vision. In this case, it is likely that segmentation activity shifts to another more conspicuous group of entities. The multiple event units formed as a result of the processing activity at a local level during Phase One provide the raw material for Phase Two processing.

Piano example: When manipulating pieces of the piano replica to demonstrate what they observed in the animation, participants typically began by moving just the hammer. Of all the piano mechanism's components, the hammer is the most distinctive in terms of both appearance (due its large size and irregular shape) and behaviour (due to its rapid motion and extensive sweep). This is particularly the case during the first part of the piano's operational cycle where the hammer dominates the perceptibility profile. In choosing to move the hammer first, participants by-pass the correct operational order of the mechanism. The sequence of actions in a piano once a key is depressed progresses via the whippen and the jack to the hammer (which is actually the *last* component of the causal chain involved in making the string sound). However, the undistinguished appearance and behaviour of the whippen and jack make them relatively inconspicuous, despite their crucial role in the piano's functioning. The high perceptual salience of the hammer compared with other parts of the mechanism dominated and made it the most likely to be noticed to the neglect of components such as the whippen and jack. Nevertheless, early perceptions of the hammer's dynamics appear to have been holistic rather than analytical. Initially, participants tended to move the hammer in a rather generalized and somewhat inaccurate fashion. For example, the hammer head either failed to reach the string or else overshot it considerably. Further, despite being invited to move pieces in pairs, participants began by moving them one at a time. This suggests that their early processing of the animation was highly localized. After another one or two viewings of the animation, participants began to move other components (such as the damper) but with little sign of connection even at the local level.

3 Phase 2: Regional Structure Formation

In Phase Two, broader scale regional structures emerge due to the formation of relationships between the individual event units generated during Phase One. There are various possibilities for establishing the connections upon which these structures are built, such as those consistent with the principles of Gestalt psychology [21]. For example, linkages between event units may be made on the basis of visuospatial characteristics (such as *Proximity*), or on the basis of their behaviour (such as *Common Fate*). This coordination process condenses the localised results of initial perceptions into more extensive dynamic micro chunks. However, these regional fragments would probably constitute isolated islands of activity across the display rather than being interlinked into a continuous structure [see 6]. It seems unlikely that the dynamic micro chunks formed in the processing of complex animations would be interconnected by coherent higher level structuring of the type found for the simple, continuous movement of a single point traveling through a static array [22].

It is possible that the boundaries defining these chunks are established perceptually via a mechanism somewhat similar to that proposed by Agam and Sekuler [23]. For example, a moving entity could sweep out a coherent 'shape' in one particular direction then stop or change direction to sweep out another shape. The intervening discontinuity would mark the boundary between one unified chunk and the next. Once these chunks are formed, some form of abstraction (such as assignment of a verbal tag) may be applied so that they can be mentally represented in a more parsimonious fashion. On the analysis suggested here, the various regional fragments established during Phase Two could be separate from each other in space and time. More specifically, some of these dynamic micro chunks may occur at the same time but in different parts of the display whereas others would occur sequentially. Clearly, important ramifications for ease of processing arise from the extent to which two or more dynamic micro chunks are spatially and temporally coincident or separated.

In this second phase of processing, dynamic conjunctions that are pre-cursors to limited attribution of domain-general causality within these regional event units begin to occur. For example, ampliation conditions [24] such as the interactions between moving objects that are thought to be the basis for inferring individual causal links become possible when multiple event units are chunked.

However, at this stage these candidates for causal linkage are likely to be limited in scope and essentially based on everyday world knowledge.

Piano example: Once participants had demonstrated the movements of the hammer and the damper as discrete components, they began to link their individual actions into a neighborhood pattern. For example, as the hammer was moved towards the string, the damper was moved away from the string. In the animation, these two actions not only occur in the same general region but also take place in a coordinated manner, both of which make them appear related to one another. In this first stage of the mechanism's operational cycle, both the hammer and the damper sweep out coherent arc-shaped areas as they swing around their pivots in a concerted fashion. This pair of regular, related movements is readily chunked to form an abstract and parsimonious representation. Tags could provide the condensed relational information that captures the dynamic basis for this chunk (e.g. 'pivot together but in opposite directions' and 'hammer sweeps out larger area than damper'). Although such chunk formation would link the behaviours of the hammer and damper, it is important to note that it is based on correlation of movements only. The causal role of the whippen, jack and spoon in synchronizing and staging these behaviours is absent from such a representation. Once this most perceptually conspicuous hammer-damper dynamic micro chunk was demonstrated, participants generally went on to demonstrate a less conspicuous set of coordinated regional movements. For example, they then showed the relationship between the rocking of the whippen and the raising of the jack. However, the whippen-jack micro chunk tended to be treated quite separately from the hammer-damper micro chunk. That is, in showing these two micro chunks within the same demonstration, the hammer-damper association typically preceded the whippen-jack association (reversing the actual sequence of activity). It therefore appears that in this phase, participants did not bridge these isolated islands of activity to form superordinate dynamic structures that are the basis for extended causal chains. However, there was some evidence that participants inferred limited, highly localized causality concerning the interaction of the whippen and the jack.

4 Phase 3: Global Characterization

During the third phase of processing, the learner develops a comprehensive internal characterization of the animation in terms of its component operations across space and time. Successful completion of Phase Three results in a mental representation that is a well structured spatio-temporal array specifying the behaviour of all essential operational entities depicted in the animation. The construction of this internal characterization during Phase Three involves confirmation of the individual causal links prefigured or identified in Phase Two, and their bridging across space and time. This concatenation allows the individual islands of activity to be integrated into overall sequences of actions that characterize the referent system at a more global level. These sequences are the basis for formation of the comprehensive causal chains that can later be interrelated to explain functional aspects of the system. However, without appropriate domain specific knowledge, causality may be wrongly characterized in everyday domain general terms [5].

The characterization produced in Phase Three provides a more detailed description of aspects such as the onset, sequencing, direction, relative magnitude, and duration of dynamic changes. In principle, Phase Three allows for inclusion of some aspects that were neglected in earlier processing because of their relatively low perceptual salience. However, prior knowledge and experience in the referent's domain can be crucial here [25]. A lack of relevant domain specific background knowledge may impose practical limits on what information the learner extracts. Without such knowledge, the learner is more likely to be influenced by the perceptual attributes of the animation and subject to the negative effects that display change can have on the capacity to extract information [c.f. 26]. Visually subtle aspects that are nevertheless crucial to the building of a high quality mental model tend to be missed by domain novices [6, 26]. Even with repeated free interrogation of a complex animation, domain novices can fail to locate key aspects because they lack the domain specific conceptual basis required to detect this high relevance information [27].

The further processing that takes place during Phase Three contributes to establishing the main operational episodes necessary for building a high quality dynamic mental model. If the learner has sufficient domain specific background knowledge, repeated careful interrogation of even a complex animation should ultimately yield the required information. The lack of such knowledge can compromise the learner's subsequent processing. However, even high knowledge learners do not finalise the mental model building process in Phase Three because the information extracted is yet to be organized around a central functional purpose with respect to the depicted referent. The ideal outcome of this phase is therefore a comprehensive and coherent account of the main physical events that occur across the entire time course of the animation. Although it specifies the 'what' of the animation, it does not set the activity in the context of the subject matter's intended overall functionality.

Piano example: Participants' piano replica demonstrations suggested that their processing tended to reach a plateau before or during Phase 3. In their earlier demonstrations, participants typically dealt with only the first of the three stages comprising the piano's complete operational cycle (i.e. *Strike*). However, in their later demonstrations most participants included something resembling the third (*Reset*) stage in which all components return into their original positions as the piano key is released. This extension of the demonstration to cover a broader time period is consistent with Phase 3 processing in which the participant's internal characterization of the animation becomes more comprehensive. There was also an elaboration of what occurred within each stage, with the interlinking of previously unconnected components being demonstrated. For example, the movement of the whippen with its attached spoon was shown as being associated with the removal of the damper from the string. As a result of these bridges being developed between actions that had formerly been treated in isolation, longer sequences of causality were being demonstrated. Further, the sets of related actions identified earlier were increasingly demonstrated in an order that corresponded with their actual sequence (rather than according to their relative perceptibility). This suggests development of a more accurately structured internal representation of the externally presented animated information. Despite participants' demonstrations becoming more comprehensive, interlinked, and appropriately structured with successive animation viewings, some functionally crucial information was still neglected. For example, *Rebound* in which only the hammer and jack move was rarely demonstrated as a distinct stage. Rather, it seemed that *Rebound* and *Reset* were blurred together into a single final stage that was essentially a simple reversal of the *Strike* stage. From a perceptual point of view, this could be explained in terms of *Rebound* having much more restrained and subtle dynamics than the *Strike* stage. Without the necessary domain specific background knowledge, it seems that participants tended to miss the functionally crucial transitory conjunction of the balance and backcheck because of its low perceptual salience. This period where the mechanism is 'frozen' with the damper separated from the string is the reason why a played note will not be cut short if the piano key continues to be held down. However, upon the key's release, the damper returns to its original position on the string and the note is stopped. Without background knowledge about this aspect of a piano's functionality to provide top-down direction of attention, its low perceptual salience makes it likely to be missed.

5 Phase 4: Functional Differentiation

Established models of multimedia learning consider the role played by prior knowledge in the comprehension of pictures. However, they tend to deal with picture processing generally rather than animations in particular and treat this processing at a broad rather than a detailed level. The generality of such models reflects their focus on the joint processing of verbal and visual information rather than more specific issues regarding picture processing *per se*. For example, the pictorial component of the Cognitive Theory of Multimedia Learning [1] is confined to general aspects of picture processing issues, especially the selection, organization, and integration of image information. Schnotz [4] also takes a broad view of picture processing and characterizes mental models as products of interplay between knowledge embodied in cognitive schemas and pictorial information from visual working memory. Cognitive load approaches tend to have a focus that targets the role of working memory constraints in their consideration of animation processing [e.g. 28]).

Although these existing approaches acknowledge the importance of prior knowledge in picture processing, none of them concerns itself with the minutiae of how it contributes to the building of high quality mental models. In particular, there is no direct consideration of the issue of functionality and how the key functional aspects that are required in these mental models are related to learning with animations. This may be because most of the educational research associated with these theoretical frameworks has involved domain novices who are given a single exposure to relatively straightforward content. However, functional aspects of mental models have been explored in the field of ergonomics. In that research, it was found that in their work with complex information systems, experts develop highly specific and efficient functional mental models [e.g. 29]. In the fourth phase in our animation processing model, the emphasis moves from perceptual processing to more top-down aspects that concern the functionality of the depicted system. Because this phase is heavily reliant on the possession of specialised knowledge, it is unlikely to be undertaken effectively by learners who are domain novices [c.f. 26]. Although a full domain general operational description of the animation may be generated during Phase Three, this description is functionally undifferentiated. The separate contributions made by the various events characterized to this point still need to be correlated with their contributions to the desired overall performance of the referent system.

In Phase Four, the events that were identified in previous phases of processing are interpreted in terms of the referent's central purpose and characterized as functionally distinct action sets. Collectively, they comprise the different subsystems that contribute to the overall functioning of the system. Within these subsystems, actions propagated along causal chains and framed by functionally important contingencies combine to produce the required performance. The role of each individual subsystem is coordinated with the roles performed by the other subsystems to produce the overall functionality. In complex systems, functional episodes can be temporally related in various ways. Some occur in a strict sequence where each episode is clearly delineated from its precedents and antecedents. Others occur as a rapid cascade in which it is difficult to perceive clear boundaries. Episodes may also overlap in time to varying extents. In some cases they occur in parallel across different parts of the display while in others, a short episode is embedded within an episode with a larger time scale. Understanding the functional purpose of these dynamics is an essential precursor to building a mental model of sufficient quality to cover the full range of performance exigencies.

Piano example: Overall, there was little indication of Phase 4 processing in participants' demonstrations with the replica piano mechanism. Considering that none of the participants had professional expertise in the subject domain, this lack of evidence for top-down processing of the animation is to be expected. The extent of participants' understanding of piano functioning appeared to be confined to very superficial aspects (e.g. when a piano key is pressed, a sound is produced). Their demonstrations did not reflect an appreciation of deeper issues concerning the role of subsystem functions in contributing to the piano's overall musical purpose. Even those who produced a reasonably comprehensive mechanical characterization of the events depicted in the animation failed to demonstrate the operation of the mechanism in terms of music-related functions of the various subsystems. The closest any of the participants came to such functional differentiation occurred in the demonstrations of a single individual who reported having once looked inside a piano out of curiosity about its inner workings. This participant knew several isolated facts about a piano's internal mechanism, such as that the damper is covered with sound-absorbent felt (not indicated in the animation or the manipulable replica) so that it can stop sound when in contact with the string. However, possession of a few disconnected fragments of appropriate background knowledge did not allow that participant to produce a coherent demonstration of the piano's subsystems in terms of musical functionality. For example, there was no evidence in her demonstrations of the subtle functional contingencies that apply in the correct operation of the hammer and damper action sets. The animation shows that the damper is actually retracted just before the hammer strikes the string. From a musical point of view, the timing of these two actions is crucial in order to obtain a clear, pure musical note. If the damper is retracted too early, the note being played will be muddied by unwanted resonance. If it is retracted too late, the note will be muffled by the damper's felt still being in contact with the string as the hammer strikes. When participants demonstrated this pair of actions on the piano replica, none of them showed this cascade effect involving tightly sequenced coordination of the damper and hammer movements. Instead, they synchronized the movements of these two components, both with the rocking of the whippen and with each other (something that would be identified by a piano technician as a serious malfunction because of its devastating effect on the quality of the sound produced). This example suggests that participants' lack of domain specific background knowledge prevented them from extracting from the animation temporal relations that are of low perceptual salience but of crucial functional relevance. It seems that even repeated viewing of the animation did not of itself allow them to obtain the essential functional information about subsystems that is a precursor to building a high quality mental model of the piano mechanism.

6 Phase 5: Mental Model Consolidation

A high quality dynamic mental model of a system must equip a learner with the capacity to understand the system's behaviour not simply in a single situation, but rather across a wide variety of circumstances (both usual and unusual). For example, it should allow the learner to predict how the system will cope with different operational requirements or to identify causes of a range of performance faults. The mental model must therefore be informed by appropriate background knowledge of not only the overall domain specific purpose that the system is required to fulfill, but also the different types of performance that it would typically be expected to produce [c.f. 30]. It therefore needs to specify both the nature of tasks likely to be encountered and how such performance variations are achieved within the constraints dictated by the components, structure, and dynamics of the system. Unfortunately, animations are by their very nature unable to represent events with the ambiguity required to allow for a high degree of performance variability [7]. Because it depicts a specific instantiation of a set of events, a single animation cannot cover a range of possible (and legitimate) alternative temporal structures. For subject matter with aspects that are not fully determined with respect to time, animation may over-constrain the inferences that can be made by learners who are domain novices.

At the present time, Phase Five remains the least elaborated aspect of our proposal. However, we are currently exploring the potential of theoretical concepts from Work Domain Analysis [31] to contribute ideas for its further development. This powerful approach for dealing with complex dynamic task environments is used to model work contexts and worker problem spaces. It characterizes these contexts and problem spaces along the two orthogonal dimensions of abstraction (five levels) and decomposition (three levels). The levels of abstraction range from physical objects at the lowest level to functional purposes, while decomposition deals with components, through subsystems, to the whole system. Preliminary work with the piano system suggests that this approach holds considerable promise to help identify the likely requirements for developing a high quality mental model from an animated diagram of a complex dynamic system. We are currently examining its applicability to natural dynamic systems that are less straightforward than mechanical devices in terms of their decomposition because their structures tend to be less readily apparent. For example, it appears that with animations of changing weather maps or hopping kangaroos, the parsing of an animated display into relevant graphic entities is strongly influenced by the localized dynamics that occur within different regions.

Piano example: Our piano animation portrays the pattern of actions that result from a particular and very standard type of key press. It shows how the piano mechanism would respond in just one specific situation, that is, when the pianist wishes to play a single note of moderate duration at moderate volume. However, it does not explicitly depict how the mechanism responds in a whole variety of other situations, such as where the requirement is for a series of very short loud notes or for one very long, soft note. These two contrasting styles of playing rely on different aspects of the piano mechanism's functionality. For the first style, the aspect of functionality required is the rapid resetting of the mechanism ready for the next key press (Stage 3 emphasis). In contrast, the second style involves the key being held down for an extended period during which time the mechanism rests in the position reached at the end of the Rebound action (Stage 2 emphasis). In this case, the event in which the hammer's rebound is halted by the backcheck on the whippen is all important to the desired functionality. These examples show that the flexibility required of the mechanism in order to produce a wide range of performances is achieved by variations in the ways the mechanism's constituent subsystems behave. These often subtle local functional variations involve changes in the spatial and temporal structuring of the mechanism's operation. Another aspect not dealt with in the animation is the types of operational adjustments that piano technicians must make during servicing in order to maintain an instrument's capacity to handle such varied demands. Nor does the animation address the types of malfunction that can seriously compromise the mechanism's basic functioning. Although the piano animation's specificity may help learners in making correct inferences about the particular performance condition it depicts, this same specificity would probably work against them dealing with a broader range of possibilities. However, these semantic constraints would not be a problem for those with expertise in the domain who could use abstractions about functionality to generate other possibilities from the specifics of this example.

7 Conclusion

The model presented here is preliminary and inevitably reflects its very specific origins in our detailed studies with the animated piano diagram. Further consideration of content differences, such as those found between mechanical and natural systems, is required to elaborate and refine the model for greater generalisability. For example, consideration needs to be given to the effect on processing of the extent to which component entities and subsystems of an animation are visually discrete (as in the piano) or continuous (as in the kangaroo). Another aspect of the model requiring considerable future development concerns the learner's knowledge of required functionality and how that interacts with the perceptual processes emphasized in the first three phases described above. Individuals who have this domain specific information available to supplement bottom up information from the animation are likely to considerably modify their perceptual processing by way of top down intervention. Our current program of research aims to explore the influence of these expertise related effects on the efficiency of animation processing and the quality of mental models that can be built as a result.

References

1. Mayer, R.E.: A Cognitive Theory of Multimedia Learning. In: Mayer, R.E. (ed.) Cambridge Handbook of Multimedia Learning. Cambridge University Press, New York (2005)
2. Mayer, R.E., Sims, V.K.: For Whom is a Picture Worth a Thousand Words? Extensions of Dual-coding Theory of Multimedia Learning. Journal of Educational Psychology 86, 389–401 (1994)
3. Narayanan, N.H., Hegarty, M.: On designing Comprehensible Interactive Hypermedia Manuals. International Journal of Human-Computer Studies 48, 267–301 (1998)
4. Schnotz, W.: An Integrated Model of Text and Picture Comprehension. In: Mayer, R.E. (ed.) Cambridge Handbook of Multimedia Learning, pp. 49–70. Cambridge University Press, New York (2005)
5. Lowe, R.K.: Extracting Information From an Animation During Complex Visual Learning. European Journal of Psychology of Education 14, 225–244 (1999)
6. Lowe, R.K.: Animation and learning: Selective Processing of Information in Dynamic Graphics. Learning and Instruction 13, 157–176 (2003)
7. Stenning, K.: Distinguishing Semantic from Processing Explanations of Usability of Representations: Applying Expressiveness Analysis to Animations. In: Lee, J. (ed.) Intelligence and Multimodality in Multimedia Interfaces: Research and Applications. AAAI Press, Menlo Park (1998)

8. Lowe, R.K.: Components of Expertise in the Perception and Interpretation of Meteorological Charts. In: Hoffman, R.R., Markman, A.B. (eds.) Interpreting Remote Sensing Imagery, pp. 185–206. Lewis, Boca Raton (2001)
9. Lowe, R.K.: Multimedia Learning of Meteorology. In: Mayer, R.E. (ed.) The Cambridge Handbook of Multimedia Learning, pp. 429–446. Cambridge University Press, New York (2005)
10. Boucheix, J.-M., Lowe, R.K., Soirat, A.: On-line Processing of a Complex Technical Animation: Eye Tracking Investigation During Verbal Description. Paper presented at the Text and Graphics Comprehension conference, University of Nottingham, UK (2006)
11. Lowe, R.K., Schnotz, W.: Animations and Temporal Manipulations: Supporting Comprehension of Complex Dynamic Information. In: Paper presented at the 12th European Conference for Research on Learning and Instruction, Budapest, Hungary (2007)
12. Lowe, R.K.: Learning with Animation: Lessons from Static Graphics. In: Paper presented at the 12th European Conference for Research on Learning and Instruction, Budapest, Hungary (2007)
13. Johnson-Laird, P.N.: Mental Models: Towards a Cogntive Science of Language, Inference and Consciousness. Cambridge University Press, Cambridge (1983)
14. Lowe, R.K.: Animated Documentation: A Way of Handling Complex Procedural Tasks? In: Alamargot, D., Terrier, P., Cellier, J.-M. (eds.) Written Documents in the Workplace, pp. 231–242. Elsevier, Amsterdam (2007)
15. Zacks, J.M., Tversky, B.: Event Structure in Perception and Conception. Psychological Bulletin 27, 3–21 (2001)
16. Zacks, J.M., Speer, N.K., Swallow, K.M., Braver, T.S., Reynolds, J.R.: Event Perception: A Mind/Brain Perspective. Psychological Bulletin 127, 273–293 (2007)
17. Wolfe, J.M., Horowitz, T.S.: What Attributes Guide the Deployment of Visual Attention and How Do They Do It? Nature Reviews Neuroscience 5, 1–7 (2004)
18. Lowe, R.K., Boucheix, J.-M.: Eye Tracking as a Basis for Improving Animation Design. In: Paper presented at the 12th European Conference for Research on Learning and Instruction, Budapest, Hungary (2007)
19. Alvarez, G.A., Franconeri, S.L.: How Many Objects Can You Track?: Evidence for a Resource-limited Attentive Tracking Mechanism. Journal of Vision 7, 1–10 (2007)
20. Wolfe, J.M., Place, S.S., Horowitz, T.S.: Multiple Object Juggling: Changing What is Tracked During Extended Multiple Object Tracking. Psychonomic Bulletin & Review 14, 344–349 (2007)
21. Wertheimer, M.: Laws of Organization in Perceptual Forms. In: Ellis, W. (ed.) A Source Book of Gestalt Psychology, pp. 71–88. Routledge, London (1938)
22. Lee, P., Klippel, A., Tappe, H.: The Effect of Motion in Graphical User Interfaces. In: Butz, A., Krüger, A., Olivier, P. (eds.) SG 2003. LNCS, vol. 2733, pp. 12–21. Springer, Heidelberg (2003)
23. Agam, Y., Sekuler, R.: Geometric Structure and Chunking in Reproduction of Motion Sequences. Journal of Vision 8, 1–12 (2008)
24. Michotte, A.: The Perception of Causality. Methuen, London (1963)
25. Kriz, S., Hegarty, M.: Top-down and Bottom-up Influences on Learning from Animations. International Journal of Man-Machine Studies 65, 911–930 (2007)
26. Lowe, R.K.: Interrogation of a Dynamic Visualisation During Learning. Learning & Instruction 14, 257–274 (2004)

27. Lowe, R.K.: Learning from animation: Where to Look, When to Look. In: Lowe, R.K., Schnotz, W. (eds.) Learning with Animation: Research Implications for Design, pp. 49–68. Cambridge University Press, New York (2008)
28. Ayres, P., Paas, F.: Can the Cognitive Load Approach Make Instructional Animations More Effective? Applied Cognitive Psychology 21, 811–820 (2007)
29. Burns, C.M.: Navigation Strategies with Ecological Displays. International Journal of Human-Computer Studies 52, 111–129 (2000)
30. Norman, D.A.: The Design of Everyday Things. Doubleday, New York (1991)
31. Naikar, N., Hopcroft, R., Moylan, A.: Work Domain Analysis: Theoretical Concepts and Methodology. Australian Government Department of Defence, Victoria (2005)

Diagrams for the Masses: Raising Public Awareness – From Neurath to Gapminder and Google Earth

Raul Niño Zambrano and Yuri Engelhardt

Department of Media Studies, University of Amsterdam, Turfdraagsterpad 9, 1012 XT Amsterdam, The Netherlands
Raul.NinoZambrano@student.uva.nl, Engelhardt@uva.nl

Abstract. In the beginning of the 1920s Austrian philosopher and sociologist Otto Neurath developed a diagrammatic language (*Isotype*) that uses simplified pictures and composition rules to convey social and economic statistical data to a general public. A current trend in the information visualization community is the presentation and sharing of information graphics via the World Wide Web (e.g. *Gapminder*, *Google Earth Outreach*). In 2007, the Nobel Peace Prize was awarded to work in which statistical diagrams played a key role in raising public awareness (Al Gore's campaign concerning the Inconvenient Truth about global warming). In this paper we explore what these recent uses of graphic representations share with Neurath's idealism about visual education in terms of access to information, empowerment and the assumed benefits of the visual. We also address how digital media have amplified these concepts, and we propose three different categories of 'diagrams for the masses'.

Keywords: public awareness, statistical graphics, Otto Neurath, online graphics, information visualization.

1 Introduction

Especially since the success of the documentary *An Inconvenient Truth* (USA: Davis Guggenheim, 2006) there has been a tremendous interest in the potential of using diagrams for getting the attention of big audiences regarding a particular issue. In Al Gore's presentation in the film, a major role is played by diagrams illustrating the relationship between temperature and levels of carbon dioxide. According to Womack (2006), Gore's effective use of these diagrams has shown that "information visualization is able to communicate the intricacies of global warming in a way no other discipline can. Its messages can be immediate and powerful, without sacrificing the level of detail necessary to represent the complex subject accurately".

Besides Al Gore's accomplishment in spreading information about climate change, a number of initiatives that deal with other issues and that also use graphics to reach a broad public have recently appeared on the Internet (e.g. *Gapminder*[1], *Google Earth Outreach*[2]). This trend, together with the United Nations' recent decision to provide free access to their statistical databases[3] and a couple of efforts in compiling useful

[1] http://www.gapminder.org
[2] http://earth.google.com/outreach/index.html
[3] http://data.un.org

G. Stapleton, J. Howse, and J. Lee (Eds.): Diagrams 2008, LNAI 5223, pp. 282–292, 2008.

guidelines for using visual information for advocacy (e.g. Emerson 2008), has been responsible for bringing the topic of 'diagrams for the masses' into discussion.

Some experts talk about "the beginning of a new era in visualization" (Kosara 2007) and situate the new trends as part of "the emergent field of social data analysis" (Viégas 2007). However, this discussion has not addressed the similarities that these trends share with the work started by Austrian philosopher and sociologist Otto Neurath in the beginning of the 1920s. In this paper we address the correspondences between Neurath's idealism and these recent trends in information visualization with regard to ideas about data accessibility, the empowerment that this access may produce and the assumed benefits of using visuals. In addition, we look at the influence of digital media concerning the roles played by diagram users and makers, and we distinguish three categories of 'diagrams for the masses'.

2 Two Examples of 'Diagrams for the Masses'

2.1 Neurath's Pictorial Statistics

Otto Neurath, leading member of the Vienna Circle, was "one of the most formidable, if controversial, intellectuals of the interwar period" (Vossoughian 2006: 48). His work included important contributions in diverse fields such as political economy, social philosophy and the theory of science. In addition, Neurath is well known for his investigations on visual communication. He believed that "Visual education is related to the extension of intellectual democracy within single communities and within mankind" (Neurath, O. 1973: 247). Therefore, he aimed for a new way to convey information that could be comprehensible for a wide range of people, and put it in practice in the different museums he directed.

After directing the Museum of War Economy in Leipzig in 1918, Neurath helped to establish the Museum of Housing and City Planning in Vienna in 1923 and later reopened it as the Museum of Society and Economy. These museums were unique institutions that attempted to bring and explain socio-economic facts to local citizens. Neurath's colleague and future wife Marie Reidemeister comments on their work for the Museum of Society and Economy:

> "Our object was to make the general public acquainted with the problems the community of Vienna had to tackle (the housing shortage, the amenities needed for children and mothers, the high infant mortality and the tuberculosis), how they were dealt with, and with what success. In colourful charts, which were like simple puzzles which everybody could solve, such problems were brought nearer to general understanding than would have been possible with just words and numbers." (Neurath, M. 1974: 130).

While working for the museum's exhibitions, Neurath and his team, among them graphic designer Gerd Arntz, developed an iconic language of pictorial statistics first known as 'The Vienna Method' but later renamed as the International System of Typographic Picture Education (*Isotype*). This iconic language uses simplified pictures and specific composition rules to "present some worthwhile information, show up some relationship or development in a striking manner, to arouse interest, direct the attention and present a visual argument which stimulates the onlooker to active participation" (Neurath, M. 1974: 146). This resulted in comprehensible diagrams that encouraged people to think about themselves and the world around them in terms of patterns, relationships and systems of organizations (see Fig. 1).

Today, the influence of Neurath's *Isotype* language can be found in the international pictograms we encounter at airports and railway stations, and also in newspaper infographics. The increasing use of infographics since the 1970s, in newspapers and magazines, and even on television, has been an important step in bringing diagrams to 'the masses'.

"The output of the *Isotype* movement as a whole" says Twyman, "draws attention to two things which are of special interest to many designers today. First, it demonstrates that successful designing depends to a large degree on clarity of thinking; secondly, it provides support for the view that the graphic designer's primary role is to serve the needs of society." (Twyman 1975/1981: 17).

2.2 Gapminder's Moving Charts

A recent example of information visualization based on "clarity of thinking" and with the aim "to serve the needs of society" can be found in the work of the *Gapminder* Foundation, a non-profit venture that promotes sustainable global development by increasing the use and understanding of statistics. Created in 2005 by Swedish doctor and researcher Hans Rosling, his son and his daughter-in-law, the foundation is committed to bringing statistical data in pleasant formats to a large audience.

After noticing problems in the distribution, access and understanding of statistical data among Swedish university students, Rosling and his team decided to develop software that "enables interactive animation of development statistics in enjoyable and understandable graphic interfaces" (Rosling et al. 2004: 2). Since 2006 *Gapminder*'s software is available for online use at www.gapminder.org. Thanks to the collaboration of the United Nations Statistics Division, it currently presents statistical data such as *infant mortality rate* and *income per capita* for all countries of the world in interactive maps and charts that turn out to be very successful in attracting the attention even of people who usually are "allergic" for charts and statistics (see Fig. 2).

Recently, *Google* has acquired *Gapminder*'s software and has made it available as part of a tool that anyone can use to visualize their own data, the *Google visualization API*[4]. In addition, the *Gapminder* site does not only offer the possibility to view and explore statistical data, but also offers printable charts and a series of videos known as'gapcasts' in which Rosling himself narrates the graphic animations. Rosling's enthusiastic performance in the videos has been described as "moving" and "touching", making of him a popular figure on the World Wide Web. As Arthur (2007) has signaled, "Rosling excels at bringing potentially dry statistics to life." Rosling's presentations, especially the ones at the *TED* conferences in 2006[5] and 2007[6], have been watched by thousands of Internet users and his gapcasts are now also available in *iPod* versions and on *YouTube*.

---

[4] http://code.google.com/apis/visualization/

[5] http://www.ted.com/index.php/talks/view/id/92

[6] http://www.ted.com/index.php/talks/view/id/140

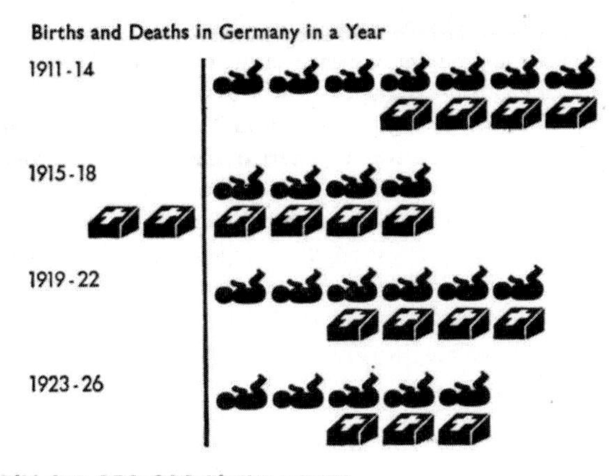

Fig. 1. *Isotype* diagram by Neurath, showing births and deaths in Germany between 1911 and 1926. The diagram shows that between 1915 and 1918 (World War I) there were more deaths than births, indicated by the two coffins extending out to the left (from Neurath, M. 1974: 132).

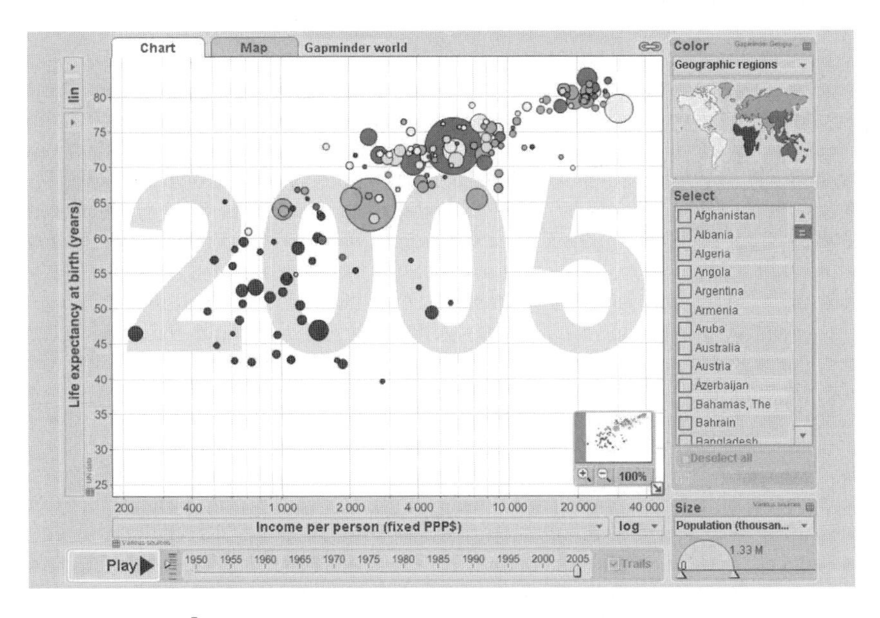

Fig. 2. *Gapminder*'s[7] chart showing the relationship between *Life expectancy* and *Income per person* for all countries of the world. Each country is represented by a colored bubble, the size of which is proportional to its population.

---

[7] http://www.gapminder.org/world

3 Seventy Years Apart – Same Spirit

Since the first *Isotype* experiments in the beginning of the 1920s up to the current access to interactive graphics as in *Gapminder*, there have been a lot of developments in the distribution and presentation of statistical data. Nevertheless, Neurath and Rosling's efforts still share a spirit that is independent of the technological means that they use. First of all, both projects were created out of the idea that statistical information must be accessible to everyone. Secondly, both key figures, Neurath and Rosling, seem to share the conviction that bringing statistical data to 'the masses' could actually lead to empowerment. Finally, *Isotype* and *Gapminder* have chosen to work with diagrams, based on the assumption that the best way to make scientific facts understandable for the general public is through visualizations.

3.1 Information Access for the Masses

In terms of giving access to statistical data, Neurath believed that "The ordinary citizen ought to be able to get information freely about all subjects in which he is interested, just as he can get geographical knowledge from maps and atlases. There is no field where humanization of knowledge through the eye would not be possible." (Neurath, O. 1939: 3). After realizing that most Austrian citizens were not aware of national and world socio-economic issues, he hoped to bring such information to the masses through museum exhibitions, so that "more than a few people will experience Austria as a unity with connections abroad" (Neurath, O., quoted in Vossoughian 2008: 59).

Today, Rosling and his team maintain on their website that "all people, independently of their political agenda, should get free access to already existing statistics about global development to easily improve their understanding about the complex society"[8]. Similar to Neurath's observations, Rosling noticed that despite the availability of data on the Internet, most Swedish medical students have wrong preconceived ideas about the health situation in different countries: "Swedish students, even those that graduated from secondary school with the highest grades, were unaware about the considerable decrease in child mortality that has occurred in the last generation in most countries in Asia, Middle East and South America." (Rosling et al. 2004: 2). For Rosling, this was the starting point for developing *Gapminder*, so that more people could get access to and a better understanding of statistical data.

3.2 Raising Awareness Towards Change

Both Neurath and Rosling have the conviction that disseminating data and making it comprehensible for a big public could play a role in social transformation. Neurath argues that "with visual aids, one could create something that is common to all, we could educate children in various countries [...] in a way which gives them the feeling of having knowledge in common for human brotherhood. Because I believe that visual aids have this peculiarity, I wish to promote visual education, as an element of human brotherhood." (Neurath, O. 1973: 248).

Along similar lines, Rosling argues that "We can change how the young generation understands the state of the world if we provide IT-tools that give them more complex

[8] http://www.gapminder.org/about/about/faq---frequently-asked-questions.html

and relevant moving images of world development in the form of animated statistics." (Rosling et al. 2004: 4).

3.3 Enthusiasm about the Visual

A final point shared by both initiatives deals with their belief in the benefits of using diagrams. While probably neither Neurath nor Rosling have much expertise regarding the cognitive aspects of diagrams use, they are both passionate advocates of the visual above the verbal. As Lupton notes, "Neurath held that vision is the saving link between language and nature, and that, hence, pictorial signs would provide a universal bridge between symbolic, generic language and direct, empirical experience." (Lupton 1986: 47). Following in the footsteps of other advocates of the visual such as Comenius[9] and Brinton[10], Neurath believes that "Words make division, pictures make connection." (Neurath, O. 1936: 18). Similarly, Rosling addresses visualization and animation as services that "unveil the beauty of statistics" (Rosling 2007: 104). For Rosling, it is *Gapminder*'s effective use of visuals that makes development statistics enjoyable and understandable.

The initiatives of the *Gapminder* Foundation aim "to create animated graphics that not only reach and please the eyes, but that also transform statistics into understanding, i.e. goes beyond the eye to hit the brain" (Rosling et al. 2004: 2), which corresponds with Neurath's enthusiasm as well. As Hartmann has noted: "While the graphic designer wants to catch the eye, Neurath wanted to catch the mind. When, in 1936, Neurath talked about an 'education by the eye', his quest was not for better visuals but for getting 'the full picture' – through perception to imagination." (Hartmann 2006).

4 Digital Media: Taking It a Step Further

Neurath believed that "communication of knowledge through pictures will play an increasingly large part in the future" (Neurath, O., quoted in Rotha 1946: 92). Indeed, popular recent projects that have reached large audiences, such as *An Inconvenient Truth* and *Gapminder,* are using pictures to make statistical data more accessible and comprehensible. Moreover, the use of new technologies in the digital era brings possibilities to the masses that Neurath could only dream of. Neurath could not experience the shift in perspective that digital media would produce, but some academics such as Nyíri have assertively signaled that "the emergence of computer graphics and multimedia computer networking might lead to a fulfillment of Neurath's vision" (Nyíri 2003: 45).

While in Neurath's diagrams data was only presented for viewing purposes, digital media now enable users to explore, produce and even share data and possible corresponding visualizations. Taking these new roles of the user into account, we propose to distinguish three categories of 'diagrams for the masses': 'View', 'Interact and Explore', and 'Create and Share' (see Table 1). These three categories are not disjoint but incremental: 'Create and Share' includes the possibility to 'Interact and Explore', and 'Interact and Explore' includes the possibility to 'View'.

---

[9] John Amos Comenius (1592-1670) is considered to be the first to create an educational picture book for children: the *Orbis Sensualium Pictus* (1658).

[10] Willard Cope Brinton (1880-1957) wrote *Graphic Methods for Presenting Facts* (1914), a book entirely devoted to the discussion of graphic representations.

Table 1. Three categories of 'Diagrams for the Masses'

	View	**Interact and Explore**	**Create and Share**
Examples	e.g.: Neurath's diagrams[11] (see Fig. 1), *An Inconvenient Truth*[12]	e.g.: *Gapminder*[13] (see Fig. 2), *EPA Emission Sources*[14], *Crisis in Darfur*[15]	e.g.: *Ushahidi*[16] (see Fig. 3), *Swivel*[17], *ManyEyes*[18] (see Fig. 4)
Data that is visualized	has been chosen by the makers	can be chosen by the user from a larger set of data that is provided by the makers	can be freely chosen by the user from any source (including the user's own data)
Type of visualization	chosen by the makers	sometimes chosen by the user, from possibilities that are provided by the makers	
Role of 'ordinary people'	passive role in raising own awareness	active role in raising own awareness	active role in raising other people's awareness

[11] http://www.fulltable.com/iso/
[12] http:www.climatecrisis.net
[13] http://www.gapminder.org
[14] http://www.epa.gov/air/emissions/where.htm
[15] http://www.ushmm.org/maps/projects/darfur/
[16] http://www.ushahidi.com
[17] http://www.swivel.com
[18] http://services.alphaworks.ibm.com/manyeyes/home

4.1 View

Neurath's *Isotype* charts (such as Fig. 1), and Al Gore's diagrams in *An Inconvenient Truth* belong to the 'View' category (see Table 1). Here, both the data and the type of visualization have been chosen by the makers of the diagrams, leaving the public only with the option of passively viewing them. Digital media have taken this a step further, enabling the creation of visuals in which users have a more active role. We have grouped such active roles of the user into 'Interact and Explore' and 'Create and Share'.

4.2 Interact and Explore

The 'Interact and Explore' category (see Table 1) includes diagrams such as the ones created by *Gapminder* (see Fig. 2), the *EPA Emission Sources* project and *Crisis in Darfur*. *EPA Emission Sources*[19] is a project by the U.S. Environmental Protection Agency that uses *Google Earth*[20] to enable the interactive exploration of air pollution sources, for example in your own neighborhood. *Crisis in Darfur*[21] is a project of the United States Holocaust Memorial Museum that also uses *Google Earth*, in this case to make compelling visual evidence available of the refugees' situation in Darfur, Sudan. The materials that were collected include the exact locations of damaged and destroyed villages and refugee camps, geo-tagged photos, videos, and personal testimonies from survivors of the genocide. *Crisis in Darfur* is one of the projects within *Google Earth Outreach*[22], a service for non-profit organizations where custom content can be created and shown as an overlay in *Google Earth*. Currently, there are more than a dozen of these projects that appear by default in *Google Earth,* appropriately labeled "Global Awareness layers."

In all these cases users are able to choose the data that is visualized from a larger set that is provided by the creators of the application. In addition, the type of visualization can sometimes be chosen by the user, from the possibilities that are provided. While in the 'View' category of 'diagrams for the masses' the user plays a passive role, in this 'Interact and Explore' category the user plays an active role in raising his own awareness about the topics at hand.

4.3 Create and Share

The 'Create and Share' category (see Table 1) includes visuals such as the ones found at *Ushahidi, Swivel* and *ManyEyes*. *Ushahidi*[23] is a website for reporting acts of violence in Kenya following the 2007 elections. *Ushahidi* uses *Google Maps*[24] to display where exactly acts of violence such as riots, property loss or rapes have been taking place (see Fig. 3). Anybody who witnesses such events can submit this information via Internet or by sending a text message with a mobile phone. *Swivel*[25] and

[19] http://www.epa.gov/air/emissions/where.htm
[20] http://earth.google.com
[21] http://www.ushmm.org/maps/projects/darfur/
[22] http://earth.google.com/outreach/index.html
[23] http://www.ushahidi.com
[24] http://maps.google.com/
[25] http://www.swivel.com

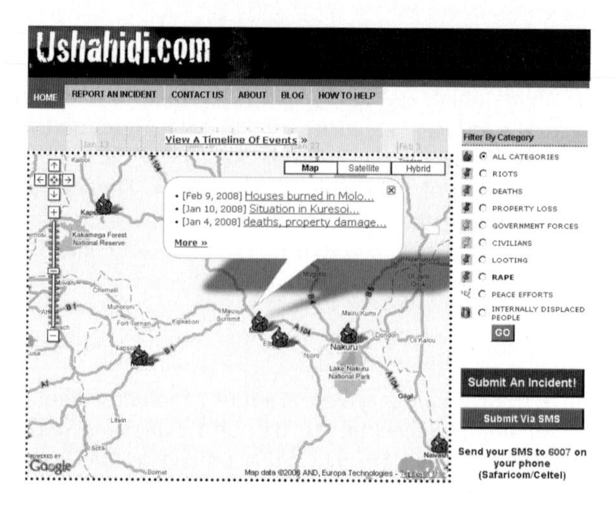

Fig. 3. The *Ushahidi*[26] project, showing citizen-reported cases of incidents such as riots, deaths and property loss in Kenya as overlay in *Google Maps*

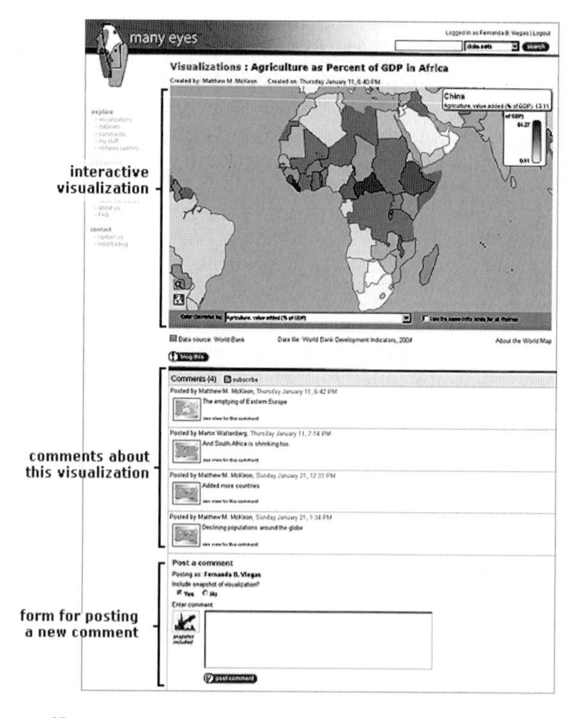

Fig. 4. *ManyEyes*[27] visualization of data concerning agriculture as percentage of GDP in Africa. Annotated in this screenshot are: the interactive visualization, the comments made by the creator and other users, and the form for posting new comments.

---

[26] http://www.ushahidi.com

[27] http://services.alphaworks.ibm.com/manyeyes/page/Tour.html

ManyEyes[28] are websites on which everybody (with access to Internet) can upload a data set, choose a visualization format for that data, add comments, and publish the result on the site, sharing it with all other users (see Fig. 4). Moreover, both *Swivel* and *ManyEyes* allow users to comment on other users' diagrams and even to download their data, in this way enabling the creation and exchange of 'diagrams for many eyes'. The *ManyEyes* website states that their goal "is to 'democratize' visualization and to enable a new social kind of data analysis"[29].

In all these examples the users can freely choose data from any source, or even provide their own data. In some cases, the type of visualization can be chosen by the user, from possibilities that are provided by the creators of the application. The powerful aspect of these initiatives is that in addition to participating in the creation of the visuals, the users can share their contributions with other users. This 'Create and Share' category goes further than 'Interact and Explore' – it offers the user the possibility of playing an active role not only in raising his own awareness about a particular issue, but also in pointing other people's attention towards it. Such peer-to-peer activity using web 2.0 technologies could be regarded as 'diagrams for and *by* the masses', or as 'diagrams for the masses 2.0'.

Even though exploration, creation and sharing of visuals by ordinary people were not real possibilities in Neurath's work, it is interesting that Neurath's vision in the 1920s already pointed towards this direction. *Isotype*'s diagrams were made by cutting and pasting individual graphic signs, often created out of woodcut models. This method was timesaving and easy to learn, and hints at possible self-production and sharing. According to Vossoughian, "Neurath felt that cut-outs allowed the masses to feel as though they were participating in the production of knowledge, which was central to his philosophy of reform in general." (Vossoughian 2008: 65)[30].

5 Conclusions

More than 70 years after Neurath's *Isotype* movement, projects such as Al Gore's *Inconvenient Truth* and the work of the *Gapminder* Foundation, share the same spirit: Data accessibility for 'ordinary people', a conscious choice of visuals above words, and a conviction that a better comprehension of statistical data through visuals could lead to desirable social change is clearly found both then and now.

Moreover, digital media have enabled 'do-it-yourself graphics' and their online peer-to-peer distribution, shifting the role of ordinary people from the passive viewer to the active creator and disseminator of 'diagrams for the masses'. What Nyíri (2003: 45) has contemplated as a future possibility has become a reality: digital media have enabled, on a previously unimaginable scale, the fulfillment of Neurath's vision.

References

Arthur, C.: Uncovering Global Inequalities through Innovative Statistics. The Guardian (January 11, 2007) (Retrieved 20-3-08), http://www.guardian.co.uk/technology/2007/jan/11/insideit.guardianweeklytechnologysection

[28] http://services.alphaworks.ibm.com/manyeyes/home

[29] http://services.alphaworks.ibm.com/manyeyes/page/About_Many_Eyes.html

[30] Robin Kinross has criticized this claim by Vossoughian for generalizing from a particular remark in which "Neurath was just talking about schools and children, not about 'masses'." (Kinross 2008).

Brinton, W.C.: Graphic Methods for Presenting Facts. McGraw-Hill Book Company / The Engineering Magazine Company, New York (1914)

Emerson, J.: Visualizing Information for Advocacy: An Introduction to Information Design (2008) (Retrieved 15-3-2008), http://apperceptive.com/infodesign.pdf

Hartmann, F.: After Neurath: The Quest for an Inclusive Form of the Icon. Stroom, The Hague, 31 (October 2006) (Retrieved 15-3-2008), http://www.medienphilosophie.net/texte/neurath.html

Kinross, R.: Isotype: recent publications. Hyphen Press Journal (Blog) (May 12, 2008) (Retrieved 13-6-2008), http://www.hyphenpress.co.uk/journal/2008/05/12/isotype_recent_publications

Kosara, R.: InfoVis 2007: InfoVis for the Masses. EagerEyes.org (November 24, 2007) (Retrieved 15-3-2008), http://eagereyes.org/blog/infovis-2007-infovis-for-the-masses.html

Lupton, E.: Reading Isotype. Design Issues 3(2), 47–58 (1986)

Neurath, M.: Isotype. Instructional Science 3, 127–150 (1974)

Neurath, O.: International Picture Language, pp. 1–49. Kegan Paul, Trench, Trubner & Co., London (1936), http://imaginarymuseum.org/MHV/PZImhv/NeurathPictureLanguage.html

Neurath, O.: Modern Man in the Making. A.A. Knopf, New York (1939)

Neurath, O.: Empiricism and Sociology. In: Neurath, M., Cohen, R.S. (eds.) Vienna Circle Collection, vol. 1. Reidel, Dordrecht (1973)

Nyíri, K.: From Texts to Pictures: The New Unity of Science. In: Nyíri, K. (ed.) Mobile Learning: Essays on Philosophy, Psychology and Education, pp. 45–67. Passagen Verlag, Vienna (2003)

Rosling, H., Rönnlund, A.R., Rosling, O.: New Software Brings Statistics Beyond the Eye. In: OECD World Forum on Key Indicators, Teatro Massimo, Palermo, Italia (November 10-13, 2004) (Retrieved 15-3-2008), http://www.oecd.org/dataoecd/39/48/33843977.doc

Rosling, H.: Visual Technology Unveils the Beauty of Statistics and Swaps Policy from Dissemination to Access. Statistical Journal of the AIOS 24, 103–104 (2007)

Rotha, P.: From Hieroglyphics to Isotypes. In: Neurath, O. (ed.) Future Books No. III, p. 92 (1946) (Retrieved 20-3-08), http://www.fulltable.com/iso/01.jpg

Twyman, M.: The Significance of Isotype. Graphic Communication Through Isotype, pp. 7–17. University of Reading, Reading (1975/1981)

Viégas, F.: Impact of Social Data Visualization. Infosthetics.com (October 31, 2007) (Retrieved 15-3-2008), http://infosthetics.com/archives/2007/10/the_impact_of_social_data_visualization_infovis_workshop.html#extended

Vossoughian, N.: Mapping the Modern City: Otto Neurath, the International Congress of Modern Architecture (CIAM), and the Politics of Information Design. Design Issues 22(3), 48–65 (2006)

Vossoughian, N.: Otto Neurath. The Language of the Global Polis. NAi Publishers, Rotterdam (2008)

Womack, D.: Seeing is Believing: Information Visualization and the Debate over Global Warming. Adobe Design Center Think Tank (October 18, 2006) (Retrieved 20-3-2008), http://www.adobe.com/designcenter/thinktank/womack_print.html

Detection of Sample Differences from Dot Plot Displays

Lisa A. Best[1], Laurence D. Smith[2], and D. Alan Stubbs[2]

[1] University of New Brunswick
Department of Psychology
P.O. Box 5050
Saint John, NB E2L 4L5 Canada
lbest@unbsj.ca
[2] University of Maine
Department of Psychology
301 Little Hall
Orono, ME 04469-5782, USA
larry.smith@umit.maine.edu, alan.stubbs@umit.maine.edu

Abstract. Cleveland and McGill [10] concluded that dot plots are effective when one judges position along a common scale. We assessed the ability of graph readers to detect sample mean differences in multipanel dot plots. In Experiment 1, plots containing vertically arranged panels with different sample sizes and levels of variability were presented. Sensitivity was greater with large samples and low variability. In Experiment 2, sensitivity depended on the location of the comparison sample, with vertical and superimposed arrays yielding greater sensitivity than horizontal or diagonal arrays. Horizontal arrays also produced a bias to judge data in right-most panels as having higher means. Experiment 3 showed that ordering of data had little effect on sensitivity or bias. The results suggest that good graph design requires attention to how the specific features of a graphical format influence perceptual judgments of data

1 Introduction

Graphs are commonly used as a powerful data analytic tool and many researchers recognize that the use of graphs is a fundamental part of data analysis. Graphical analyses allow one to easily extract information from complex data sets and this promotes scientific discovery. They are used in education, industry, and science to convey patterns in data that would be difficult to express in words. Research has shown that graphs play an important role in science [21], [22] and have figured centrally in a number of major scientific discoveries [15], [17], [33]. Graphs occupy almost 20% of the page space in natural-science journals [8], [34] and are common in science textbooks [2], [27]. The importance of graphs has led some to speak of “graphicacy” (competence in reading, interpreting, and constructing graphs) as a set of cognitive skills on par with literacy or numeracy [38]. The ability to read and interpret graphs is becoming crucial as the number of graph formats and computerized displays increases [39].

Graphs are an important means of representing data for a number of reasons. Well-designed graphs serve an integrative function, inviting comparisons and inferences

G. Stapleton, J. Howse, and J. Lee (Eds.): Diagrams 2008, LNAI 5223, pp. 293–307, 2008.

about data by converting statistical effects into similarities and differences in visual patterns, which the human visual system is well-equipped to discern. Depending on the intended functions of particular graphs, relevant comparisons can be between individual cases or experimental conditions within a single data set, between data sets from diverse sources, or between data sets and theoretical curves. Latour [22] has linked the integrative functions of graphs to their centrality in science, arguing that graphs allow scientists to draw things together by literally *drawing things together* (p. 60). From a perceptual standpoint, drawing things together encourages the seeing of things together, a property of graphs that has been called their synoptic power [34]. Not surprisingly, graph designers have advocated the use of multiple-panel graphs as a way to enhance the synoptic qualities of visual displays [37], [4], [5], [19]. Others have found that multipanel graphs are especially common in the harder branches of science [34] and that their use in physics has increased over time [1].

Graphs can be thought of as communication channels that transmit the information contained in a data set to the visual system of the perceiver. An ideal graph would be one whose format supports the drawing of sound inferences by conveying all of the statistical information available in the data and such a graph could be said to bring the perceiver into full contact with the world of codified phenomena. In the case of synoptic graphs, this would entail the optimal use of the visual system's ability to integrate across cases and conditions by invoking the pattern recognition abilities that underlie multiple comparisons of subsets of the data. Synoptic graphs must be designed to allow the visual system to integrate information and include enough information to maintain sensitivity to subset differences and statistical effects inherent in the data.

One type of multipanel graph that has received increasing attention is the *dot plot*, developed by the statistician William Cleveland [4], [7], [29], [36]. In this format, dots representing individual cases are arrayed in rows with the horizontal position of each dot representing the data value (see examples in Figure 1). Dot plots have the advantage of coding data as position along a common scale, which supports relatively accurate graph-based judgments [11]; but see [16]. Dot plots are superior to bar graphs in certain respects [7] and are applicable to a wide range of data sets [29].

Multiway dot plots are matrices of dot plots in which different samples are displayed in separate panels. Cleveland [4] argued that these graphs support rapid judgments of similarities and differences between samples and allow for the quick detection of errors in data-coding and other problems associated with data integrity. As a way to facilitate information extraction, data presented in a dot plot can be sorted by individual values, and the panels in a matrix can be sorted by sample means or medians. This sorting allows one to easily locate important data points and make comparisons between the panels [4], [5], [6]. Two samples can be presented in a single panel by superposition of distinctive plotting symbols. Because Cleveland's advocacy of dot plots is based explicitly on perceptual theory (and in some cases empirical research), dot plots represent fertile candidates for experimental study.

The use of signal-detection methods [26] for graphical perception lends itself to graph research by virtue of the parallels between signal-detection theory and null-hypothesis statistical testing [12], [14]. Assuming a normal distribution, optimal statistical tests–such as the t test applied to sample pairs–can be regarded as *ideal observers*, that is, as representing performance in which optimal use is made of all the information in a data set. Performance of human observers compared to an ideal observer is known as *relative efficiency*, which can be expressed as the ratio of performances (the squared ratio of the performances can also be used but in this article, we use the unsquared formulation; see [25]).

Legge, Gu, and Luebker [23] argued that relative efficiency provides an absolute scale by which observers' performances can be compared with those of ideal observers or statistical tests. A relative efficiency of 1 means that the perceiver's performance matched the performance of the statistical test and, in terms of signal-detection theory, that the distributions of sensory events governing the perceiver's judgments are as far apart (in standard deviation units) as the populations of data. Relative efficiencies of less than 1, as are typically found, would reflect performances falling short of the ideal observer's, and indicate that the populations of sensory events governing judgments are less separated than the original data populations.

1.1 Purpose of the Current Studies

The overall purpose of these studies was to evaluate the effectiveness of the dot plot and we were interested in comparing the performance of typical graph readers to that of common statistical tests. In statistical inference, the accuracy of a test (its ability to discriminate between genuine and spurious effects) depends on sample size and the variability in the data. To determine how sensitivity depends on these factors in graphical inference, they were varied systematically in Experiment 1. The goal was to determine if graph readers are able to (a) integrate information from increasing sample sizes and (b) overcome the noise represented by data variability and discriminate between sample differences arising from chance from those arising from actual population differences.

Experiment 2 investigated the challenges posed by the multiway dot plot format. Accordingly, the location of the comparison sample was varied across vertical, horizontal, and diagonal positions, relative to the standard sample. The aim was to assess observers' ability to integrate across samples displayed in different panels of synoptic graphs. Experiment 3 investigated whether the barriers to accurate graphical inference found in Experiment 2 could be reduced by ordering the data within panels, as recommended by Cleveland [4], [7]. It was hypothesized that the ordering of data would provide stimulus patterning that would improve performance by reducing the demands on perceivers' ability to perceptually organize the arrays.

2 General Method

2.1 Stimuli

The stimuli were based on numbers drawn randomly from two normal distributions with equal variances. The mean of the signal distribution was always 10 units higher than the mean of the noise-only (i.e., non-signal). The strength of the signal was sometimes varied across conditions (Experiment 1) by adjusting the standard deviations of both parent distributions, and sample size could be manipulated across a representative range of sizes (n = 8, 16, or 32).

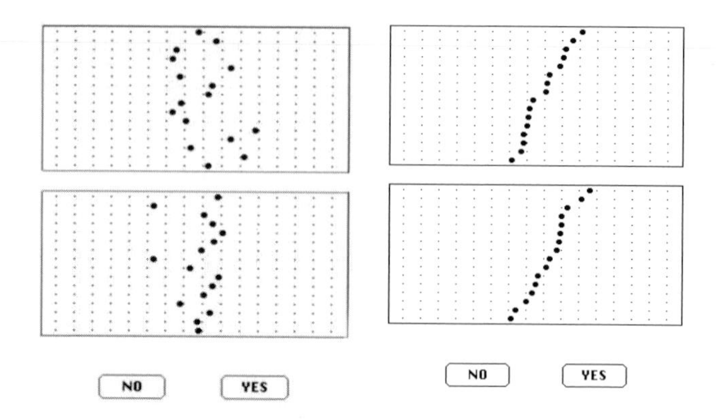

Fig. 1. Dot plots representing a sample stimulus from Experiment 1 (left) and Experiment 3 (right), each showing a standard sample (top) and a comparison sample (bottom)

For the displays, the generated data values for each trial were coded as the horizontal positions of dots in a dot plot and each sample was displayed in its own panel (see Figure 1). The panels were framed by rectangles that measured 8.0 cm across, subtending a visual angle of 7.6° at the viewing distance of 60 cm. The frames were 2.0 cm high (1.9°) when the sample size was 8, 4.2 cm (4.0°) with a sample size of 16, and 8.3 cm (7.9°) with a sample size of 32. The dots were 1.4 mm in diameter, subtending .13° of visual angle. Panels were separated by a space of 6 mm (.57°) between adjacent frames. The panels contained light grid lines consisting of vertical dotted lines spaced 5 mm (.47°) apart.

2.2 Materials

A Macintosh LC475 computer was used to present the stimuli to the participants, provide feedback and record responses, response latencies, and trial-by-trial statistical information about the samples. The monitor was a Radius Pivot monochrome monitor operated in portrait mode.

2.3 General Procedure

At the beginning of each session, participants were informed of the sample size, the level of variability (effect size) that would be in effect during the session, and the corresponding value of d' for the ideal observer. Each trial began with the presentation of a dot plot and response buttons for "yes" and "no" responses. The graph remained on the screen until the participant made a response by clicking one of the buttons with a mouse. On a random half of the trials, the comparison sample came from the parent population with the higher mean. Participants were instructed to respond "yes" if they judged that sample to come from the population with the higher mean and "no" if they judged the samples to come from populations with equal means. Participants were asked to judge the position of the comparison sample *relative to* that of the standard sample rather than by its absolute position within its own panel frame. To further reduce the possibility of reliance on absolute position, a constant (ranging in value from 0 – 20 units) that varied randomly from trial to trial was added to the values in both samples, so that the samples' horizontal positions within the frames changed unpredictably across trials.

After a response was made, the computer gave feedback in the form of a brief tone for a correct response or no tone for an incorrect response. The trial-by-trial feedback was provided to participants to enhance their interest and *challenge* them as they moved through the trials. In general, psychophysics researchers have reported that trial feedback does not improve accuracy. The participant then clicked an "OK" button to proceed to the next trial. Response latencies were recorded for each trial, timed from the onset of the stimulus to the occurrence of the response. Sessions lasted for 50 trials.

3 Experiment 1

3.1 Participants

Seven participants served in the experiment. Four were undergraduate psychology students, two had doctorates in psychology, and one was a graduate student in psychology. All were well-practiced on the task and had normal or corrected-to-normal vision. To ensure that there were no practice effects, performance on the early (Sessions 1 – 6) and late (Sessions 30 – 36) trials were compared and results showed no statistically significant differences in accuracy.

3.2 Method

Participants were presented with vertically arrayed two-panel dot plots (Figure 1) and were asked to judge whether the comparison sample in the bottom panel was drawn from a population with a higher mean. The strength of the signal was varied across conditions by adjusting the standard deviations of both parent distributions (SD = 14.28, 20, or 33.33) so that the resulting statistical effect sizes (Cohen's d) were 0.7, 0.5, or 0.3. These values were chosen so as to represent effect sizes that are typical in psychological research. Sample size was also manipulated across conditions, using a range of sizes commonly found in psychological research (n = 8, 16, or 32). The three levels of sample size and variability (effect size) were factorially crossed and varied across blocks of sessions. Participants completed 4 sessions with each of the 9 combinations of sample size and variability. The resulting 36 sessions, each containing 50 trials, yielded 1800 trials per participant.

As noted above, this task is formally analogous to performing a statistical test for a one-tailed two-sample independent group design. The optimal strategy for the task is to compute the two sample means, subtract the mean of the standard sample from the mean of the comparison sample, and respond "yes" if the difference exceeds a criterion value and "no" if it does not. With a sample size of n, the performance of the ideal observer is given by

$$d'_{\text{ideal}} = \text{ES} \,/\, (2/n)^{1/2},$$

where ES is the statistical effect size, Cohen's d,

$$d = (M_2 - M_1) / SD,$$

where M_2 and M_1 are the means of the comparison and standard populations and SD is the standard deviation of the parent normal distributions. With $n = 8$, the resulting values of d'_{ideal} were 1.40 (low SD), 1.00 (medium SD), and 0.60 (high SD). When $n = 16$ the values were 1.98, 1.41, and 0.85, and were 2.80, 2.00, and 1.20 when $n = 32$.

3.3 Results

To assess the effects of variability and sample size on the sensitivity of participants, a 3 x 3 repeated measures analysis of variance was conducted on d' and η^2, a widely used measure of effect size, was calculated. There were significant main effects for both sample size, $F(2, 12) = 6.30$, $p = .013$, $\eta^2 = .51$, and variability, $F(2, 12) = 21.00$, $p = .0001$, $\eta^2 = .78$. There was no interaction between variability and sample size, $F(4, 24) = .32$, $p = .86$, $\eta^2 = .05$. Sensitivity varied directly with sample size and inversely with variability, as would be expected on the basis of parallels with statistical tests. Sensitivity ranged from 1.53 when variability was low to 1.08 when variability was intermediate and .77 when variability was high. Post hoc tests revealed significant differences between each of the variability levels. Mean sensitivity was 1.32 with 32 data points, 1.11 with 16 points and .95 with 8 points. Post hoc tests revealed that accuracy was significantly greater with 32 data points than with 8 points; the performance differences between sample sizes of 32 and 16 and between 16 and 8 fell short of significance.

A separate 3 x 3 repeated measures ANOVA was performed on relative efficiency. The analysis indicated that performance relative to the ideal observer's declined as a function of increasing sample size, $F(2, 12) = 7.46$, $p = .008$, $\eta^2 = .55$, but that changes in variability did not affect relative efficiency, $F(2, 12) = 1.02$, $p = .39$, $\eta^2 = .15$. The interaction between variability and sample size was not significant, $F(4, 24) = .23$, $p = .92$, $\eta^2 = .04$. Figure 2 shows that relative efficiency decreased monotonically as a function of sample size at all three levels of variability. Averaged across variability levels, the mean efficiency was close to 1 ($M = .96$) when the sample size was 8 and fell to .80 and .68 with sample sizes of 16 and 32. Thus, even though sensitivity increased with sample size, efficiency dropped because participant sensitivity grew at a slower rate than that of the ideal observer. In contrast with the effect of sample size on efficiency, there was no effect of variability, suggesting that the performances of both human and ideal observers deteriorated at the same rate as variability increased.

3.4 Discussion

Averaged across conditions, the mean sensitivity of participants was 1.13, indicating a substantial ability of the visual system to discriminate between genuine and spurious sample differences presented on dot plots. Using scatterplot displays in which data values were coded as vertical positions of dots, Legge and his colleagues [23] also found average d′ values of well over 1, supporting the claim of Cleveland [4], [11] that graphical coding of data as position along a common scale supports accurate graph reading. Legge and his colleagues [23] also compared graph- based performance

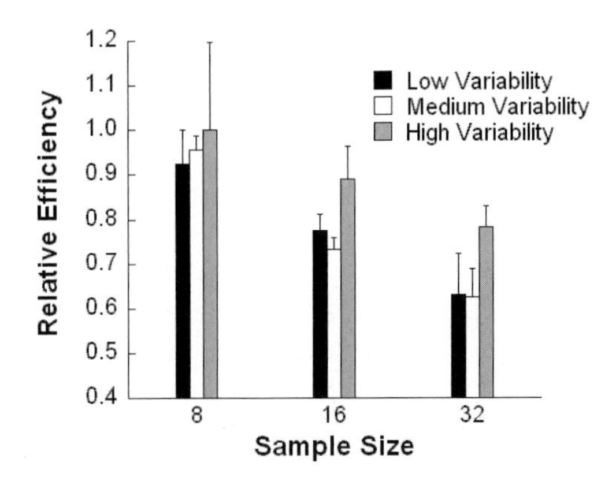

Fig. 2. Mean relative efficiency as a function of sample size and variability. Error bars represent the standard error of the mean.

to performance based on numerical tables and luminance displays and found that sensitivity associated with these displays was lower, providing overall support for the claim that graphs are superior to tables (see also [3]).

The present finding that graph-based performance in detecting sample differences varies inversely with data variability and directly with sample size indicates that visual inference parallels the performance of statistical tests, whose accuracy also declines with variability and increases with sample size. In both cases, increases in variability reduce signal strength, defined in terms of the difference between sample means relative to standard deviation, and increases in sample size provide more information relevant to the task of discriminating signal from noise. Legge and his colleagues [23] and Lewandowsky and Spence [24] also reported that performance improves with larger sample sizes. Overall, these results suggest that graph perception involves at least some degree of parallel processing (see discussion in [23]) and because this type of processing is rapid it may give graphs an advantage over other presentation formats [5], [18], [19], [20].

Although sensitivity varied in ways that paralleled the performance of a *t* test, the overall level of sensitivity of participants was lower than that of the inferential test. This shortfall of performance relative to the ideal observer depended on sample size. The fact that efficiency decreased with sample size even though sensitivity increased indicates that the participants made some use of the additional information contained in the larger samples but did so less and less effectively relative to the total available information in the samples. Such drop-offs in efficiency as a function of sample size were also reported by Legge et al. [23] and Smith, Boynton, and Stubbs [31].

Increases in variability affected the human and ideal observers equally, suggesting that perceivers were able to "see through" the noise as effectively as the corresponding numerical methods. This similarity in performance suggests that observers are able to overcome the obscuring effects of noise when making graph-based decisions about whether variable data sets reflect mean differences. The finding that graph-based judgments keep pace with statistical tests when confronted with increasing variability is somewhat surprising, especially in light of the common view that the major benefit of inferential statistics is their ability to detect signals against a background of noise.

4 Experiment 2

The overall goal of this study was to examine the effects of panel placement on the accuracy of judgements. The location of the comparison sample was systematically varied, such that it was positioned vertically (Figure 1), horizontally, or diagonally, with respect to the standard sample. In a fourth condition, the two samples were superimposed , in a single panel, with data values coded by open dots for the standard sample and filled (black) dots for the comparison sample. The inclusion of this condition was motivated by Cleveland's [7] recommendation of using superimposed data for dot plots with smaller sample sizes. A fifth condition, called the *horizontal-left* condition, was a variant on the horizontal condition in which the locations of the standard and comparison samples were reversed so that the comparison sample was in the left panel; in this case, participants were instructed to respond "yes" if the *left* sample was judged to come from the population with the higher mean. This condition was included as a control because of pilot data suggesting that horizontal arrays may induce response biases.

4.1 Participants

Six participants served in the experiment. Three were undergraduate students, two had doctorates in psychology, and one was a graduate student. The undergraduate participants were not the same as those who served in Experiment 1. All participants were well-practiced on the task and had normal or corrected-to-normal vision.

4.2 Methods

The basic procedure for this experiment was similar to that of Experiment 1 in that participants were shown dot plots containing two samples and had to decide whether the comparison sample came from a population with a higher mean. In this experiment, the sample size was held constant at $n = 16$ and variability was fixed at the intermediate level (population *SD*s = 20, effect size = 0.5). Each participant completed a block of 18 sessions under each of the five conditions; at 50 trials per session, there were a total of 4500 trials per participant.

4.3 Results

A one-way repeated measures ANOVA performed on d' showed that panel placement significantly affected the sensitivity of participants, $F(4, 20) = 10.93$, $p = .0001$, $\eta^2 = .69$. Figure 3 shows that sensitivity was higher for superimposed data ($d' = 1.16$) than for vertical ($d' = 1.10$), diagonal ($d' = .72$), right ($d' = .58$), or left ($d' = .70$) panels. Post hoc tests revealed that sensitivity in the superimposed condition differed significantly from all other conditions. Sensitivity was also significantly higher for

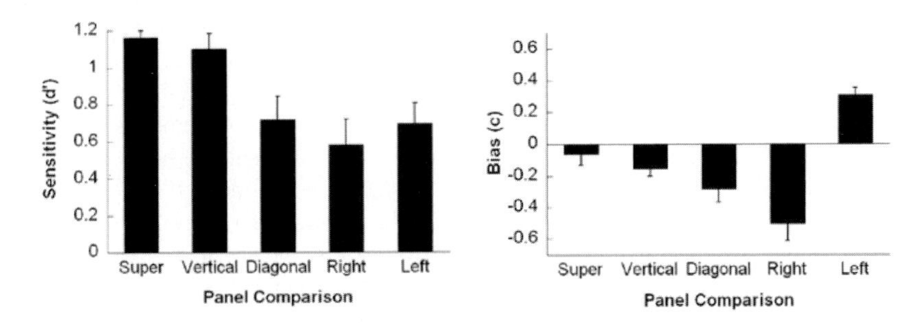

Fig. 3. The sensitivity (left) and response bias (right) as a function of panel placement. For response bias, negative values indicate a *yes* bias and positive values indicate a *no* bias.

vertical comparisons than for the diagonal and horizontal comparisons, but there were no differences among the diagonal and horizontal conditions.

The decision criterion, or response bias, of an observer refers to whether there is a tendency to respond in a certain way, in this case to respond "yes" versus "no." An ANOVA performed on c showed that bias was affected by the placement of the dot plot panels, $F(4, 20) = 32.74$, $p = .0001$, $\eta^2 = .87$. As can be seen in Figure 3, bias was least (i.e., closest to 0) when participants made superimposed (c = -.06) and vertical (c = -.16) comparisons. Participants adopted a more liberal criterion for diagonal (c = -.29) and horizontal-right (c = -.51) comparisons. In contrast, when horizontal-left comparisons were made, the response criterion became more conservative, shifting to a "no" bias (c = +.31). Post hoc tests showed that all pair-wise differences between the five conditions were significant, except for the difference between vertical and superimposed and the difference between vertical and diagonal. In sum, the results from the two horizontal conditions show that there was a tendency to judge whichever sample was located to the right as having a higher mean. A similar but lesser tendency was found when judging samples located down and to the right (diagonal condition).

4.4 Discussion

Multiway graphs are thought to represent a highly effective way to display data because they permit visual comparisons between multiple data sets and encourage synoptic viewing. However, in the case of dot plots, the conclusions drawn from viewing multipanel graphs are influenced by the location of the panels. Panel placement affected both discrimination accuracy and decision criterion. Perceivers' sensitivity to subtle differences in sample means was relatively good when panels were arranged vertically and when samples were superimposed on a single panel, but decreased when the panels were arranged diagonally or horizontally. This finding is consistent with the claim of Cleveland and McGill [4], [10] that coding data by position along a common aligned scale is superior to coding data by position on common but nonaligned scales (on the advantages of superposition in line graphs, see [30]). Although irrelevant to the information contained in the data, panel placement thus has perceptual consequences that can affect conclusions drawn from data. This is an important consideration that should be taken into account when constructing dot plots [4], [39].

The value of multiway plots derives largely from their potential for supporting judgments of various pair-wise comparisons, but differential criterions based on panel placement would likely compromise their consistency. As shown here, samples in panels displaced rightward from the standard sample were more readily judged to have higher means, especially when the displacement was directly to the right (horizontal) but also when it was down and to the right (diagonal). A side-to-side reversal of the standard and comparison samples (horizontal-left condition) produced a symmetrical reversal in the tendency to say "yes", showing that the right-most sample tended to be seen as having a higher mean regardless of its status as the standard or comparison. Because participants reported that they generally inspected the standard sample before the comparison sample (although not explicitly instructed to do so), this finding implies that the shift in decision criterion is due to the rightward displacement itself rather than to the direction of eye movements while scanning the graph or other factors arising from the arbitrary designation of samples as standard or comparison. It may be that the overall task demand of judging data values by their horizontal position, with larger values coded as positions farther to the right, led participants to perceptually incorporate some of the rightward distance between panels into their judgments of dot position within panels. In this way, the format of the multiway dot plots may inadvertently induce reliance on factors other than the critical factor of within-frame position of dots.

The right-bias phenomenon proved quite durable in the experiment, remaining strong over hundreds of trials despite the trial-by-trial feedback given to participants and despite the presence in the displays of grid lines that could have been used to aid the accurate estimation of sample means. The effect was phenomenologically powerful to observers, being experienced as a compelling illusion. Participants continued to experience the phenomenon even after being informed of its existence and informal tests showed that it occurred when the displays were printed on paper as well as when presented on the computer screen. The effect has also been confirmed, and its magnitude measured, using a method-of-adjustment procedure [32]. All in all, the right-bias effect appears to constitute a type of "graphical illusion," a class of perceptual aberrations that can arise when graphs are viewed and that compromise the accurate extraction of information from them. The phenomenon of graphical illusions was discussed by Graham [13] and explored by Poulton [28]; see also [9], and[19], pp. 276, 282).

5 Experiment 3

The chief purpose of this experiment was to test Cleveland's [4], [5], [6] assertion that the extraction of information from dot plots is facilitated by ordering the data according to their values. A second purpose was to determine whether this sorting of data, aside from any possible enhancement of sensitivity, would eliminate or reduce the bias-inducing effects of horizontal panel placement that were demonstrated in Experiment 2.

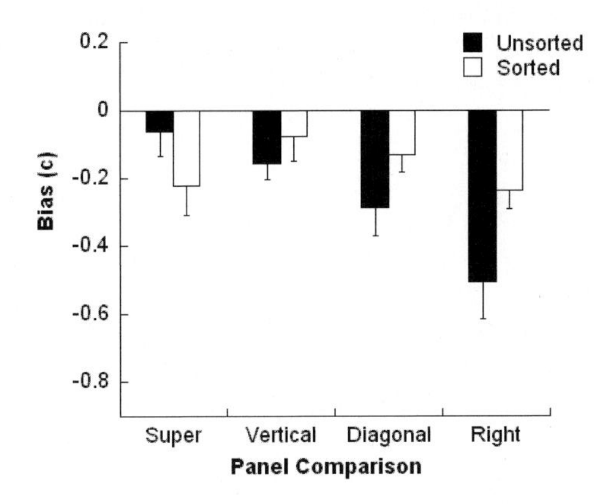

Fig. 4. Mean bias as a function of panel placement and sorting condition

5.1 Participants

The six participants who completed Experiment 2 served in the experiment.

5.2 Procedure

The procedure was the same as that of Experiment 2, except that the data were sorted and displayed in increasing order, bottom to top, within the panels. Because the samples were drawn from normal parent distributions, this sorting resulted in dot patterns that approximated normal ogives (see example in Figure 1). Panel placement varied across blocks of sessions, with the samples arrayed in the vertical, horizontal-right, diagonal, and superimposed positions (the horizontal-left condition was omitted).

5.3 Results

To assess the effects of panel location on sensitivity with the sorted data, a one-way repeated measures ANOVA was performed on the sorted data. As in the previous experiment, sensitivity depended on panel location, $F(3, 15) = 7.86$, $p = .002$, $\eta^2 = .61$. Post hoc tests revealed that sensitivity was again significantly higher with superimposed panels ($d' = 1.29$) than with vertical ($d' = .98$), horizontal ($d' = .73$), or diagonal ($d' = .77$) panels. Although the mean d' for the vertical condition was higher than for the diagonal or horizontal-right conditions, these differences failed to reach significance, $.10 > ps > .05$. The diagonal and horizontal conditions did not differ significantly.

The effects of panel location on response bias with the sorted data are shown in Figure 4 (unfilled bars). As can be seen, sorting tended to reduce the differences in response bias as a function of panel location, in part by reducing bias in the diagonal and right conditions. However, a one-way ANOVA performed on the bias measure c for the sorted data revealed that significant differences remained, $F\ (3, 15) = 3.76$, $p = .034$, $\eta^2 = .43$. Post hoc tests showed that vertical panels produced significantly less bias than superimposed and right panels, with no difference between the vertical and diagonal conditions. Thus, despite some reduction in the amount of bias in the right and diagonal conditions, sorting failed to eliminate the effects of panel placement.

5.4 Discussion

In terms of sensitivity, the results with the sorted data closely matched the results with unsorted data from Experiment 2. Overall sensitivity was unaffected by sorting, and the effects of panel position on sensitivity for sorted data mirrored the pattern of effects for unsorted data. Thus, sorting neither enhanced the average sensitivity to sample differences nor reduced the differential sensitivity attributable to variations in panel location.

It remains unclear why sorting fails to enhance sensitivity in these graph-based tasks. Cleveland [4] argued that the process of *assembly*, a stage of graph perception in which individual data points are grouped into patterns, is facilitated by the sorting of data. Legge et al. [23] noted that sorting would be expected to facilitate the detection of sample medians (because median values could be quickly picked out from the center of the sorted samples), and they showed through computer simulations that a strategy of comparing sample medians would support highly efficient performance when discriminating differences in population means. However, fits of their participants' performance to simulated performance with a median-based strategy were poor, suggesting that participants were not adopting such a strategy.

Although the sorting of data failed to improve sensitivity, the results hint at the possibility of beneficial effects of sorting on bias. Sorting tended to reduce overall response bias (although with only marginal statistical significance), and it appeared especially to reduce bias in the horizontal-right condition where it had been greatest with unsorted data. As a result, there was a tendency for sorting to equalize differences in bias across panel placements. If confirmed in further research, such an effect could be helpful in situations involving multiple comparisons because sorting would encourage the use of similar criteria for judging differences across various comparison pairs regardless of their placement in a graph (even if location-related differences in sensitivity remain).

6 General Discussion

Taken as a whole, the current results show that dot plots function fairly efficiently to convey mean differences between samples. The plots supported perceptual judgments that preserved much of the signal strength represented in the underlying population differences. Compared to the corresponding statistical tests, however, dot plot-based judgments suffered a loss of sensitivity when samples sizes were relatively large, as observers failed to use all of the information contained in the additional cases (Experiment 1). There was no comparable effect of increasing variability, as perceivers' sensitivity kept pace with that of the tests as variability increased.

An overall impression of how well dot plots support visual judgments of sample differences can be gained by comparing the judgments with decisions made by t tests applied to the same samples across a range of variations in signal strength. Figure 5

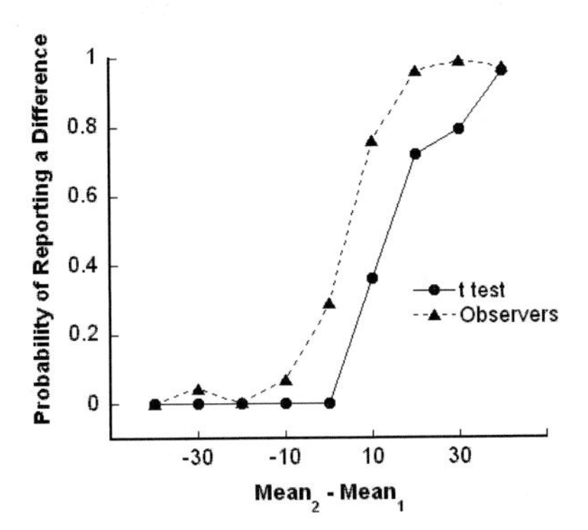

Fig. 5. The probability of judging that the comparison sample came from a population with a higher mean, as a function of the difference between sample means

shows the likelihood that the comparison sample was judged to come from the higher-mean population as a function of the difference between sample means. The curves are based on all 12,600 trials in Experiment 1, pooled across conditions and participants, and the data for the *t* test represent trial-by-trial determinations of significance using one-tailed tests at the .05 level of significance.

Three features of the data are noteworthy. First, the function for the *t* test lies to the right of the function for the observers, a difference that represents the conservative bias of the statistical test. Because the t-test is constrained to make decisions that produce false alarms on only 5% of trials, it makes fewer "yes" responses than do human observers. Second, as the curves rise through the middle portion of the graph, they remain fairly parallel, indicating good sensitivity of the observers in the crucial range where sample differences are especially informative. Third, despite this good sensitivity, the function for the observers begins to rise substantially before the mean difference reaches 0, indicating that comparison samples were sometimes erroneously judged to lie above the standard samples. These nonoptimal responses likely reflect the difficulty of estimating means from graphical displays [35] and constitute a major reason why performance fell short of the *t* test's. Subsequent analysis showed that the majority of such responses occurred when the sample size was large, suggesting the difficulty of integrating information across large numbers of cases.

The potential of multipanel plots to facilitate synoptic viewing was compromised when panels did not share aligned scales (Experiment 2). First, judgments suffered reduced sensitivity when comparison panels were displaced to the right, either horizontally or diagonally, from the standard panel. Second, there was a bias to judge the right-displaced samples as having higher means. The sorting of data failed to reverse the position-related sensitivity decrements and had only marginal effects on the placement-induced bias differences (Experiment 3). These problematic effects, which compromise the accuracy and consistency of graph-based comparisons, are perceptual phenomena, and so do not affect the decisions made by statistical tests (or ideal observers). The bias of statistical tests, no less than their sensitivity, is immune to panel placement effects. In setting strict decision criteria that control error rates, tests also adopt *uniform* criteria that are applied consistently to all sample pairs, something that perceivers in these experiments were unable to do in making their visual judgments. Comparisons of visual inference with normative numerical inference thus show that the benefits of synopticality that have stimulated the interest of graph designers in multipanel displays come with associated risks, risks arising from perceptual properties that appear to be inherent.

References

1. Bazerman, C.: Theoretical integration in experimental reports in twentieth-century physics: Spectroscopic articles in physical review, 1893-1980. In: Bazerman, C. (ed.) Shaping Written Knowledge, pp. 153–186. University of Wisconsin Press, Madison (1988)
2. Butler, D.L.: Graphics in psychology: Pictures, data, and especially concepts. Behav. Res. Meth. Instrum. Comput. 25, 81–92 (1993)
3. Chambers, J.M., Cleveland, W.S., Kleiner, B., et al.: Graphical methods for data analysis. Wadsworth & Brooks/Cole, Pacific Grove (1983)
4. Cleveland, W.S.: The elements of graphing data (rev. ed.). AT&T Bell Laboratories, Murray Hill (1994)
5. Cleveland, W.S.: Visualizing data. AT&T Bell Laboratories, Murray Hill (1993)
6. Cleveland, W.S.: The elements of graphing data. Hobart Press, Summit (1985)
7. Cleveland, W.S.: Graphical methods for data presentation: Full scale breaks, dot charts, and multibased logging. Am Stat. 38, 270–280 (1984)
8. Cleveland, W.S.: Graphs in Scientific Publications. Am Stat. 38, 261–269 (1984)
9. Cleveland, W.S., Diaconis, P., McGill, R.: Variables on scatterplots look more highly correlated when the scales are increased. Science 216, 1138–1141 (1982)
10. Cleveland, W.S., McGill, R.: Graphical perception and graphical methods for analyzing scientific data. Science 229, 828–833 (1985)
11. Cleveland, W.S., McGill, R.: An experiment in graphical perception. Int. J. Man Mach. Stud. 25, 491–500 (1986)
12. Gigerenzer, G., Murray, D.J.: Cognition as intuitive statistics. Erlbaum, Hillsdale (1987)
13. Graham, J.L.: Illusory trends in the observations of bar graphs. J. Exp. Psychol. 20, 597–608 (1937)
14. Green, D.M., Swets, J.A.: Signal Detection Theory and Psychophysics. Wiley, New York (1966)
15. Hankins, T.K.: Blood, dirt, and nomograms: A particular history of graphs, pp. 50–80. ISIS (1999)
16. Hollands, J.G., Spence, I.: Judgments of change and proportion in graphical perception. Hum. Factors 34, 313–334 (1992)
17. Holmes, F.L., Olesko, K.M.: The images of precision: Helmholtz and the graphical method in physiology. In: Wise, M.N. (ed.) The Values of Precision, pp. 198–221. Princeton University Press, Princeton (1995)
18. Kosslyn, S.M.: Graph design for the eye and mind. Oxford University Press, New York (2006)
19. Kosslyn, S.M.: Elements of graph design. W. H. Freeman, New York (1994)
20. Kosslyn, S.M.: Cognitive neuroscience and the human self. In: Harrington, A. (ed.) So Human a Brain: Knowledge and Values in the Neurosciences, pp. 37–56. Birkhaeuser, Boston (1992)
21. Krohn, R.: Why are Graphs so Central in Science? Biol. Philos. 6, 181–203 (1996)
22. Latour, B.: Drawing things together. In: Lynch, M., Woolgar, S. (eds.) Representation in Scientific Practice, pp. 19–68. MIT Press, Cambridge (1990)

23. Legge, G.E., Gu, Y., Luebker, A.: Efficiency of graphical perception. Percept Psychophys. 46, 365–374 (1989)
24. Lewandowsky, S., Spence, I.: Discriminating strata in scatterplots. J. Am. Stat. Assoc. 84, 682–688 (1989)
25. Licklider, J.C.R.: Theory of signal detection. In: Swets, J.A. (ed.) Signal Detection and Recognition by Human Observers: Contemporary Readings, pp. 95–121. Wiley, New York (1964)
26. Macmillan, N.A., Creelman, C.D.: Detection Theory: A User's Guide. Cambridge University Press, New York (1991)
27. Peden, B.F., Hausmann, S.E.: Data graphs in introductory and upper level psychology textbooks: A content analysis. Teach. Psychol. 27, 93–97 (2000)
28. Poulton, E.C.: Geometric illusions in reading graphs. Percept. Psychophys. 37, 543–548 (1985)
29. Sasieni, P.D., Royston, P.: Dotplots. 45, 219–234 (1996)
30. Schutz, H.G.: An evaluation of formats for graphic trend displays: Experiment II. Hum. Factors 3, 99–107 (1961)
31. Smith, L.D., Boynton, D.M., Stubbs, D.A.: Intuitive statistics as signal detection: Perceptual judgments of sample differences. Poster presented at the meeting of the American Psychological Association, Boston, MA (1990)
32. Smith, L.D., Stubbs, D.A., Best, L.A.: Graphical illusion in multipanel dot plots: Method-of-adjustment analysis. Paper presented at the American Psychological Association Annual Meeting, Honolulu (2004)
33. Smith, L.D., Best, L.A., Cylke, V.A., et al.: Psychology without p Values: Data Analysis at the turn of the 19th century. Am. Psychol. 55, 260–263 (2000)
34. Smith, L.D., Best, L.A., Stubbs, D.A., et al.: Constructing knowledge: The role of graphs and tables in hard and soft psychology. Am. Psychol. 57, 749–761 (2002)
35. Spencer, J.: Estimating averages. Ergonomics 4, 317–328 (1961)
36. Sternberg, R.J.: The psychologist's companion: A guide to scientific writing for students and researchers, 3rd edn. Cambridge University Press, Cambridge (1993)
37. Tukey, J.W.: Data-based graphics: Visual display in the decades to come. Stat. Sci. 5, 327–339
38. Wainer, H.: Visual Revelations: Graphical Tales of Fate and Deception from Napoleon Bonaparte to Ross Perot. Copernicus, New York (1997)
39. Wainer, H., Velleman, P.F.: Statistical graphics: Mapping the pathways of science. Annu. Rev. Psychol. 52, 305–335 (2001)

Visualizing Non-subordination and Multidominance in Tree Diagrams: Testing Five Syntax Tree Variants

Leonie Bosveld-de Smet[1] and Mark de Vries[2]

[1] Department of Information Science
[2] Department of Theoretical Linguistics
University of Groningen, The Netherlands
{l.m.bosveld,mark.de.vries}@rug.nl

Abstract. In linguistics, it is quite common to use tree diagrams for immediate constituent analysis of sentences. Traditionally, these trees are binary and two-dimensional. However, phenomena such as coordination and right node raising, have led to the view that a simple hierarchical approach of sentences is inadequate: some linguistic phenomena rather seem to involve non-subordination and multiple dependencies. The central question of the present research is this: what are workable alternative tree-like diagrams that can accommodate to this view? An experiment has been set up to test five different types of tree visualizations, including three-dimensional trees. Subjects were asked to respond to various questions concerning coordination and (non-constituent) right node raising constructions, and to mark their preference for each tree visualization. This paper will discuss the representation problems, and present the experiment and its results. It turned out that the tree most rich in information was the least usable one, whereas the tree, most close to the traditional syntax tree, but with colour enrichment, performed best.

Keywords: tree diagrams, non-subordination, multidominance, usability.

1 Introduction

In this paper we test representations of information that can mostly be ordered in a hierarchical way, but, for some part, goes beyond hierarchy. Five tree-like diagrams have been designed to represent these data. An empirical test has been set up in order to see which tree visualization would come out as the most usable one.

Tree diagrams, also called hierarchies, are a way to visualize hierarchically ordered information, not only in non-scientific, but also in scientific contexts. As a scientific diagram, the tree diagram belongs to the category of schematic diagrams [11, 15]. Schematic diagrams are abstract diagrams that provide an overview of the components and their organization of a set of raw data according to some model. Not only in science education, but also in scientific research, they are an important tool for thinking and reasoning. They often serve as an external memory device for multilevel hierarchical information, such as class inclusion or componential analysis. Being schematic diagrams, they rely on learned conventions [15, 14]. They are not restricted

G. Stapleton, J. Howse, and J. Lee (Eds.): Diagrams 2008, LNAI 5223, pp. 308–320, 2008.

to one specific application domain. They are so-called domain-general diagrams. As a consequence, most people are familiar with them. They know how to create and use them, and reason with them.

Tree diagrams also occur in the domain of linguistic theory. Harleman Stewart [10] points out that the tree in linguistics bears at least four meanings: genesis, taxonomy, componential analysis, and constituent analysis. The type of data represented and the context of use of the tree determines which meaning should be assigned to it. If the data is a sequence of word tokens constituting a natural language sentence and the context of use is the linguistic subdomain of generative syntax, the meaning of the tree diagram is the immediate constituent analysis of the sentence. The tree depicts the sentence in terms of its constituents. It shows which word token units are constituents of which constructions at which level. An example is the tree in Fig 1.

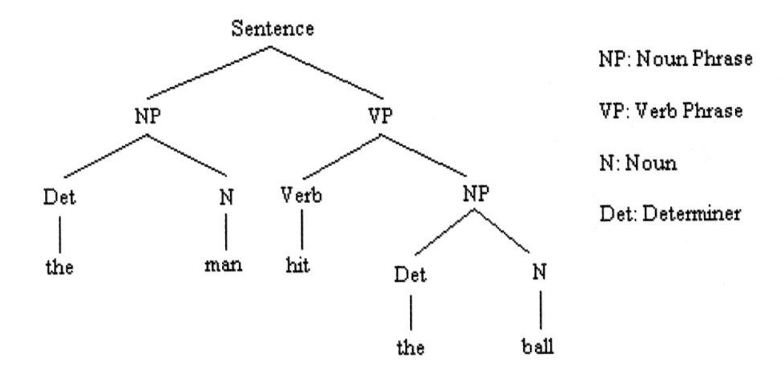

Fig. 1. Constituent analysis of the sentence *The man hit the ball* [6][1]

In the context of generative syntax, tree diagrams are also called syntax trees. It should be pointed out that the process by which diagrams in linguistics are interpreted is a process of circular reasoning. Diagrams are used to make statements about linguistic data for which the questions raised have already been answered. So diagrams in linguistics are merely used to convey and idea, and are not designed for helping to solve a linguistic problem. In terms of information visualization, they visualize the data according to a given, predetermined view. They are not intended to 'amplify cognition', i.e. to solicit visual thinking in order to get new ideas [4].

Graphically, the syntax tree is a set of nodes and branches. As for their arrangement, it follows the definition of an ordered binary two-dimensional tree. The vertical dimension expresses succession, not in time or space, but in the sense of substitution. In the example above, NP, for instance, gives way to Det and N, VP is substitutable by Verb and NP, and so on. In linguistic terms, the branches represent relations of dominance or subordination. The horizontal dimension expresses a relation between sentence parts that are not substituble one for another. In the example above, Det and N form one constituent, verb and NP another. Node and branch also express constituent function. The difference between Subject (*the man*) and Object (*the ball*), for instance, is conveyed in the example above, by a left branching NP from Sentence versus a right branching NP from Sentence via VP. Whereas most tree diagrams

[1] Note that Chomsky used the symbol T instead of Det for determiners.

require units to appear at the nodes, in immediate constituent trees, no units need to appear at their nodes. The expression of immediate constituent analysis is unhampered by the lack of category symbols like NP, VP, and so on. The terminal units form the basic data of the tree. At the terminal level, the horizontal dimension expresses the order in which the terminal units, i.e. the word tokens, appear in the sentence. In linguistic terms, the horizontal axis expresses the relation of precedence. Both vertical and horizontal relations are asymmetric.

Since Chomsky [6], things have changed a lot: new models have appeared, new ideas have been proposed. For quite a long time, the traditional syntax tree could accommodate fairly well to all these changes, until syntax got interest in so-called paratactic phenomena[2] and new theories were produced about these, starting with McCawley [13] (see also [8, 23, 9, 19, 12]). Generative syntax has indeed been preoccupied by hypotaxis for a long time. Hypotactic relations are relations of subordination. Sentence (1) is an example.

(1) He couldn't come because he was ill.

The sentence part *he was ill* is subordinated to *he couldn't come*. However, natural language sentences are not restricted to hypotactic relations between sentence parts, but also manifest paratactic relations. Coordination and parenthesis, for instance, exemplify parataxis. (2) is an example of coordination, (3) of parenthesis.

(2) The man was sitting and the woman was standing.
(3) I told them, mistakenly, it turned out, that she had already left.

In (2), *the man was sitting* is coordinated to *the woman was standing*. These two sentence parts are called conjuncts of the coordination. In (3), *mistakenly, it turned out* is a comment clause inserted in the main clause. Coordination often gives rise to a special construction, the so-called 'right node raising' construction, as exemplified by (4). 'Right node raising' is also known as 'backward conjunction reduction'.

(4) Mary wrote and John signed the letter.

In (4), where the intonation pattern is such that the verbs *wrote* and *signed* are accentuated, the object NP *the letter*, which is a full constituent, is shared by both conjuncts of the coordination, viz. *Mary wrote NP* and *John signed NP*. Sharing does not always involve whole constituents, as demonstrated by (5).

(5) John offered, and Mary actually gave a gold Cadillac to Billy Schwartz.

In (5), the two conjuncts *John offered NP PP*[3] and *Mary actually gave NP PP* share a sequence of phrases and words, viz. *a gold Cadillac to Billy Schwartz*, that does not form a constituent.

Linguists do not quite agree on the way how to account for paratactic relations and how to deal with conjuncts sharing constituent or non-constituent parts of a sentence. In this paper, it is assumed that sentences with coordination imply a third asymmetric relation, alongside those of dominance and precedence. This relation is called non-subordination. It is further assumed that right node raising constructions involve multidominance. Non-subordination and multidominance are views put forward by de Vries (see [20, 21, 22]) in order to account for the connection between conjuncts of a

[2] In grammar, parataxis refers to phrases and clauses arranged independently.
[3] PP stands for prepositional phrase.

coordination. When a node N_1 is in a relation of non-subordination with another node N_2, N_2 can be viewed as lying behind N_1. Subordination constitutes the first representation problem for the traditional syntax tree.

The other representation problem is the relation of multidominance, i.e. the sharing of nodes. In an ordered binary tree, one node is replaced by one or more nodes at a lower level. Dominance is indeed a one-many relation. Ordered binary trees do not allow for several nodes being replaced by one node at a lower level.

So, non-subordination and multidominance challenge the representational properties of the traditional syntax tree.The two representation problems raised first led to the question how to modify the traditional and among linguists commonly accepted and well-known tree diagrams in order to convey the new theoretical views on paratactic phenomena in natural language. With the renewed and expanding interest in visualizations of all types, design theories offer interesting visualization techniques for optimizing representations of all kinds of data [2, 18, 4]. We have made use of some of these to enrich the traditional tree diagrams in order to represent visually the new syntactic views. The enrichments led to five alternative solutions. These various options then raised a second question, namely the queston which enrichment would be the most usable one. In order to answer this question, an experiment has been set up to test the usability of the five alternatives designed. Performance of the task, which consisted in answering a series of questions on syntactic configurations, was measured by the registration of error rates and response times. User preference for the various tree diagrams was measured as well.

2 The Five Visual Enrichments of the Traditional Syntax Tree

The preliminary condition for all alternative tree diagrams was to maintain the hypotactic information as visualized by the traditional syntax tree following X-bar theory for Dutch syntax, see [1]). The five alternatives differed with respect to the visualization of non-subordination and multidominance. Below we show the five tree diagrams proposed. All trees have been composed in Blender, the open source, cross-platform suite of tools for 3D creation.

All five trees visualize the structure of five different, but comparable, Dutch sentences illustrating each coordination with non-constituent right node raising. The structure of these sentences all involve non-subordination and multidominance.

The tree in Fig. 2 is called the 2D variant. It is most close to the traditional two-dimensional tree diagram. It has only a special accommodation for multidominance relations. These are marked by red-colored branches. The tree in Fig. 3 is labelled 2D+. It is like 2D in that it has color coding for multidominance. Moreover, it marks the different syntactic categories at the internal nodes as well by different colors. These also determine the final colors of the multidominance relations. Fig. 4 shows a three-dimensional tree, called 3D. It is like 2D, augmented with a special encoding for non-subordination. The two conjuncts of the coordination are represented as two subtrees, one behind the other at a third dimension, and connected to each other by a pink-colored branch.Figure 5 shows the 3D+ variant, which is like 3D, but adds to it the color marking of 2D+ of the internal nodes. 3D+ is the most rich tree from a graphical and a semantic point of view.

D_planes, illustrated by Fig. 6, overlaps with 2D, but differs from it by representing the conjuncts as subtrees in different planes, thus creating a pseudo three-dimensional picture. Finally, it should be noted that an important concern for the composition of each of the above trees was clutter. Clutter has been avoided as much as possible.

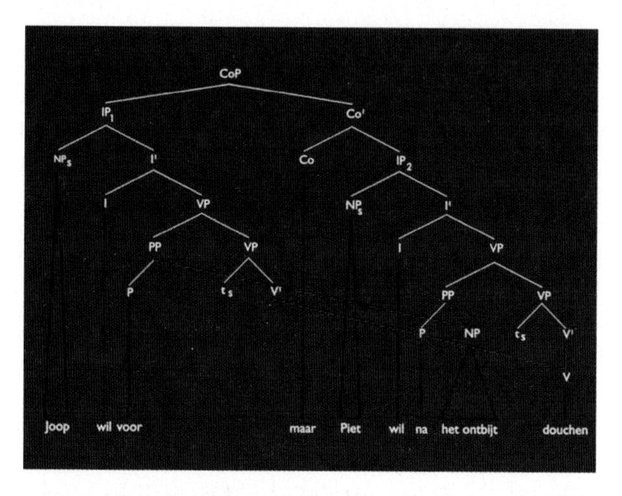

Fig. 2. 2D tree visualizing the structure of Dutch *Jan wil voor maar Piet wil na het ontbijt douchen* (Eng. *John wants before but Peter wants after breakfast to take a shower*)

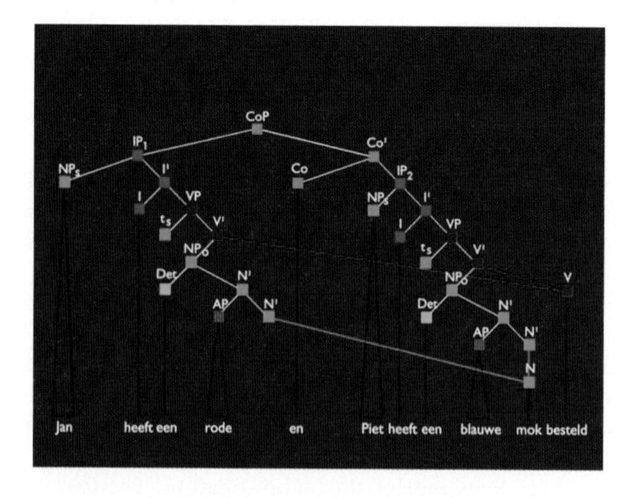

Fig. 3. 2D+ tree visualizing the structure of Dutch *Jan heeft een rode en Piet heeft een blauwe mok besteld* (Eng. *John has a red and Peter has a blue cup ordered*)

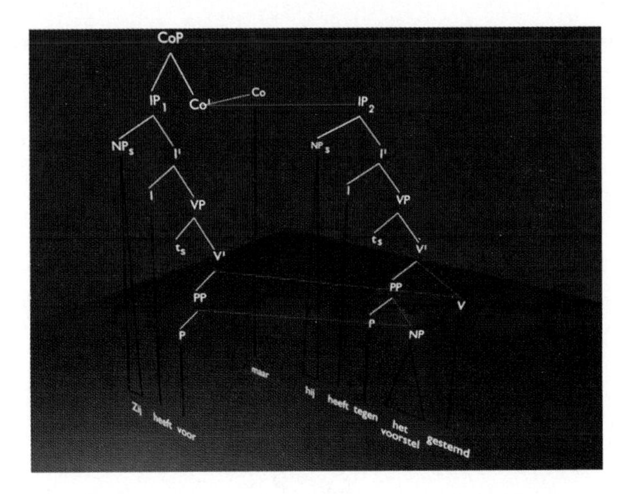

Fig. 4. 3D tree visualizing the structure of Dutch *Zij heeft voor maar hij heeft tegen het voorstel gestemd* (Eng. *She has for but he has against the proposal voted*)

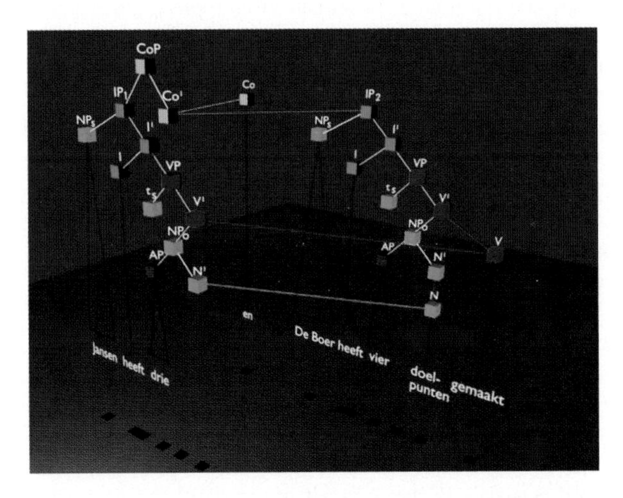

Fig. 5. 3D+ tree visualizing the structure of Dutch *Jansen heeft drie en De Boer heeft vier doelpunten gemaakt* (Eng. *Jansen has three and De Boer has four goals made*)

3 The Experiment

An experiment has been set up to test the usability of the above enriched syntax trees. What did we consider to be the likely outcome? From an information visualization point of view, one might be tempted to think that the richer the tree in structure and color, the more informative the representation is, and the more usable it will be for the illustration and demonstration of, and reasoning with non-subordination and multidominance.

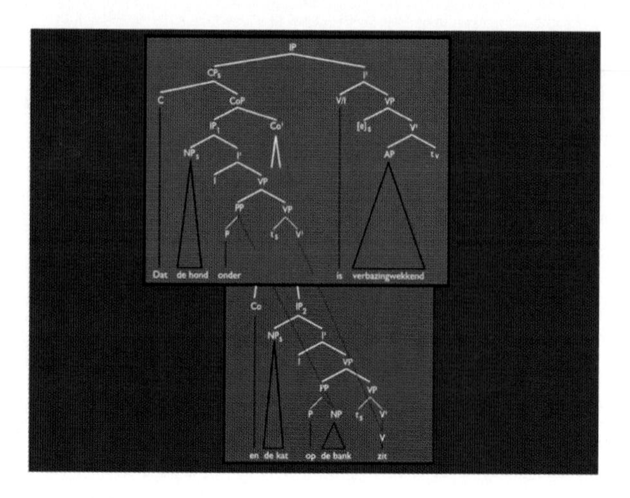

Fig. 6. D_planes tree visualizing the structure of Dutch *Dat de hond onder en de kat op de bank zit is verbazingwekkend* (Eng. *That the dog under and the cat on the bench sits is surprising*)

Three-dimensionality would signal coordination and its conjuncts, the color coding of the internal nodes would make it easier to recognize and identify phrases belonging to a certain syntactic category, and the color coding on the branches would help to see the deviant types of links, viz. those indicating non-subordination on the one hand, and those indicating multidominance on the other hand. However, this positive view on semantically rich representations might be tempered by sceptical cognitive views (see [16, 5] among others). Moreover, in the literature on the use of graphical representations in instruction and user interfaces, one can also find observations that run counter to the expectation that informationally rich diagrams will be appreciated by their users. In [3] for instance, it is pointed out that the more familiar the diagram, the easier it is to process. New diagrams have to be learned in order to be effective in the sense of facilitating a task. Results from empirical research show that three-dimensional visualizations in user interfaces are not necessarily easier to process than their two-dimensional variants (see [17, 6]).

Forty-six persons served as subjects in this study. This group contained 24 females and 22 males. All of them had native or near-native knowledge of the Dutch language. Ages varied between 18 and 65. All of them either followed or had finished a university education. With respect to the background knowledge in the specific application domain of generative syntax, three different groups could be distinguished. One group of 17 subjects (8 males, 9 females) with ages varying from 18 to 65, did not have any education in syntax. For us this was the group with no background knowledge. The second group, a group of 18 students of the faculty of Arts of the University of Groningen (9 males, 9 females, with ages varying from 18 to 30) got some introductory course in generative syntax. They could be seen as the group with weak background knowledge. The third and last group contained 11 experts (5 males and 6 females) in the field of generative syntax. They had finished a linguistic education at the university and some of them have got a Ph.D.. It was the group with strong background knowledge. Evidently, the expert group was most familiar with the

traditional syntax trees, and their conventions, the laymen were least familiar with them. For all subjects, the enriched trees were new. The two-dimensional tree diagram as domain-general diagram could be considered common knowledge of all subjects.
The five enriched syntax trees were tested in a web-based application. The application consisted of five parts:

1. The web-based test started with a page where the subjects were asked to enter age, sex, education and background knowledge.
2. A short introduction and help was provided explaining and illustrating the syntactic category symbols as used in X-bar theory, and the notions of constituency, dominance (or subordination), and coordination. The notions and symbols were illustrated with the help of the traditional syntax tree.
3. The actual test showed the five enriched trees visualizing three sentence constructions in Dutch, viz. coordination, constituent right node raising and non-constituent right node raising. Fifteen different pictures were shown to each subject, who had to answer three different type of questions for each picture. Each subject was thus solicited to inspect each of the five tree variants on the basis of nine questions. The actual test covered forty-five pages.
4. After the questions, a page summarized the five enriched syntax tree variants and asked the subject to mark his preference for each tree on a Likert scale with five values, ranging from very good to very bad.
5. The test was closed by thanking the subject. The subject could also give his or her comments on this page.

The independent variables of the experiment were the type of tree diagram (five levels), the degree of syntactic background knowledg (three levels), the syntactic construction type (three levels), and the question type (five levels). The dependent variables were error rate, performance time, and user preference. The experiment was conducted as a within-subjects design.The main interest of the study was in the manipulation of the tree diagram. Transfer of learning effects were lessened by varying the order of the five trees between the subjects, and by varying the content of the trees. The three construction types were illustrated in each visualization type by different sentences, as shown by Fig. 2-6 above. The questions were formulated as yes-no questions and in domain knowledge terms of the notions explained in the introductory and help page. For instance, the questions asked with respect to the 2D tree in Fig. 2 were the following (below, they are given in Eng.):

1. Is *voor het ontbijt douchen* a constituent?
2. Are IP_1 and Co dominated by the same nodes?
3. Is V dominated by PP in the first conjunct?

From the diagrammatic view proposed by Novick and Hurley [15], who distinguish three categories of properties of hierarchies, the questions required the subject to consider the basic structure of the tree, as well as details about the nodes and links in the tree, and potential movement of information through the tree.

The subjects were invited to take the web-based test in their own time at a quiet place.

4 Results

The question which tree is the most usable one for the representation of non-subordination and multidominance relations is answered by looking at the scores on the three different usability aspects measured for each tree. The global results are given in Fig. 7, 8 and 9 below. Fig. 7 gives the results in percentages for the mean number of errors made for each visualization type. A proportion test revealed that the difference between 2D+ and D_planes, on the one hand, and 2D, 3D and 3D+, on the other hand , is significant ($p < 0.05$). In Fig. 8 , the mean response times are given for each visualization type. The subjects took most time for answering the questions with the 3D+ variant. The difference of 3D+ with each of the other visualization types turned out to be significant, as indicated by a MANOVA test (Bonferroni post hoc, $p<0.05$). Fig. 9, at last, shows the preference results. These are the subjective judgments made by the subjects on each visualization type. It is clear that the tree the subjects liked best was the 2D variant. The differences are significant for 2D versus the other visualization types, for 2D+ versus 3D+ and D_planes, and for 3D versus 3D+ and D_planes, as shown by a One-way ANOVA test (Bonferroni Post Hoc, $p<0.05$).

When we look at the overall results of the experiment, we can conclude that none of the enriched syntax trees has the best score on all three usability aspects measured. As for error rate, 2D+ scored best, as for response times, 2D+ scored best as well, but 2D scored best with respect to preference. 3D+ scored second worst on error rate, and worst on response time. Together with D_planes, it is the worst appreciated tree diagram. If all usability aspects are assigned equal weights, one can say that 2D+ is the overall winner, and, surprisingly, that 3D+ is the overall loser.

Worth mentioning are also some more local results, especially those differentiated between the three subject groups. In Table 1 and 2 below, we see that the group with

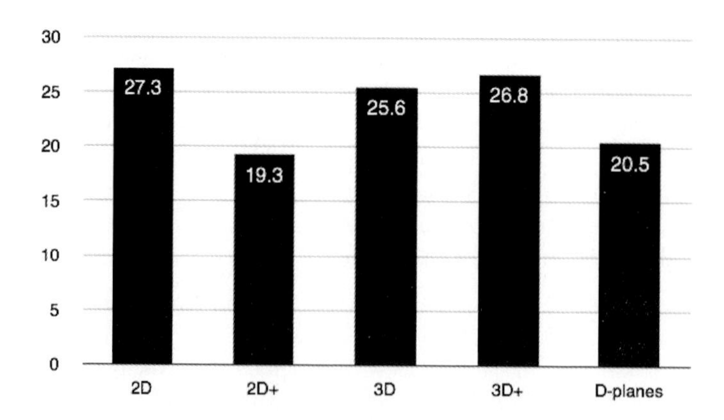

Fig. 7. Mean error rate per visualization type (in %)

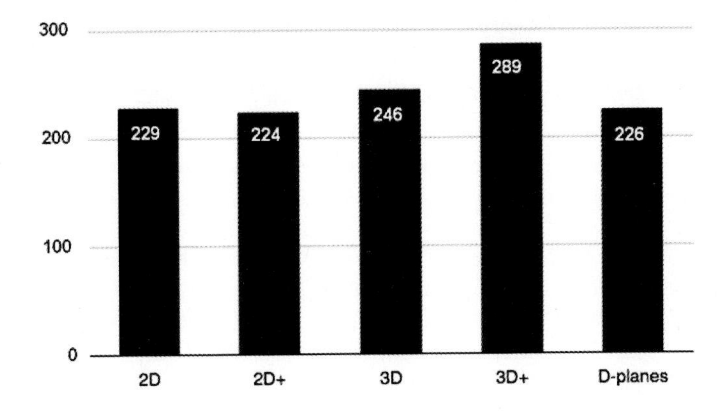

Fig. 8. Mean response times per visualization type (in sec.)

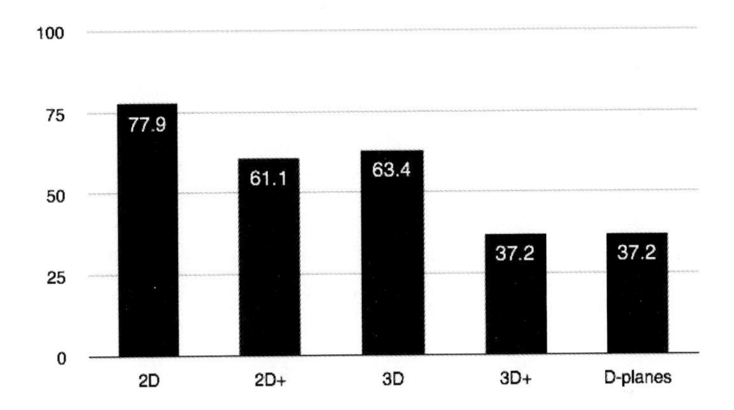

Fig. 9. Subjects' preferences per visualization type (100% = very good; 0%= very bad)

strong domain knowledge performed best in the least time. The group with no domain knowledge performed worst, but, as for error rate, did quite well in comparison with the weak domain knowledge group. Strikingly, all groups were unanimous in their preference for 2D. 3D+ was disliked by both strong and weak domain knowledge-groups, while the laymen were more positive about it, and disliked the D_planes version. Notably, the latter scored nearly as well as 2D+. It can further be noted that the two 3D variants (3D and 3D+) are neither popular, nor lead to optimal performance.

When we look at the construction types (see Table 3 and 4 below), it can be observed that for constituent right node raising (const. RNR), most errors were made with the 2D and 3D+ diagrams. This was a pattern demonstrated by all groups. Also, 3D+ took most time for the subjects of all three groups to answer the questions about coordination (coord.) and constituent right node raising. Globally, non-constituent right node raising (non-const. RNR) turned out to be the least problematic construction, although D_planes was here the least facilitative tree diagram.

Table 1. Total test mean error rate (in %) and mean response time (in min.) per subject group

Domain knowledge	*mean error rate (%)*	*mean response time (min.)*
strong	12	17
weak	26	18
no	30	26

Table 2. Interesting visualization type results per subject group

Domain knowledge	*most preferred*	*least preferred*	*least errors*	*least response time*
strong	2D	3D+	2D+	2D
weak	2D	3D+	D_planes	2D+
no	2D	D_planes	2D+	2D+

Table 3. Error rate for each visualization type per construction type

Construction type	*2D*	*2D+*	*3D*	*3D+*	*D+planes*
coord.	30	32	36	27	27
const. RNR	56	30	39	59	25
non-const. RNR	27	18	31	25	33

Table 4. Response time (sec.) for each visualization type per construction type

Construction type	*2D*	*2D+*	*3D*	*3D+*	*D+planes*
coord.	215	227	276	325	215
const. RNR	234	247	243	337	195
non-const. RNR	237	199	218	204	269

5 Conclusion

The main goal of the experiment was to test the usability of five variants of tree diagrams augmented with extra graphical elements to encode non-subordination and multidominance relations, which are assumed to be implied by syntactic constructions such as coordination and right node raising. It turns out to be difficult to answer this question unambiguously. 2D+ does a good job, and 3D+ does not, suggesting that color coding influences understanding of the new relations introduced positively, and that the addition of a third dimension is not facilitative at all. A two-dimensional syntax tree seems to be easier to process than a three-dimensional one. Familiarity seems to influence user preference strongly. 2D trees are familiar, and people like to use them, probably because they know how to interpret them. Apparently, subjects did not prefer the tree diagram which facilitated most the task performance. Which tree diagram can best be used to convey the views on parataxis as illustrated by coordination depends on the values attached to the different usability aspects. If preference is more important than correct and fast understanding, then we can stick to the variant most close to the traditional syntax trees. If not, we can opt for the 2D+ variant. It has also been shown that domain-specific knowledge plays an important role in task performance with the syntax trees, supporting research findings on schematic diagrams in other domains (Novick, 2006). The experiment relied on tree diagrams composed in a particular way in Blender. It might be the case that three-dimensional trees designed in another way will be more facilitative. We think that the results of this experiment show that visualization techniques should be applied with care, and that learning and cognitive aspects should be taken into account in the design of novel diagrams, even when these rely on familiar ones.

Acknowledgement

The authors would like to thank Edwin Wildeboer for making the tree visualizations in Blender, for implementing the web-based test, and for help with the gathering of the data. They are grateful to Wilbert Heeringa for his statistical advice.

References

1. Bennis, H.: Syntaxis van het Nederlands. Amsterdam University Press, Amsterdam (2000)
2. Bertin, J.: Semiologie Graphique. Les Diagrammes, les Réseaux, les Cartes. Paris: Éditions Gauthier-Villars; Paris-La Haye: Éditions Mouton & Cie (1967)
3. Brna, P., Cox, R., Good, J.: Learning to Think and Communicate with Diagrams: 14 Questions to Consider. In: Blackwell, A.F. (ed.) Thinking with Diagrams, pp. 115–134. Kluwer Academic Publishers, Dordrecht (2001)
4. Card, S.K., Mackinlay, J.D., Shneiderman, B. (eds.): Readings in Information Visualization. Using Vision to Think. Morgan Kaufmann Publishers, San Francisco (1999)
5. Cheng, C.-H., Lowe, R.K., Scaife, M.: Cognitive science approaches to understanding diagrammatic representations. Artificial Intelligence Review 15, 97–94 (2001)
6. Chomsky, N.: Syntactic Structures. Mouton, The Hague (1957)
7. Cockburn, A., McKenzie, B.: 3D or not 3D? Evaluating the effect of the third dimension in a document management system. CHI 3(1), 434–441 (2001)
8. Goodall, G.: Parallel structures in syntax: Coordination, causatives and restructuring. Cambridge University Press, Cambridge (1987)
9. Grootveld, M.: Parsing coordination generatively. Ph.D Dissertation, Leiden University (1994)
10. Harleman Stewart, A.: Graphic Representation of Models Linguistic Theory. Indiana University press, Bloomington, London (1976)
11. Hegarty, M., Carpenter, P.A., Just, M.A.: Diagrams in the comprehension of scientific texts. In: Ban, R., Kamil, M.L., Mosenthal, P., Pearson, P.D. (eds.) Handbook of Reading Research, vol. 2, pp. 641–668. Longman, NY (1991)
12. Kluck, M.: The perspective of external remerge on right node raising. In: Proceedings of CamLing 2007, pp. 130–137 (2007)
13. McCawley, J.: Parentheticals and Discontinuous Constituent Structure. Linguistic Inquiry 13, 91–106 (1982)
14. Novick, L.R.: The importance of both diagrammatic conventions and domain-specific knowledge. In: Barker-Plummer, D., Cox, R., Swoboda, N. (eds.) Proceedings of 4th International Conference, Diagrams 2006, Diagrammatic Representation and Inference, Stanford, CA, USA, June 2006, pp. 1–11. Springer, Heidelberg (2006)

15. Novick, L.R., Hurley, S.M.: To matrix, network, or hierarchy: That is the question. Cognitive Psychology 42, 158–216 (2001)
16. Scaife, M., Rogers, Y.: External cognition: how do graphical representations work? International Journal of Human-Computer Studies 45, 185–213 (1996)
17. Sebrechts, M.M., Vasilakis, J., Miller, M.S., Cugini, J.V., Laskowski, S.J.: Visualization of Search Results: A Comparartive Evaluation of Text, 2D, and 3D Interfaces. In: Proceedings of the 22nd Annual International ACM/SIGIR Conference, Berkley, CA USA (August 1999)
18. Tufte, E.R.: The Visual Display of Quantitative Information. Graphics Press, Cheshire (1983)
19. de Vries, M.: Coordination and Syntactic Hierarchy. Studia Linguistica 59, 83–105 (2005a)
20. de Vries, M.: Ellipsis in nevenschikking: voorwaarts deleren maar achterwaarts delen. TABU 34, 13–46 (2005b)
21. de Vries, M.: Invisible Constituents? Parentheses as B-Merged Adverbial Phrases. In: Deh, N., Kavalova, Y. (eds.) Parentheticals, pp. 203–234. John Benjamins, Amsterdam (2007a)
22. de Vries, M.: Internal and External Remerge: On Movement, Multidominance, and the Linearization of Syntactic Objects. University of Groningen. Revised version (manuscript, 2007b)
23. Williams, E.: Across-the-board Rule Application. Linguistic Inquiry 9, 31–43 (1978)

The Effects of Users' Background Diagram Knowledge and Task Characteristics upon Information Display Selection

Beate Grawemeyer and Richard Cox

Representation & Cognition Group
Department of Informatics, University of Sussex, Falmer, Brighton BN1 9QH, UK
b.grawemeyer@sussex.ac.uk, richc@sussex.ac.uk

Abstract. This paper explores factors associated with effective external representation (ER) use. We describe an information-processing approach to the assessment of ER knowledge. We also present findings from a study that examined the effects of users' background knowledge of ERs upon performance and their preferences for particular information display forms across a range of database query types that differed in their representational specificity. A representationally specific task is one which can only be performed effectively with one type of representation (or a narrow range of representations). On highly representationally specific tasks, optimal ER selection is crucial. Both ER selection performance and reasoning performance are, in turn, predicted by an individual's prior knowledge of ERs. On representationally nonspecific tasks, participants performed well with any of several different ER types regardless of their level of prior ER knowledge. It is argued that ER effectiveness crucially depends upon a three-way interaction between user characteristics (*e.g.* prior knowledge), the cognitive properties of an ER, and task characteristics.

1 Introduction

This paper presents an investigation of the interactions between a user, certain types of database query tasks and external representations (ER). It explores factors that are associated with effective ER use.

Within mathematical problem solving Polya [1] describes processes that are involved in solving particular problems: problem comprehension and interpretation; formulating of a plan to find a solution; using the devised plan to find a solution to the problem; and examining the plan according to its effectiveness. In analytical reasoning with ERs, Cox [2] developed a five stage approach, which loosely follows the stages described by [1]. The components that have been identified in reasoning with ERs are: 1) reading and comprehending the reasoning task; 2) selecting an ER; 3) constructing an ER; 4) using the ER to read off solutions and 5) responding to the task (answering).

However, individuals differ in their familiarity with particular forms of ER, as well as in their cognitive style. This paper extends earlier work (*e.g.* [3]) by

G. Stapleton, J. Howse, and J. Lee (Eds.): Diagrams 2008, LNAI 5223, pp. 321–334, 2008.

examining the relationships between subjects' background knowledge of ERs and their ability to select appropriate information displays in the course of responding to various types of database query tasks that differ, *inter alia*, in terms of their representational specificity[1]. In the study, a prototype automatic information visualization engine (AIVE) was used to present a series of questions about the information in a database. In earlier work [3], we investigated the representation selection and reasoning behaviour of participants who were offered a choice of information-equivalent data displays (*e.g.* tables, bar charts) across a range of database query tasks. Some of the tasks required the identification of unique entities, whereas others required the detection of clusters of similar entities, and others involved the qualitative comparison of values. This paper focusses on the effect of users' background knowledge of ERs upon information display selection in tasks that differed widely in terms of representational specificity.

2 Experimental Study

2.1 Hypotheses

The hypotheses were that different degrees of representation specificity of various tasks would influence participants' performance with respect to a) selecting an appropriate ER, b) the correctness of their response on the database query task and c) the time to answer the database query (latency). It was predicted that participants with high ER knowledge would select more appropriate representations on highly representation specific tasks, compared to participants with lower ER knowledge. Specifically, it was predicted that the gap in reasoning performance and time to answer the query between participants with high and low ER knowledge would be wider on more representationally specific tasks.

Performance was predicted to differ in terms of answer accuracy and latency, where participants with high ER knowledge (in contrast to low ER knowledge) would spend less time answering the question and would have a higher task response accuracy on high representation specific tasks, due to better knowledge of an ERs' applicability conditions [4] (where the applicability conditions specify which type of diagram best fits the structure of a particular problem). In contrast, the gap in reasoning performance and latencies between high and low ER knowledge participants was predicted to be narrower (even absent) on low representationally specific tasks, *i.e.* those tasks on which a wide range of different types of ERs can be used with equal effectiveness.

2.2 Participants

Twenty participants were recruited: 5 software engineers, 1 graphic designer, 1 html programmer, 2 IT business/project managers, 7 postgraduate students, and 4 research officers/fellows (6 female/14 male). Twelve out of the twenty participants had a strong maths, physics or computer science background. Five had a psychology background and three participants had an arts background.

[1] Representationally specific tasks are those for which only a few, specialised, representational forms are useful.

2.3 Procedure

In order to assess participants' background knowledge of ERs, 4 pre-experimental cognitive tasks were administered over the computer online. This was followed by the AIVE database query problem solving session. The pre-experiment session took between 90-115 minutes to complete. 30-40 minutes were spent on the AIVE tasks.

ER Knowledge Tasks. The ER knowledge pre-tests were devised by Cox [5] and consist of a series of online cognitive tasks designed to assess ER knowledge at the perceptual, semantic and output levels of the cognitive system. A large corpus of ERs was used as stimuli. The corpus contained 112 ERs including maps, set diagrams, text, lists, tables, graphs & charts, trees, node & arc, plans, notations & symbols, pictures, illustrations, scientific diagrams and icons. Figure 1 shows 12 examples from the 112 ER corpus.

The ER knowledge test consists of the following 4 different types of cognitive tasks, as described in [5]:

- *decision task*, which consists of a visual recognition task requiring real/fake decisions[2].
- *categorisation task* that assessed semantic knowledge of ERs - participants categorised each representation as 'graph or chart', or 'icon/logo', 'map', etc.
- *functional knowledge task*, here participants were asked *'What is this ER's function'?*. An example of one of the (12) response options for functional knowledge items was *'Shows patterns and/or relationships of data at a point in time'*.
- *naming task*, for each ER, participants chose a name from a list. For example, 'Euler's circle diagram', 'timetable', 'scatterplot', 'Gantt chart', 'entity relation (ER) diagram', etc.

The four tasks were designed to assess ER knowledge representation using an approach informed by picture and object recognition and naming research as described by [6]. The cognitive levels ranged from the perceptual level (real/fake decision task) through to 'deeper' cognitive tasks involving language production (ER naming) and semantics (functional knowledge of ERs).

AIVE Database Query Task. Following the ER knowledge pre-test, participants performed the AIVE database query tasks. Participants were asked to make judgements and comparisons between cars and car features based on information in a database. The database contained information about different cars including manufacturer, model, purchase price, average cost per month, CO_2 emission, engine size, horsepower, fuel consumption, insurance group, number of doors and safety features, such as the inclusion of driver, passenger or side airbag. Each participant responded to 30 database questions.The question types were based on the low level visual task taxonomy from [7]. In comparison to the earlier study, a new and different type of task set was used within this experiment.

[2] Some items in the corpus are invented or chimeric ERs.

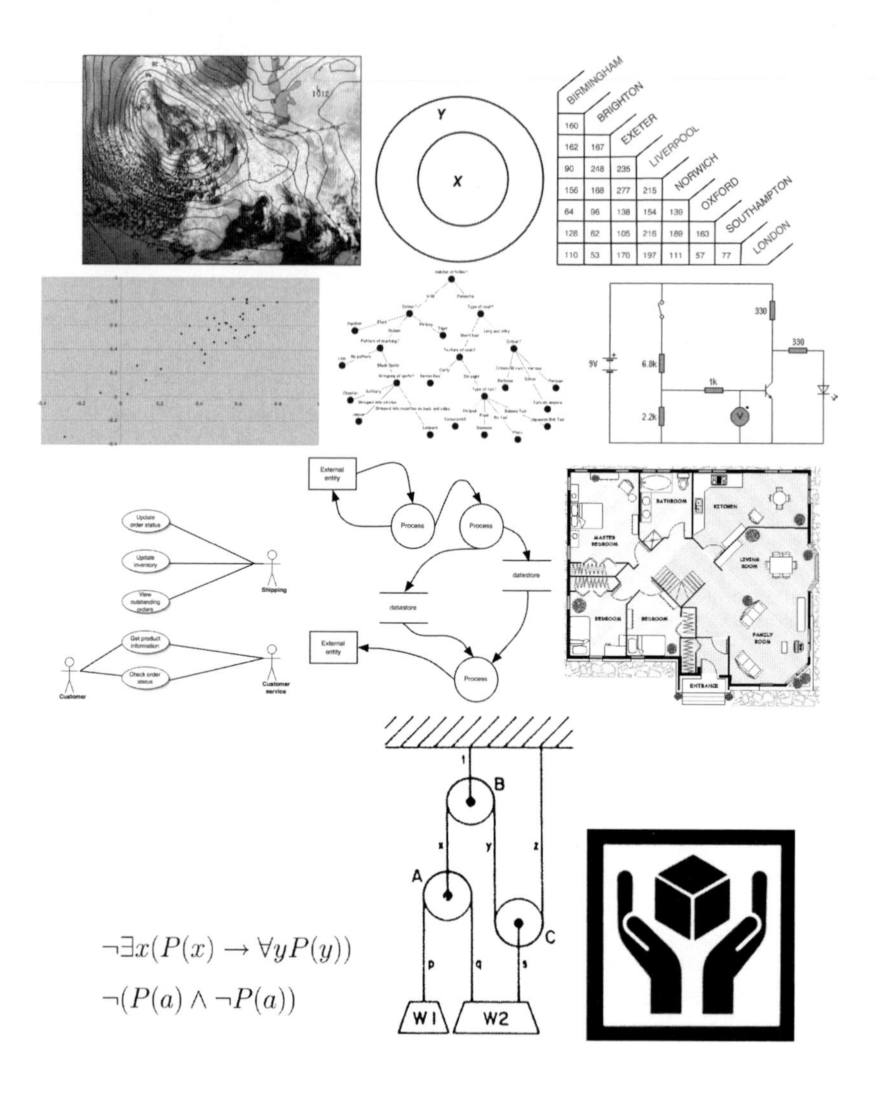

Fig. 1. 12 Examples from the 112 ER corpus

The database questions were of the following 6 types: compare (5 questions), correlate (5), distribution (5), identify (5), locate (5) and quantifier-set (5). Table 1 shows an example query for each different type of task. Participants were informed that to help them answer the questions, the system (AIVE) would supply the appropriate data from the database and instantiate their chosen representation with it. AIVE also offered participants a choice of representations. They could choose between various types of ERs. The set of representational options provided from AIVE included bar chart, pie chart, scatter plot, sector graph, set diagram, and table. Figure 2 shows an example of AIVE's representation selection interface. All representations were offered by the system

Table 1. Example queries for the different task types used in this AIVE experiment

Task type	Query example
compare neg.	If you wanted to buy a car with a horsepower over 70 hp which cars would you ignore?
correlate	Which two cars are most similar with respect to their CO_2 emission and cost per month
distribution	Which of the following statements is true? A: Insurance group and engine size increase together. B: Insurance group increases and engine size decreases. C: Neither A nor B
identify	Which car costs 12700 pounds?
locate	Where would you place a Fiat Panda with an engine size of 1200 cc inside the display?
quantifier set	Is it true that some cars which have an average cost over £350 per month have a fuel consumption of below 7.5 l/km?

for any query at all times. Some previous research (*e.g.* [8,9,10,11] and [12]) has implied that particular 'optimal' ERs for tasks can be identified. However, in this study, all of the different types of database query tasks could be answered with any of the representations offered by the system with the exception of set diagrams, which could only be used on tasks involving quantifiers (some, all). If the set diagram representation was chosen on a non-quantifier task, an error message was displayed and participants were asked to choose another representation.

The task sequence was as follows. AIVE first displayed the database query question and offered the representation selection buttons as described above. The spatial layout of the buttons was randomized across the 30 query tasks in order to prevent participants from developing a response 'set'. After the participant made their representation choice, AIVE generated and displayed the representation with

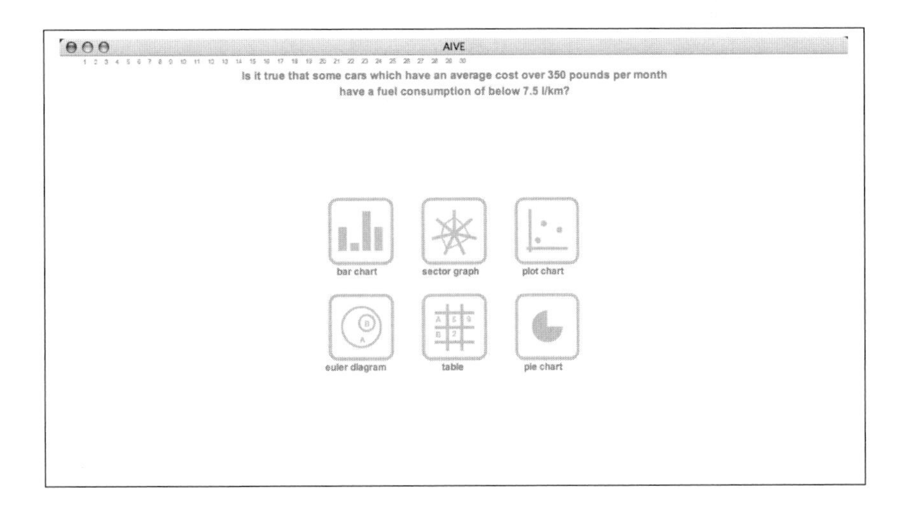

Fig. 2. Example of AIVE's representation selection interface

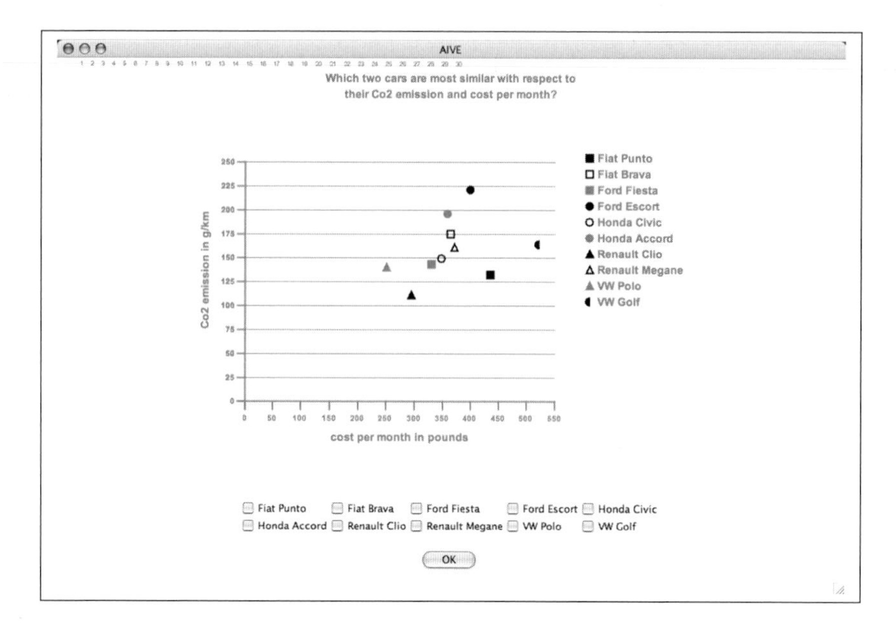

Fig. 3. Example of AIVE's information display interface with plot representation for two-term questions

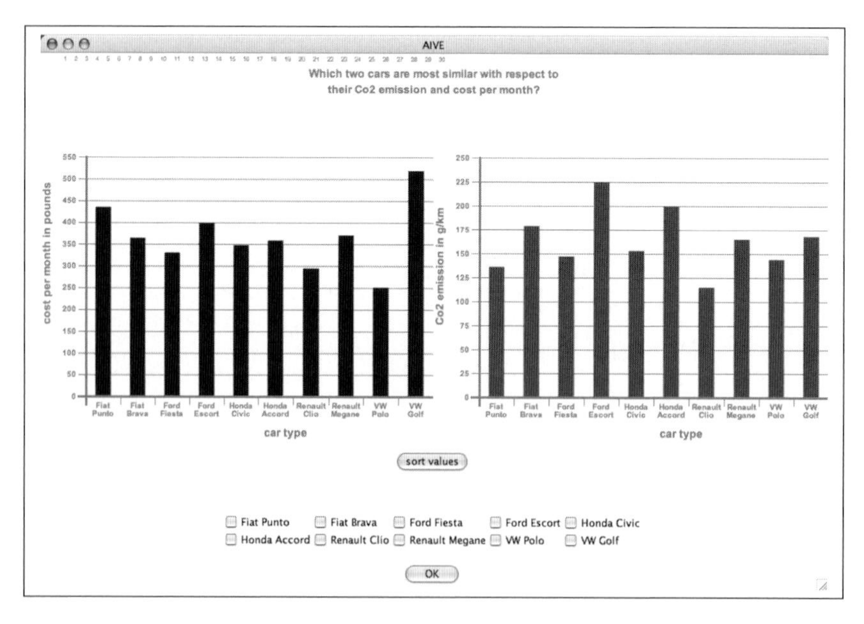

Fig. 4. Example of AIVE's information display interface with bar chart representations for two-term questions

the data required for answering the question. Participants then answered the question using the chosen visualization. The representations that were offered by AIVE were information equivalent[3], but not computationally equivalent, in terms of [13]. The database query task questions differed in terms of the number of attribute values they included: some were one-term questions; others two-term. For example, a typical one-term question was the identify task 'Which car has a CO_2 emission of 147 g/km?'. Here, only one attribute value, CO_2 was involved. A typical two-term questions was *e.g.* the correlate task 'Which two cars are most similar with respect to their fuel consumption and engine size?', which included two attribute values, *fuel consumption* and *engine size.* While some representations that AIVE created were able to represent two-term questions (such as the scatter plot), others were only able to represent one-term questions, including bar chart, pie chart and sector graph. If the user chose a representation that could handle one-term questions only for a two-term task, then AIVE displayed two representations next to each other, each depicting one attribute value. Figure 3 shows an example of a two-term question using a scatter-plot. In comparison, Figure 4 shows an example of AIVE's information display interface if a bar chart was selected for a two-term question.Participants were not permitted to change the representation following their initial selection. Following a completed response, participants were presented with the next task and the sequence was repeated. AIVE recorded users' representation choices, the time to read a question and select a representation, participants' responses to the questions, time to answer the question using the chosen representation, and the randomized position of each representation icon from trial to trial.

3 Results

3.1 ER Knowledge Task

For each participant 4 scores were derived for each of the 4 different ER knowledge tasks. The scores were assigned according to participants' accuracy on the decision, categorisation, naming and functional knowledge tasks. Table 2 shows the mean percent correct responses across participants for each ER knowledge task plus standard deviation (SD).Participants performed best on the decision task with 77.20% correct responses, which was followed by the ER categorisation task with 70.55%. Most errors were made on the naming and functional knowledge task with only 53.15% and 62.75% correct responses. Correlations between the cognitive tasks showed that decision and functional knowledge tasks were significantly and positively correlated ($r=.55$, $p<.001$). The categorisation task was significantly and positively correlated with the decision task ($r=.28$, $p<.001$) a well as the functional knowledge task ($r=.67$, $p<.001$) and also showed a significant positive correlation with the naming task ($r=.60$, $p<.001$). The naming task and the functional knowledge task were significantly and positively correlated too ($r=.74$,

[3] Except for set diagrams in the case of tasks that were non-quantifier tasks, as described earlier.

Table 2. Means and standard deviations for the cognitive ER knowledge tasks

ER knowledge task	mean % correct	SD
decision	77.20	13.61
categorisation	70.55	7.52
naming	53.15	8.66
functional knowledge	62.75	8.38

$p<.001$). The tasks that correlated significantly with all the other tasks were functional knowledge and categorisation tasks. The highest correlation was between the naming and functional knowledge tasks ($r=.74$, $p<.001$), which were also those that showed the lowest response accuracy rates (53.15% and 62.75% correct database query task responses, respectively).

3.2 AIVE Database Query Task

Table 3 shows the mean performance of all participants in percentage for display selection accuracy (based on an 'optimal' ER match[4]) and database query accuracy with standard deviation and the mean latency (sec) for the time to select a representation (display selection latency) and the time spent in answering the task (database query latency). To recapitulate, each of 20 participants performed 30 AIVE tasks (600 data points in total). The simple bivariate correlations, which included Bonferroni correction [14], across all AIVE tasks for display selection accuracy, database query answering accuracy, display selection latency and database query answering latency, were as follows:

- Display selection accuracy was correlated significantly and positively with database query answering accuracy ($r=.30$, $p<.001$)
- Display selection accuracy was significantly and negatively correlated with display selection latency ($r=-.17$, $p<.001$)
- Display selection accuracy and database query answering latency were significantly and negatively correlated ($r=-.32$, $p<.001$)
- There was a significant and negative correlation between database query answering accuracy and database query answering latency ($r=-.28$, $p<.001$)
- Display selection latency and database query answering latency were significantly and positively correlated ($r=.30$, $p<.001$)

Across all AIVE task types, the results show that good display-type selection leads to better query answering performance. The selection latency results indicate that rapid selection of a display representation is associated with good display-type choices. Less time spent responding to the database query question is associated with a good display-type choice and correct query response. Additionally, less time spent on selecting a representation is associated with fast query responding.

[4] For example, the 'optimal' ER for comparison tasks was the bar chart, scatter plot for correlate/distribution tasks, table for identify and locate tasks, and set diagram for quantifier set tasks.

Table 3. Participants' overall performance on AIVE's database query tasks (display selection accuracy and database query accuracy mean in %; display selection latency and database query answering latency mean in sec)

	mean %	SD
display selection accuracy	55.65	20.20
database query answering accuracy	82.80	11.13
display selection latency	12.65	10.37
database query answering latency	23.20	19.15

Figure 5 shows the frequency with which the representations were chosen by participants by task type for correct response (in %). The results show that the most used representations were tables, scatter plots and bar charts, which were also, over all tasks, the most effective ones. However, looking at the different types of tasks, it can be seen that the most popular representations, such as tables, were not necessarily effective on *e.g.* correlate or distribution tasks. The set diagram could only be chosen on quantifier set tasks and when it was selected it had a very high response accuracy (m=.93; SD=.26). The most difficult task type was the correlate task (m=.62; SD=.34 for correct response) followed by the distribution task (m=.77; SD=.42 for correct response). Correlate and distribution tasks were more representation specific than others. Here only one representation could be used effectively (scatter plot) in order to complete the task. This can be contrasted with the compare (m=.97; SD=.17 for correct response) and locate (m=.94; SD=.24 for correct response) tasks, where a range of representations were used effectively to answer the task.

3.3 Relationship between ER Knowledge and AIVE Performance

Correlations (with Bonferroni correction [14]) between the cognitive ER knowledge and AIVE tasks revealed a significant and positive correlation between ER functional knowledge and database query accuracy (r=.66, p<.001), and display selection accuracy was correlated significantly and positively with database query accuracy (r=.30, p<.001). The results indicate that high ER functional knowledge is associated with good query performance. Noticeably, task performance on the functional knowledge task is a better predictor of database query performance than display selection accuracy.

The examination of bivariate correlation matrices can be fraught, as the numerous partial inter-correlations, suppressor effects can make interpretation by visual inspection rather difficult. So to further pursue the relationship between ER knowledge and AIVE task variables, two stepwise multiple linear regression (MLR) analyses were also performed, using the 4 ER knowledge tasks as predictor variables. Stepwise methods in regression analysis provide a useful tool for exploratory research in new areas [15], where the focus is upon initially descriptive model-building. Display selection accuracy was the dependent variable in the first analysis and database query answer accuracy was the dependent variable in the second analysis. For both variables, the inclusion of more than one predictor

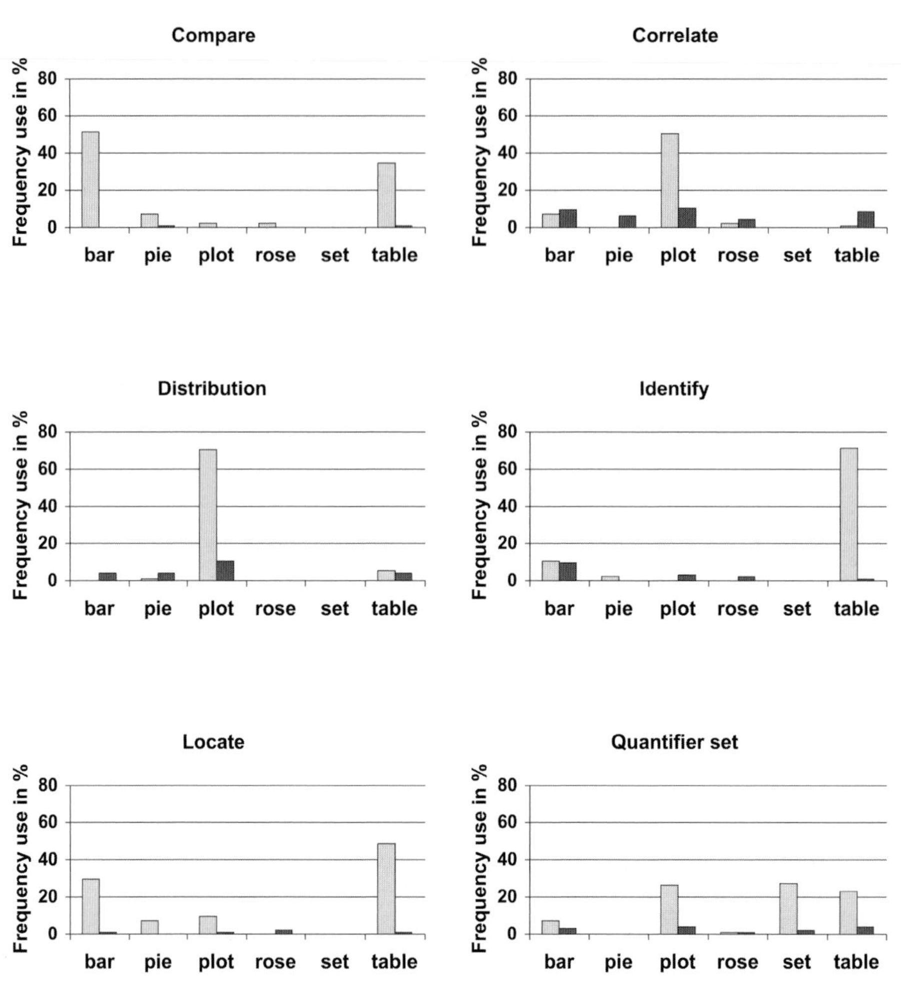

Fig. 5. Frequency with which each type of representation was used for each type of database query task type in % (all participants' data, n=20)

variable did not significantly improve the MLR models. For display selection accuracy the significant predictor was ER categorisation knowledge ($r=.57$, $p<.01$). For database query answer accuracy the predictor was functional ER knowledge ($r=.66$, $p<.01$).

It can be seen that conceptual (classificatory) knowledge of ERs predicts success at appropriate information display selection on the AIVE tasks but deeper semantic (functional) knowledge of ERs is associated with success at *using* the selected ER *i.e.* reading-off information and using it to respond correctly to the database query. However, it should be borne in mind that the correlations between display selection and database query answer accuracy reported above are computed across all AIVE task types - these differ extensively in terms of their representational specificity. Two contrasting task types were selected below to elucidate this issue.

3.4 Comparing a High Representation Specific AIVE Task and a Less Representation Specific AIVE Task.

To recapitulate, the AIVE tasks differed in terms of their representation specificity. Here, two extreme cases, the AIVE distribution task (highly representation specific) with the AIVE locate task (much less representation specific) were compared.

The AIVE Distribution Task - High ER Specificity. As shown in Figure 5 the most appropriate representation was the scatter plot. If it was selected it was effectively used in 87% of cases (SD=.34). The correlations between ER knowledge and AIVE's distribution tasks were:

- ER classification was correlated significantly and positively with display selection accuracy (ER classification $r=.71, p<.001$)
- Display selection accuracy and database query accuracy were significantly and positively correlated ($r=.44$, $p<.001$)
- Display selection latency and database query answering latency were significantly and positively correlated ($r=.41$, $p<.001$)

The results show that performance on one ER knowledge task - ER classification knowledge - predicts good display selection performance. Moderate, but statistically significant, correlations were found between two particular subset of ERs. Classificatory knowledge of *set diagrams* was significantly and positively correlated with display selection accuracy ($r=.58$, $p<.01$), as well as functional knowledge of *graphs and charts* ($r=.60$, $p<.01$). The ER knowledge task variables were entered as predictor variables into two MLR stepwise regression analyses, one with display selection accuracy as the dependent variable and the other with database query answer accuracy as the dependent variable. Again, single predictor models resulted from the procedure - display selection accuracy was strongly predicted by ER classification knowledge ($r=.71, p<.001$) and database query answer accuracy was also significantly predicted by ER classification knowledge ($r=.47$, $p<.05$).

The results show that good display-type selection is associated with high query response accuracy. Longer display selection latency is associated with longer time spent responding to the database query question. Hence, there does not seem to a be a speed/accuracy trade-off in display selection - participants either know which ER to choose and continue with the task, or they do not.

The AIVE Locate Task - Low ER Specificity. Figure 5 shows that 4 different data displays are effective for the locate task. Overall, participants answered queries with a high degree of accuracy (m=.94; SD=.24). However, participants chose the 'right' representation (table or matrix) in only 51% of cases. A range of other AIVE display forms were also effective (bar and pie charts, and scatterplots).

Database query answer accuracy and display selection performance were not significantly correlated. This would be expected on a task in which accurate responding to database queries can be achieved by using any one of 4 different information displays. ER knowledge and database query answer accuracy were not significantly correlated. This implies that less 'graphical literacy' on the part of participants is required for this task compared to the distribution task, for example. Stepwise regression analyses were performed but no ER knowledge task predictors reached significance for inclusion in the models for either display selection performance or database query answer accuracy.

4 Discussion

Participants with high ER knowledge demonstrate more appropriate representation selection behaviour compared to low ER knowledge participants. The prediction that the representational specificity of the task would affect participants' performance was supported on the whole. The results show that participants' reasoning performance was lowest on correlate and distribution tasks, where it was crucial to select an appropriate representation (scatter plot). In contrast, on low representationally specific tasks, such as compare and locate, reasoning performance was very high. Here, more than one representation could be used effectively (*e.g.* bar chart, table, pie chart and scatter plot).

The prediction that participants with high ER knowledge would show superior reasoning performance in high representation specific tasks, in comparison to participants with low ER knowledge, was supported by the results too. The high representation specific distribution task showed clearly that good display-type selection will lead to high query answering performance. ER classification knowledge (especially of set diagrams) predicts good display selection performance. Database query answer accuracy was also significantly predicted by ER classification knowledge. In contrast, the low representation specific locate task showed no significant correlation between display selection accuracy and database query performance, as would be expected on a task in which accurate responding to database queries can be achieved by using any one of 4 different information displays. ER knowledge and database query answer accuracy were not significantly correlated which implies that less 'graphical literacy' on the part of participants is required for this task compared to *e.g.* the distribution task.

The selection latency results showed that speedy selection of a display type in AIVE is associated with a good display-type choice. This implies that users either recognise the 'right' representation and proceed with the task or they procrastinate and hesitate because of uncertainty about which display form to choose. If, as suggested by [4] (p.201), representational knowledge (and applicability condition knowledge) 'might be to some extent implicit' then representation selection decisions would be expected to be automatic and fast. Less time spent responding to the database query question was associated with a good display-type choice and correct query response.

Task performance on the cognitive functional ER knowledge task was a better predictor of database query performance than display selection accuracy. This suggests that for predicting query response accuracy, it is as useful to know a participant's functional ER background knowledge as it is to know what type of information display he or she prefers.

As Cox describes in [2], selecting an appropriate representation depends upon the interpretation and comprehension of the problem solving task and matching the characteristics of the problem to an appropriate ER. Problem interpretation and devising a plan for solving the problem depends upon familiarity with the task. For example, a person who is not familiar with set theory might not identify quantifiers such as *for all* as keywords in a quantifier reasoning problem. Additionally, matching a representation to the characteristics of the task depends upon a person's ER repertoire as well as on their cognitive style, and knowledge of applicability conditions [4].

Previous studies ([8,9,10,11] and [12]) have tended to imply that 'optimal' ERs for tasks can be identified. For example [9] proposed 'cognitive fit' theory, which argues that if certain characteristics of a) the ER, b) the task and c) the underlying cognitive processes fit together then this will lead to quick and accurate problem solving. Our results suggest that caveats to [9] are needed - a task's representational specificity must also be considered. The degree to which a problem is representation specific might be determined by characteristics such as its degree of determinacy (extent to which it is possible to build a single model of the information given in the problem). As described in [16] a task can be characterized by (1) the existence of multiple ways or paths to solve a particular problem, (2) the presence of multiple desired solution outcomes, (3) the existence of conflicting interdependence among solution paths to multiple outcomes and (4) the presence of uncertain links among the solution paths and solution outcomes. Applying these task characteristics to the findings of the experimental study, it appears that multiple solution paths might be used effectively within low representation specific tasks, whereas in the case of highly representationally specific tasks only one solution path leads to a solution.

5 Conclusion

An individual might be more or less familiar with certain types of tasks and might apply different problem-solving plans to find a solution for a particular problem. Based on how the problem is comprehended and interpreted, individuals differ in terms of the ERs they match to problems. Selecting an appropriate ER depends upon an individual's representational repertoire *i.e.* the range of ERs with which he or she is familiar.

Tasks differ quite widely in terms of their representational specificity. The extent to which a problem is representation specific is determined, at least in part, by level of determinacy (the extent to which it is possible to build a single model of the information in the problem). In highly representationally specific tasks only a few specialised representational forms are effective, whereas in low representations specific tasks more than one representation can be applied successfully.

This study highlighted the differences in reasoning performance between participants with high or low ER knowledge. Participants with high ER knowledge outperformed participants with low ER knowledge on high ER representation specific tasks, where it was crucial to select an appropriate representation - one which matches the cognitive demands of the task. In contrast, no difference in reasoning performance were found between the two groups on tasks that had low representational specificity.

Hence, the effectiveness of an ER depends upon a 3-way interaction between user characteristics (*e.g.* prior knowledge), the cognitive properties of an ER, and task characteristics (*e.g.* representational specificity). For example, the effectiveness of Euler's circles for solving set problems involving quantifiers might vary according to an individual's prior knowledge of such tasks as well as his or her representational repertoire. In future work we plan to investigate how an individuals prior knowledge of *tasks* interacts with his or her prior knowledge of external representations.

References

1. Pólya, G. (ed.): How to Solve It. Penguin books, London (1945)
2. Cox, R.: Analytical reasoning with external representations. Ph.D thesis, Department of Artificial Intelligence, University of Edinburgh (1996)
3. Grawemeyer, B., Cox, R.: The effect of knowledge-of-external-representations upon performance and representational choice in a database query task. In: [17], pp. 351–354 (2004)
4. Novick, L., Hurley, S.: To matrix, network, or hierarchy, that is the question. Cognitive Psychology 42, 158–216 (2001)
5. Cox, R., Romero, P., du Boulay, B., Lutz, R.: A cognitive processing perspective on student programmers 'graphicacy'. [17], 344–346
6. Humphreys, G., Riddoch, M.: Visual object processing: A cognitive neuropsychological approach. Lawrence Erlbaum Associates, Hillsdale (1987)
7. Wehrend, S., Lewis, C.: A problem-oriented classification of visualization techniques. In: Proceedings IEEE Visualization, vol. 90, pp. 139–143 (1990)
8. Gilmore, D.J., Green, T.R.: Comprehension and recall of miniature programs. Int. J. Man-Mach. Stud. 21, 31–48 (1984)
9. Vessey, I.: Cognitive fit: A theory-based analysis of the graphs versus tables literature. Decision Sciences 22, 219–241 (1991)
10. Cheng, P.H.: Functional roles for the cognitive analysis of diagrams in problem solving. Cognitive Science, 207–212 (1996)
11. Day, R.S.: Alternative representations. In: Bower, G.H. (ed.) The Psychology of Learning and Motivation, vol. 22, pp. 261–305. Academic Press, New York (1988)
12. Norman, D.A. (ed.): Things that make us smart. Addison-Wesley, Reading (1993)
13. Larkin, J.H., Simon, H.A.: Why a diagram is (sometimes) worth ten thousand words. Cognitive Science 11, 65–100 (1987)
14. Bonferroni, C.E.: Teoria statistica delle classi e calcolo delle probabilità. Pubblicazioni del R Istituto Superiore di Scienze Economiche e Commerciali di Firenze 8, 3–62 (1936)
15. Hosmer, D.W., Lemeshow, S. (eds.): Applied logistic regression. Wiley and Sons, New York (1989)
16. Campbell, D.J.: Task complexity: A review and analysis. The Academy of Management Review 13, 40–52 (1988)
17. Blackwell, A.F., Marriott, K., Shimojima, A. (eds.): Diagrams 2004. LNCS (LNAI), vol. 2980. Springer, Heidelberg (2004)

Multimodal Comprehension of Graphics with Textual Annotations: The Role of Graphical Means Relating Annotations and Graph Lines

Cengiz Acarturk[1,*], Christopher Habel[1], and Kursat Cagiltay[2]

[1] University of Hamburg, Computer Science, Vogt-Koelln-Str. 30, 22527 Hamburg, Germany
{acarturk,habel}@informatik.uni-hamburg.de
[2] Middle East Technical University, Computer Education and Instructional Technology 06531 Ankara, Turkey
kursat@metu.edu.tr

Abstract. Graphs are often accompanied by text, i.e. linguistically coded information, augmenting the information presented diagrammatically. Thus, graph comprehension by humans often constitutes comprehension and integration of information provided by different representational modalities, namely graphical elements and verbal constituents. In this study we focus on textual annotations to line graphs providing information about events, processes and their temporal properties as well as temporal relations about the events and processes in question. We present results of an experimental investigation on parameters which influence subject's interpretations concerning the temporal properties of the annotated events and on eye movement behavior. In particular, we discuss the role of graph shape and the role of graphical means for relating textual annotations and determined parts of the graph line.

Keywords: text-graphics comprehension; annotations; line graphs; temporal relations.

1 Introduction

Multimodal documents combining text and pictorial representation such as newspaper articles, educational material and scientific papers are wide-spread in print media as well as in electronic media.[1] Comprehension of multimodal documents is based on almost automatically performed cognitive processes underlying the integration of information provided by the different modalities. Researchers from different disciplines investigated multimodal documents of different types in different domains, [2], [3], [4] among many others. Nevertheless, the research on cognitive mechanisms underlying multimodal integration is currently in a premature state due to abundant

* The research reported in this paper has been partially supported by DFG (German Science Foundation) in ITRG 1247 'Cross-modal Interaction in Natural and Artificial Cognitive Systems' (CINACS). We thank three anonymous reviewers for their helpful comments.

[1] In this paper, we use the term 'modality' as shorthand for 'representational modality' [1].

G. Stapleton, J. Howse, and J. Lee (Eds.): Diagrams 2008, LNAI 5223, pp. 335–343, 2008.

possible variations of the external representations and the modes of communication (e.g., speech vs. written text). In the present paper we focus on a specific type of pictorial representations, namely on *diagrams*—in particular *line graphs*—and on comprehension of diagrams in the context of text. The research on *cognitive* (rather than *perceptual*) processes in graph comprehension is scarce [5] in the last decades with some exceptions [6], [7], [8]. Furthermore, multimodal comprehension of graph-text elements has seldom been in the main focus of the research so far.

In contrast to *pictures* (or *images*) diagrams possess internal syntactic structures in the sense of representational formats [9]. Thus the syntactic analysis of a graph can be exploited by succeeding processes of semantic and pragmatic analyses in graph comprehension [7], [10], [11]. From a linguistic point of view, the process of *referring*, which is constituted by a *referential expression* that refers to an *entity* of the domain of discourse, is the core of comprehension. Based on this, *co-reference*, the backbone of text coherence has to be established by speaker and hearer employing internal—conceptual—representations, which mediate between language and the domain of discourse. In processing text-diagram documents, in which both modalities have to contribute to a common conceptual representation, additional types of reference and co-reference relations have to be distinguished. Foremost, there exist corresponding referential relations (reference links) between graphical entities and entities in the domain of discourse. Furthermore, there exist referential links between linguistic and graphical entities.[2] Beyond this—traditionally discussed—type of text-diagram multimodality there exists a second kind, the diagram-internal multimodality of graphical representations and diagram-internal text, e.g. labeling of axes, annotation to graph lines etc. Figure 1 shows that both types of multimodality can be involved in multimodal comprehension. Therefore it is important to investigate the role of annotating textual elements in the graph region of a text-diagram document.

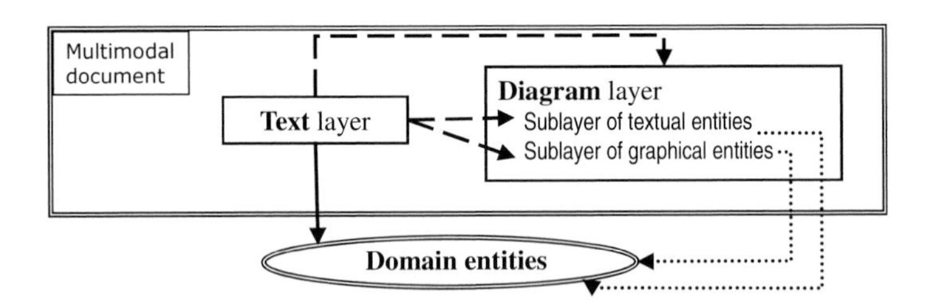

Fig. 1. Reference links between layers and sub-layers of text-diagram documents

The graph region includes the graph (proper) and graph-related text information (graph title, annotations etc.), and usually is separated by a frame from the rest of the document (e.g. paragraphs). We propose that annotated textual elements serve the purpose of bridging the two representational modalities in graph-text documents [13].

[2] In [12] we discussed the concept of *text-graphic coherence* based on an analysis of different types of referential links. Furthermore, we described how the interaction between information graphics and language is mediated by common conceptual representations.

If annotations are available contiguous to the graph, separately constructed representations of the text and the graph can be connected via these constituents. If not, integration of modalities is achieved with further cognitive effort of encoding spatially represented information on the graph and constructing co-reference relations between the paragraphs and the graph. Our purpose in the present study is to investigate the diagram-internal multimodality of graphical elements and diagram-internal text, i.e. textual annotations. We exemplify this topic focusing on line graphs concerning sequences of events/processes and describing temporal properties of these events and processes (see Figure 2). For annotated line graphs, the content provided distributively by graphical elements and verbal annotations has to be integrated via co-reference relations to reach coherent conceptual representations. The construction of co-reference links can be induced by spatial contiguity between annotations and the graph line or prominent parts of the graph line (e.g. "Famine 1930" or "1948 lake freeze-over") or by explicit 'pointers', called 'annotation icons', (e.g. "Fire 1936")[3].

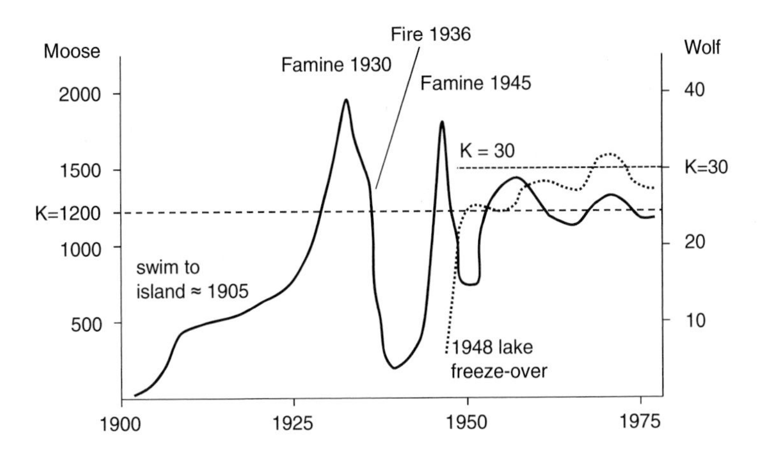

Fig. 2. A sample annotated line graph: Isles Royale Moose/Wolf Progression[4]

Accordingly, the following research questions were central to the present study: How do graph format (i.e. straight- vs. curved-shape) and annotation icon affect interpretations concerning the temporal properties of the annotated event? How do the salient points change under the presence of annotation text and annotation icon?

Methodologically, compared to research on eye movement control in reading, there are few studies investigating eye movement characteristics in multimodal documents

[3] In this study, we use the term 'annotation' to mean the combination of textual elements (e.g. phrases) and a connecting icon, for example a vertical line which on the one end points the textual element, on the other end points a specific location on the graph. The term 'annotation icon' is used to mean this vertical line. The term 'annotation text' is used to mean the verbal constituents of the annotation.

[4] Figure 2 is based on a line graph used in N.C. Heywood's course material in Biogeography (http://www.uwsp.edu/geo/faculty/heywood/Geog358/Population/Populate2.htm). Fig. 2 goes back to Harris, A. and Tuttle, E. Geology of National Parks. ISBN 0-8403-2810-9.

[14]. The underlying assumptions in these studies, as well as in the present study, are based on the *eye-mind hypothesis* [15]; see [16] for objections to the eye-mind hypothesis in its proposed form.

2 The Experiment

Subjects. A total of 36 subjects (mean age 22.8, SD = 2.49) were paid to participate in the experiment. The subjects were either undergraduate or graduate students in different general academic areas at the Middle East Technical University. All subjects were native speakers of Turkish, which was the language of the experiment.

Materials and Design. Simple line graphs that present the change of a domain value in time were prepared by the experimenter. The graphs included an annotation that includes information about an event's occurrence (see Figure 3).

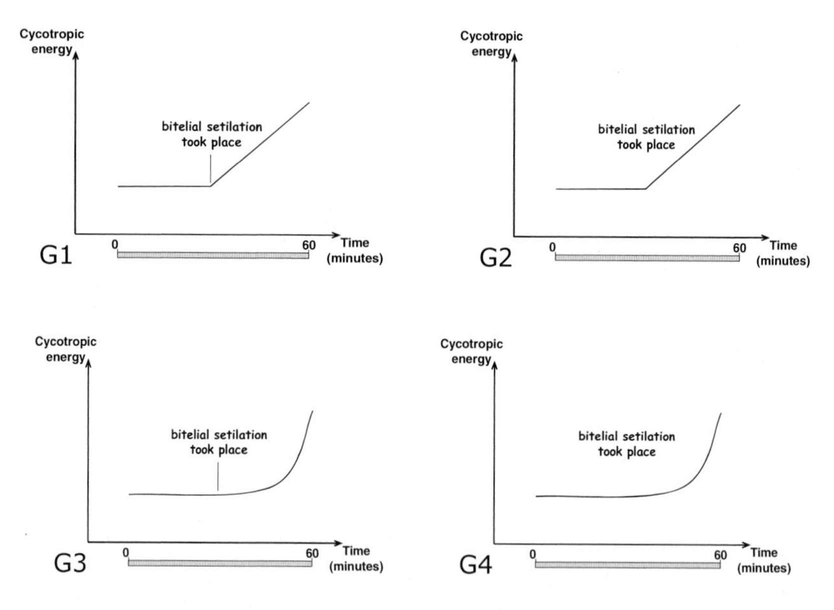

Fig. 3. The four graphs show sample material for the four experimental conditions G1 to G4 (translated to English by the first author; fonts in the figure were changed for better visibility)

The design was 2x2, with two independent parameters (two within-subjects parameters), and one dependent parameter. The first independent parameter was the *graph format*. The graph was either a straight-shaped line graph (G1 and G2) or curve-shaped line graph (G3 and G4). The second independent parameter was the *annotation icon*. The graph either included an annotation icon (G1 and G3) or did not include an annotation icon (G2 and G4). As a result, each subject was presented four experimental conditions. The dependent parameter was subjects' reports of interpretations concerning the *duration* of the annotated event, explained below. The order of presentation of the conditions was randomized. In addition to the four experimental conditions, two graphs with straight- and curve-shaped lines without annotations were presented to the subjects without a specific given task (C1 and C2). They were presented for the purpose of comparison with the experimental conditions (G1 to G4). The domain was not evaluated as an independent parameter in this study. The domain value labels, as well as the annotated events were prepared for four different fictional domains. Subjects were informed that the graphs were excerpted from lecture notes in medicine, which was an unfamiliar domain for the participants of the study. Eye-tracking data were recorded by a 50 Hz. Tobii 1750 EyeTracker.

Procedure. The subjects attended the experiment in single sessions. After the preliminary information about statistical information graphics with samples, a practice session was presented to explain the task and the use of the response scale. In the experiment session, in each screen, subjects investigated the graph and reported their interpretation by clicking on the response scale, which was given below the x-axis in the same screen (see Figure 3). The response scale was a colored horizontal bar extending from the beginning to the end of the horizontal time axis of the graph. If the subject interpreted the event as point-like, i.e. the event occurred in a specific point in time then he/she clicked the time when the event happened, on the response scale with the mouse (i.e. point interpretation). If the subject interpreted the event as durative, i.e. the event occurred in a time interval rather than a specific point in time then he/she clicked the time when the event started and the time when the event ended, on the response scale (i.e. interval interpretation). The experiment was self-paced, and took a total of approximately 10 minutes to complete.

3 Results

The distribution of the number of subjects who made the *point* and *interval* interpretations for the temporal properties of the annotated event showed that the main determinant was the *graph format*. The presence or absence of the *annotation icon* had a marginal effect. A Cochran test was conducted to evaluate differences between related proportions. The test was significant, χ^2 (3, N = 36) = 20.47, $p < .01$, Kendall coefficient of concordance was .19. Follow-up pairwise comparisons were conducted using a McNemar's test. The results showed that the number of subjects who made point interpretation was significantly higher in conditions G1 and G2 than the number of subjects who made point interpretation in conditions G3 and G4. Correspondingly, the number of subjects who made interval interpretation (almost half of the subjects) was significantly higher in conditions G3 and G4 than the number of subjects who made interval interpretation in conditions G1 and G2.

Further analysis of the effect of the annotation icon as well as the graph format was investigated with the analysis of eye movement parameters, namely fixation count, gaze time and fixation duration. For the analysis, the region covered by the graph line was divided into 15 rectangle AOIs (Area of Interest), namely *AOI 1* to *AOI 15* (Figure 4). The same AOI template was used to evaluate all experimental conditions.

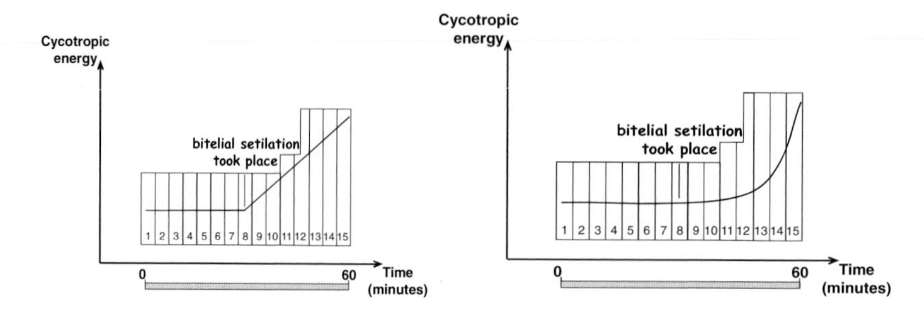

Fig. 4. The specified AOIs (Area of Interest) for the analysis of eye movement parameters

Mean fixation counts on the AOIs were calculated, and z-score normalization was used for the analysis. A within-subjects repeated measures ANOVA was conducted with the factors being the conditions and mean fixation counts on the fifteen AOIs. The results indicated a significant condition effect, Wilks's $\Lambda = .18$, $F(5, 31) = 28.95$, $p < .01$, a significant AOI effect, Wilks's $\Lambda = .04$, $F(14, 22) = 35.53$, $p < .01$, and a significant interaction between the conditions and the AOIs. The distribution of the mean fixation counts for C1 (the straight graph without annotations) and C2 (the curved graph without annotations), shown in the left part of Figure 5, reveals information about visually/informationally salient regions on the graph lines. In the C1 graph the salient region was AOI 8, whereas in the C2 it was AOI 12.

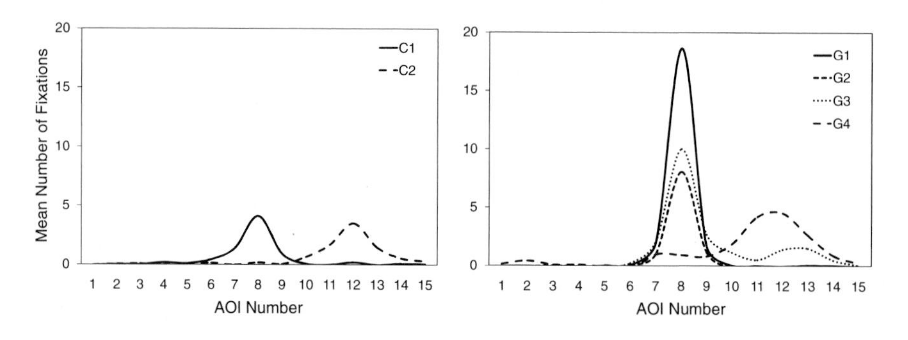

Fig. 5. The distribution of mean number of fixations (i.e. fixation counts) on the AOI 1 to AOI 15 in C1 and C2 conditions (on the left) and G1, G2, G3, and G4 conditions (on the right)

How does the distribution change in the presence of annotation text? This corresponds to the conditions G2 and G4, in other words the two annotated graphs with different graph format and without annotation icon. The results (the right part of Figure 5) revealed a similar distribution to the C1-C2 distribution. The salient region in the G2 condition was AOI 8, whereas the salient regions in the G4 condition were AOI 11 and AOI 12. Furthermore, the comparison of mean fixation counts for the C1-C2 conditions and the G2-G4 conditions shows that in G2 and G4, the addition of the annotation text resulted in an overall increase in the number of eye fixations on the AOI 8, which was the region below the annotation text.

How does the distribution change in the presence of an annotation icon, as well as the annotation text? This corresponds to the conditions G1 and G3, in other words the two annotated graphs with different graph format and with the annotation icon. The results showed that the distribution of mean fixation counts in the G1 and G3 conditions was different than the distributions in the previous conditions. The comparison of mean fixation count distributions for the conditions G1 and G3 shows that the salient region was AOI 8 for both conditions. In other words, especially for the G3 condition, the visually/informationally salient region was shifted from AOI 11 and AOI 12 (in the condition G4) to AOI 8 (in the condition G3) with the addition of the annotation icon.[5]

The analysis of gaze time values on the previously specified AOIs revealed similar distributions to the ones for mean fixation counts.

The results for mean fixation durations were calculated for the AOIs that had an average number of fixations of one or greater than one. Accordingly, mean fixation durations were calculated for AOI 7 to AOI 9 in G1 and G2 conditions and for AOI 7 to AOI 13 in G3 and G4. A univariate ANOVA test was conducted for the analysis of the differences between conditions G3 and G4. The results for the ANOVA indicated a significant condition effect, $F(1, 296) = 6.09$, $p < .05$ and a significant AOI effect, $F(6, 296) = 3.15$, $p < .01$. Important from the focus of this study is that mean fixations in the G4 condition were longer than the ones in the G3 condition.

4 Discussion

The subject's reports of interpretations concerning the temporal properties of the annotated events show that the graph format, rather than the presence or absence of the annotation icon, is the main determinant for the temporal properties of the annotated events. Most of the subjects reported that the annotated event took place at a specific point in time in the straight-graph conditions. On the other hand, almost half of the subjects reported that the annotated event took place in a time interval in the curved graph conditions. Nevertheless, the role of the annotation icon on subjects' reports is not significant between the conditions.

Further analysis of eye movement parameters reveals more detailed information for the differences between the conditions and the effect of the presence or absence of an annotation icon. The results show that on the one hand, the addition of the annotation texts to the graphs does not reveal major changes in the distribution of average number of fixations and gaze time values on the previously specified AOIs, compared to the distributions on the non-annotated graphs. This implies that the annotation text, without the annotation icon does not strongly affect the visually/informationally salient regions on the graphs. On the other hand, mean fixation counts and gaze time values increase with the addition of the annotation text. The increase in fixation counts and gaze time values on the graph line points to subjects' effort for the integration of information provided by the graphical elements and the annotation text.

[5] Whether this shift is stable, or is an artifact due to the very slight change at the beginning of the curves (*inflections*) used in this experiment—as suspected by two reviewers—will be investigated in a future study. *Inflections* have been found to be the second mostly used segmentation points after *negative minima* [17], [18]. We argue here, that annotation icons belong to those *top down factors* which interact with geometric factors in segmentation [18].

Furthermore, the addition of an annotation icon (together with the addition of the annotation text) results in major changes in the distribution of mean fixation counts and in the distribution of average gaze time values in the curved graphs. In other words, the presence of annotation icons shifts the visually/informationally salient points in the curved graphs. This is expected, since the addition of the new graphical element, namely the annotation icon, attracts subjects' attention to this region. In addition, average number of fixations and gaze time values further increase with the addition of the annotation icon. This increase may imply subjects' further effort to integrate the information provided by the annotation text and the annotation icon, as well as the annotation icon and the *relevant part of the graph.*

More important from the perspective of this study is that the subjects experienced difficulties in determining the *relevant part of the graph* in the absence of the annotation icon in the curved graphs. The results of the analysis for mean fixation durations support this idea showing that the absence of the annotation icon in the curved graphs results in longer fixations on the salient regions of the graph.

5 Conclusions and Future Work

Comprehension of multimodal documents includes the construction of co-reference relations between the graph (proper) and text elements (e.g. annotations) in the graph region, as well as the construction of co-reference relations between the graph region and the main text (e.g. paragraphs) of the document. From the perspective of automated generation of graph-text documents, investigation of multimodal integration at both levels is necessary for the design of easily comprehended documents and graphs by humans. Poorly designed diagrams may create misconceptions, deceive or confuse an issue, do not ease the comprehender's task and hinder comprehension and learning. Nevertheless, in the current state of the art, the design and use of annotations is based on the experience and practice of the designers of graphs and multimodal documents, rather than theory, guidelines or systematic empirical research. Furthermore, on the application side, recent data visualization components of popular statistical and mathematical software programs offer limited capacity for annotation design and generation for graphs. Preparing effective diagrams requires both practice and also evidences from empirical research studies.

In this study, we investigated how the graphical elements (i.e. the graph format, as well as the absence or presence of an annotation icon) affect subjects' interpretations concerning the temporal properties of events annotated by textual elements, and eye movement characteristics during multimodal graph comprehension. In the future, we will investigate the role of additional graphical means (e.g. arrows as well as lines, textbox etc.) for annotations by experiments with human subjects, as well as by corpus studies.

References

1. Bernsen, N.O.: Foundations of Multimodal Representations: A Taxonomy of Representational Modalities. Interacting with Computers 6, 347–371 (1994)
2. Hegarty, M.: The Mechanics of Comprehension and Comprehension of Mechanics. In: Rayner, K. (ed.) Eye Movements and Visual Cognition: Scene Perception and Reading, pp. 428–443. Springer, New York (1992)

3. Butcher, K.R.: Learning from Text with Diagrams: Promoting Mental Model Development and Inference Generation. Journal of Educational Psychology 98, 182–197 (2006)
4. Narayanan, N.H., Hegarty, M.: On Designing Comprehensible Interactive Hypermedia Manuals. International Journal of Human Computer Studies 48, 267–301 (1998)
5. Scaife, M., Rogers, Y.: External Cognition: How Do Graphical Representations Work? International Journal of Human Computer Studies 45, 185–213 (1996)
6. Peebles, D.J., Cheng, P.C.-H.: Extending Task Analytic Models of Graph-based Reasoning: A Cognitive Model of Problem Solving with Cartesian Graphs in ACT-R/PM. Cognitive Systems research 3, 77–86 (2002)
7. Pinker, S.: A Theory of Graph Comprehension. In: Freedle, R. (ed.) Artificial intelligence and the Future of Testing, pp. 73–126. Erlbaum, Hillsdale (1990)
8. Mautone, P.D., Mayer, R.E.: Cognitive Aids for Guiding Graph Comprehension. Journal of Educational Psychology 99, 640–652 (2007)
9. Kosslyn, S.M.: Image and Mind. Harvard University Press, Cambridge (1980)
10. Kosslyn, S.M.: Understanding Charts and Graphs. Applied Cognitive Psychology 3, 185–226 (1989)
11. Tversky, B.: Semantics, Syntax, and Pragmatics of Graphics. In: Holmqvist, K., Ericsson, Y. (eds.) Language and Visualisation, pp. 141–158. Lund University Press, Lund (2004)
12. Habel, C., Acarturk, C.: On Reciprocal Improvement in Multimodal Generation: Co-reference by Text and Information Graphics. In: van der Sluis, I., Theune, M., Reiter, E., Krahmer, E. (eds.) Workshop on Multimodal Output Generation (MOG 2007), Aberdeen, United Kingdom, pp. 69–80 (2007)
13. Acarturk, C., Habel, C., Cagiltay, K., Alacam, O.: Multimodal Comprehension of Language and Graphics: Graphs with and without Annotations (in press, accepted for publication in Journal of Eye Movement Research)
14. Underwood, G., Jebbett, L., Roberts, K.: Inspecting Pictures for Information to Verify a Sentence: Eye Movements in General Encoding and in Focused search. The Quarterly Journal of Experimental Psychology 57A, 165–182 (2004)
15. Just, M.A., Carpenter, P.A.: A Theory of Reading: From Eye Fixations to Comprehension. Psychological Review 87, 329–354 (1980)
16. Anderson, J.R., Bothell, D., Douglass, S.: Eye Movements Do Not Reflect Retrieval Processes: Limits of the Eye-mind Hypothesis. Psychological Science 15, 225–231 (2004)
17. Cohen, E., Singh, M.: Geometric Determinants of Shape Segmentation: Tests Using Segment Identification. Vision Research 47, 2825–2840 (2007)
18. De Winter, J., Wagemans, J.: Segmentation of Object Outlines into Parts: A Large-scale Integrative Study. Cognition 99, 275–325 (2006)

Talk to the Hand: An Agenda for Further Research on Tactile Graphics

Frances Aldrich

Reginald Phillips Research Programme
www.lifesci.sussex.ac.uk/reginald-phillips
Dept. of Psychology, University of Sussex, Brighton BN1 9QG, UK
F.K.Aldrich@sussex.ac.uk

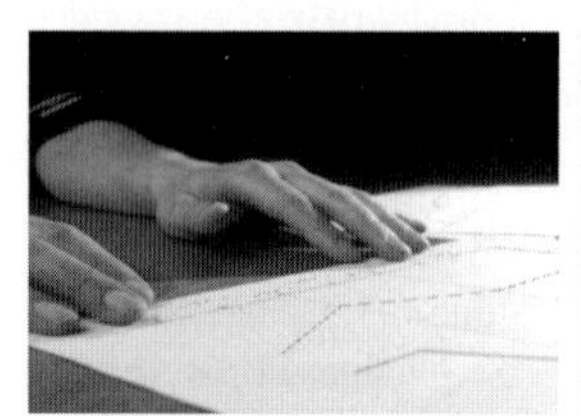
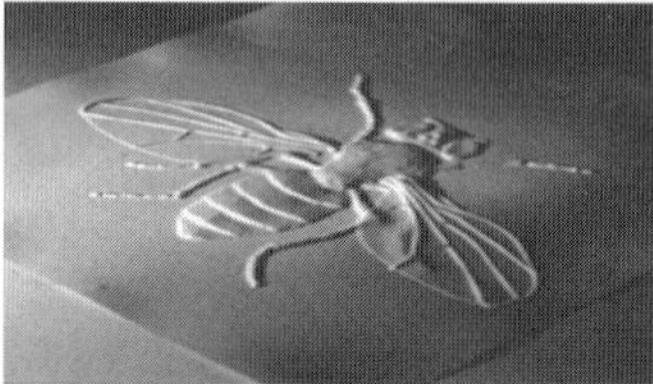
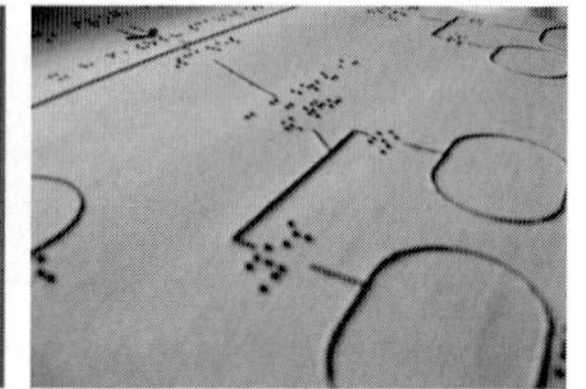

© 2008 RNIB (with permission)

1 Rationale

Tactile graphics are the primary means by which blind people access maps, graphs, diagrams and other graphical representations. Tactile graphics are made up of raised lines, areas, textures and symbols, and are intended to be felt rather than seen [1], [2] and [3].

Major obstacles to the successful use of tactile graphics are that:

- touch cannot discriminate the fine detail that sight can;
- extracting information through a sequence of touches, then re-integrating it, imposes a heavy memory load; and
- many graphical representations need visual experience for interpretation.

A variety of technologies for the production of tactile graphics have been explored and refined. To date this has been the main thrust of research in the field of tactile graphics, see [4]. Although this endeavour has been essential, it is not sufficient. Now that a range of acceptable production methods exists, it is time to shift emphasis away from 'the medium' and onto 'the message' – and to look harder at what can be communicated and how [5].

The people active in advancing tactile graphics are predominantly engineers, technicians, educators, and service providers. How tactile graphics can be designed and used as effective representations is receiving little attention from researchers (although see [6] for some notable exceptions). Mainstream graphics researchers could make a valuable contribution.

G. Stapleton, J. Howse, and J. Lee (Eds.): Diagrams 2008, LNAI 5223, pp. 344–346, 2008.

My intention here is to put forward an agenda that will orient mainstream researchers to the principle challenges in the field of tactile graphics. Set out below is a 'smorgasbord' of research opportunities in a very worthwhile area! The principles of good design, the prerequisites for use, the pedagogy, and the possibilities, all largely remain to be investigated, preferably using a level of analysis that will inform practice effectively.

2 Agenda

With the aim of focusing endeavour where it will be of most practical benefit, I propose the following agenda:

1. Effective Design
Tactile graphics are usually simplifications of visual graphics. What would they be like if we designed them from scratch?

There are design guidelines and kits for making tactile graphics, embodying accepted practice in simplifying and translating visual graphics, and aimed at ensuring tactile legibility (see [3] for examples). As well as designing for the limitations of touch, another key issue is eliminating visual references such as occlusion and vanishing perspective. Designing new forms of graphical representation, specifically for the tactile medium, is a possibility [7]. However we need more research to find out whether, and how, the design of tactile graphics might differ profitably from that of visual graphics.

2. Teaching and Learning
There is little guidance for teachers on how and when to introduce tactile graphics, what concepts to introduce or materials to use. What should we tell them?

Blind children generally grow up in a graphically impoverished world, with almost no incidental exposure to graphical representations, and little opportunity or incentive for 'mark-making' or drawing. Tactile graphics can be very effective, but teachers may be reluctant to bother with them, and pupils sometimes lose confidence in working with them [8] and [9]. Teaching the use of tactile graphics needs to encompass tactile discrimination, tactile graphic reading strategies, and knowledge of graphic conventions, as well as domain knowledge. Ideally it should begin well before formal schooling. Although the route to tactile graphicacy ought perhaps to differ from the route to visual graphicacy, this has not been debated. We need more research to understand what skills and knowledge are needed to use tactile graphics successfully, and to aid in developing effective teaching and materials.

3. Cognitive Processing
We know little about the cognitive benefits of using tactile graphics. But are people right to consider graphical representations to be essentially visual?

The argument for provision of tactile graphics usually focuses on blind people's access to the information contained within them. If the information can be put into words, this is generally seen as a perfectly acceptable substitute. However, research into sighted people's thinking and problem solving suggests that the form of representation plays a crucial role in success (e.g. [10] and [11]). This may also be the case for blind people and is a strong but rarely heard argument for the right of blind people to have access to graphics [12]. More general cognitive advantages to sighted people have also been put forward for graphical representations [13]. We do not know which of these hold for blind people using tactile graphics. For example, the heavily sequential nature of extracting tactile information is likely to diminish the 'computational offloading' that visual graphics allow. Whether or not a person has any sight experience to relate to tactile graphics is also likely to have some effect. Systematic research is needed into what cognitive leverage blind people can gain from tactile graphics.

3 Conclusion

This is not an exhaustive list of the challenges that lie in the field of tactile graphics. But hopefully it will succeed in attracting the attention of mainstream researchers, stimulating their interest, and providing sufficient orientation for them to see how they could contribute to this worthwhile field.

References

1. Sheppard, L., Aldrich, F.K.: Tactile graphics: A Beginner's Guide to Graphics for Visually Impaired Children. Primary Science Review 65, 29–30 (2000)
2. RNIB National Centre for Tactile Diagrams, `http://www.nctd.org.uk`
3. The Tactile Graphics Website, `http://www.tactilegraphics.org/`
4. 'Tactile Graphics' conference series, `http://www.nctd.org.uk/Conference`
5. Aldrich, F.K., Sheppard, L., Hindle, Y.: First Steps towards a Model of Tactile Graphicacy. British Journal of Visual Impairment 20(2), 62–67 (2002)
6. RNIB's Scientific Research Unit's website, `http://www.tiresias.org/research/index.htm`
7. Challis, B.P., Edwards, A.D.N.: Good Tactile Diagrams Can Look Bad. In: International Conference on Tactile Diagrams, Maps and Pictures (2000)
8. Sheppard, L., Aldrich, F.K.: Tactile Graphics in School Education: Perspectives from Teachers. British Journal of Visual Impairment 19(3), 93–97 (2001)
9. Aldrich, F.K., Sheppard, L.: Tactile Graphics in School Education: Perspectives from Pupils. British Journal of Visual Impairment 19(2), 69–73 (2001)
10. Cheng, P., Pitt, N.G.: Diagrams for Difficult Problems in Probability. Mathematical Gazette 87(508), 86–97 (2003)
11. Barone, R., Cheng, P.: Representations for Problem Solving: on the Benefits of Integrated Structure. In: Banissi, E., Borner, K., Chen, C., et al. (eds.) 8th International Conference on Information Visualisation, pp. 575–580. IEEE Press, Los Alamitos (2004)
12. Aldrich, F.K., Hindle, Y., Morley Wilkins, S., Gunn, D.: Tools of thought – towards independence in thinking and problem-solving. Visability 34 (2004)
13. Scaife, M., Rogers, Y.: External Cognition: How do Graphical Representations Work? International Journal of Human Computer Studies 45, 185–214 (1996)

Openproof - A Flexible Framework for Heterogeneous Reasoning

Dave Barker-Plummer[1], John Etchemendy[1], Albert Liu[1], Michael Murray[1], and Nik Swoboda[2]

[1] Stanford University
Stanford, CA, 94305-4101, USA
[2] Universidad Politécnica de Madrid
Boadilla del Monte, Madrid, 28660, Spain

Abstract. In this paper we describe the Openproof heterogeneous reasoning framework. The Openproof framework provides support for the implementation of heterogeneous reasoning environments, i.e., environments for writing arguments or proofs involving a number of different kinds of representation. The resulting environments are in a similar spirit to our *Hyperproof* program, though the Openproof framework goes beyond *Hyperproof* by providing facilities for the inclusion of a variety of representation systems in the same environment. The framework serves as the core of a number of widely used educational programs including *Fitch*.

1 Introduction

In [1,2,3] our group pioneered the notion of *formal heterogeneous deduction*: formally specified inference systems in which different representations are used in concert to reach conclusions. This work resulted in the implementation of *Hyperproof* [4], the first proof checker for a heterogeneous logic.

Hyperproof employs two representations: the sentential representation of first-order logic (FOL), and a diagrammatic representation consisting of blocks on a checkerboard. Our theory, however, is general and does not depend on these two specific representations. The Openproof framework abstracts over the specific representations that are employed in the deduction. The framework can be used to create heterogeneous (and homogeneous) proof environments for any combination of representations by "plugging in" implementations of the specific representations. Aspects of this general architecture have been described in [5,6,7,8] though this is the first description of the software implementation.

The Openproof framework is in use as the core of the *Fitch* application for homogeneous deduction in FOL [9]. Additional Openproof modules are currently being built for Block's World, Euler/Venn [10], Position, and Coincidence Grid diagrams [11], and for sentences in plain text. Here we will briefly describe the different components of the Openproof framework.

G. Stapleton, J. Howse, and J. Lee (Eds.): Diagrams 2008, LNAI 5223, pp. 347–349, 2008.

2 Framework

The Openproof framework consists of three tiers:

- **Kernel** – the Openproof kernel provides services including the loading of components, saving and restoring files, and interfacing with the operating system and Java environment.
- **Tools** – the second tier consists of tool kits which can be used to build modules and provide intra-module and inter-module communication. Examples of tools provided by the framework include a tool kit to build basic diagrammatic editors, and facilities for integrating changes in representations into a proof.
- **Modules** – the third tier provides component-level interfaces for specifying the particular kinds of modules that can be combined into a heterogeneous proof environment. There are two kinds of modules, representation modules and proof modules. Representation modules have a common structure consisting of:
  - Representations in one of two flavors: sentential and diagrammatic.
  - Editors to enable users to construct and manipulate those representations.
  - Inference engines to support two kinds of reasoning: homogeneous and heterogeneous. (Homogeneous engines involve only one representation, while heterogeneous engines involve more than one.)

  Proof modules also have a common structure consisting of:
  - Editors to allow users to build and change proofs.
  - Proof engines to support the logic behind the proof system.

Thus, the Openproof framework provides core services to allow the automatic generation of heterogeneous reasoning systems. At the moment the framework does not provide support for the construction of single proof environments involving more than one proof system at the same time, but this is a possibility which opens a number of interesting theoretical and practical questions.

2.1 The Role of *interlingua*

By design, the framework does not have a single common language into which all representations in the system must be translated in order to perform or check inferences, i.e., an *interlingua*. As discussed in [2], it is our belief that heterogeneous rules of inference need not be defined upon the basis of a translation into an interlingua. Following this philosophy gives greater latitude to module designers when thinking about the design of new components. In general, not requiring an interlingua simplifies the design of homogeneous rules of inference (as they can be defined directly from one instance of a representation system to another without having to pass through a second kind of representation), and allows the possibility for a range of heterogeneous rules to co-exist simultaneously.

While we do not require the use of interlingua, the framework supports the partial or total use of *implicit* and *explicit* interlingua in particular heterogeneous systems. By implicit, we mean the use of an interlingua to define heterogeneous rules of inference (a use which is transparent to the reasoner). For example, when defining rules of inference in an Euler diagram module, internally the rules could translate the Euler diagram into a Venn diagram and then rely upon already defined Venn rules. By explicit, we mean the use of a single representation as the "go-between" for a number of others. An example of this would be a system in which exchanging information between any two different representations requires translating the first into FOL and then translating from FOL into the second.

3 Conclusion

The Openproof framework will serve as a tool to free researchers interested in heterogeneous reasoning systems from many of the mundane tasks involved in developing such systems and thereby allow them to focus on the design of individual representations and heterogeneous relations between those modules.

References

1. Barwise, J., Etchemendy, J.: Information, infons and inference. In: Situation Theory and Its Applications, pp. 33–78. CSLI Publications, Stanford (1990)
2. Barwise, J., Etchemendy, J.: Heterogeneous logic. In: Logical reasoning with diagrams, pp. 179–200. Oxford University Press, New York (1996)
3. Barwise, J., Etchemendy, J.: Visual information and valid reasoning. In: Logical reasoning with diagrams, pp. 3–25. Oxford University Press, New York (1996)
4. Barwise, J., Etchemendy, J.: Hyperproof. CSLI Publications, Stanford (1994)
5. Barker-Plummer, D., Etchemendy, J.: Visual decision making: A computational architecture for heterogeneous reasoning. In: Kovalerchuk, B., Schwing, J. (eds.) Visual and Spatial Analysis: Advances in Data Mining, Reasoning and Problem Solving, pp. 79–109. Springer, Berlin (2004)
6. Barker-Plummer, D., Etchemendy, J.: A computational architecture for heterogeneous reasoning. Journal of Theoretical and Experimental Artificial Intelligence 19(3), 195–225 (2007)
7. Barker-Plummer, D., Etchemendy, J.: Applications of heterogeneous reasoning in design. Machine Graphics and Vision 12(1), 39–54 (2003)
8. Swoboda, N., Allwein, G.: Modeling heterogeneous systems. In: Hegarty, M., Meyer, B., Narayanan, N.H. (eds.) Diagrams 2002. LNCS (LNAI), vol. 2317, pp. 131–145. Springer, Heidelberg (2002)
9. Barwise, J., Etchemendy, J., Allwein, G., Barker-Plummer, D., Liu, A.: Language Proof and Logic. CSLI Publications, University of Chicago Press, Stanford (1999)
10. Swoboda, N., Allwein, G.: Heterogeneous reasoning with euler/venn diagrams containing named constants and fol. Electronic Notes in Theoretical Computer Science 134, 153–187 (2005)
11. Barker-Plummer, D., Swoboda, N.: A sequent based logic for coincidence grid. In: CEUR Workshop Proceedings, vol. 274, pp. 1–12 (2007)

Cognitive and Semantic Perspectives of Token Representation in Diagrams

Rossano Barone and Peter C-H. Cheng

Representation and Cognition Research Group,
Department of Informatics, University of Sussex. Falmer, UK
{r.barone,p.c.h.cheng}@sussex.ac.uk

Abstract. The article considers a perspective on token representation in diagrams and its relation to other properties typical of diagrammatic external representations (ERs). We distinguish between direct and indirect cognitive modes of token reference based on whether a referential attribute of ER tokens, represented by a cognitive system, is used as a surrogate for represented tokens. We then consider how this characterization sheds light on particular cognitive properties of representing objects and relations in diagrams and diagrammatic kinds of engagement with sentential classes of ER.

1 Token Representation

A common property of diagrammatic external representations (ERs) is that its tokens are used to stand in for the tokens being represented [1]. This kind of token reference can be contrasted with ERs in which tokens are identified by types of symbols [2]. To illustrate this distinction consider a semantic network representing a cognitive model of a person's knowledge in which each node is intended to stand in for a particular knowledge structure in a hypothetical memory system. In contrast consider algebraic expressions in which types of symbols for example R1 in expressions like $\mathbf{V_{R1}} = \mathbf{V_{R2}}$ and $\mathbf{V_{R1}} < \mathbf{V_{R3}}$, rather than the tokens themselves denote particular token resistors in an electric circuit. This distinction has been discussed by Stenning and colleagues from a semantic perspective [2]. Here we outline a cognitive perspective which appeals to the internal representation of tokens and how tokens are referred to in ERs. We will then consider cases that reveal dependencies between token reference and other properties of diagrams and diagrammatic forms of cognitive engagement.

We assume that when a cognitive system internally represents particular tokens rather than just classes of them (e.g., mental models, problems state representations, mental imagery, perceptual object representations) it must also create a type of internal reference that identifies the continuity of the token being internally represented. We will call such representations *token identities* (in contrast to type identity) which we think of as a kind of demonstrative representation that represents or simulates external reference. A form of token identity is assumed in theories of perceptual object representation sometimes referred to as numerical identity. We also consider the theory of visual indexing proposed by Pylyshyn [3] as an account of token identity within the visual system. In cognitive models of token representation,

G. Stapleton, J. Howse, and J. Lee (Eds.): Diagrams 2008, LNAI 5223, pp. 350–352, 2008.

token identity may be implicit in internal representing tokens or modeled by pointers to them. With respect to the cognitive representation of tokens we believe that the separation of token identity from selected bindings (e.g., attributes and classifications) is a more coherent characterization as the bindings linked to a token identity may change over the course of an interpretation/mental simulation.

We will use token identity to more clearly specify the cognitive mode of token reference typically afforded by diagrammatic forms of ERs which we call *direct token reference*. Direct token reference in an ER occurs whenever a cognitive system invokes semantically interpreted propositions that refer to the token identity of the ER tokens. Hence, if t1 and t2 are names that we give to a cognitive representation of token identities of two nodes in a semantic network then semantically interpreted propositions such as *spreads-activation-to (t1, t2)* involve direct token reference. When direct token reference is employed one can think of the token identity of ER tokens as standing in for the token identity of represented tokens. This can be contrasted with *indirect token reference* which occurs whenever a cognitive system derives semantic propositions from an ER that do not refer to the token identities in the ER. In such cases an interpretation of imagined token identities may need to be formulated offline.

2 Some Implications

The mode of token reference appears to be an important concept in the characterization of other cognitive and semantic properties of diagrams. We outline three examples below: object conceptualization, acquiring relations and the use of direct token reference in sentential ERs.

Diagrams often afford different ways of conceptualizing representing objects depending on the conceptual requirements of the task. For example, the arcs of a graph may be conceptualized as directed paths in tracing a course of activation between nodes but only as objects when being enumerated. A node representing Bill may be referred to as the Bill node or more abstractly as the male node. In comparison, symbols in sentential expressions are normally much more specific about the conceptualizations they designate of represented objects [4]. In direct token reference the cognitive systems treats the token identity as invariant to different conceptualization of its perceptual attributes akin to non-representational forms of object conceptualization. In indirect token reference as typically employed in sentential ERs there are no representing token identities to support such invariance assumptions.

The mode of token reference characterizes a difference in the representation of relations between diagrammatic and sentential classes of ERs. In diagrams, represented relations are normally represented by actual relations even when symbols are involved (e.g., semantic networks use arc symbols to semantically relate token identities of nodes). This can be contrasted with sentences that use symbols that identify types of relations. The former class of relations are direct token reference relations because the token identity of the ER tokens are the referents of semantic propositions about them. This perhaps provides a useful operational distinction of the difference between the representation of relations in diagrammatic and sentential ERs at least where symbols of expressions do not represent token values (e.g. **2 + 2 =**).

Direct token reference is conventionally employed with many sentential ERs. Some cases involve individuating a symbol grouping and assigning a semantic interpretation to that individual based on an indirect token reference interpretation of its subsumed sentential structures. For example, in meaningfully reordering blocks of written text in a document a user will have formulated one or more propositions about its content. This mundane case involves direct token reference because its token identity is the subject of propositions about its semantic content. There are many other cases to consider in which notational structures are purposely designed to support direct token reference such as in programming languages (e.g., class, frame, production declarations) and mathematical notations (e.g., right hand side of an equation) which may all be ascribed meaning via direct token reference. Such cases have similarities with derivative meaning in diagrams [5] for which direct token reference seems to be an important condition [4]. Our analysis suggests that opportunities for direct token reference in sentential ERs is a common factor in blurring their boundaries with diagrams.

3 Conclusion

Types of ERs and their structures differ in the mode of token reference they afford. Direct token reference is a way of processing tokens typically afforded by diagrammatic ERs and has dependencies with several cognitive and semantic properties of diagrams. The mode of token reference implicates a kind of demonstrative representation possessed by a cognitive system which can be viewed as a semantic attribute of representing tokens in diagrams. It is suggested that this characterization may be of use in theories of external representation.

References

1. Barwise, J., Etchemedy, J.: Heterogeneous Logic. In: Glasgow, J., Narayanan, N.H., Chandrasekaran, B. (eds.) Diagrammatic Reasoning: Cognitive and Computational Perspectives, pp. 211–234. AAAI Press, Menlo Park (1995)
2. Stenning, K., Inder, R., Neilson, I.: Applying semantic concepts to analyzing media and modalities. In: Glasgow, J., Narayanan, N.H., Chandrasekaran, B. (eds.) Diagrammatic Reasoning: Cognitive and Computational Perspectives, pp. 303–338. AAAI Press, Menlo Park (1995)
3. Pylyshyn, Z.: Seeing and Visualizing: It's not what you think. MIT Press, Cambridge (2004)
4. Barone, R., Cheng, P.C.-H.: Conditions for Selection and Conceptualization in Diagrams and Sentences. In: Proceedings of the 30th Annual Conference of the Cognitive Science Society (in press)
5. Shimojima, A.: Derivative meaning in graphical representations. In: Proceedings of 1999 IEEE Symposium on Visual Languages, pp. 212–219 (1999)

Estimating Effort for Trend Messages in Grouped Bar Charts*

Richard Burns[1], Stephanie Elzer[2], and Sandra Carberry[1]

[1] Dept of Computer Science, Univ. of Delaware, Newark, DE 19716 USA
{burns,carberry}@cis.udel.edu

[2] Dept of Computer Science, Millersville Univ., Millersville, PA 17551 USA
elzer@cs.millersville.edu

Abstract. Information graphics found in popular media contain communicative signals which help the viewer infer the graphic designer's intended message. One signal is the relative effort required for different recognition tasks. This paper presents a model of the effort required to recognize a trend in a *grouped bar chart*. The model is developed using the ACT-R cognitive framework and validated via eye tracking experiments.

1 Introduction

Information graphics, ranging from bar charts and line graphs to grouped, composite and more complex graphic representations, appear in popular media and contain an intended message. The goal of our research is to identify the message conveyed by an information graphic. Our work has two applications that we are pursuing: 1) better summarization of documents in digital libraries, and 2) access to multimodal documents for individuals with impaired eyesight. Currently, research on summarization has focused on an article's text. Our goal is to integrate the messages conveyed by an article's information graphics into a richer summary of a multimodal document. In the case of individuals with sight impairments, we want to provide access to an article's information graphics by providing a brief summary (rendered via speech) of the high-level content of the graphic, centered on its overall message, and then responding to followup questions from the user.

2 Message Recognition from Information Graphics

In addition to words and phrases, language includes any deliberate action or signal (or lack of) that has a communicative intent.[1] Under this broad view, information graphics in popular media are a form of language. A graphic designer has an intended message to communicate to the viewer, and thus will use specific signals in attempting to convey that message.

* This material is based upon work supported by the National Science Foundation under Grant No. IIS-0534948.

G. Stapleton, J. Howse, and J. Lee (Eds.): Diagrams 2008, LNAI 5223, pp. 353–356, 2008.

Our research group was the first to apply language understanding techniques to information graphics. We developed a system[2] which recognizes the intended message of a *simple bar chart*. The system extracts communicative signals from a simple bar chart and uses them as evidence in a Bayesian network that hypothesizes the message conveyed by the graphic. The system was shown to have a success rate of 79.1% on a corpus of bar charts whose messages had been previously identified by human annotators.

Our current work is concerned with recognizing the intended message of a more complex kind of information graphic, *grouped bar charts* which consist of two or more visually distinguishable groups of bars.

2.1 Message Categories

Messages conveyed by grouped bar charts are more complex than simple bar chart messages. Our analysis of a corpus consisting of approximately 100 grouped bar charts has identified 14 categories of possible messages. On occasion, it appears that multiple messages can exist. Two examples of messages are: *Rising-Trend-All* (trend of values is increasing for every series of bars) and *Gap-Decreasing* (gap between entities in a group decreases across groups).

2.2 Communicative Signals

Captions contain communicative signals which can be identified via simple shallow processing. For example, nouns in the caption of a grouped bar chart serve to make a set of bars salient. This might be the set of bars comprising one group in the chart, or might be the same bar depicted over each group. Also, sets of bars can become salient via design choices, such as positioning, made by the graphic designer.

In their work on generating information graphics, the AutoBrief group contended that graphic designers construct a graphic that facilitates as much as possible the tasks that the viewer will need to perform in understanding the graphic's message.[3] Thus, from a recognition viewpoint, the relative effort required for different tasks is a communicative signal about which tasks the graphic designer intended the viewer to perform. Consider, for example, Fig. 1; the two grouped bar charts were constructed from exactly the same data[1]. The message conveyed by the graphic on the left is that salaries are lower for women than men in all disciplines shown. Note that this graphic makes it easy to compare female and male salaries in each discipline. However, extracting that same message from the graphic on the right would require much more effort. In fact, a different message would most likely be inferred from the right graphic: that salaries are greatest in engineering and physical science for both men and women.

In our experiments with simple bar charts, we found that relative effort was the communicative signal that had the greatest impact on our system's

[1] The grouped bar chart on the left appeared in the 2000 Report of the NSF Committee on Equal Opportunities in Science and Engineering.

Fig. 1. A graphic which is informationally but not computationally equivalent

performance.[4] Thus, we hypothesize that relative effort will also be very important in recognizing the intended message of a grouped bar chart.

3 Model of Task Effort

To gain insight into the features that impact task effort in grouped bar charts, we performed a preliminary set of experiments with human subjects and an eye tracker, to identify fixations, measure fixation durations, and compute the time required to process different graphics. In particular, we examined the properties of grouped bar chart trends which affect task effort. Based on the results, we found that 1) the number of groups in a graphic, 2) the size of a graphic, 3) the density of a graphic, 4) the presence of visual clutter, 5) and the presence of exceptions, all affect the effort required to recognize a trend. Using ACT-R[5] and the EMMA add-on[6], we used these observations in designing a cognitive model to estimate relative effort for recognizing a trend in different graphics.

We ranked a set of graphics in terms of the median effort required by subjects to recognize a trend and the effort required by our developed model to process a trend. Here, effort was defined as time required to recognize a trend. The rankings for the test set are presented in Table 2. As shown, the graphics varied in number of bars per group, number of groups, size, presence of exceptions, and existence of increasing and decreasing trends. The *crossover* column is only applicable in graphics with opposite trends, and reflects where the trends intersected. An *exception* is any bar whose bar value does not follow the strict increasing or decreasing bar value ordering of preceding or following bars. The Spearman Rank Order Correlation Coefficient (ρ) is a statistical test which measures the correlation between two sets of ordinal data, such as our rankings. For our data, $\rho = 0.7874$, indicating a strong correlation between the two sets of rankings ($p < 0.0001$). Thus we conclude that our model produces a very good estimate of the relative effort required to recognize trends in a grouped bar chart.

Table 1. Rank-order correlation of model and validation experiment

Graphic						Rank	
Number of Bars per Group	Graphic Size	Trends	Crossover group if present	Number of Exceptions	Number of Groups	Model	Experiment
2	small	2 inc.		0	10	1	1
2	small	2 inc.		0	5	2	2
2	small	2 inc.		0	7	3	3
2	large	2 inc.		0	12	4	8
2	large	1 inc. & 1 dec.	5th	0	8	5	9
2	large	2 inc.		0	10	6	4
2	large	1 inc. & 1 dec.	6th	0	8	7	7
2	large	1 inc. & 1 dec.	7th	0	8	8	6
2	small	1 inc. & 1 dec.	6th	0	8	9	5
2	small	2 inc.		1	5	10	18
2	small	2 inc.		1	10	11	13
2	small	2 dec.		2	10	12	15
2	large	2 inc.		1	7	13	16
2	large	2 inc.		1	10	14	11
2	large	2 dec.		2	10	15	12
3	small	2 inc.		0	7	16	10
3	large	2 inc.		0	4	17	14
3	large	2 inc. & 1 dec.		0	4	18	17

4 Future Work

Future work on this project involves expanding our effort estimation to handle all types of grouped bar chart messages. We also need to create and implement a Bayesian network, which will use the communicative signals present in a grouped bar chart as evidence about the graphic's intended message.

References

1. Clark, H.: Using Language. Cambridge University Press, Cambridge (1996)
2. Elzer, S., Carberry, S., Zukerman, I., Chester, D., Green, N., Demir, S.: A probabilistic framework for recognizing intention in information graphics. In: Proceedings of the International Joint Conference on Artificial Intelligence (2005)
3. Green, N.L., Carenini, G., Kerpedjiev, S., Mattis, J., Moore, J.D., Roth, S.F.: Autobrief: an experimental system for the automatic generation of briefings in integrated text and information graphics. International Journal of Human-Computer Studies 1, 32–70 (2004)
4. Carberry, S., Elzer, S.: Exploiting evidence analysis in plan recognition. In: International Conference on User Modeling (2007)
5. Anderson, J.R., Matessa, M., Lebiere, C.: Act-r: A theory of higher level cognition and its relation to visual attenion. Human-Computer Interaction 12, 439–462 (1997)
6. Salvucci, D.D.: An integrated model of eye movements and visual encoding. Cognitive Systems Research 1, 201–220 (2001)

Types and Programs from Euler Diagrams

James Burton*

Visual Modelling Group, University of Brighton, UK
j.burton@brighton.ac.uk

Abstract. Type theory provides a formal basis for programming languages and can also be used to model reasoning systems such as Euler diagrams. We present part of a simple type theory of Euler diagrams. Expressing a system of reasoning with Euler diagrams as a collection of types and operations on types (which correspond to diagrams and reasoning rules) is a first step towards embedding visually modelled constraints directly into the type system of a programming language.

1 Introduction

Constraint diagrams [3] are a sound and complete diagrammatic logic for expressing constraints on software systems, of which Euler diagrams [4] are a sound and complete fragment. Type theory allows formal abstractions such as Euler diagrams to be implemented directly in the type systems of suitably expressive programming languages. In this paper we present part of a simple type theory of Euler diagrams as a step towards our goal of embedding visually modelled specifications directly into a programming language. Since existing tools for specifying software are external to the implementation environment and require specialised training, we believe our approach has the potential to provide a more direct way of creating software from formal specifications.

2 Euler Diagrams

A sound and complete logic for reasoning with Euler diagrams is presented in [4]. A unitary Euler diagram is a collection of closed curves called *contours* which represent sets, within an enclosing rectangle. Figure 1 shows two examples which each have three contours, labelled A, B and C. Containment, intersection and disjointness are represented by the relative placement of contours. A *zone* is the maximal set of points in the diagram that can be described as being inside certain contours and outside all others. The region outside of all contours is also a zone. Shading within a zone asserts the emptiness of the set represented by that zone. So, in Figure 1, d' asserts $A \cap B = \emptyset$. Compound Euler diagrams are formed by conjunction, disjunction and negation, so if d_1 and d_2 are diagrams then so are $d_1 \wedge d_2$, $d_1 \vee d_2$, $\neg d_1$ and so on. Reasoning is carried out by the application

* James Burton is supported by EPSRC Grant EP/P501318/1.

G. Stapleton, J. Howse, and J. Lee (Eds.): Diagrams 2008, LNAI 5223, pp. 357–359, 2008.

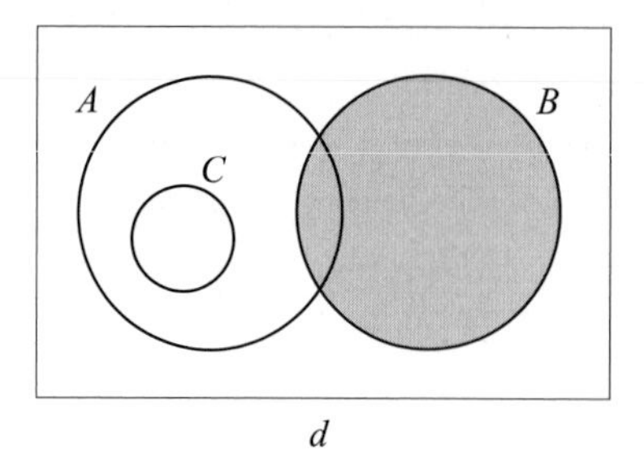

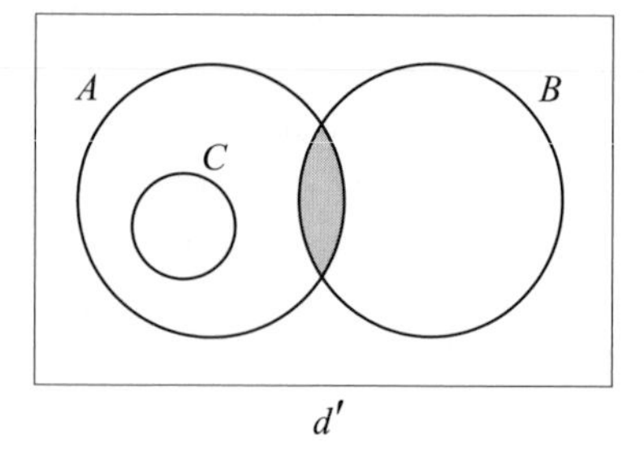

Fig. 1. Two Euler diagrams

of rules which transform one diagram into another, such as *Add Contour* and *Remove Shading*. In Figure 1 d can be transformed into d' by a single application of the rule *Remove Shading*. An abstract (symbolic) syntax is used to represent diagrams and reasoning rules in definitions and meta-level results.

3 Type Theory

Modern type theory originated in the work of Martin-Löf [2] and formalises the behaviour of systems such as the simply typed lambda calculus. It leads to many notions of more complex types such as *dependent types*. Dependent types are families of types indexed by a value, such as *Vector* a n, the type of vectors with length n containing elements of type a. To construct a vector, the constructor *Vector* is supplied with a type for its input a (e.g. the type of booleans) and a natural number for its input n. Because such types carry more information than those found in widely used languages, more aspects of correctness can be checked mechanically by the compiler. This provides the opportunity to rule out certain errors which would not normally be detected until runtime, such as an attempt to extract an element from an empty vector. In general, dependent types offer the programmer many opportunities for increased safety and expressiveness [1]. Entire diagrammatic reasoning systems may be specified as a theory of types, with the abstract syntax translated to rules for forming and eliminating types. Implementing this type theory in a suitable language delegates the task of choosing whether a given diagram can be formed or whether a transformation can be applied to the typechecker of that language.

4 The Types of Euler Diagrams and Reasoning Rules

This section gives a brief flavour of a type theory of Euler diagrams. Three kinds of element appear in the judgment rules: typing judgements, e.g. $x : T$ (x has type T), type constraints, e.g. *Diagram* d type (meaning that this type can be formed), and functions from types to types such as *union* x y. In each case this takes place within a context Γ, which we omit for clarity.

We assume a type level library of sets such as given in [2] and types for unitary diagrams, zones and invariants between them, created using the abstract syntax as a model. In the abstract syntax a zone is represented by a pair of sets of contour labels, those inside and those outside the zone. A diagram is represented by the tuple of sets of labels, zones and shaded zones. This gives us the types $Z\ lin\ lout$ (zones) and $D\ l\ z\ z^*$ (diagrams). The semantic properties of unitary Euler diagrams are captured as typing judgements, not shown for reasons of space, which guarantee that only well-formed diagrams can be constructed. We then define compound diagrams by induction:

$$\frac{\mathit{Diagram}\ d\ \text{type}}{\mathit{Diagram}\ \neg d\ \text{type}}$$

$$\frac{\mathit{Diagram}\ d_1\ \text{type} \quad \mathit{Diagram}\ d_2\ \text{type}}{\mathit{Diagram}\ (d_1 \vee d_2)\ \text{type}} \quad \frac{\mathit{Diagram}\ d_1\ \text{type} \quad \mathit{Diagram}\ d_2\ \text{type}}{\mathit{Diagram}\ (d_1 \wedge d_2)\ \text{type}}$$

As an example, the reasoning rule *Remove Shading* mentioned previously can be captured as an application of a type function *transform*:

$$\frac{\mathit{Member}\ z_1\ z^*\ \text{type} \quad z' : \mathit{delete}\ z_1\ z \quad z^{*\prime} : \mathit{delete}\ z_1\ z^*}{\mathit{transform}\ \mathit{RemoveShadedZone}\ z_1\ (D\ l\ z\ z^*) : (D\ l\ z'\ z^{*\prime})}$$

5 Conclusions and Further Work

To represent Euler diagrams as a theory of types we have extended the approach of the previous section to a complete set of reasoning rules. The resulting types will, in turn, be implemented as part of a Domain Specific Embedded Language (DSEL). This DSEL will combine strong type-level assurances with a runtime (term-level) interface for users and a graphical representation of concrete diagrams. The motivation for doing this is to build a foundation for more complex systems. We intend to provide a DSEL for the system of generalized constraint diagrams, which is expressive enough to be used as the basis of tools for software modelling [3]. The types of diagrams created using such tools will form constraints that we aim to use directly in finished software. The question of which non-trivial constraints can be effectively modelled and exported as types will form one of the main subjects of the ongoing research.

References

1. Altenkirch, T., Mcbride, C., Mckinna, J.: Why dependent types matter (2005) (accessed 01/02/08), `http://www.cs.nott.ac.uk/~txa/publ/ydtm.pdf`
2. Nordstrom, B., Petersson, K., Smith, J.M.: Programming in Martin-Löf's Type Theory. OUP (1990)
3. Stapleton, G., Delaney, A.: Evaluating and generalizing constraint diagrams. Journal of Visual Languages and Computing (JVLC 2008)
4. Stapleton, G., Masthoff, J., Flower, J., Fish, A., Southern, J.: Automated theorem proving in Euler diagrams systems. Journal of Automated Reasoning 39, 431–470 (2007)

Diagrams in the UK National School Curriculum

Grecia Garcia Garcia and Richard Cox*

Representation and Cognition Group, University of Sussex, Brighton, UK
{G.Garcia-Garcia,richc}@sussex.ac.uk

Abstract. We review the use of diagrams in the UK National School Curriculum (NC) and assess it in the light of current research on childrens' graph comprehension. It is noted that some diagrams are very ubiquitous (e.g. illustrations and bar graphs) whereas other representations which are potentially appropriate for younger children (e.g. hierarchies) are relatively rarely used. We also query some of the teaching practices advocated in the NC in that, for example, diagrammatic representations are conflated with figurative illustrations (pictographs). Suggestions for improving the NC are offered in the conclusion.

1 Diagrams in the UK National Curriculum

Despite the calls for a graphicacy curriculum and the growing evidence on children's diagrammatic abilities (e.g. [2], [6], [7]), it seems that it has yet to influence the UK National Curriculum (NC). We discuss this by presenting (1) results from a survey of the use of different external representations[1] (ERs) in the NC activities, and (2) recommendations from the literature about students' use of some ERs. These two points are discussed and some recommendations to educators are made about how ERs should be used in the UK school curriculum.

The NC sets out the subjects[2] that children should study at each of four key stages (KS) across ages ranging from 5 to 16 years. Each KS identifies a band of school years, with children's performance assessed at the end of each band. The assessment includes a national test (SAT, Standard Assessment Task) for KS 2 and KS 3.

The attainments, knowledge and skills for each subject are outlined in the "Program of Study". However, the content of each subject is not overly specified in terms of detail. This allows educators to follow other NC guidelines[3] and to freely decide which teaching methods are appropriate for their students. Our focus is upon the National Curriculum in Action[4] [5] because it explicitly presents ER-related classroom activities.

* Corresponding author.

[1] For the purpose of this paper, we define them as those graphical representations listed in Table 1 and based on Harris' diagrammatic taxonomy [3].

[2] See: `http://www.qca.org.uk/qca_7124.aspx`

[3] Such as the "Schemes of Work": `http://www.standards.dfes.gov.uk/schemes3/`

[4] `http://curriculum.qca.org.uk/index.aspx`

G. Stapleton, J. Howse, and J. Lee (Eds.): Diagrams 2008, LNAI 5223, pp. 360–363, 2008.

Classification of Data. Activities were selected and categorized by subject and year. This resulted in a list of activity examples together with descriptions and, in most cases, an example of pupils' work. The activity examples and their associated sets of ERs were examined, and their representations were categorized and named based on [3] (see Table 1). KS 4 data is not included because our focus is on children below 14 years. Additionally, examples from the subjects 'Modern Foreign Languages' and 'Citizenship' were excluded because they do not appear in all of the NC key stages.

Table 1 shows the ERs found across school years 1-9. The gray boxes indicate that at least one of that table cell's representations was conflated with pictograms. The practice of conflating diagrammatic and figurative illustrations could be implicated as a cause of misconceptions such as 'graph-as-picture' (discussed below).

Table 1. External representations found in key stages 1, 2 and 3. 'Other graphs' refers to 'frequency and distribution graph', 'radar graph' and 'scatter graph' as in [3]. KS stands for Key Stage.

	Total	KS 1		KS 2				KS 3		
Age:		5-6	6-7	7-8	8-9	9-10	10-11	11-12	12-13	13-14
Representation vs Year		1	2	3	4	5	6	7	8	9
Illustration chart	**137**	15	24	7	10	13	20	13	17	18
Table	**95**	3	7	7	8	11	15	15	11	18
List	**32**	1	9	3	2		2	3	7	5
Bar graph	**29**	5	3	4	3	1	6	2	1	4
Map	**27**	2	3	1	4	2	2	1	4	8
Line graph	**15**					3	5	1		6
Conceptual diagram	**11**			1	1		1	1	5	2
Block diagram	**10**	1		1			1	1	4	2
Flow chart	**8**		2					2	3	1
Tree diagram	**7**						2		1	4
Other graphs	**5**					1			1	3
Tally chart	**5**		1	2			1		1	
Venn diagram	**5**		2		1		2			
Network diagram	**3**				1			2		
Pie chart	**3**								1	2
Relationship diagram	**3**									3
Time line chart	**3**		2					1		
Calculation chart/tree	**1**									1
Cartesian coordinates	**1**					1				
Graph theory	**1**					1				
TOTAL	**401**	27	**53**	26	30	33	**57**	42	56	**77**

Table 1 shows which ERs are ubiquitous across the 3 key stages. It is interesting to note that there is an increase in the number of different ERs used at the end of each KS (when SAT-type assessment tests are administered). It seems that children are exposed to a wider variety of ERs during those periods. Additionally, mixtures of pictorial and diagrammatic representations were found for several representational forms in the early years. Figure 1 presents an example of a 'pure' diagrammatic form and an example of a conflated pictorial-diagrammatic representation. They are representative of those found throughout the NC in Action.

Considering the number of ERs in each subject area, it is apparent that illustration charts, tables and lists represent more than 56% of the representations in all subjects except in Geography and Physical Education. It would be interesting

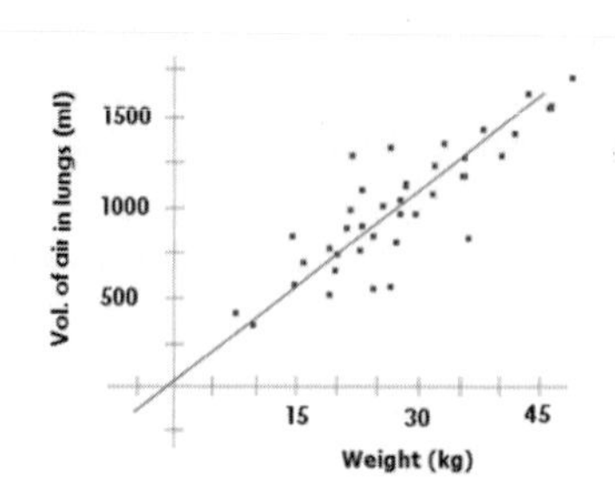

traffic survey

type of vehicle	number of vehicles in 5 minutes
cars	3
vans	1
bicycles	2
others	3

Fig. 1. A scatter graph and a tally chart. Note tally chart includes pictograms.

to establish whether the ubiquity of illustration charts, tables and lists makes the use of representations in the NC too homogeneous. An increased variety of representations in the NC might improve it. For example, [2] has shown that hierarchies can be effectively used in educational contexts by children as young as 7 years.

'Graphical literacy' is directly targeted in some subjects. For example, the ICT Schemes of Work (Unit 1E - Representing information graphically: pictograms) states[5]: "In this unit children learn how to use ICT to represent information graphically. They learn how to create pictograms and how to answer simple questions on the data shown in their pictograms." However, it is not clear what is meant precisely by the terms 'pictogram' and 'represent information graphically'.

2 Discussion - Use of ERs by Primary Students

An issue with curriculum implications concerns children's **graphicacy readiness**. Research on children's ability to use bar graphs indicates that 9 year old pupils have difficulties with comprehension [6]. It seems that children can identify labels in a graph, but they experience difficulties when doing operations and drawing inferences [6]. Nevertheless, first graders are encouraged to use them in the UK NC (see Table 1). In the case of line graphs, 7 and 8 year olds are able to interpret them [7]. However, examples of the use of that ER are not found at that stage (Table 1). Similarly, hierarchies are relegated to the very end of primary school, despite research evidence which shows that children aged 7, 9 and 11 years are able to understand and construct them [2]. As far as we know, the NC contains no clear suggestions to teachers for using hierarchies in teaching and learning contexts. Another area of relevant research is that on children's **misconceptions**. In the case of line graphs, it has been noted that some of the students' difficulties include slope/height confusion (graphs), interpreting intervals, and the graph-as-picture misconception, among others [1,4,7]. In particular, the graph-as-picture misconception seems quite prevalent [1]. Children

[5] http://www.standards.dfes.gov.uk/schemes2/it/itx1e/?view=get

who have this misconception confuse an abstract, diagrammatic ER with another kind of representation (e.g. confusing a line graph with a map, or a line graph with picture of a mountain range). We suspect that the graph-as-picture misconception may be partially due to the frequent use of pictorial representations in conjunction with 'purer' diagram forms such as line graphs. Thus, in attempt to introduce formal diagrammatic forms via the inclusion of familiar pictorial images, it is possible that teachers might inadvertently seed pictorial/diagrammatic representational confusions in students' minds by confounding two entirely different forms of representations (Table 1 - gray boxes).

In the survey of representations used in the NC there is a disparity between the very prevalent use of illustration charts and bar graphs and other potentially useful alternative forms. Ubiquitous forms potentially 'displace' alternative appropriate forms. The NC's attempt to 'bridge' pictorial and diagrammatic representations by combining them might be counterproductive and may engender misconceptions such as the graph-as-picture misconception. The results also suggests that students would probably benefit from earlier instruction in the use of hierarchies and line graphs. A systematic trend was for the variety of representations used in the NC to increase markedly at the end of each key stage (when national achievement tests are administered).

To develop their graphical skills, children need broad experience with a varied mix of representations across the curriculum. Research-based 'graphicacy development' guidelines could be produced to help instructors. These might include information about methods of teaching representational skills and the matching of representations to tasks. They might also help to prevent the development of graphical misconceptions by students.

References

1. Bell, A., Brekke, G., Swan, M.: Diagnostic teaching: 4 graphical interpretation. Mathematics Teaching 119, 56–59 (1987)
2. Greene, T.R.: Children's understanding of class inclusion hierarchies: The relationship between external representation and task performance. Journal of Experimental Child Psychology 48, 62–89 (1989)
3. Harris, R.L.: Information graphics: A comprehensive illustrated reference. Oxford University Press, New York (1999)
4. Leinhardt, G., Zaslavsky, O., Stein, M.K.: Functions, graphs, and graphing: Tasks, learning, and teaching. Review of Educational Research 60(1), 1–64 (1990)
5. National Curriculum in Action: Pupil's work online (Retrieved May 30, 2007) (2001), `http://www.ncaction.org.uk/`
6. Pereira-Mendoza, L., Mellor, J.: Students' concepts of bar graphs - Some preliminary findings. In: Vere-Jones, D. (ed.) Proceedings of the Third International Conference on Teaching Statistics, vol. 1, pp. 150–157. International Statistical Institute, Voorburg, The Netherlands (1991)
7. Phillips, R.J.: Can juniors read graphs? A review and analysis of some computer-based activities. Journal of Information Technology for Teacher Education 6(1), 49–58 (1997)

LePUS3: An Object-Oriented Design Description Language

Epameinondas Gasparis[1], Jonathan Nicholson[1], and Amnon H. Eden[1,2]

[1] The Two-Tier Programming Project, Department of Computing & Electronic Systems, University of Essex, United Kingdom
[2] Centre for Inquiry, Amherst, NY, USA

1 Introduction

LePUS3 [1] (lepus.org.uk) is a logic, visual, object-oriented Design Description Language: a formal specification language designed to capture and convey the building-blocks of object-oriented design. LePUS3 minimal vocabulary constitutes of abstraction mechanisms that can specify effectively and precisely design patterns and the design of Java™ (C++, Smalltalk, etc.) programs at any level of abstraction.

LePUS3 was tailored to integrate the strength of other specification and modelling notations, most notably UML, but it is unique in addressing the combination of the following concerns:

- **Rigour**. LePUS3 is a logic visual language: a chart stands for a formula in an axiomatized theory in the classical first-order predicate calculus.
- **Parsimony & scalability**. LePUS3 offers powerful abstractions: charts scale well and do not clutter with the size of the program.
- **Minimality**. LePUS3 vocabulary (Figure 1) is minimal, consisting of 15 tokens.
- **Decidability & verifiability** [2]. Consistency between a given specification (a chart) and an implementation (a Java program) can be verified by a button-click.
- **Program Visualization**. Charts modelling Java programs can be reverse-engineered from source code.

To emphasize practicalities, we focus on tool support [3][4] in specifying, (automatically) verifying, and visualizing Java programs in LePUS3.

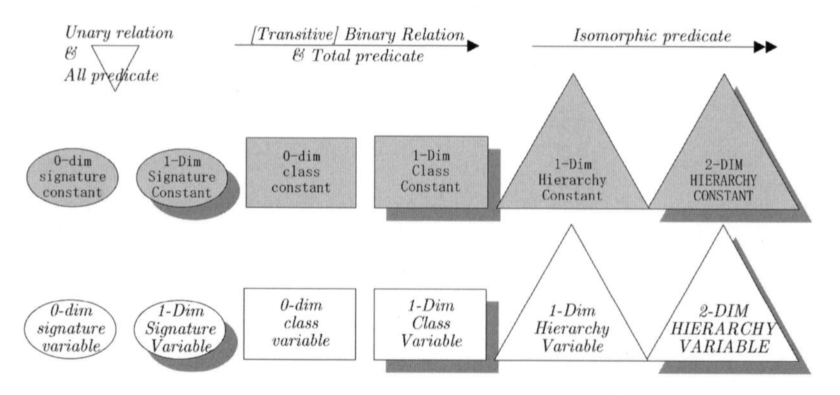

Fig. 1. LePUS3 vocabulary

G. Stapleton, J. Howse, and J. Lee (Eds.): Diagrams 2008, LNAI 5223, pp. 364–367, 2008.

2 Visualizing Programs

We take program visualization to be a tool-assisted process of discovering some of the building-blocks in the design of programs, and charting them at the appropriate level of abstraction. The motivation is usually understanding and re-engineering large, complex, and inadequately documented programs. We call the approach we take "Design Navigation" [5]: a user-guided process of reverse-engineering LePUS3 charts from the source code of arbitrarily-large Java™ programs. Design Navigation in package `java.io` (Java™ Software Development Kit 1.6) is demonstrated below using the Two-Tier Programming Toolkit [3][4].

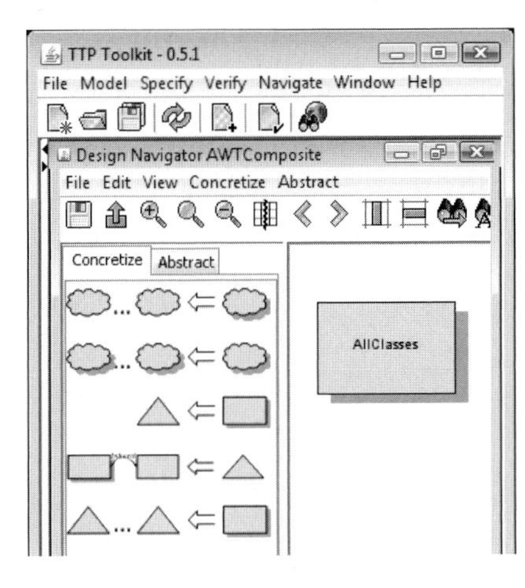

Chart 1. All the classes in `java.io`

After analyzing package `java.io`, Design Navigation commences from the Top Chart (Chart 1), the most abstract representation of any Java program. Chart 1 depicts the set of all static types (classes, interfaces, etc.) in `java.io` as a '1-dimensional class'. From Chart 1, Design Navigation proceeds by a user-guided, tool-assisted step-wise application of concretization ('zoom-in') and abstraction ('zoom-out') operators (left panel, Chart 1). At each step, the Two-Tier Programming Toolkit discovers inheritance class hierarchies (triangles), sets of classes (shaded rectangles), sets of dynamically-bound methods (superimposed ellipses), and correlations amongst them (arrows), visualized using LePUS3 terms and predicates (Figure 1).

For example, Chart 2 offers a birds-eye view of the `Closeable` class inheritance hierarchy in `java.io`, generated in a short sequence of concretizations of Chart 1.

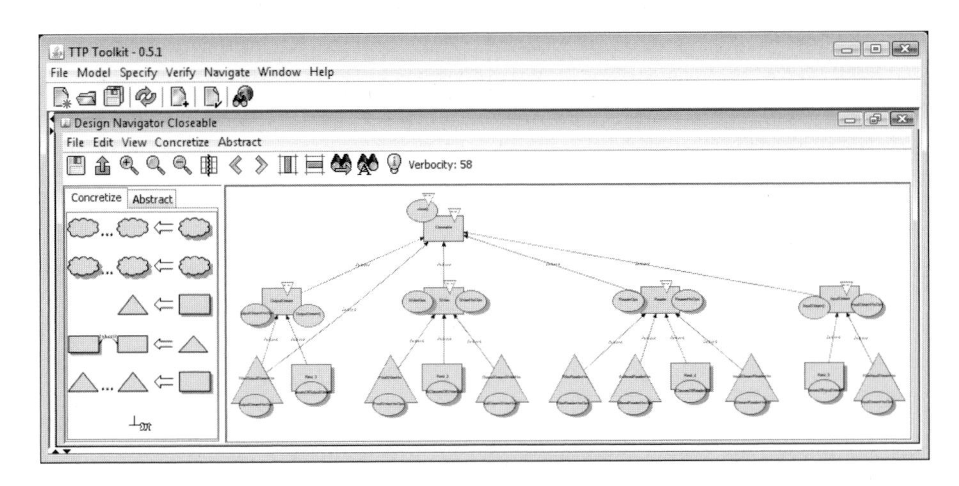

Chart 2. The `Closeable` hierarchy in `java.io`

3 Specifying Programs and Patterns

We take the goal of visual specification to be that of articulating non-functional requirements on object-oriented design in a precise visual language *at the appropriate level of abstraction*. LePUS3 can be used to specify generic design motifs, such as the Composite design pattern (Chart 3), as well as the overall design of a specific implementation, such as package `java.awt` (Chart 4).

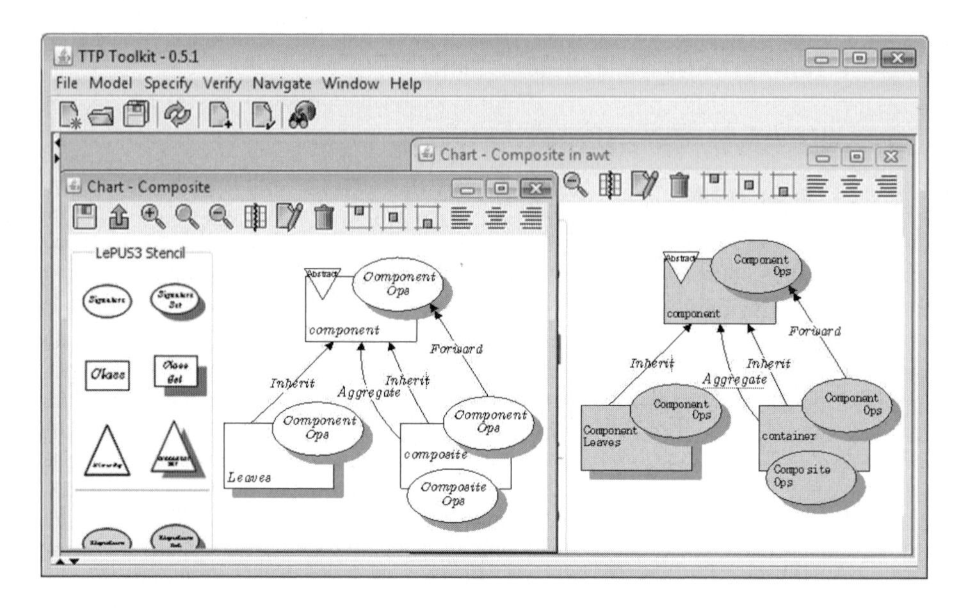

Chart 3. The Composite pattern **Chart 4.** `java.awt` (selected elements from)

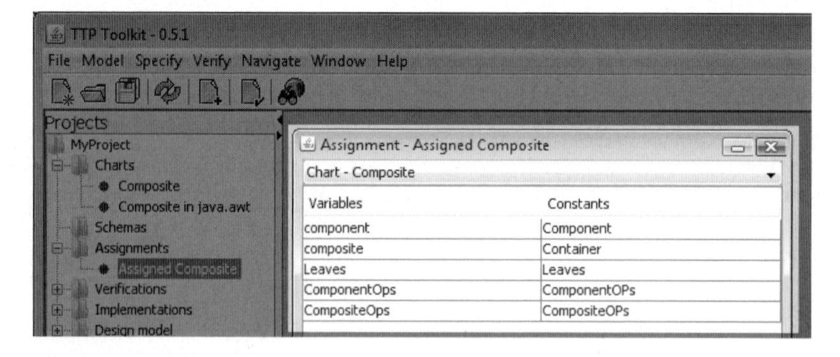

Fig. 2. Assignment from the Composite design pattern (Chart 3) to `java.awt` (Chart 4)

The similarity between Chart 3 and Chart 4 is not accidental: package `java.awt` was designed to implement the Composite design pattern [6]. In LePUS3, this intent is made explicit by a logic *assignment* (Figure 2): a mapping which indicates where (and how often) the design pattern is supposed to be implemented. The Two-Tier

Programming Toolkit can be used not only to specify assignments but, as we show in the next section, also verify them at a click of a button.

4 Automatically Verifying LePUS3 Specifications

By *verification* [2][7] we refer to the rigorous, conclusive, and decidable process of establishing or refuting consistency between a specification (chart) and a program. Fully automated verification is possible in principle since LePUS3 is decidable. It is also possible in practice, as proven by the Two-Tier Programming Toolkit, which implements the verification algorithm. For example, the toolkit can verify by a button-click that package `java.awt` *satisfies* the Composite design pattern as specified in Chart 3, delivering its results in seconds (Figure 3). If verification fails, the toolkit indicates clearly which parts of the specification were violated to allow programmers to fix such inconsistencies.

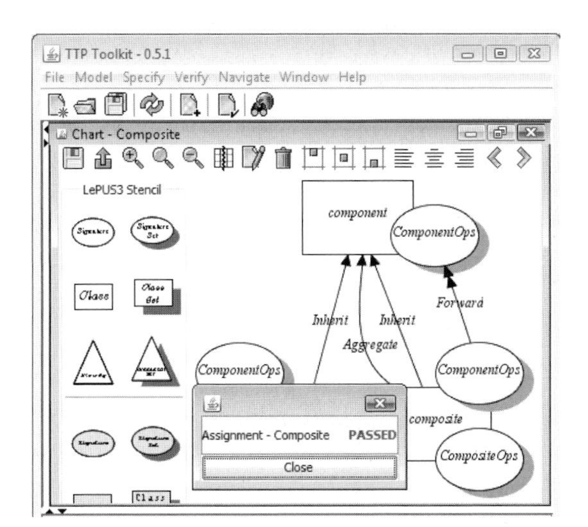

Fig. 3. The TTP Toolkit verification results

Acknowledgments. This work was partially funded by EPSRC. The authors wish to thank Ray Turner and Rick Kazman for their numerous contributions to this project.

References

[1] Eden, A.H., Gasparis, E., Nicholson, J.: LePUS3 and Class-Z Reference Manual. University of Essex, Tech. Rep. CSM-474, ISSN 1744-8050 (2007)
[2] Nicholson, J., Eden, A.H., Gasparis, E.: Verification of LePUS3/Class-Z Specifications: Sample models and Abstract Semantics for Java 1.4. University of Essex, Tech. Rep. CSM-471, ISSN 1744-8050 (2007)
[3] Gasparis, E., Eden, A.H., Nicholson, J., Kazman, R.: The Design Navigator: Charting Java Programs. In: 30th Int'l Conf. Software Engineering, Leipzig, Germany, May 10–18 (2008)
[4] `http://ttp.essex.ac.uk/`
[5] Gasparis, E., Eden, A.H.: Design mining in LePUS3/Class-Z: search space and abstraction/concretization operators. University of Essex, Tech. Rep. CSM-473, ISSN 1744-8050 (2007)
[6] Gamma, E., et al.: Design patterns: elements of reusable object-oriented software. Addison-Wesley, Boston (1995)
[7] Wing, J.M.: A Specifier's Introduction to Formal Methods. Computer 23(8), 8–24 (1990)

Utilizing Feature Diagrams to Assess the Capabilities of Tools That Support the Model Driven Architecture

Benjamin Gorry

BAE Systems Rapid Engineering
W374B, Warton Aerodrome, Warton, Lancashire, UK, PR4 1AX
ben.gorry@baesystems.com

Abstract. The Model Driven Architecture (MDA) is being increasingly adopted as part of the systems development life cycle. A plethora of tools which support the MDA are currently available. The time it takes to gain knowledge of the capabilities of a particular tool can be costly. This poster outlines how *Feature Diagrams* have been used to provide a graphical representation of the capabilities of a set of given MDA tools.

Keywords: Feature Diagrams, OMG, MDA.

1 Syntax and Semantics of Feature Diagrams

The Object Management Group (OMG) [1] has championed an initiative called the Model Driven Architecture (MDA) [2] approach. The model driven approach to system development facilitates better understanding of systems requirements capture, design, construction, and generation. Several tools which support the development for systems using the MDA are available; many of these tools support use of the Unified Modeling Language [3]. The MDA provides supports for standard specifications such as the UML, while other examples include [4, 5].

Each MDA tool supports the various features of the MDA to a differing extent.Feature Diagrams are used to communicate the MDA characteristics of each tool and provide assistance when selecting a tool for use in a systems development project.

Feature Diagrams [6] are used in Domain Analysis to capture commonalities and variabilities of systems in a domain. Each diagram represents a hierarchical decomposition of features (in this case, the features are taken from an understanding of the MDA; discussed in section 3), giving an indication of whether or not a feature is:

- *Mandatory* – represented as an edge ending with a filled circle.
- *Alternative* – only one of a type of feature may be supported; represented as edges connected by a clear arc.
- *Optional* – represented as an edge ending with an empty circle.
- *Or-features* – one or more of a type of feature may be supported; represented as edges connected by a filled arc.

The elements of Feature Diagrams are displayed in Figure 1. By taking the elements of Feature Diagrams, and distilling a number of MDA features from [2], a Feature Diagram structure for the MDA capabilities of a tool was produced.

G. Stapleton, J. Howse, and J. Lee (Eds.): Diagrams 2008, LNAI 5223, pp. 368–370, 2008.

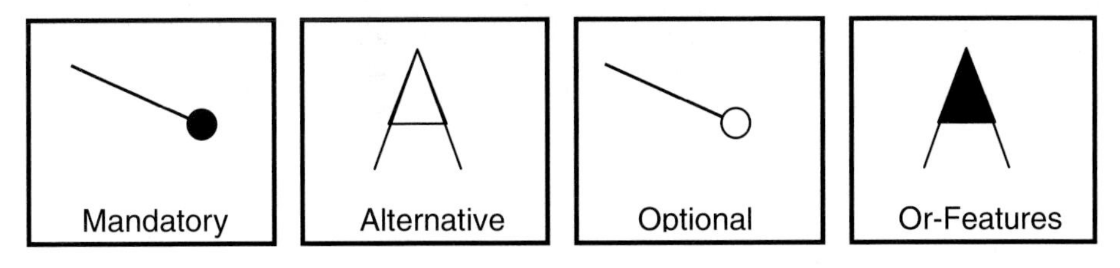

Fig. 1. Elements of Feature Diagrams

2 Feature Diagram Representation for an MDA Tool

From [2], several features which an MDA tool should support were extracted; these are displayed in Figure 2 and are now discussed. *Metamodeling* – it is not a requirement of MDA that a metamodeling approach is supported – but it is important in the domain of model portability; therefore this is identified as being optional. The features marked *Testable Transformations*, supported by an *OMG Metamodeling Architecture*, *Transformations Defined Between Metamodels*, and *Testable Metamodels* have been included as mandatory features. Even though *Metamodeling* is represented as being optional, if we are to support metamodeling these four features are required.

Testing– even though testing is an essential part of the software life cycle, it is not a required feature of the MDA; therefore it is identified as being optional. The features marked *Support Testing of the Models* and *Provide Testing Documentation* have

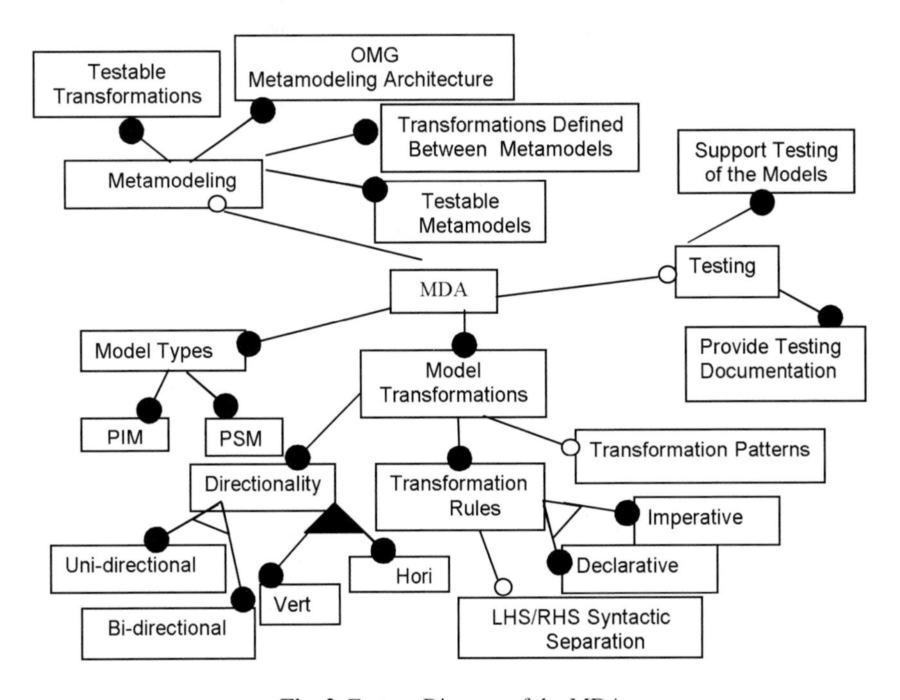

Fig. 2. Feature Diagram of the MDA

been included as mandatory features. Though *Testing* is represented as being optional, if we are to adopt testing these two features are mandatory. *Model types* in the MDA are mandatory, with *PIM* (Platform Independent Models) and *PSM* (Platform Specific Models) being mandatory.

Model Transformations in the MDA are mandatory. There are three features of *Model Transformations*; *Directionality* and *Transformation Rules* (which are mandatory), and *Transformation Patterns* (which are optional).

Within *Directionality*, model transformations must have a direction which is either *Uni-directional* or *Bi-directional*, but not both. This choice is represented using the alternative representation. Also within Directionality, model transformations can be *Vertical*, *Horizontal*, or both. Within *Transformation Rules*, model transformation rules may have an optional *LHS* (Left Hand Side)/*RHS* (Right Hand Side) *Syntactic Separation* and transformation rules must be either *Imperative* or *Declarative* (this is represented in Figure 2 using the alternative representation).

3 Conclusions

The Feature Diagram displayed in Figure 2 has been used to assess the capabilities of 21 different MDA tools. If a tool supports a particular feature, the feature is shaded on the diagram. By adopting a diagrammatic approach to analyzing the capabilities of a set of tools we can view the diagrams and quickly make an informed choice on the best tool to use in a given situation. Feature Diagrams provide a flexible method for analyzing the capabilities of a software development tool. We hope to expand our use of Feature Diagrams to tools in other application domains.

References

1. Object Management Group, HYPERLINK, `http://www.omg.org/`
2. Object Management Group.: MDA Guide Version 1.0.1, Technical Guide, Object Management Group (June 2003)
3. Object Management Group.: Unified Modeling Language: Superstructure Version 2.1.1 (2007)
4. Object Management Group.: OMG Systems Modeling Language (SysML) Version 1.0 (2007)
5. Object Management Group. Meta Object Facility (MOF) Version 2.0 Core Specification (2003)
6. Czarnecki, K., Eisenecker, U.W.: Generative Programming – Methods, Tools, and Applications. Addison-Wesley, Reading (2000)

Diagrammatic Knowledge-Based Tools for Complex Multi-dynamic Processes

Ronald R. Grau and Peter C.-H. Cheng

University of Sussex
Representational Systems Laboratory
Falmer, Brighton BN1 9QH, United Kingdom
r.r.grau@sussex.ac.uk
p.c.h.cheng@sussex.ac.uk

Keywords: Knowledge Acquisition, Complex Processes, Process Modeling, Diagrammatic Knowledge-based Tools, Heterogeneous Knowledge.

1 Introduction

This paper describes the initial work in a program of research that aims to create diagrammatic tools for the modeling and discovery of an extremely challenging domain: Complex multi-dynamic (CMD) processes. Diagrams play a central role in the approach because they provide a range of representational, cognitive and computational properties that may yield potential solutions to the challenges posed by CMD processes. Motivation for this work comes from previous research on the use of diagrams in computational scientific discovery [2], [3] and the potential of diagrams for conducting knowledge acquisition [4].

2 CMD Processes

We have coined the term *complex multi-dynamic processes* to address a particular class of knowledge that combines structural, behavioral and functional types of complexity. CMD processes are found in the biological and medical sciences and play a central role in various industrial sectors including chemical and materials production, pharmaceuticals and food manufacturing. CMD processes are a major challenge to product designers because the processes involved in the production process are fundamental to the structure and target characteristics of the final product. This is in stark contrast to most types of mechanical or electronic products, in which the manufacturing procedures (e.g., machining, component assembly) are substantially independent of the final properties, structure or function of the product (e.g., generating mechanical power, amplifying electrical signals).

Due to the different kinds of processes involved, available expert knowledge of CMD processes is usually both qualitative and quantitative in nature, contains many uncertain elements and is expressed using different formal notations, types of diagrams or informal statements in natural language. Drawn from many different disciplines, the knowledge appears fragmented and heterogeneous, and is often encoded as

G. Stapleton, J. Howse, and J. Lee (Eds.): Diagrams 2008, LNAI 5223, pp. 371–373, 2008.

many types of information, such as causal relations, simple facts, limitations, or contexts. Further, the tacit nature of much of the expert's knowledge of CMD processes makes it problematic to elicit.

Some previous approaches in the fields of AI and knowledge engineering have developed techniques that address selected problems of knowledge acquisition, modeling and discovery of particular facets of CMD processes, e.g. [1], [5], [6], [7] or [9]. Although they all address important aspects, the lack of a systematic integrative framework that draws them together has been a major obstacle to progress on capturing and modeling knowledge about CMD processes.

3 The CMD Framework

The CMD framework uses diagrams to integrate different knowledge representations and methods within knowledge-based tools to facilitate and support the representation and elicitation of knowledge about CMD processes. It uses flexible taxonomies for organizing the conceptual model components and their properties. Different diagrammatic tools (Two of which are shortly mentioned below) provide domain experts with a notation system that is capable of jointly representing process knowledge which exists in the context of different scientific paradigms. The representation of this knowledge requires the use of different abstraction levels, granularities and perspectives and may often involve elements that are vague, incomplete, or hard to quantify. Parts of the framework have been implemented as a software prototype, *CMD SUITE*, which is currently being applied and evaluated using sets of physical, chemical and bio-chemical processes from the domain of industrial bakery product manufacturing.

An overall compositional approach to the creation of a model structure is used for the specification and interconnection of fragments of process knowledge. Interactions are represented using general formalisms that apply to the physical, chemical and biochemical interactions occurring in the domain. Apart from this basic model paradigm, few constraints are imposed on the domain experts in order to minimize biases in knowledge acquisition.

Central to our approach is the hierarchical and temporal decomposition of a CMD process into *model components* and component processes, represented by *process fragments*. Experts are provided with diagrammatic tools that allow them to carry out general knowledge operations which are necessary to perform the higher-level tasks of acquiring and modeling CMD knowledge. At the start, a simple algorithm is run on an available production recipe in order to decompose and extract important knowledge objects. The aim of this recipe decomposition is to create a time ordered description of substances and activities involved in the making of a product as a basis for modeling. A *Compound Diagram* (CD) can then be created to determine the boundaries of subordinate process contexts on several levels of structural abstraction and identify substances that have a potential to interact with each other. For this purpose, a *CD* generates a diagrammatic matrix of material substances available for interaction in subsequent temporal steps of the production process. This constrains the size of initial interaction systems, and provides an expert with starting pointers for knowledge acquisition about component processes. For the specification of this kind of knowledge, *Process Scenario Diagrams* (*PSD*) form the basis for complex tools that combine stacks of line graphs with drawing and graph-manipulation methods to record changes to the structural elements involved in a component process.

The diagrams and underlying notations are fundamental to our approach as they provide the common integrative representation for modeling the many types of knowledge that characterize CMD processes. Our aim is to capture a more general but distinctive notion of behavior and function in these kinds of models, which may provide hints about tacit knowledge encoded within the system. Later, we want to inspect CMD processes along their conceptual dimensions and align, manipulate and evaluate the model fragments that constitute these processes, and track directions of physical change. Shrager [8] highlighted the need to go beyond simulation models towards comprehensive knowledge models in finding new approaches to complex problems in biology. Our framework presents an attempt to realize this aspiration for CMD processes with the goal of supporting knowledge acquisition, modeling and discovery.

Acknowledgements

The research was funded by the EPSRC ICASE Food Processing Faraday Partnership with the Campden & Chorleywood Food Research Association, Chipping Campden, GL55 6LD, UK (10/2004 - 09/2007). Bakery expert knowledge has been provided by Stanley Cauvain and Linda Young from BakeTran, UK.

References

1. Bechtel, W., Richardson, R.C.: Discovering complexity: decomposition and localization as strategies in scientific research (1993)
2. Cheng, P.C.-H., Simon, H.A.: Scientific discovery and creative reasoning with dia-grams. In: Smith, S., Ward, T., Finke, R. (eds.) The Creative Cognition Approach, pp. 205–228. MIT Press, Cambridge (1995)
3. Cheng, P.C.-H.: Scientific discovery with law encoding diagrams. Creativity Research Journal 9(2&3), 145–162 (1996)
4. Cheng, P.C.-H.: Diagrammatic knowledge acquisition: Elicitation, analysis and issues. In: Shadbolt, N.R., O'Hara, H., Schreiber, G. (eds.) Advances in knowledge acquisition: 9th european knowledge acquisition workshop, pp. 179–194. Springer, Berlin (1996)
5. Iwasaki, Y.: Reasoning with Multiple Abstraction Models. In: Faltings, B., Struss, P. (eds.) Recent Advances in Qualitative Physics. MIT Press, Cambridge (1992)
6. Kulkarni, D., Simon, H.A.: The processes of scientific discovery: The strategy of experimentation. Cognitive Science 12, 139–176 (1988)
7. Langley, P.: Lessons for the computational discovery of scientific knowledge. In: Proceedings of the First International Workshop on Data Mining Lessons Learned, Sydney (2002)
8. Shrager, J.: Tools for thought in the age of biological knowledge. In: Workshop on Quantita-tive Education in Biological Science, University of Maryland (2004)
9. Van Paassen, M.M., Wieringa, P.A.: Reasoning with multilevel flow models. Reliability Engineering and System Safety 64, 151–165 (1999)

Supporting Reasoning and Problem-Solving in Mathematical Generalisation with Dependency Graphs*

Sergio Gutiérrez[1], Darren Pearce[1], Eirini Geraniou[2], and Manolis Mavrikis[2]

[1] Birkbeck College
London Knowledge Lab
{darrenp,sergut}@dcs.bbk.ac.uk
[2] Institute of Education
London Knowledge Lab
{m.mavrikis,e.geraniou}@ioe.ac.uk

Abstract. We present a brief description of the design of a diagram-based system that supports the development of thinking about mathematical generalisation. Within the software, the user constructs a dependency graph that explicitly shows the relationships between components of a task. Using this dependency graph, the user manipulates graphical visualisations of component attributes which helps them move from the specific case to the general rule. These visualisations provide the user with an intermediate representation of generality and facilitate movement between the specific details of the task, the appropriate generalisations, verbal descriptions of their understanding and various algebraic representations of the solutions.

1 Introduction

The need to recognise, express and justify generality is at the core of mathematical thinking and scientific enquiry. However, a voluminous body of mathematics education research [1] suggests that expressing generality, recognising and analysing patterns and, in particular, articulating structure is complex and problematic for students [2].

We are addressing this problem by developing a mathematical microworld [3] that aims to promote the learning of mathematical generalisation by encouraging students to connect their actions during activities with the need to express generality. The microworld will be part of a learning environment which will also contain components that are (a) able to provide personalised support adapted to students' construction processes and (b) foster and sustain an effective online learning community by advising learners and teachers as to which constructions of other students to view, compare, critique and build upon.

A first version of this microworld has been presented in [4]. Several pilot studies using the software have shown that although learners are able to use the software effectively and that it facilitates their reasoning processes, there

* Work supported by project MiGen (TLRP e-Learning Phase-II, RES-139-25-0381).

G. Stapleton, J. Howse, and J. Lee (Eds.): Diagrams 2008, LNAI 5223, pp. 374–377, 2008.

is a need to highlight the dependencies between shapes and the expressions manipulated. Since diagrammatic representations have been shown to support the reasoning process [5] and they hold potential for facilitating transfer of skills across subject matter [6], the remainder of this paper presents a brief description of proposed improvements to the user interface that use dependency diagrams to address the need described above.

2 Dependency Graphs

This section presents the new on-screen representation for the relationships between objects in the system. Building on lessons learnt from previous versions [4], the goals are to make the relationships between shapes explicit, direct and unambiguous. A graph explicitly relates expressions and shapes. In this *dependency graph* both expressions and shapes are represented as nodes, while the edges of the graph represent the relationships between them. Expressions use icon-variables [4] to represent concepts such as 'width of a shape' (see Figure 1). This is a compelling alternative to using letters which has been shown to be confusing when learning algebra [7].

There are three types of nodes: *shapes*, *shape attributes* (e.g. width), and *expressions* (that define shapes: attributes have to be described by means of an expression, e.g. '4' or 'width of that shape + 2'). Accordingly, there are three types of edges: *shape-parameter* (these edges link a shape with its parameters; therefore, they are *declarative* links, as they show a relationship that is defined by the system and not the learner), *expression-parameter* (that link a parameter with an expression; they are *assignment* links, they give a value — either concrete or variable — to a parameter), and *parameter-expression* (similar to the former ones, they are also *assignment* links; in this case the parameter gives the value to the expression, but the expression can only be an icon-variable). A simple example can be seen in Figure 1. The height and the width of the rectangle on the right are based on the dimensions of the rectangle on the left; they are both two units larger. It can be seen that the links that connect the parameters with icon-variables (e.g. dotted line connecting the left shape's parameters with the icon-variables used in the expressions that define the right shape) 'cross' the boundaries of the expression. This effect distinguishes these links from the other two types (apart from the different drawing style) and highlights their importance in the generalisation process.

A major concern in explicitly displaying the dependency graph is that of cognitive overload and generally making the system difficult to use. This is not only a performance issue, but it has been shown that learners are able to focus only on a small number of objects on the screen at the same time [8]. To address this, we propose that shapes are always shown but the graph is not shown in full on the screen; only those parts of the graph relevant to the student are shown. This could be inferred from the position of the cursor so that the user chooses which nodes are important by moving the mouse cursor over them, and the rest of the graph is hidden until requested. The user can additionally select which

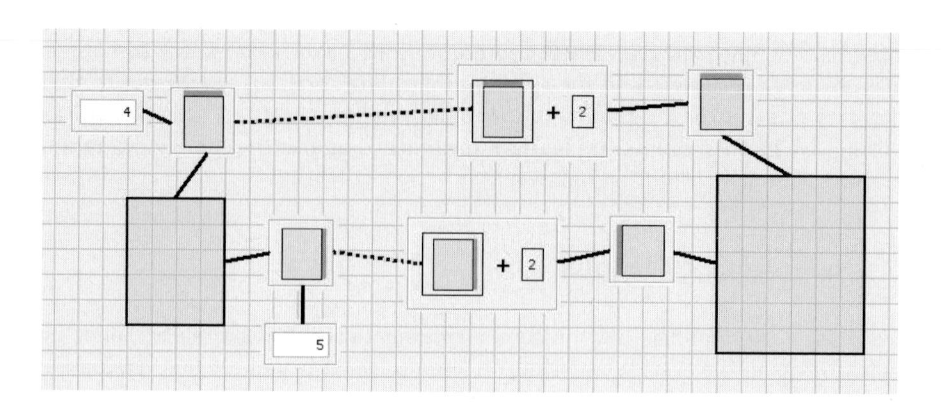

Fig. 1. A shape whose height and width depend on the dimensions of another shape

parts of the graph they want to be able to see at all times. This affords the user full control over the degree to which the dependency graph is displayed.

An important feature of dependency graphs is not only the use of multiple representations for the students, but the visual links between them. When students construct new shapes based on their original shape, they can refer at any time to the connections between them. It must be noted that icon-variables are visually linked to the shapes from which they originate in a meaningful way (e.g. 'width' variables are connected to one of the horizontal lines of a rectangle).

From a pedagogical perspective, the importance of the use of a dependency graph is in giving students the opportunity to reason within their constructions. Students can see how their shapes are linked (or not) together and possibly be motivated to construct related shapes rather than looking for specific solutions. These graphs can be used as stepping stones in students' thinking before a general solution is reached, but also as justifications for their actions to themselves or their peers. By explaining their way of thinking, justifying how they reached their solutions and answering questions, they can demonstrate their understanding. Such actions could also be considered as a first approach to formal proof, since students try with such visual help to express generality and achieve reasoning 'in general' rather than 'in particular' cases.

The importance of this representation is that it helps students to make the shift from specific properties (i.e. numbers) to general relationships (by using icon-variables). Showing students creative representations (as well as symbolic) reinforces connections between them [9].

3 Conclusions and Future Work

We have presented a feature of a system that helps in the development of mathematical generalisation thinking. It clearly shows the relations that exist between several shapes. Studies using a former version of the software suggested that a more explicit relation between the attributes would help the learner to understand them and think in general terms (i.e. using variables instead of numbers). Dependency graphs are used to justify general rules and not just to describe them. Therefore, the system supports learners engaging in the process of mathematical generalisation by providing them with diagrams and representations that can be used as a resource for reasoning about generalisation. The system is diverted from the usual teaching method of algebra, in which students are introduced to letters that represent an unknown value and scaffolds the leap from the concrete to the general through icons that represent variables (e.g. 'width' of a shape).

Next step in the research involves performing studies with learners in order to assess the impact of the new approach as an assistance for the development of generalisational thinking. Different students learn in different ways and this multiplicity of representations can accommodate different kinds of learners. For example, visualisers might have an advantage when using this system. The impact of this representation on different types of learners demands further investigation.

References

1. Noss, R., Healy, L., Hoyles, C.: The construction of mathematical meanings: Connecting the visual with the symbolic. Educational Studies in Mathematics 33(2), 203–233 (1997)
2. Mason, J., Graham, A., Johnston-Wilder, S.: Developing Thinking in Algebra. Paul Chapman Publishing, Boca Raton (2005)
3. Balacheff, N., Kaput, J.: Computer-based learning environments in mathematics. In: Bishop, A.J., Clements, K., Keitel, C., Kilpatrick, J., Laborde, C. (eds.) International Handbook on Mathematics Education. Kluwer, Dordrecht (1996)
4. Pearce, D., Mavrikis, M., Geraniou, E., Gutiérrez, S.: Issues in the design of an environment to support the learning of mathematical generalisation. In: Proc. of the European Conference on Technology-Enhanced Learning (2008)
5. Larkin, J., Simon, H.: Why a diagram is (sometimes) worth ten thousand words. Cognitive Science 11, 65–99 (1987)
6. Novick, L., Hurley, S., Francis, F.: Evidence for abstract, schematic knowledge of three spatial diagram representations. Memory and Cognition 27, 288–308 (1999)
7. Küchemann, D.: Algebra. In: Hart, K.M. (ed.) Children's Understanding of Mathematics, pp. 102–119. Antony Rowe Publishing Services (1981)
8. Moreno, R., Mayer, R.E.: Visual presentations in multimedia learning: Conditions that overload visual working memory. In: Huijsmans, D.P., Smeulders, A.W.M. (eds.) Visual Information and Information Systems, pp. 793–800. Springer, Heidelberg (1999)
9. Warren, E., Cooper, T.: The effect of different representations on year 3 to 5 students ability to generalise. ZDM Mathematics Education 40, 23–37 (2008)

A Concept Mapping Tool for Nursing Education

Norio Ishii and Saori Sakuma

Aichi Kiwami College of Nursing, Jogan-dori 5-4-1, Ichinomiya, Aichi, 491-0063 Japan
{n.ishii.t,s.sakuma.t}@aichi-kiwami.ac.jp

Abstract. We have developed a concept mapping tool for nursing education. We first conducted a student attitude survey on concept mapping to identify points that the students felt were difficult. The three main problem areas were 1) thinking about correlations, 2) drawing maps, and 3) collecting reference materials. We therefore designed a tool that would help students overcome these difficulties. The tool, which makes use of Microsoft Excel VBA, enables students to easily draw, review, and revise concept maps.

Keywords: nursing education, learning support tool, concept map.

1 Introduction

The drawing of a concept map is one of the integral activities in the nursing practice [1-2]. In the field of nursing, the concept map has the general advantages of 1) promoting patient understanding, 2) making the nursing process more efficient, and 3) enhancing the critical thinking ability of the medical care professional.

Previous research has focused on strategies [3-4] and evaluation standards [5-6] related to drawing concept maps in nursing practice. However, because Japanese instructors use different methods to teach students how to draw concept maps, these earlier R&D results are difficult to incorporate into a Japanese educational environment.

The aim of our research is thus to develop a student support tool for concept mapping based on current conditions in nursing education. Specifically, we first conduct a student attitude survey on concept mapping to identify points that students feel are difficult. The support tool is then designed from the viewpoint of alleviating these difficulties.

2 Attitude Survey on Concept Mapping

We first conducted a student attitude survey on concept mapping in nursing practice. The purpose of this survey was to clarify how students feel about concept mapping and to grasp their difficulties in doing so.

2.1 Method

A total of 51 third-year students from Aichi Kiwami College of Nursing participated.

The participants responded to questions in a paper survey: Is drawing a concept map onerous (difficult)? The questions were composed of two parts. In the first part,

G. Stapleton, J. Howse, and J. Lee (Eds.): Diagrams 2008, LNAI 5223, pp. 378 – 381, 2008.

participants evaluated whether drawing a concept map is difficult according to five stages. In the second part, they explained the reasons for their responses in the five stage evaluation.

2.2 Results

The average value for the five stage evaluation was 3.82 (I think so (12 participants), I rather think so (23), I can't say either way (11), I don't really think so (4), and I don't think so (1)). Thus nearly 70% of the students felt that a concept map is difficult to draw.

In the free response part, participants gave a total of 63 reasons why they felt the concept mapping is difficult to draw. The free responses were divided into six categories according to the KJ method: Understanding correlations (17 participants), arranging and revising diagrams (13), time-consuming (10), examining reference materials (8), establishing the work flow (6), and understanding patient pathology and conditions (5).

The survey results indicate that many students feel that drawing a concept map is difficult. Specifically, students have trouble with examining correlations, arranging correlations into a diagram, making revisions, and searching for appropriate reference materials. Because of these difficulties, students do not have enough time to draw a map.

Based on the responses from this survey, we set about to design a student support tool for concept mapping. Our aim was to develop a tool that supports three specific activities: 1) thinking about correlations, 2) drawing maps, and 3) collecting reference materials.

3 Development of Concept Mapping Tool

Based on the results of the attitude survey on concept mapping, we designed a concept mapping tool for nursing students. This section provides an overview of the tool and its development going forward.

3.1 Tool Overview

The tool, which is based on Microsoft Excel VBA, was developed for the purpose of alleviating the difficulty of drawing maps (Fig.1). Using the tool involves the following steps.

Preparing Information List. The first step is to prepare an information list. The student fills in basic information about the patient, covering several categories (e.g. physical, mental, and social state). It is also possible to enter other nursing-related matters at this time.

Arranging Nodes. There is a button beside each item list column that when pressed records the content of the respective item and creates a node. The student can arrange nodes freely. By pressing the line type button, it is possible to switch the node line between solid and dotted.

Connecting Links. Pressing the link button creates an arrow connection between two nodes. There are three types of links: straight line, curved line, and hooked line. Similar to a node, by pressing the line type button, it is possible to switch the link line between solid and dotted.

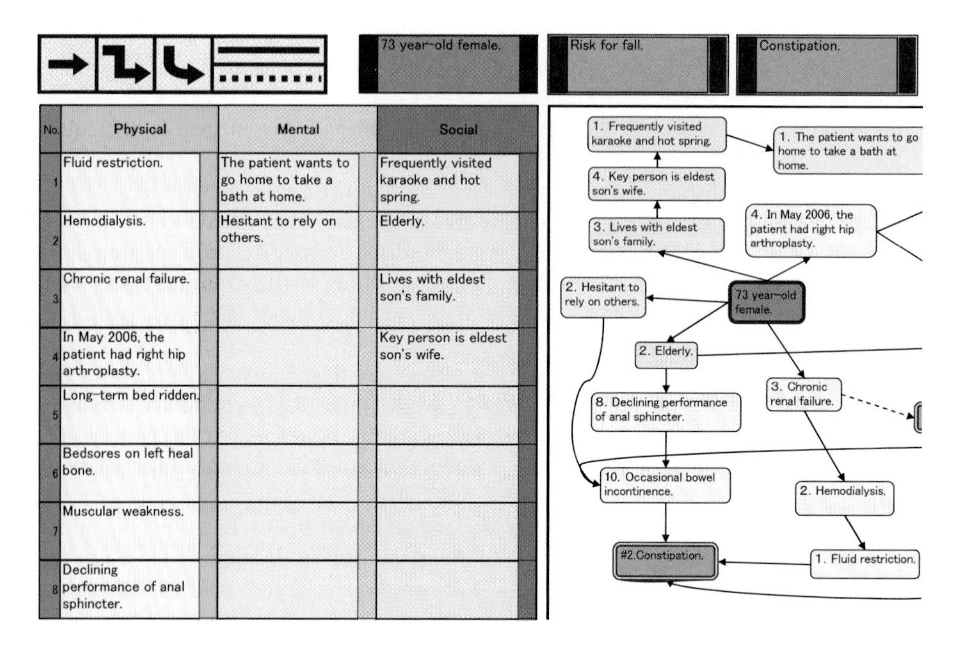

Fig. 1. The concept mapping tool

3.2 Development Going Forward

Going forward, we plan to develop the tool further for educational use in the following ways.

Adding Data Search and Collection Features. To make it easier to search for and collect materials from textbooks and case studies for drawing concept maps, future versions of the tool will incorporate enhanced functionality. Specifically, the tool will generate a database of essential information. The user will then be able to search the database using keywords.

Examining Tool Effectiveness. The tool will undergo evaluation testing to ensure its educational effectiveness and ease of use for nursing college teachers and nursing students. In the following fiscal year, the tool will be introduced into the classroom where its practical effectiveness will be evaluated.

Examining the Characteristics of Concept Maps Created by Using the Tool. By using the tool, it is easy to analyze characteristics of concept maps such as how students create content for items and links between items. In addition, it is possible to automatically record and keep track of the process by which students learn through trial and error to create concept maps. Analyzing the creation processes enables us to examine more deeply how to develop a more effective tool.

Acknowledgments. This research was partially supported by the Ministry of Education, Science, Sports and Culture, Grant-in-Aid for Young Scientists (B), 18791649, 2006-2008.

References

1. Akinsanya, C., Williams, M.: Concept Mapping for Meaningful Learning. Nurse Education Today 13(1), 41–46 (2004)
2. Hsu, L., Hsieh, S.: Concept Maps as Assessment Tool in a Nursing Course. J. Professional Nursing 21(3), 141–149 (2005)
3. All, A., Havens, R.: Cognitive/Concept Mapping: A Teaching Strategy in Nursing. Journal of Advanced Nursing 25(6), 1210–1219 (1997)
4. Schuster, P.M.: Concept Mapping: A Critical Thinking Approach to Care Planning. F.A.Davis, Philadelphia (2002)
5. Castellino, A.R., Schuster, P.M.: Evaluation of Outcomes in Nursing Students Using Clinical Concept Map Care Plan. Nurse Educator 27(4), 149–150 (2002)
6. Toyoshima, Y., Itou, F., Hagi, Y., Nishibori, Y., Kazaoka, T., Kishita, S., Itou, S.: Evaluation of Student Learning about the Nursing Process Using Written Simulation in Adult Nursing Course (Part 3): Analysis of Student Self-evaluation in Learning with Sequence of Events. Bulletin of Department of Nursing Seirei Christopher College 13, 81–90 (2005)

Cognitive Methods for Visualizing Space, Time, and Agents

Angela M. Kessell[1,*] and Barbara Tversky[1,2]

[1] Stanford University, Stanford, CA
[2] Columbia Teachers College, New York, NY
{akessell,btversky}@stanford.edu

Abstract. Visualizations of space, time, and agents (or objects) are ubiquitous in science, business, and everyday life, from weather maps to scheduling meetings. Effective communications, including visual ones, emerge from use in the field, but no conventional visualization form has yet emerged for this confluence of information. The real-world spiral of production, comprehension, and use that fine-tunes communications can be accelerated in the laboratory. Here we do so in search of effective visualizations of space, time, and agents. Users' production, preference, and performance aligned to favor matrix representations with time as rows or columns and space and agents as entries. Both the diagram type and the technique have broader applications.

Keywords: diagram, space, time, agent, production, comprehension, preference.

Visual communications are some of the oldest as well as the newest form of communication, from local maps inscribed in stone to glitzy graphs in daily papers. They have been found in diverse cultures and preceded written language. Despite their prevalence, visualizations of the non-visible, notably data, are a recent phenomenon. Although many, such as the periodic table or Minard's depiction of Napoleon's unsuccessful campaign against Russia, are praised for their clarity, others are opaque and confusing. Designing good graphics is a challenge. Many common graphics, such as route maps, developed through practice in a community of users, a process that fine-tunes the graphics and improves communicative efficacy. That process can be accelerated in the laboratory, and used to reveal design principles in specific domains [1]. Users produce visualizations, and those visualizations are tested for performance and preference in other users. When there is convergence, that is, when the same visualizations prevail in all three tasks, design principles can be extracted.

This user-as-designer method was applied here to a common visualization problem, simultaneously communicating information about space, time, and agent, using the paradigmatic task of visualizing the changing locations of agents in time. Users' spontaneous graphic productions yielded strong consensus on a tabular or matrix representation of that information. Moreover, respondents put time on an axis, but placed people and locations equally on an axis or in the cells.

The performance and preference tasks assessed two main variants and two subvariants of matrices. The performance task was verification of many kinds of assertions

* Corresponding author.

G. Stapleton, J. Howse, and J. Lee (Eds.): Diagrams 2008, LNAI 5223, pp. 382–384, 2008.

possible from the visualizations, the kinds of information that might naturally be ascertained. Were certain people in the same place at the same time? Did a person go from one specific place to another? Did people congregate in a particular place? In the main variant of the visualization, table entries were color-coded dots; in the other, they were color-coded lines connecting cell entries (Fig. 1). In both cases, cell entries represented people. The visualizations also varied on the orientation of time, vertical or horizontal. To assess performance, participants verified whether a large set of relations among space, time, and agent were true or not, using the different variants of the visualization. Afterwards, they were asked which visualization they preferred for verifying a range of assertions of space, time, agent relations.

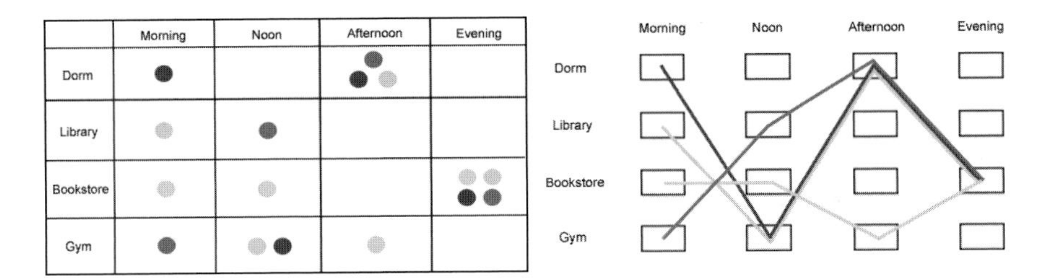

Fig. 1. Example stimuli

One expectation was that the lines would facilitate assessment of the relations of the people, specifically, their temporal paths. This prediction derived from previous work where lines encouraged trend interpretations [2]. The lines connected the people across time and place. Lines, however, clutter, making it more difficult to discern cell entries. Hence, other comparisons should be facilitated by unconnected dots as cell entries. The predictions for the orientation of time are less clear. On the one hand, earlier work on spontaneous graphic productions in children and adults found that they arrayed time horizontally more frequently than vertically [3]. This is, of course, the standard in graphs. Moreover, horizontal is often preferred for neutral dimensions, such as time, and the vertical for evaluative dimensions, such as preference, perhaps because the horizontal dimension is more neutral than the vertical.

Performance was better for matrices with dots than for matrices with lines ($F(1,126) = 52.29$, $p < .001$; Fig. 2) for nearly all statement types. The exception was statements of temporal sequence, such as "Alex went directly from the dorm to the

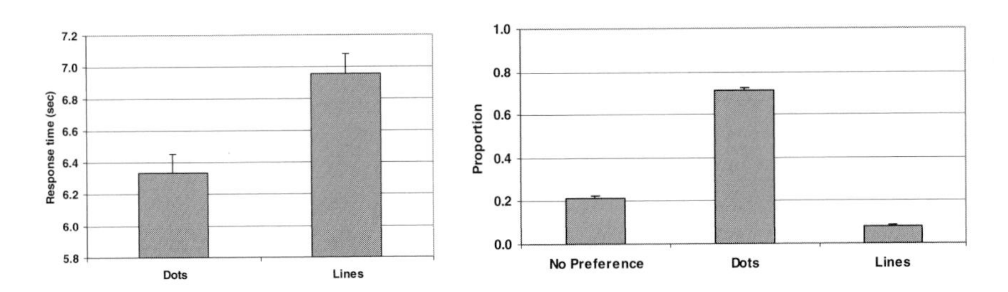

Fig. 2. Response times and preferences. Error bars represent standard errors of the means.

gym." As for the orientation of time, verification performance was equally good when time was vertical as when it was horizontal. Participants' preferences were in alignment with their performance. That is, they preferred matrices with dots for all question types (χ_2^2s > 34.33; *ps* < .001) except for temporal sequence, and these preferences held whether time was oriented vertically or horizontally.

Thus, production, performance, and preference for visualizing time, space, and agents converged on a tabular representation with dots for entries, except for temporal sequences, where lines may be better. There was also convergence that time should be an axis, though there was no preference for the orientation of the axis, vertical or horizontal. This may be because there are strong correspondences of time to both vertical and horizontal in common visualizations. Graphs typically plot time horizontally, as do time lines. However, calendars and date books display time vertically, with earlier times at the top and later ones at the bottom. Because participants have undoubtedly used both tools, they are used to imagining time both vertically and horizontally.

The users-as-designers paradigm produced consensus on visualizations of time, space, and agent. This paradigm has been used successfully in other domains, namely diagrams of routes and assembly [1]. Together, the projects provide support for this set of procedures as a general tool for designing tools for human use. There are undoubtedly limitations to this design tool. In all of these situations, users had some familiarity with the task and with visualizations of the task. Thus, there is reason to believe that users can serve as effective designers only when they have some domain expertise, that is, past experience in comprehending and perhaps even in producing relevant visualizations, and in thinking more abstractly about the information. With that caveat, co-opting users to be designers and tweaking and testing their designs looks promising as part of a program for creating effective designs.

Acknowledgments. We are grateful for the support of the Stanford Regional Visualization and Analysis Center, National Science Foundation grant IIS-0725223 and to NSF REC-0440103.

References

1. Heiser, J., Phan, D., Agrawala, M., Tversky, B., Hanrahan, P.: Identification and validation of cognitive design principles for automated generation of assembly instructions. In: Proceedings of the working conference on Advanced Visual Interfaces, Gallipoli, Italy (2004)
2. Zacks, J., Tversky, B.: Bars and lines: A study of graphic communication. Memory & Cognition 27(6), 1073–1079 (1999)
3. Tversky, B., Kugelmass, S., Winter, A.: Cross-cultural and developmental trends in graphic productions. Cognitive Psychology 23, 515–557 (1991)

Benefits of Constrained Interactivity in Using a Three-Dimensional Diagram*

Peter Khooshabeh[1], Mary Hegarty[1], Madeleine Keehner[2], and Cheryl Cohen[1]

[1] University of California, Department of Psychology, Santa Barbara, California 93106-9660 USA
{khooshabeh,hegarty,c_cohen}@psych.ucsb.edu
[2] University of Dundee, School of Psychology, Dundee, DD1 4HN, Scotland, UK
m.m.keehner@dundee.ac.uk

Abstract. In four experiments participants were allowed to manipulate a virtual 3-D object in order to infer and draw 2-D cross sections of it. Key differences between the experiments were the interface and degree of interactivity available. Two experiments used a three degrees-of-freedom inertia tracking device allowing unconstrained interactions and the other two experiments used a slider bar that allowed only one degree-of-freedom movement at a time. Somewhat counter-intuitively, we found that the constrained interface allowed people to access task-relevant information more effectively and resulted in better performance on the task.

Keywords: constrained interactivity, spatial cognition, input devices, interaction techniques, medical visualization.

1 Introduction

As computers become more pervasive in classrooms and work environments, diagrams that were once displayed only in print are increasingly being presented in digital format. This representation allows for new possibilities of accessing information from diagrams. Among these new possibilities is interactive control with multiple degrees of freedom (DOF), as well as more realistic rendering in three-dimensions (3-D), either with stereopsis or compelling perspective rendering with texture and shading providing depth cues.

Individuals need to appropriately control the interface in order to benefit from it, and graphical aspects of direct manipulation are an important aspect of interactivity. Graphical representations are special not simply because they facilitate more direct control methods, but because they aid inferential processing. For example, Scaife and Rogers [1] argue that graphical elements "are able to constrain the kinds of inferences that can be made about the underlying represented world (p. 189)."

* This research was supported by National Science Foundation grant 0313237 and by a U.S. Department of Homeland Security (DHS) Graduate Fellowship to Peter Khooshabeh under DOE contract number DE-AC05-00OR22750.

G. Stapleton, J. Howse, and J. Lee (Eds.): Diagrams 2008, LNAI 5223, pp. 385–387, 2008.

The task studied in our experiments involved inferring and drawing a cross section of an unfamiliar anatomy-like 3-D object (see Figure 1). A superimposed line on 2-D printed images indicated where participants should imagine the object had been sliced, and an arrow showed the orientation from which participants were to imagine the cross section (see the sample trial in Figure 1a). The task was to infer and draw what the cross section would look like from the viewpoint of the arrow; Figure 1c shows the viewpoint of the arrow (which we will call the arrow-view) for the sample trial, which provides the most task-relevant informative view—an individual can access this view by a mental spatial transformation or by rotating a 3-D interactive computer diagram of the structure to this view. The correct cross section is shown in Figure 1d and examples of participants' drawings are shown in Figure 1e-g.

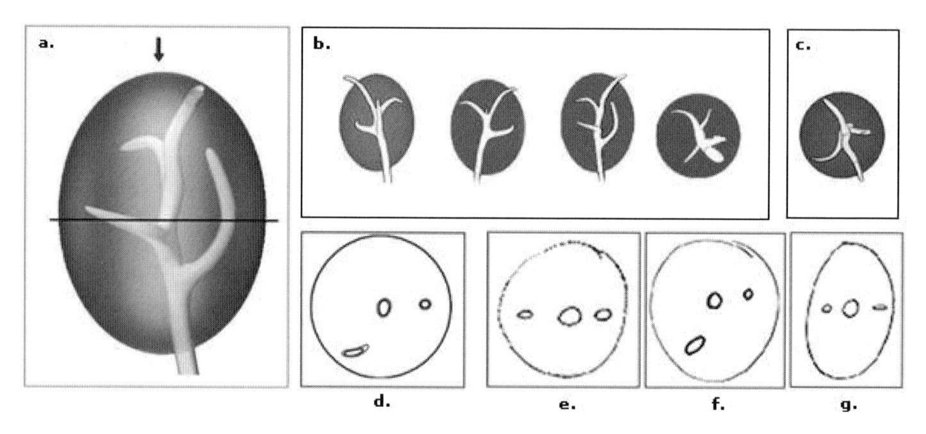

Fig. 1. a) Printed stimulus of 3-D object showing the cutting plane and viewing-direction arrow for one trial. b) The object in various orientations. c) Arrow-view for the trial d) Correct cross section for the trial shown. e–g) Sample drawings by participants for the trial shown.

2 Experimental Evaluation of Interfaces

Method. In the *free manipulation* condition [2, 3] participants rotated the virtual 3-D model in any direction using a 3 degrees-of-freedom motion sensor, the InterSense Intertiacube2. For the *slider control interface* condition [3] participants used a mouse to move a horizontal slider in two QuickTime media players to control yaw and pitch. Participants could operate only one of these at a time and one rotation in a QuickTime player did not affect the other.

3 Results

Patterns of Interactivity. The mean proportion of trials on which participants accessed the arrow-view in the different conditions are shown in Figure 2a. A univariate ANOVA indicated a significant effect of experiment on this proportion, $F(3,113) = 3.5$, $p < .05$. Post-hoc (LSD) tests indicated that participants in the slider interface experiments (Experiments 3 & 4) accessed the arrow-view on significantly more trials than those who used the inertia-cube interface (Experiments 1 and 2), but there were no differences in access between Experiments 1 and 2 or between Experiments 3 and 4.

Task Performance. Mean proportions of correct drawings in the four experiments are shown in Figure 2b. There was a significant effect of experiment, $F(3, 115) = 3.5$, $p < .05$. Post-hoc (LSD) tests indicated that participants in Experiment 4, who used the slider interface, performed better than those in the other three experiments ($p < .05$), but performance on those three experiments did not differ significantly.

Across the four experiments, accessing the arrow view was significantly correlated with the proportion of trials on which the arrow-view was accessed ($r = .58$).

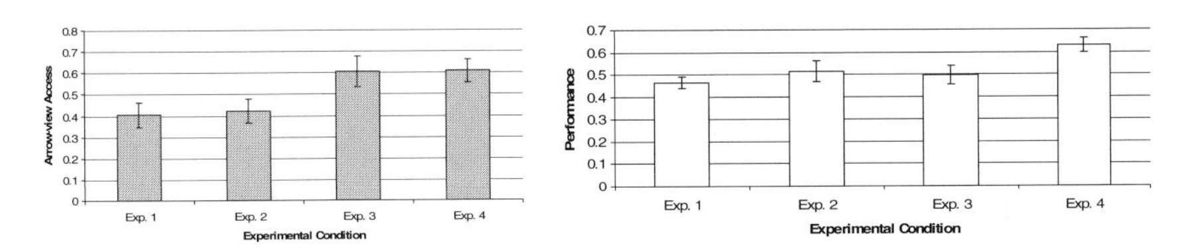

Fig. 2. Participants in the slider control (constrained interactivity [Exp. 3, 4]) conditions accessed the arrow-view on more trials (A.) and performed better on the experimental task (B.)

4 Discussion

We found that performance in a task of inferring a 2-D cross section from a 3-D object was better when individuals used a 2 degrees-of-freedom slider interface than when they used a 3 degrees-of-freedom free manipulation interface. In general, participants were more likely to access the arrow-view with the slider interface and performed better using it. This set of experiments suggests that interactive control and information elicited by a constrained slider interface can be just as useful, if not more useful than an unconstrained interface, under some conditions.

References

1. Scaife, M., Rogers, Y.: External cognition: how do graphical representations work? International Journal of Human-Computer Studies 45 (1996)
2. Cohen, C.A., Hegarty, M.: Individual differences in use of an external visualization to perform an internal visualization task. Applied Cognitive Psychology 21, 701–711 (2007)
3. Keehner, M., Hegarty, M., Cohen, C., Khooshabeh, P., Montello, D.R.: Spatial Reasoning with External Visualizations: The Role of Individual Differences in Distributed Cognition. Cognitive Science (in press)

A Strategy for Drawing a Conceptual Neighborhood Diagram Schematically

Yohei Kurata

SFB/TR8 Spatial Cognition, Universität Bremen
Postfach 330 440, 28334 Bremen, Germany
ykurata@informatik.uni-bremen.de

Abstract. Conceptual neighborhood diagram is a network diagram that schematizes a set of spatial/temporal relations. This paper proposes a strategy for arranging the relations in a diagrammatic space such that the graph highlights the symmetric structure of the relation set.

1 Introduction

Conceptual neighborhood diagram (*CNG*) [1] is a diagram in which spatial/temporal relations (or potentially any set of concepts in a certain domain) are networked based on their similarity. A number of CNGs has been developed for various relation sets (e.g., [1-6]) and its usefulness for schematizing relations has been reported repeatedly, because well-designed CNGs highlight the symmetric structures of the relation sets. However, how to design such schematic CNGs is not well discussed except the definitions of neighbors. Thus, this paper proposes a strategy for arranging the relations in a diagrammatic space such that the CNG becomes schematic.

2 A Drawing Strategy

In a CNG, each node corresponds to a relation and each link indicates a pair of *conceptual neighbors*. Typically, two relations are considered conceptual neighbors if an instance of one relation can be transformed into an instance of another relation by a continuous transformation (e.g., sliding a part of one object). The nodes (i.e., relations) are arranged in a diagrammatic space under the following conventions:

- the number of crossing links becomes as small as possible;
- the nodes are arranged in a linear or reticular pattern as much as possible; and
- the pairs of symmetric relations are located at symmetric locations.

As for the third convention, for instance, in the CNG for topological relations between two simple regions embedded in a spherical surface $\mathbf{C}^1$ (Fig. 1d) [4], every pair of symmetric relations, derived from each other by exchanging the interior and exterior of one region (e.g., Fig. 1a), is located symmetrically with respect to the horizontal axis. Similarly, in the CNG for topological relations between a directed line (arrow) and a region [6], every pair of relations, derived from each other by exchanging the line's direction, is located symmetrically with respect to the middle-height plane. If the

G. Stapleton, J. Howse, and J. Lee (Eds.): Diagrams 2008, LNAI 5223, pp. 388–390, 2008.

relations are modeled by the 9-intersection [7] or its relative models (e.g., [3, 5, 6]), such symmetric pairs of relations are represented by similar patterns of icons with two exchanged rows or columns (Fig. 1a) and thus the pairs are easily identified.

Based on this observation, we propose the following strategy for drawing a CNG:

Step 1: Determine neighbors among the given set of relations R (Fig. 1b). Under the 9-intersection or its relative models, the candidates for such neighbors are determined computationally based on the similarity of the relations' iconic patterns [2, 5, 6].

Step 2: Determine one or two concepts of symmetry C_i (e.g., exchanging the interior and exterior of one region, exchanging the line's direction, etc.).

Step 3: For each C_i, identify R's subset R_i that is self-symmetric with respect to C_i. Then, among the relations in $\bigcap R_i$, identify the relation r^* that has the largest number of neighbors, and put r^* at (0, 0, 0). In our example, r^* is *overlap* relation (Fig. 1c), since this relation holds even if we exchange the interior and exterior of each region.

Step 4: Locate the relations in $R_1 \backslash \{r^*\}$, if they exist, on the x-axis at $(a, 0, 0)$ ($a \in \mathbf{Z}$), such that the length of each link between the relations in R_1 becomes two. Leave the relations without any link. In a similar way, locate the relations in $R_2 \backslash \{r^*\}$ on the y-axis, if they exist. In our example, both $R_1 \backslash \{r^*\}$ and $R_2 \backslash \{r^*\}$ are empty.

Step 5: Locate all other relations at $(a, b, 0)$ ($a, b \in \mathbf{Z}$), such that:

- each relation is located at equal distance from its neighbors, whenever possible;
- the remaining relations in R_1 and R_2 are located on x- and y- axes, respectively;
- symmetric relation pairs with respect to C_1 and C_2 are located symmetrically with respect to x- and y-axes, respectively (e.g., *meet* and *covers* relations in Fig. 1c).

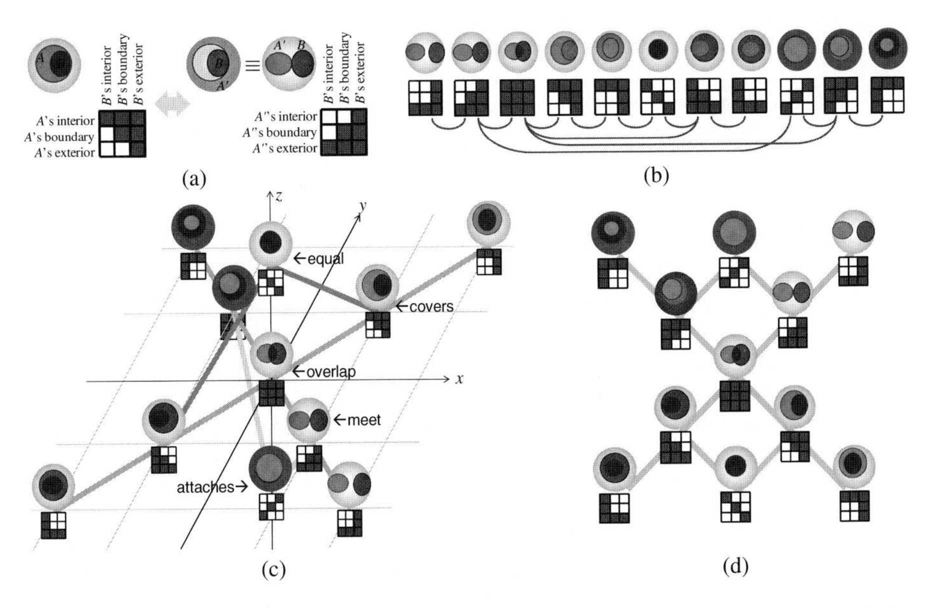

Fig. 1. (a) Icons of two symmetric relations, (b) neighboring pairs of relations, and (a-b) CNGs for topological relations between two simple regions in $\mathbf{C}^2$ before and after Step 7

Step 6: If two or more relations are located at the same point, then relocate them to (a, b, c) ($c \in \mathbf{Z}$) such that the links do not intersect with each other and the links' total length becomes the smallest. In our example, *equal* and *attaches* relations, which are placed originally at (0, 0, 0) in Step 5, are relocated to (0, 0, ±1) by this rule (Fig. 1c).

Step 7: If preferable, reduce the dimension of the CNG by continuous transformation while avoiding the creation of intersecting links as much as possible. In our example, we obtain the two-dimensional CNG in Fig. 1a from the CNG in Fig. 1b by rotating its right half ($x > 0$) by 180 degree and its middle vertical plane ($x = 0$) by 90 degree.

3 Conclusions

This paper proposed a strategy for arranging relations in a CNG by which symmetric properties of the relation set are highlighted. We already confirmed the applicability of this strategy to various sets of topological relations, including 13 interval relations in a linear time frame [8], 16 interval relations in a cyclic time frame [3], 8 region-region relations in $\mathbf{R}^2$ [7], 11 region-region relations in $\mathbf{C}^2$ [4], 26 arrow-region relations in $\mathbf{R}^2$ [6], and 80 arrow-arrow relations in $\mathbf{R}^2$ [5] (a slight technique necessary), as demonstrated in our poster. It is left for a future research to examine the applicability of this strategy to other sets of relations, including more complicated topological relations (e.g., those between two regions with holes) and non-topological relations (e.g., directional relations). In addition, given an arbitrary set of relations, their neighborhood information, and all symmetric pairs among them, how to design a schematic CNG in a fully automated way is an interesting future challenge.

References

1. Freksa, C.: Temporal Reasoning Based on Semi-Intervals. Artificial Intelligence 54, 199–227 (1992)
2. Egenhofer, M., Al-Taha, K.: Reasoning about Gradual Changes of Topological Relationships. In: Frank, A., Campari, I., Formentini, U. (eds.) GIS 1992. LNCS, vol. 639, pp. 196–219. Springer, Heidelberg (1992)
3. Hornsby, K., Egenhofer, M., Hayes, P.: Modeling Cyclic Change. In: Chen, P., Embley, D., Kouloumdjian, J., Liddle, S., Roddick, J. (eds.) Advances in Conceptual Modeling. LNCS, vol. 1227, pp. 98–109. Springer, Heidelberg (1999)
4. Egenhofer, M.: Spherical Topological Relations. Journal on Data Semantics III, 25–49 (2005)
5. Kurata, Y., Egenhofer, M.: The Head-Body-Tail Intersection for Spatial Relations between Directed Line Segments. In: Raubal, M., Miller, H., Frank, A., Goodchild, M. (eds.) GIScience 2006. LNCS, vol. 4197, pp. 269–286. Springer, Heidelberg (2006)
6. Kurata, Y., Egenhofer, M.: The 9+-Intersection for Topological Relations between a Directed Line Segment and a Region. In: Gottfried, B. (ed.) 1st International Symposium for Behavioral Monitoring and Interpretation, pp. 62–76 (2007)
7. Egenhofer, M., Herring, J.: Categorizing Binary Topological Relationships between Regions, Lines and Points in Geographic Databases. In: Egenhofer, M., Herring, J., Smith, T., Park, K. (eds.) NCGIA Technical Reports 91-7, pp. 91–97. NCGIA, Santa Barbara (1991)
8. Allen, J.: Maintaining Knowledge about Temporal Intervals. Communications of the ACM 26, 832–843 (1983)

Supporting Relational Processing in Complex Animated Diagrams

Richard Lowe[1] and Jean-Michel Boucheix[2]

[1] Curtin University, Australia
r.k.lowe@curtin.edu.au
[2] University of Burgundy, France
J-M Boucheix@u-Bourgogne.fr

Abstract. The psychological utility of different approaches to cueing key information within complex animated diagrams remains relatively unexplored. Because of the varied processing demands required to comprehend such representations, it is likely that providing a range of different cue types would be best for supporting effective user information extraction. Traditional cueing approaches used with static diagrams may need to be reconsidered in order to meet the new design challenges associated with their animated counterparts.

Keywords: Complex animated diagrams, relational processing, dynamic cues.

1 Introduction

Animated diagrams are increasingly used to present complex dynamic information in both professional and educational settings. However, the complexity of these animations means that they can be extremely demanding for users to process [1]. Contributors to the complexity of animated diagrams range from more obvious characteristics such as the sheer number of graphic entities shown in the display space, to quite subtle aspects such as highly transitory temporal relationships that occur between those entities. Because of the extreme processing demands imposed by these animations, there is an ever present danger of crucial information being neglected or misinterpreted. Although display simplification is a possible way to reduce these demands, there are many situations where this is neither feasible nor appropriate. An alternative approach is to retain the animated diagram's complexity but provide users with additional processing support that helps them to cope with its inherent demands. It is this second approach that is addressed here.

2 Effective Direction of Attention

Users can cope with a complex dynamic display if they are able direct attention strategically and assign their limited information processing resources efficiently

G. Stapleton, J. Howse, and J. Lee (Eds.): Diagrams 2008, LNAI 5223, pp. 391–394, 2008.

to extraction of the most task-appropriate information. In general, different aspects of such a display are not all equally important at any one time. Typically, a subset only of the displayed entities have high relevance to the depicted situation's central theme with the remainder serving a largely contextual role. In order to extract the key information from an animated depiction, users need to direct their attention to the display's most thematically relevant entities by looking in the right place, at the right time, for the right aspects [2]. If users already possess extensive background knowledge in the domain of the depicted subject matter, this type of highly selective attention direction can be implemented in a top down manner. However, if the user is a domain novice or if the depicted situation is highly unusual, another form of attention guidance is required. Visual guidance in the form of arrows and highlighting has long been used to direct attention in static graphics. These approaches tend to offer more precision in their directional function than non-visual alternatives (such as accompanying text). Unfortunately, they can also increase the display's visual clutter and (in the case of arrows) may even distract attention from the aspects they are meant to cue by becoming a focus of attention themselves [3].

Years of trial and error development by design practitioners have demonstrated that the judicious application of highlighting to a static display's thematically relevant aspects can provide an embedded form of cueing that reduces the types of problems referred to above. However, care is needed in applying the lessons learned from static graphics regarding such cueing to animations because of the added temporal dimension that distinguishes animated depictions from their static cousins [c.f. 4]. In order to process a complex animation effectively, the viewer's direction of attention must be continually adjusted to align with thematically relevant changes in the display. Unfortunately, these key display changes are not always particularly conspicuous and so may be neglected if the viewer is largely reliant on bottom-up processing due to a lack of appropriate background knowledge. Fixed highlight cues of the type long used in static diagrams (such as the colouring of key entities in an otherwise black and white depiction) do not take account of the need to shift the focus of attention amongst different regions of the display as an animation progresses. What was thematically relevant early in the animation's time course may not be so at some later point in time. This raises the possibility that a fixed cue could actually prejudice processing effectiveness by continuing to direct attention to a particular region at a time when it should rightly be directed elsewhere in the display.

3 Dynamic Cueing

One way of adapting traditional static highlight cueing approaches to the requirements of processing complex animations is to synchronize the operation of the cueing effect with the changes in thematic relevance that occur as the presentation progresses. This synchronization may be implemented at a number of levels, depending on the phase of processing that is being targeted. At a broad level, dynamic highlighting can direct the user's attention in a general

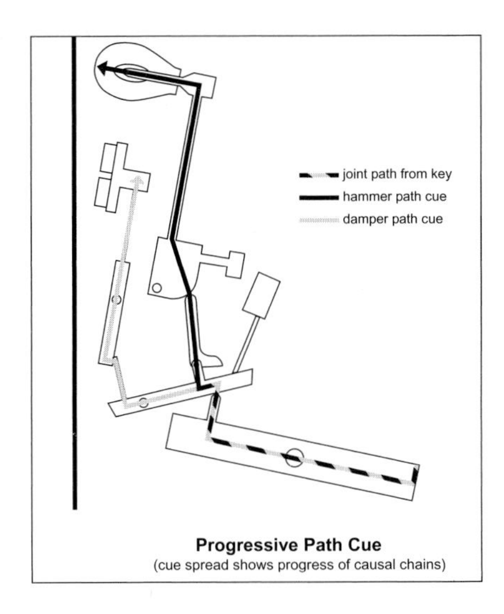

Fig. 1. Progressive Path Cue: Spreading bands of contrasting colours/shades guide attention along progress of causal chains through piano mechanism's functionally related components

sense by indicating the path along which it should progress as the animation runs its course. This *progressive path* cueing (Figure 1) has been found to be effective in helping viewers to notice aspects of a complex animation that are high in thematic relevance but would otherwise tend to be missed because of their low perceptual salience. In particular, cueing in the form of a colour band that spreads through the sequence of graphic entities involved in a causal chain allows viewers to follow the line of action through a mechanism [3].

These spreading colour cues target the broad alignment of viewer attention shifts with linear changes in the locus of thematically relevant activity. However, they do not directly tackle the issue of drawing attention to thematically relevant associations between graphic elements that are not connected in a continuous chain. Comprehension of many types of complex animated content depends on the viewer's capacity to appreciate relationships between spatially distal elements that occur at specific instants during the animation's time course. When closely related behaviors occur between elements that are widely separated in a display, the likelihood of the relationship being missed by viewers rises considerably. Under these circumstances, effective guidance of attention is especially crucial. A type of highlight that appears to hold considerable promise for dealing with this demanding situation is *relational intensity* cueing (Figure 2). In this approach, the cue strength is modulated over time to align with the thematic relevance of the relationship between the target graphic entities during the animation's time course. As the relationship becomes more important over time, the cue strength (e.g. colour intensity) increases then decreases as its importance wanes. One important aspect of relational intensity cueing is its potential to help viewers form

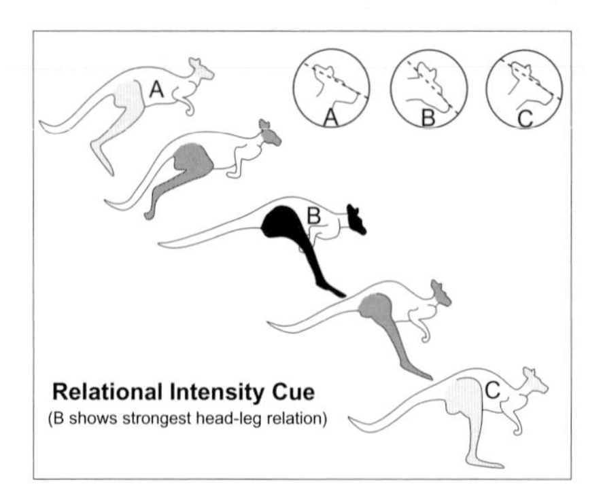

Fig. 2. Relational Intensity Cue: Intensity of colour/shade cues on successive frames of kangaroo animation indicates changing strength of relation between head angle and leg position

high thematic relevance information into spatio-temporal chunks that capture and represent key dynamic relationships in a parsimonious manner [5].

4 Conclusion

To date, the comparison of different cueing approaches in terms of their distinctive potentials for supporting the processing of animated diagrams has been neglected. Different types of cueing can address the varied demands of the hierarchical processing that is required to dealing effectively with complex animations. The support offered by these different cue types can range from generalized guidance along the continuous route via which a causal chain is propagated, to far more specific direction of attention targeting thematically relevant multiple events that have close but inconspicuous conceptual relationships due to their separation in space and time.

References

1. Lowe, R.K.: Interrogation of a Dynamic Visualisation During Learning. Learning & Instruction 14, 257–274 (2004)
2. Lowe, R.K.: Learning from animation: Where to Look, When to Look. In: Lowe, R.K., Schnotz, W. (eds.) Learning with Animation: Research Implications for Design, pp. 49–68. Cambridge University Press, New York (2008)
3. Boucheix, J.-M., Lowe, R.K.: Eye Tracking as a Basis for Improving Animation Design. Learning and Instruction (submitted)
4. Tversky, B., Heiser, J., Mackenzie, R., Lozano, S., Morrison, J.: Enriching Animations. In: Lowe, R.K., Schnotz, W. (eds.) Learning with Animation: Research Implications for Design, pp. 263–285. Cambridge University Press, New York (2008)
5. Schnotz, W., Lowe, R.K.: A Unified View of Learning from Animated and Static Graphics. In: Lowe, R.K., Schnotz, W. (eds.) Learning with Animation: Research Implications for Design, pp. 304–356. Cambridge University Press, New York (2008)

Animated Cladograms: Interpreting Evolution from Diagrams

Camillia Matuk

Northwestern University, Learning Sciences, School of Education and Social Policy
2120 Campus Drive, Evanston, IL 60208, U.S.A
cmatuk@northwestern.edu

Abstract. Misconceptions of evolution are well-documented and may be attributed to features of its diagrammatic representations. For example, embodied space, the tendency to structure experiences through narrative, and base perceptual processes interact in people's interpretation of cladograms, a diagram used to reason about species' evolutionary relationships. Clinical interviews with students indicate an effect of animation on the interpretation of the depicted narrative of evolution. Implications are in the design of diagrams for teaching evolution.

Keywords: animation, evolution, diagrams, interviews.

1 Introduction

Biologists use *cladograms* to make predictions about the relationships among species, and student understanding of such tools of professional practice may lead to greater public scientific literacy. However, meaning from cladograms may be constructed through various uncontrollable interactions among prior knowledge, expectations, and base perceptual processes. In this poster, I report on clinical interviews with students, in which I attempt to understand how people build understanding of evolution from a cladogram (Fig. 1). Through animation, I add to it the explicit dimension of time, and ask how this may affect the perception of the depicted narrative.

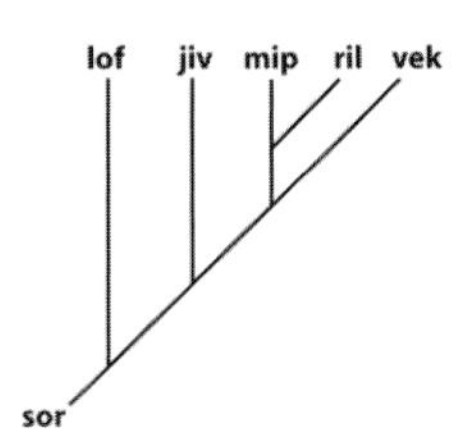

Fig. 1. This cladogram was presented in one of three conditions: static, revealed from bottom to top, or revealed from top to bottom. Nonsense words ensured that participants relied on visual features of the diagram rather than on prior knowledge of familiar organisms in giving their responses.

G. Stapleton, J. Howse, and J. Lee (Eds.): Diagrams 2008, LNAI 5223, pp. 395–397, 2008.

On hearing the word *evolution*, most people imagine a gradual march from primitive to more sophisticated organisms through time. This canonical icon, based in medieval thought, persists in current popular imagery and prevents a correct understanding of the true complexity of the history of life [1]. Interestingly, this misunderstanding can be traced to a number of sense-making processes either implicit or explicit in the modern-day ladder cladogram. By the *Gestsalt Principle of Good Continuation*, students are drawn to perceive the diagonal line as continuous, which prevents them from effectively reasoning about the relationships depicted [2]. That same line may also appeal to our tendency to read *time* into static diagrams [1], for it takes little effort to extrapolate the experience of drawing a line, which necessarily starts and ends, to thinking of the line as a timeline. The slant of the line toward the upper right further taps into the idea of *embodied space*, whereby certain graphic conventions map naturally onto bodily experiences [3], and by which people easily equate the upper right-hand corner of a ladder cladogram with a position of advanced evolution and superiority. Thus narrative, Gestalt perception, and embodied space interact and may cause us to mistakenly translate cladograms into progressions from primitive to more sophisticated organisms: A compelling story with a clear beginning and a determinate end that appeals to our penchant toward narrative for structuring and communicating our experiences [4]. Can a cladogram be animated such that it counters these powerful perceptual and interpretive tendencies? In what ways might simple variations in the presentation of a cladogram influence the resulting understanding that people construct of evolution?

2 Methods

In a between-subjects design, 32 undergraduate students viewed the same cladogram (Fig. 1) on a computer screen under one of three conditions: Revealed gradually from bottom to top (*Bottom-Top*), from top to bottom (*Top-Bottom*), or as a static graphic (*Static*). During one-on-one clinical interviews, participants were asked to describe, explain, and reason with the diagram. Information on their prior experience with similar diagrams, their science background, and their belief systems was collected in an exit survey. Interviews were qualitative coded for emergent patterns and themes.

3 Findings and Discussion

There was a strong effect of the manner of animation on students' perceptions of the cladogram. When asked to describe the image, participants in both the *Static* and the *Bottom-Top* conditions told a story of how a single primitive organism, *sor*, developed over time into a larger diversity of organisms. An overwhelming majority of them identified the apparent diagonal line from *sor* to *vek* as a continuous timeline along which *sor* underwent gradual changes. When asked to choose a likely subdivision of the cladogram, they tended to prefer the one that preserved the continuity of the line rather than one that broke it in half; responses that confirm the effect of Good Continuation on their perception of the diagram. In agreement with the explanation of embodied space, some participants further identified *vek,* the organism in the upper-right corner of the diagram, to be the most evolutionarily advanced, somehow better than the organisms to its left, and certainly better than *sor,* the organism located at the bottom left. But participants in the *Top-Bottom* condition told a different story. As one such participant described, the organisms at the top "combine and then eventually they all come together to form sor," the present-day, most highly evolved organism of all.

However, prior knowledge appeared to more greatly influence interpretations than did animation. One participant, who had recently encountered such diagrams in a biology course, described how the animation began in the present-day, then "went back through the years to one organism that broke off into all of these, which would be the *sor.*" Contrary to other students, it appeared that her stronger mental model of this more typical reading of evolution trounced the interpretation the animation would otherwise have elicited. In other cases, we see how the short-lived quality of the animation, or its *evanescence,* affects the interpretations made. One participant in the *Top-Bottom* condition, who began by describing the diagram as a progression from many organisms to a single one later said: "...because of the way the video played, I would have assumed we were progressing in a *[downward]* time line. But I guess it could just as easily have been the other way around." This participant demonstrated an interesting interaction between the external image, which in her mind appeared to flip back and forth between the brief animation and the static graphic that persisted onscreen, and the two competing internal images she constructed as a result.

Because diagrams help us build internal representations [5], their careful design is important. That animation can affect the manner in which a cladogram is interpreted has implications in how cladograms should be introduced to novices. I am currently investigating manners of animation that may counter the effects of Good Continuation, and comparing interpretations of the ladder cladogram with those of the less commonly-used tree cladogram, which lacks a conspicuous diagonal line. I am also investigating the effects of not priming participants with the idea of evolution before they view the cladogram. How those canonical icons of evolution constrain our interpretations of evolution, and how such diagrams would otherwise be interpreted could have further implications in the design of images of evolution for education.

References

1. Gould, S.J.: Ladders and cones: Constraining evolution by canonical icons. In: Silvers, R.B. (ed.) Hidden histories of science. New York Review of Books, New York (1995)
2. Novick, L.R., Catley, K.M.: Understanding phylogenies in biology: The influence of a Gestalt perceptual principle. Journal of Experimental Psychology: Applied 13(4), 197–223 (2007)
3. Tversky, B.: Some ways that graphics communicate. In: Allen, N. (ed.) Working with words and images: New steps in an old dance, Ablex Publishing Corporation, Westport (2002)
4. O'Hara, R.J.: Telling the tree: Narrative representation and the study of evolutionary history. Biology and Philosophy 7, 135–160 (1992)
5. Hegarty, M.: Dynamic visualizations and learning: getting to the difficult questions, pp. 343–351. Elsevier, Amsterdam (2004)

Automatic Diagram Drawing Based on Natural Language Text Understanding

Anirban Mukherjee[1] and Utpal Garain[2]

[1] RCC Institute of Information Technology, Beliaghata, Kolkata 700015, India
[2] Indian Statistical Institute, 203, B.T. Road, Kolkata 700108, India
attreyee@yahoo.com, utpal@isical.ac.in

Abstract. This article presents a general framework for automatic conversion of a piece of text into a diagram that is described in the given text. Such kind of text is often found in many branches of Science & Engineering. Secondary school-level geometric problems are considered as a reference in this study. A knowledge base or lexical resource named GeometryNet is used in interpreting geometric meaning of a given text to draw the corresponding diagram.

Keywords: Text to diagram conversion, Geometric problems, Lexical resource: Geometry net.

1 Introduction

In many disciplines of Science and Engineering (e.g. geometry, mechanics, engineering drawings, etc.) we often encounter text that basically describes a figure or diagram. This study attempts to find an automatic way for drawing the diagram described by such a piece of text. Since the text is in natural language, therefore the problem becomes the understanding of the natural language text and drawing the corresponding diagram. To the best of our knowledge no study so far deals with this specific problem. However, a few approaches [e.g. 1-4] are reported in the literature for solving natural language mathematical problems. Most of these studies solve limited word problems mostly in the domain of Physics, Mathematics and Mechanics.

Our proposed approach consists of following four modules: 1) a natural language processing (NLP) component that is employed for syntactic, semantic and logical analysis of the problem text, 2) a knowledge base GeometryNet [5] for technical understanding of the problem, 3) an intermediate representation of the text by consulting the GeometryNet and 4) graphics module to convert the intermediate representation into the final diagram.

2 GeometryNet

Borrowing major ideas from the WordNet [6], GeometryNet [5] is constructed as a lexical resource for representing specialized knowledge about geometric entities and concepts. The basic organization for the GeometryNet is based on relational pointers linking the generic instances of different geometric terms – entity (e.g. triangle),

G. Stapleton, J. Howse, and J. Lee (Eds.): Diagrams 2008, LNAI 5223, pp. 398–400, 2008.

attribute (e.g. equilateral) and relation (e.g. inscribe), each pertaining to a syntactic category – noun/adjective/verb. Any geometric term in these categories is generically defined with all necessary input parameters/variables, mathematical conditions and correlated terms. E.g. entity instances 'center' and 'circle' will occur in the entity (noun) file as

center	[@->point][#p circle]
circle	(xc, yc), r, [r!=0, angle = 360]

Here relational pointers @-> and #p implies 'a kind of' and 'part of', respectively. This representation conveys the concept that center is basically a point and forms a part of circle which apart from the center has other parameter (radius r) and mathematical conditions (stated within [ ]). When a particular geometric term is searched in the GeometryNet, the knowledge thus stored against its entry and also all the linked entries are retrieved which taken together constitutes the related concept.

3 Processing of a Problem

Consider a geometric word problem: *ABCD is a parallelogram and X is the midpoint of BC. The line AX produced meets DC produced at Q. CQ is extended upto P and the parallelogram ABPQ is completed. Prove that DC = CQ = QP.*

For a given problem as above, the problem statement is first POS tagged to identify the nouns implying geometric entities, proper nouns indicating entity names, adjectives for entity attributes and verbs representing relations between entities. A linguistic routine then works on the tagged output to list out number of different entities (e.g. parallelogram, line, midpoint etc.) along with their names (e.g. ABCD, BC, X etc.), attributes (e.g. produced) and relation with other entities (e.g. 'CQ' 'extended upto' 'P'). Different entities but of the same type are numbered e.g. line_1, line_2 etc. Another semantic routine relates the nouns as arguments of verbs, e.g. meets(produced(line_2=AX), produced(line_3=DC), point_1=Q) or parts of another noun, e.g. (line_1= BC).(midpoint_1 = X).

Next, corresponding to each noun, related adjectives and verbs are searched in the GeometryNet to retrieve corresponding knowledge or generic data set. For example, search for parallelogram_1 = ABCD extracts four points (e.g. vertex_1 = point_1 = A,(x1,y1)), four lines (e.g. side_1 = line_1: vertex_1, vertex_2, m1 = (y2–y1)/(x2–x1)) and four conditions (e.g. m1=m3) involving eight basic variables x1,y1 ….x4,y4. Again search for midpoint_1= X yields point_5 =X,(x5, y5), #p->line_2,[x5=(x2 + x3)/2, y5=(y2 + y3)/2]. Now the complete geometrical meaning of the given text is conceptualized in terms of parameters and conditions of different drawing elements. But the variables need to be instantiated with suitable numerical values to draw the target diagram.

The process of numerical value assignment to the variables results in a format called *internal representation* of the problem. This representation lists only the different drawing elements (e.g. parallelogram_1, line_1 etc.) with drawing arguments separated with other parameters and conditions. For example, occurrence of line_1 (a component of parallelogram_1) will be only as line_1: (x1,y1), (x2,y2) – just what is required for drawing line_1. Here (x1,y1), (x2,y2) will be assigned default numerical values stored against parallelogram vertices in GeometryNet as these variables can't be derived from any other relation pertaining to the problem. Parameters (like slope m1 for line_1) and conditions (like m1=m3 for parallelogram_1) attached to the listed entities are evaluated. All the parameters may not be evaluated immediately, e.g. the coordinates of point P initially. Eventually, such unknowns are solved from conditions occurring later in the list. The *graphics module* takes individual drawing elements with numerical arguments (e.g. line_1: (x1,y1), (x2,y2)) and calls the corresponding draw function. Sequential execution of all the draw functions ultimately yields the desired diagram (as shown in the given figure).

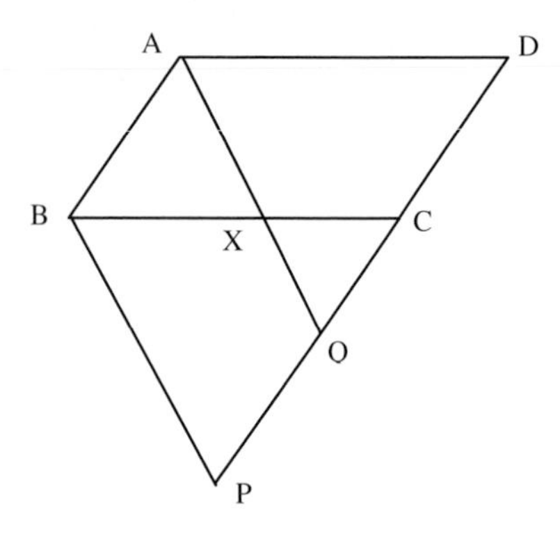

4 Conclusions

Besides serving as a Geometry Tutor in the field of intelligent computer-aided teaching, the framework would expect to find its application in computer aided engineering design and problem solving, visual languages and computing, robotics etc. However, rigorous testing of the system is required to identify the coverage and limitation of the system in terms of geometric entities and relations. A method for quantitative evaluation of the performance of such a system has also to be developed.

Reference

1. Bobrow, D.G.: Natural language input for a computer problem solving system, Ph.D. Thesis, Department of Mathematics. MIT, Cambridge (1964)
2. Novak, G.S.: Computer understanding of Physics problems stated in natural language, Ph.D. Thesis, University of Texas at Austin, USA (1976)
3. Bundy, A., Byrd, L., Luger, G., Mellish, C., Palmer, M.: Solving Mechanics Problems Using Meta-Level Inference. In: Proceedings of IJCAI 1979, Tokyo, Japan, pp. 1017–1027 (1979)
4. Wong, W.K., Hsu, S.-C., Wu, S.-H., Lee, C.-W., Hsu, W.-L.: LIM-G: Learner-initiating instruction model based on cognitive knowledge for geometry word problem comprehension. Computers and Education 48, 582–601 (2007)
5. Mukherjee, A., Garain, U., Nasipuri, M.: On Construction of a GeometryNet. In: Proceedings of IASTED International Conference on Artificial Intelligence and Applications (AIA 2007), Innsbruck, Austria, pp. 530–536 (2007)
6. WordNet: a lexical database for the English language, Cognitive Science Laboratory, Princeton University, Princeton, NJ, USA, `http://wordnet.princeton.edu/`

Texts and Graphs Elaboration: The Effect of Graphs' Examination on Recall

Gisella Paoletti and Sara Rigutti

Department of Psychology, Via S. Anastasio, 12, 34100 Trieste, Italy
paolet@units.it, rigutti@psico.units.it

Abstract. We investigated readers' tendency to ignore graphs in textual contents, the effect of graphs' visibility on recall as well as the spatial contiguity effect of texts and graphs. Eighty subjects read on a computer screen a text with graphs that depicted statistics related to the contents and were tested using a recall questionnaire. The recall performance was found to be improved for observers that examined the graphs while neither the spatial contiguity nor the graphs' visibility affected the mnestic performance. Moreover we found that people tend to examine integrated graphs rather then graphs visualized within pop–up windows.

Keywords: graphs, texts, recall, spatial contiguity effect, pop–up windows, learning.

1 Recall of Texts and Graphs

Different studies provided evidences on the positive effects of pictures within the texts on recall and comprehension of contents [1,2,3,4]. Pictures help students to comprehend hard concepts facilitating information acquisition [2,5,6,7], because the presentation of information by both a verbal and a visual format allows for a dual coding [3,4,8,9]. However, if text and graphs are separated, the cognitive load can be excessive, given that the reader should search for the corresponding figure [10]. According to the *spatial contiguity effect* [4], students learn better when corresponding words and pictures are presented near rather than far from each other on the page or screen. In contrast with these results, other studies found a tendency in readers to neglect or to elaborate superficially the iconic information, [11,12]. Even when the readers consider the iconic information, the time dedicated to the elaboration of figures is often insufficient to capture their global meaning. With the development of the world–wide–web, several web–sites show textual information within the web–pages and the related iconic information within pop–up windows. With this set up, the visibility of the iconic information is limited. The use of pop–up windows requires indeed that the reader clicks on a small icon to visualize the figure in its actual size. According to the cognitive load theory [10], visual integration reduces the reader cognitive load, while some cognitive activities, such as clicking an icon to examine a figure, can

G. Stapleton, J. Howse, and J. Lee (Eds.): Diagrams 2008, LNAI 5223, pp. 401–403, 2008.

reduce comprehension and recall. Our study sheds light on these open questions by providing empirical evidences on some effects of figures embodiment.

2 Summary of the Experiment and Expectations

We conducted an experiment in which 80 participants read a text about the occupational condition of Italian graduates that was displayed on a computer screen. The experimental material included some graphs depicting statistics that were related to the textual contents. After the reading, the participants answered to a recall questionnaire with 14 items (10 referred to the text and 4 to the graphs contents). The answers to recall questionnaire were coded assigning the score 1 to each correct answer and the score 0 to each incorrect one (maximum obtainable score: 14 points). Participants' reading behavior was also recorded by the experimenter, while sitting on the side of the participant during the experimental session. Each participant was thus categorized into two possible types, those who examined the graphs and those who did not. The aim of our experiment was to investigate the tendency of readers to ignore the figures within textual contents, through the observation of their natural behavior. Moreover, by comparing the recall performances in conditions where the graphs were either near to the text (just after the paragraph referencing it), or else, far from it (at the end of the page), we measured the effect of the spatial contiguity of text and graphs. The effect of graphs visibility was also studied through comparison of the recall performance for conditions in which figures where either integrated within the text, or else, visualized within pop-up windows. According to the literature, the recall performance was expected to be improved by the following factors

1. *graphs examination* [1,2,3,4], with a better recall for those participants who examined the graphs, in comparison to the participants that did not,
2. *proximity* between graphs and related text [4], with a better recall for the near graphs condition, rather than the far one,
3. *embodiment* of graphs within the text [10], with a better recall when the graphs were integrated, rather than visualized within a pop-up window.

2.1 Results

A large number of participants (39) neglected the iconic information neither examining the integrated graphs, nor the graphs within pop-up windows. Most of the participants who examined the graphs belonged to the integrated graphs condition rather than the pop-up condition (28 vs. 13, $\chi^2 = 11.75, p < 0.01$). The recall performance was analyzed in a 3-way ANOVA with graphs-examination, spatial contiguity and graphs visibility as between subject factors. A significant main effect of graphs-examination ($F = 9.361, p < 0.05$) was found in a direction consistent with our first expectation, with a significant improvement of the performance for the participants who examined the graphs (score = 10.9) vs. those who did not (score = 8.46). Expected effects of proximity and embodiment were not corroborated by the data, since neither the spatial contiguity nor the graphs' visibility significantly affected the recall performance.

3 Conclusions

Our pattern of data confirmed the positive role of graphs on text's recall: subjects who processed graphs had better performances. Moreover, data suggest a relationship between graphs' visibility and the likelihood readers examine the graphs. In fact reader examined graphs integrated within the text rather than those visualized within pop-up windows. Results suggest that the active processing of both sources of information cannot be taken for granted, and that it can be influenced by promoting graphs' visibility and integration.

References

1. Alesandrini, K.L.: Pictorial-verbal and analytic-holistic learning strategies in science learning. Journal of Educational Psychology 73, 358–368 (1984)
2. Peeck, J.: The role of illustrations in processing and remembering illustrated text. In: Willows, D.M., Houghton, H.A. (eds.) The psychology of illustration, Basic research, vol. 1, pp. 115–151. Springer, New York (1987)
3. Paivio, A.: Mental representations: A dual coding approach. Oxford University Press, Oxford (1986)
4. Mayer, R.E.: Multimedia Learning. Cambridge University Press, New York (2001)
5. Brody, P.J.: Affecting instructional textbooks through pictures. In: Jonassen, D.H. (ed.) The technology of text, pp. 301–316. Educational Technology Publishers, Englewood Cliffs (1982)
6. Schallert, D.L.: The role of illustrations in reading comprehension. In: Spiro, R.J., Bruce, B.C., Brewer, W.F. (eds.) Theoretical issues in reading comprehension: Perspectives from cognitive psychology, linguistics, artificial intelligence and education, pp. 503–524. Erlbaum, Hillsdale (1980)
7. Levie, W.H., Lentz, R.: Effects of text illustrations: A review of research. Educational Communication and Technology Journal 30, 195–232 (1982)
8. Hegarthy, M., Carpenter, P.A., Just, M.A.: Diagrams in the comprehension of scientific texts. In: Barr, R., Kamil, M.L., Mosenthal, P., Pearson, P.D. (eds.) Handbook of reading research. II, pp. 641–668. Erlbaum, Mahwah (1991)
9. Carney, R.N., Levin, J.R.: Pictorial illustrations still improve students' learning from text. Educational Psychology Review 14, 5–26 (2002)
10. Chandler, P., Sweller, J.: Cognitive Load Theory and the format of instruction. Cognition and Instruction 8, 472–517 (1991)
11. Weidenmann, B.: Codes of instructional pictures. In: Schnotz, W., Kulhavy, W. (eds.) Comprehension of Graphics, pp. 29–42. North-Holland, Amsterdam (1994)
12. Paoletti, G.: Writing-to-learn and graph drawing as aids for the integration of text and graphs. In: Rijlaarsdam, G., Rijlaarsdam, G., Van den Bergh, H., Couzijn, M. (eds.) Studies in writing, Effective learning and teaching of writing, 2nd edn., vol. 14, pp. 587–598. Kluwer Academic Publishers, Dordrecht (2004)

Diagrammatic Logic of Existential Graphs: A Case Study of Commands

Ahti-Veikko Pietarinen

Department of Philosophy
P.O. Box 9, FI-00014 University of Helsinki
ahti-veikko.pietarinen@helsinki.fi

Abstract. Diagrammatic logics have advantages over symbolic cousins. Peirce thought that logical diagrams (Existential Graphs, EG) are capable of "expression of all assertions", as our reason is no longer limited to the "line of speech" (MS 654). This paper points out one such value: the economy resulting from combining multi-dimensional diagrams with multi-modal features. In particular, EGs are well-suited for representing and reasoning about non-declarative assertions, such as *questions* (interrogatives, *vert*), *commands* (e.g., imperatives, *vair*) and *the compelled* (*potent*). An advantage over symbolic-logical counterparts is multi-dimensionality that entitles recognition of non-declarative moods in an instantaneous fashion: there is no need to attune to the phonetics of expressions. The paper suggest an application of diagrammatic logic of commands to the cases where (i) minimal *reaction time* to commands is of essence, (ii) a full *comprehension* of the meaning of imperatives ('search for their objects') is needed, and (iii) an effective *discrimination* of commands from other non-declarative moods is critical.

Keywords: diagrammatic logic, existential graphs, multi-dimensionality, multi-modality, tinctures, commands.

1 Diagrammatic Logic of Existential Graphs

Peirce's diagrammatic logic of Existential Graphs (EGs, [1,5]) is both visual (iconic) and formally rigorous. The expressive power goes up to higher-order modal logics. EGs can represent any assertion that has propositional content.

1.1 Multi-dimensionality

EGs are scribed on a multi-dimansional manifold, the *Sheet of Assertion.* For example, in the theory of BETA graphs, which corresponds to the theory of predicate logic with identity, the sheet is 4-dimensional.

1.2 Modality

EGs capture modalities in terms of a *broken cut* [3,5]. A broken cut does not compel the interpreter to admit that the graph P within its enclosure is true:

G. Stapleton, J. Howse, and J. Lee (Eds.): Diagrams 2008, LNAI 5223, pp. 404–407, 2008.

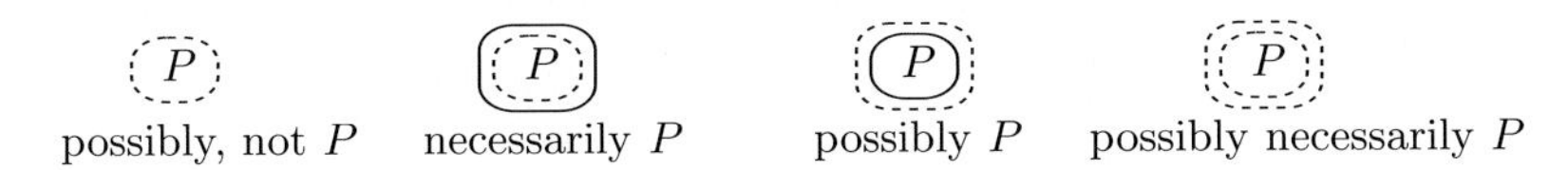

These correspond to the modal-logical formulas $\Diamond\neg P$, $\neg\Diamond\neg P$, $\Diamond\neg\neg P$ and $\Diamond\neg\Diamond\neg P$, respectively. Different modalities are represented by different colours or tinctures.

1.3 Multi-modality: Tinctured Graphs

Different *modes of being*, or different kinds of *universes* (actualities, possibilities, ignorance, power, futurity, intention, etc.) are represented as follows:

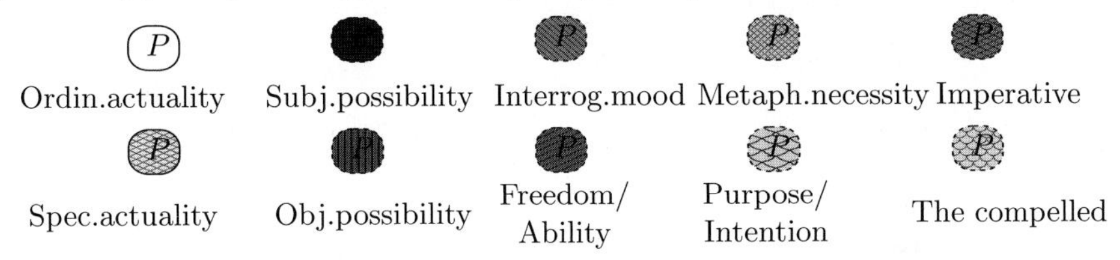

Peirce's examples was "There is a Turk who is the husband of two different persons" (CP, 1906):

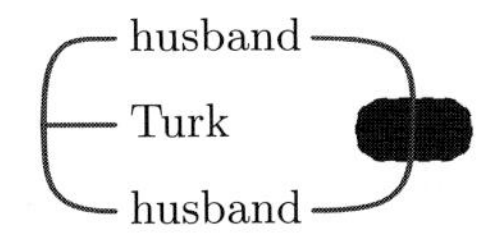

That is, it would be contrary to *what is known* (subjective, epistemic possibility), by the one who scribes the graphs, that the two individuals are identical.

Peirce confesses "that the use of the Tinctures is, in practice, perplexing" (MS 300: 40). Nevertheless, with tinctures, he anticipated alethic modal logic, epistemic logic, erotetic logic, deontic logic, belief-desire-intention logic, and the logic of commands. We study the last case.

1.4 Non-declarative Assertions

Since non-declarative assertions have propositional content, they can be represented and reasoned about by means of EGs.

2 Diagrammatic Logic of Commands

Commands are species of imperatives. Commands have propositional content, since their (immediate) objects are found in the intentions of those asserting the commands.[1] In case the command is actually complied with, the (dynamic) object is in the excecution subsequent to the command.

[1] Witness Peirce: "You may say, if you like, that the Object is in the Universe of things desired by the Commanding Captain at that moment. Or since the obedience is fully expected, it is in the Universe of his expectation" (CP 8.178).

2.1 The Logic of Commands in EGs

Rescher [4] interprets commands in a tense-logical sense ('Do now!'; 'Do always!') and proposes a notion of validity and inference.

In EGs commands may be represented using colours and tinctures (*vert*). To exclame 'If you love me, kiss me!!' is to scribe:

'Help all victims of war!!' and 'Help implies the existence of war' do not entail 'Help the existence of war!!' (Good Samaritan Paradox thus resolved):

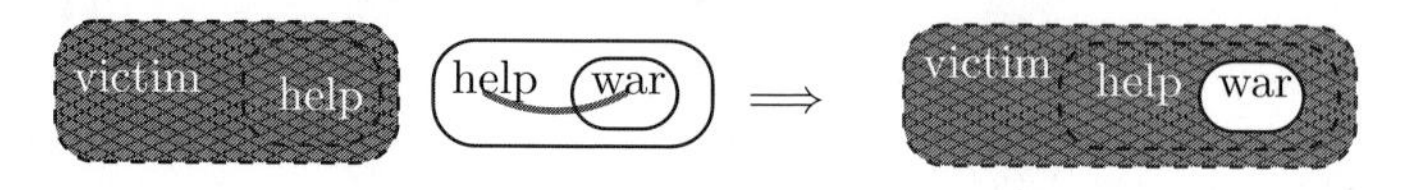

The notion of satisfiability and validity is by *possible-worlds semantics*. Tinctures denote different sets of accessible worlds corresponding to different modalities.

To reason about command graphs is to perform *illative transformations*, including cross-modal transformations (e.g., from the compelled to imperatives). Reasonings are iconic; they capture our 'Moving pictures of thought' [3].

Such graphs may be combined with *tensed modalities*, to deal with commands such as 'Do p whenever (= now and henceforward) X is the case!'[2]

3 Explanations and Applications

Such graphs are very amenable to applications. The logic is *diagrammatic*, and hence recognition of non-declarative moods is instantaneous. There is no reading or hearing sentences, vital to quick but reasoned human responses.

Reaction time to commands is of essence in numerous applications. Time span also depends on two other qualities:

Comprehension of imperatives ('search for their objects') means movement from *immediate object* of a command to its *dynamic object*. The objects are sets of possible worlds. Immediate objects are sets of possible worlds compatible with what the Commander *intends*. Dynamic objects are sets of possible worlds compatible with what the Executor *brings about*. Moreover: (i) Intersection of the two sets is larger for iconic than non-iconic representations. (ii) Movement from immediate to dynamic object is more effective when the iconic logic of commands is used than with linear symbolic sentences. (iii) Empty intersection is *bypass*, hence the logic is three-valued.

[2] The logical form of 'Whenever X happens, p happens' is beyond the expressive power of first-order logic, and so a corresponding modification to graphs is needed [2].

Discrimination An effective and quick *discrimination* of commands from other non-declarative moods is critical whenever having to decipher the line of speech is too time consuming. Because iconicity, in EGs such discrimination is more effective and less error prone than symbols.

4 Conclusions

It is to be expected that representing the logical structure of commands using iconic and diagrammatic notation of EGs comes close to the actual workings of our cognitive processes about their interpretation, reasoning, and excecution.

References

1. Peirce, C.S.: Collected Papers. Harvard University Press, Cambridge (1931-1958)
2. Pietarinen, A.: Peirce's diagram logic in IF persp. In: Blackwell, A.F., Marriott, K., Shimojima, A. (eds.) Diagrams 2004. LNCS (LNAI), vol. 2980, pp. 91–97. Springer, Heidelberg (2004)
3. Pietarinen, A.-V.: Signs of Logic. Springer, Dordrecht (2006)
4. Rescher, N.: The Logic of Commands. Routledge, London (1966)
5. Roberts, D.D.: The Existential Graphs of C.S. Peirce. Mouton, The Hague (1973)

Diagrammatic Reasoning in Separation Logic

M. Ridsdale[1], M. Jamnik[1], N. Benton[2], and J. Berdine[2]

[1] University of Cambridge
[2] Microsoft Research Cambridge

Abstract. A new method of reasoning about simple imperative programs in separation logic is proposed. Rather than proving program specifications symbolically, the hope is to model more closely human diagrammatic reasoning, and to perform automated diagrammatic reasoning in separation logic.

1 Introduction

Separation logic is used for reasoning about low-level imperative programs that manipulate pointer data structures. It enables the writing of concise proofs of correctness of the specifications of simple programs, and such proofs have been successfully automated.

When reasoning informally about separation logic, it is often useful to draw diagrams representing program states, with memory locations represented by boxes, pointers represented by arrows, etc. We aim to formalise these diagrams and implement an automated theorem prover (ATP) which makes use of this formalism. This proposal outlines a promising direction for research. The ideas on diagrammatic reasoning are drawn from [1], which is on diagrammatic theorem proving in arithmetic; we also draw on ideas from [2], which is on symbolic automated theorem proving in separation logic.An example of the kind of problem we want to be able to solve is shown in Fig. 1.

Each square represents a memory location, which can hold values (α_i) or pointers to other memory locations. x, y, k outside of the squares represent program variables, which store memory locations. The program shown reverses a linked list; this can be proved symbolically using separation logic. An informal diagrammatic proof is given in Fig. 1. It shows a partial execution trace of the list-reversal program, through two iterations of the `while` loop. On each iteration, the direction of one pointer in the list is reversed. A person following this trace can see intuitively that the entire list will be reversed when the program is complete.

Following [1], we intend to turn this into a formal proof using an ATP which makes use of *schematic proofs*. This approach allows us to avoid including abstractions such as ellipses in diagrams, and doing inductive proofs over diagrams. Informally, schematic proofs are intended to capture the notion of a 'general proof': they are functions from some set of parameters to a set of proofs, for some more specific notion of a proof. In the example in Fig. 1, the schematic

G. Stapleton, J. Howse, and J. Lee (Eds.): Diagrams 2008, LNAI 5223, pp. 408–411, 2008.

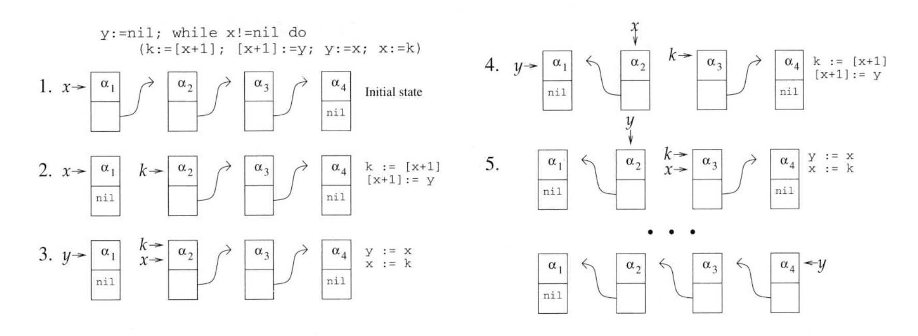

Fig. 1. An informal diagrammatic proof, that the program shown reverses a list

proof *sch-pf* would be a function of the length of the list, such that *sch-pf* (n) is a proof that lists specifically of length n are reversed by the program.

A disadvantage of this approach is that it requires at least one example proof from the user in every case. Possible future work would be to look at an alternative diagrammatic procedure which is fully automated.

In order to implement the above, we first need to define the following:

1. Diagram syntax: the arrangements of shapes on a page or screen which constitute a well-formed diagram.
2. Diagram semantics: a function mapping each diagram to a set of memory states represented by the diagram.
3. Diagram operations: a set of operations on diagrams that will permit us to perform automated reasoning. These operations must be sound with respect to the semantics, and preferably complete, particularly since the symbolic system in [2] is complete.
4. Theorems and proofs: a schematic proof will output a concrete proof for each value of parameter n. A definition is required of what would constitute such a concrete proof.

Once the appropriate definitions are in place, the ATP would work in the following stages:

1. User provides a few example proofs for a specific instance of a theorem.
2. Program generalises from the examples, using a heuristic which suggests a schematic proof.
3. Program verifies the schematic proof using a sound verification procedure.

2 Syntax and Semantics

We can define diagrammatic objects corresponding to predicates in separation logic. These predicates represent data structures and statements about data structures. For example, the two diagrams in Fig. 2 represent memory states in which the predicates `list` and `list segment` hold. The `list` is terminated by `nil`, while the `list segment` is terminated by a program variable.

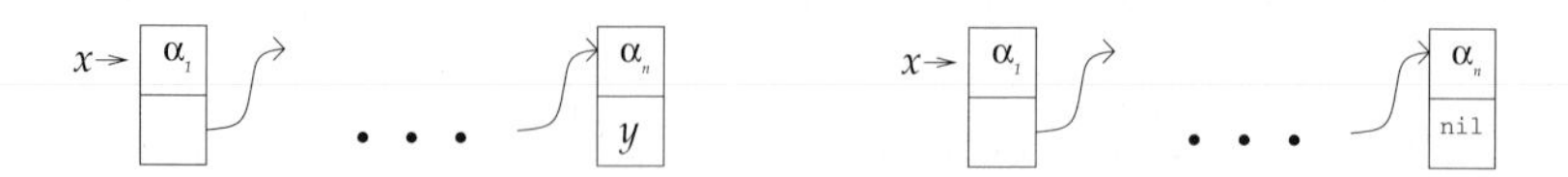

Fig. 2. Diagrams representing a list segment (left) and a list (right)

3 Theorems and Proofs

Before being able to reason about programs using the schematic proofs mentioned previously, we need to be able to reason about static memory states, computing whether two diagrams are equivalent, or whether one diagram entails another, more general one. For example, a single list might be drawn as the concatenation of any possible decomposition into sublists, and we need a way to automatically recognise these equivalences. An example from [2] is shown in Fig. 3. There is a `list segment` from x to t, and a separate `list` beginning at y. The two are connected by a single element, indexed by t. It follows that there is in fact a `list` from x terminated by `nil`.

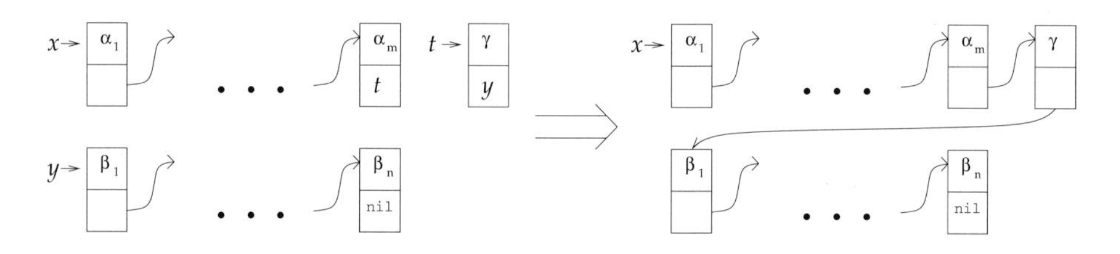

Fig. 3. A *list segment* and a *list*, which are implicitly connected

In order to perform automated reasoning, we define a set of operations on diagrams, which will be picked to allow for sound reasoning with respect to the semantics. These can be of two types: static operations which preserve the diagrams' meaning, for reasoning about static memory states, and dynamic operations corresponding to program commands. An example of the former is shown in Fig 3. It constitutes a simple diagrammatic proof of the problem described in the previous paragraph: two instances of a `replacement` operation. This operation simply replaces both instances of program variables y and t with pointers between the relevant cells. For comparison, here is the symbolic proof that the procedure in [2] would give:

$$\dfrac{\dfrac{\dfrac{\dfrac{t \neq \mathsf{nil} \mid \mathsf{ls}(y, \mathsf{nil}) \vdash \mathsf{ls}(y, \mathsf{nil})}{t \neq \mathsf{nil} \mid t \mapsto [n\!:\!y] * \mathsf{ls}(y, \mathsf{nil}) \vdash t \mapsto [n\!:\!y] * \mathsf{ls}(y, \mathsf{nil})}}{t \neq \mathsf{nil} \mid t \mapsto [n\!:\!y] * \mathsf{ls}(y, \mathsf{nil}) \vdash \mathsf{ls}(t, \mathsf{nil})}}{t \neq \mathsf{nil} \mid \mathsf{ls}(x, t) * t \mapsto [n\!:\!y] * \mathsf{ls}(y, \mathsf{nil}) \vdash \mathsf{ls}(x, \mathsf{nil})}}{\mathsf{ls}(x, t) * t \mapsto [n\!:\!y] * \mathsf{ls}(y, \mathsf{nil}) \vdash \mathsf{ls}(x, \mathsf{nil})}$$

We believe the diagrammatic proof is more human-readable.

4 Conclusion

Developers frequently use diagrams informally when discussing separation logic problems with one another. Our aim is to formalise these diagrams and implement an automated diagrammatic theorem prover for separation logic. A necessary first step to creating an ATP which can reason about entire programs is an ATP which can reason about static memory states, and this is our initial direction of research.

References

1. Jamnik, M., Bundy, A., Green, I.: On automating diagrammatic proofs of arithmetic arguments. Journal of Logic, Language and Information 8(3), 297–321 (1999)
2. Berdine, J., Calcagno, C., O'Hearn, P.: Symbolic execution with separation logic. In: Yi, K. (ed.) APLAS 2005. LNCS, vol. 3780, pp. 52–68. Springer, Heidelberg (2005)

Method of Minimal Representation: An Alternative Diagrammatic Technique to Test the Validity of Categorical Syllogisms

Sumanta Sarathi Sharma

Department of Humanities and Social Sciences, Indian Institute of Technology Kanpur, 208016 Uttar Pradesh, India
sumantas@iitk.ac.in

Abstract. Some valid Categorical Syllogisms in traditional interpretation befall invalid in the modern point of view. Euler Circles and its variants evaluate the validity as per the traditional interpretation whereas Venn Diagrams and its modifications examine the modern point of view. Hence, we fail to find any standard diagrammatic technique, which incorporates both the points of view together. The present paper explores the possibility of developing an alternative diagrammatic technique to test the validity of Categorical Syllogisms. The proposed technique also attempts to test the validity in both the formats.

Keywords: Euler Circles, Venn Diagrams, Method of Minimal Representation.

1 Introduction

There are two major diagrammatic techniques to test the validity of syllogisms: Euler circles which satisfy the traditional point of view and Venn diagrams which corroborate the modern viewpoint. The present paper attempts to devise an alternative diagrammatic technique, which can substantiate both the points of view together.

2 Method of Minimal Representation

The proposed representation method is based on the following findings. Traditional understanding seeks to find representation in the objected area, i.e. whether the diagram draws the area claimed by the conclusion. However, in case of modern viewpoint we look for specific demonstration. Therefore, in the proposed method, diagrams remain the same but the evidence we look for changes in both the perspectives. We use rectangles for universal propositions and right-angled triangles for particular propositions. When we are corroborating the traditional perspective, we seek for the representation of the conclusion in the diagrammed figure. However, when we deal with the modern interpretation, the shape of the geometric figure becomes important and necessary differences (if any) are made.

G. Stapleton, J. Howse, and J. Lee (Eds.): Diagrams 2008, LNAI 5223, pp. 412–414, 2008.

The diagrams are explained below:

1. *Universal Affirmative Proposition* (**A**)–All *S* is *P*. Here, a rectangle is drawn from the right bottom edge containing it and it is shaded. The arrow shows that the orientation or probability of finding *S* is inside *P* only.

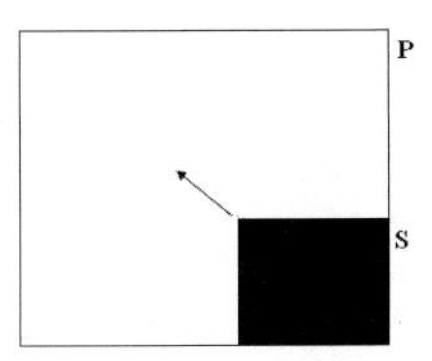

2. *Universal Negative Proposition* (**E**)–No *S* is *P*. Here, two disjoint rectangles are drawn and the arrow shows that the orientation or probability of finding *S* is outside *P* only.

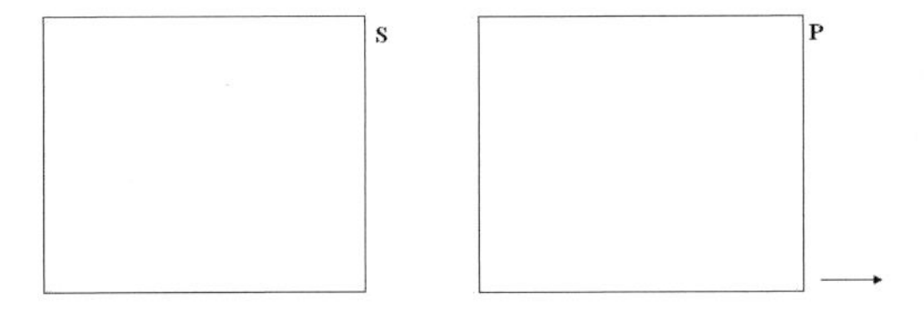

3. *Particular Affirmative Proposition* (**I**)–Some *S* is *P*. Here, a right-angled triangle is drawn from the right bottom edge containing it and is shaded. The arrows suggest that the orientation or probability of finding *S* is both inside as well as outside *P*.

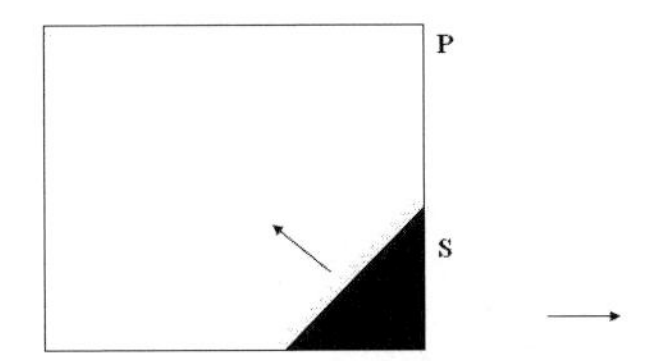

4. *Particular Negative Proposition* (**O**)–Some *S* is not *P*. Here, a right-angled triangle is drawn from the right bottom edge outside it and is shaded. The arrows suggest that the orientation or probability of finding *S* is both inside as well as outside *P*.

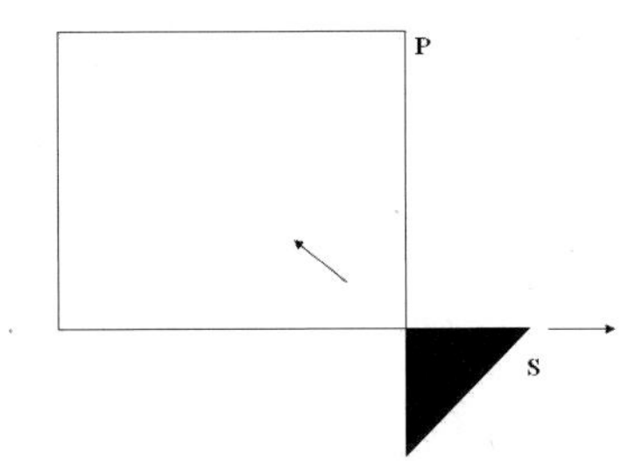

3 Working of the Diagrams

Let us take the following categorical syllogism for understanding the working of this proposed technique.

EAO – IV, which is represented as,

No **P** is **M**.
All **M** is **S.**
Some **S** is not **P**.

As per the technique, the below given diagram can be drawn:

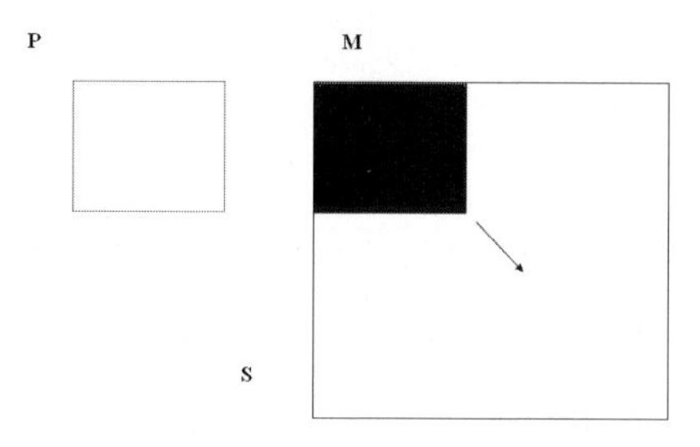

We draw two disjoint rectangles **P** and **M** and then **S** is drawn which contains **M**. The conclusion says, "Some **S** is not **P**" and we find some part of **S** is outside **P**. Therefore, **EAO – IV** is valid in the traditional viewpoint. In the modern interpretation, we search for a right-angled triangle as conclusion. Inability to find the exact geometric figure makes **EAO – IV** invalid in modern interpretation.

4 Conclusion

Valid moods have been tested with the proposed diagrammatic technique. The results were found confirming both the points of view. The scope of the paper is limited to check the efficacy of the illustrated diagramming technique to categorical syllogisms only. This paper attempts to unify the existing stalemate over representing valid syllogistic reasoning with a single diagrammatic technique in both the elucidations.

References

1. Kneale, W., Kneale, M.: The Development of Logic. Oxford Clarendon Press, London (1962)
2. Shin, S.J.: The Logical Status of Diagrams. Cambridge University Press, New York (1994)

The Relationship between Graph Comprehension and Spatial Imagery: Support for an Integrative Theory of Graph Cognition

Brandie M. Stewart, Aren C. Hunter, and Lisa A. Best

University of New Brunswick
Department of Psychology
P.O. Box 5050
Saint John, NB E2L 4L5 Canada
brandie.stewart@unb.ca,
ahunter3@connect.carleton.ca, lbest@unbsj.ca

Abstract. Trickett and Trafton [7,8] suggested an integrative model of graph comprehension that accounts for the role that spatial cognition plays in comprehension. This paper presents a preliminarily examination of spatial and object imagery abilities in relation to Trickett and Trafton's three levels of graph comprehension [8]. Results confirmed Trickett and Trafton's levels of comprehension and suggested that spatial imagers have higher levels of graph comprehension.

1 Introduction

Diagrams are important in all levels of scientific inquiry [3, 5] and it has been suggested that visual representation is central to scientific discovery. Some philosophers argue that a defining feature of science is the use of visual inscriptions to analyze and present data [5]. Graphs are prevalent in science because they allow complex relationships to be conveyed through basic perceptual patterns. Theories of graph comprehension focus on both perceptual and cognitive processes and interpretation occurs as a series of perceptual and cognitive events result in an overall understanding of the data. Perceptual processes allow the reader to decode the basic features and result in the formation of a visual description of the array [6]; the end result of these processes is the extraction of simple information [2]. Cognitive processes are instrumental in relating the basic features to the concepts and influence interpretation in a top-down fashion [6].

Trickett and Trafton [7, 8] proposed that graph comprehension involves different tasks and levels of interpretation. Extraction of explicit information, such as the value of a specific data point, is the easiest comprehension task [7, 8] and the extraction of explicit information can be accounted for by perceptual models. Tasks that require comparisons between graphical elements are more difficult and at these, more complex, levels of interpretation, implicit information must be extracted. The extraction of implicit information involves more complicated cognitive processes [6]. Trickett and Trafton [8] argued that spatial processing is central to comprehension and allows for comparisons between graphical elements and results in the extraction

G. Stapleton, J. Howse, and J. Lee (Eds.): Diagrams 2008, LNAI 5223, pp. 415–418, 2008.

of implicit information. The ability to make spatial transformations is an integral aspect of visual problem-solving and these processes are central to the comprehension of graphical displays [7, 8]. Current theories of graph cognition lack a complete description of the spatial processes that are preformed when both explicit and implicit information must be extracted.

Kozhevnikov, Hegarty, and Mayer [3] proposed that spatial imagery and object imagery are two distinct abilities. These researchers found that participants who performed well on spatial ability tasks tended to generate abstract schematic representations and those with lower spatial abilities tended to generate detailed pictorial images. Blajenkova and her colleagues [1] confirmed that the images generated by object imagers were more pictorial and those generated by spatial imagers were more schematic.

Kozhevnikov, Kossyln, and Shephard [4] developed specific tests to assess object and spatial imagery. The Object Imagery Ability Test (OIA) assesses the ability to identify objects presented in a degraded picture and the Spatial Imagery Ability Test (SIA) assesses performance on box folding, cube rotation, and block orientation tasks. These researchers found that object imagers typically have lower scores on analytical problem solving tasks, including analytic graph interpretation problems. This difficulty with analytic problem solving may be attributed to the tendency to process information globally and extract the overall meaning of a display. Abstract comprehension requires the restructuring of specific graphical elements and involves the consideration of each element; these comparisons involve local processing. Thus, the differences in comprehension could be accounted for by the use of a local or global processing strategy [4].

One goal of the present study was to further validate Trickett and Trafton's [7, 8] three levels of graph comprehension. A second goal was to preliminarily examine the association between imagery abilities and the comprehension of graphical displays.

2 Method

Forty three participants from the University of New Brunswick completed a graph comprehension test, the SIA test and the OIA test. The graph comprehension test included 36 multiple choice questions designed to measure three levels of comprehension (see Figure 1). The SIA consisted of 24 spatial ability questions and the OIA consisted of 20 object imagery questions.

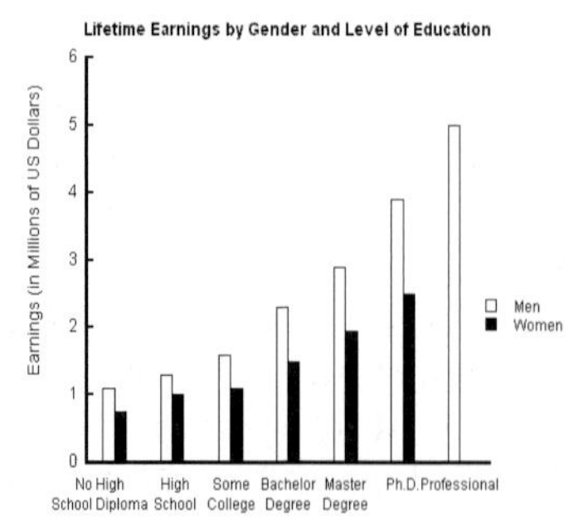

Read-off: The lifetime earning of a man who has a Bachelor's degree is _____ million.

Transformation: The difference in lifetime earning for a woman who has a master's degree as compared to a woman who has not graduated from high school is _____ million.

Interpretation: Based on the information in the graph, what do you predict would be the lifetime earnings of women with a "professional" degree?

Fig. 1. Example of a set of questions from the graph comprehension test

3 Results

The participants (21 males; 22 females) ranged in age from 17 to 45 years (M = 20.37, SD = 5.06). For ease of interpretation, SIA and OIA scores were converted to percentages. In general, SIA scores (M=35.08, SD=20.13) were higher than OIA scores (M=30.93, SD=12.06) but these differences were not statistically significant. As reported by Kozhevnikov and her colleagues [8], scores on the SIA and the OIA were independent of one another ($r = .23$, $p = .14$).

To examine the relationship between imagery and graph comprehension, the difference score between spatial and object imagery was calculated for each participant. The frequency distribution of the difference scores were examined and based on the distribution, participants were grouped according to their imagery preferences. A difference score greater than +3 indicated a preference for spatial imagery, a difference score ranging from 0 to +3 indicated a preference for neither spatial or object imagery, and a difference score less than 0 indicated a preference for object imagery. This grouping resulted in 15 participants being classified as object imagers, 16 were classified as having no preference, and 12 were classified as spatial imagers.

A 3 (graph type) x 3 (question type) x 3 (imagery group) mixed ANOVA was conducted. The main effects for graph type was statistically significant, F (2, 80) =18.38, p=.0001, η^2=.32. Accuracy was 91.67% for pie charts, 66.27% for line graphs, and 56.59% for bar charts. Post hoc tests indicated statistically significant accuracy differences between each of the graph types. Accuracy also differed according to question type, F (2, 82) =188.99, p=.0001, η^2=.83. Comprehension was highest for read-off questions (M=90.50), intermediate for transformation questions (M=57.36), and lowest for interpretation questions (M=46.71). Post hoc analyses indicated statistically significant differences between each question type.

There was a statistically significant main effect for imagery group, F (2, 40) =5.2, p=.01, η^2=.21. On average, object imagers and those with no preference had lower graph comprehension scores than those who were classified as spatial imagers. Post hoc tests showed that spatial imagers had significantly higher graph comprehension scores than the object imagers and those with no preference (see Figure 2).

Correlational analyses were conducted to examine the relationship between spatial imagery and graph comprehension. There was a statistically significant relationship between spatial imagery and graph comprehension, r (43) =.42. As expected, the

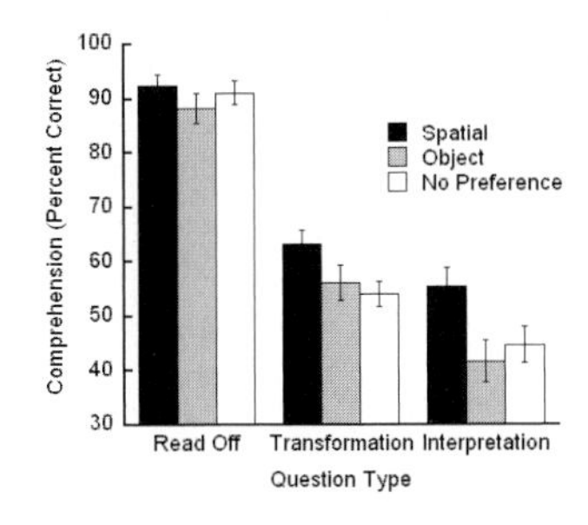

Fig. 2. Accuracy of Spatial and Object Imagers on different comprehension questions

correlation between object imagery and graph comprehension was not statistically significant, $r(43) = -.017$. These findings confirm Kovhevnikov et al.'s [7] distinction between object and spatial imagery and Trickett and Trafton's [15] argument that graph comprehension involves spatial cognition.

4 General Discussion

The current results suggest that graph comprehension is a multi-faceted skill, as evidenced by the fact that comprehension was higher when read-off judgments were made and lower for questions involving spatial transformation and overall interpretation. The results confirm Trickett and Trafton's [7, 8] assertion that there are different levels of graph comprehension.

These results also support the differentiation between object and spatial imagery [1, 3, 4]. In the current study, spatial imagers had higher graph comprehension scores than object imagers; suggesting that graph reading is, at least in part, a spatial task [see 8]. Although the interaction between imagery and question type was not statistically significant, the accuracy differences of object and spatial imagers were larger on transformation and interpretation questions. Future studies should further investigate the links between graph comprehension and preferred imagery strategy.

References

1. Blajenkova, O., Kozhevnikov, M., Motes, M.A.: Object-spatial imagery: A new self-report imagery questionnaire. Applied Cognitive Psychology 20, 239–263 (2006)
2. Cleveland, W.S., McGill, R.: Graph Perception: Theory, experimentation and application to the development of graphical methods. Journal of American Statistical Association 79, 531–554 (1984)
3. Kozhevnikov, M., Hegarty, M., Mayer, R.E.: Revising the visualizer/verbalizer dimension: Evidence for two types of visualizers. Cognition & Instruction 20, 47–77 (2002)
4. Kozhevnikov, M., Kosslyn, S., Shephard, J.: Spatial versus object visualizers: A new characterization of visual cognitive style. Memory & Cognition 33, 710–726 (2005)
5. Latour, B.: Drawing things together. In: Lynch, M., Woolgar, S. (eds.) Representation in Scientific Practice. MIT Press, Cambridge (1990)
6. Pinker, S.: A theory of graph comprehension. In: Friedle, R. (ed.) Artificial intelligence and the future of testing. Erlbaum, Hillsdale (1990)
7. Trafton, J.G., Marshall, S.P., Mintz, F., Trickett, S.B.: Extracting Explicit and Implicit Information from Complex Visualizations. In: Hegarty, M., Meyer, B., Hari Narayanan, N. (eds.) Diagrammatic representation and Inference, pp. 206–220. Springer, Berlin (2002)
8. Trickett, S.B., Trafton, J.G.: Toward a Comprehensive Model of Graph Comprehension: Making the Case for Spatial Cognition. In: Barker-Plummer, D., Cox, R., Swoboda, N. (eds.) Diagrammatic Representation and Inference, pp. 244–257. Springer, Berlin (2006)

Using MusicXML to Evaluate Accuracy of OMR Systems

Mariusz Szwoch

Gdansk University of Technology, ul.Narutowicza 11/12,
80-952 Gdansk, Poland
szwoch@eti.pg.gda.pl

Abstract. In this paper a methodology for automatic accuracy evaluation in optical music recognition (OMR) applications is proposed. Presented approach assumes using ground truth images together with digital music scores describing their content. The automatic evaluation algorithm measures differences between the tested score and the reference one, both stored in MusicXML format. Some preliminary test results of this approach are presented based on the algorithm's implementation in OMR Guido application.

Keywords: Optical Music Recognition, MusicXML, Performance of Systems.

1 Introduction

Diagrams are widely used to represent information of different kind. Musical notation is a good example of diagrams in which musical symbols represent sounds of musical instruments and the way of their interpretation and performing [1]. Because many diagrams still exist only in a paper form many pattern recognition algorithms have been developed for information retrieval from such documents. The images of musical scores are processed by Optical Music Recognition (OMR) algorithms. The research work in the OMR field has been done for over 40 years [2] and has lead to creation of several commercial (SmartScore, SharpEye, Photoscore, etc.), and even more academic (O^{3}MR, Gamera, Guido etc.), complete OMR systems.

One of the important issues in developing OMR algorithms and systems is evaluation of their accuracy or recognition rate. Because it is a very time consuming task, if done by-hand, some automation of this process should be introduced. The obvious solution to this problem is creation of the ground truth databases allowing for automatic accuracy evaluation for different OMR software. Unfortunately, at the present time, no such databases of musical scores are available. Another problem is a lack of standard format for digital scores. Though there are many different formats available only MusicXML format[1] gained the status of a *de facto* standard and is accepted by most commercial and academic applications, allowing for musical information exchange between different score-writing, music recognition and music sequencer software.

[1] MusicXML is developed by the Recordare LLC – http://www.recordare.com/xml.html (2008).

G. Stapleton, J. Howse, and J. Lee (Eds.): Diagrams 2008, LNAI 5223, pp. 419–422, 2008.

2 Using MusicXML to Evaluate Accuracy of OMR Systems

Evaluation of the recognition's accuracy or efficiency of complex documents such as musical sheets or other diagrams requires taking into account a different importance of particular classes of symbols. Also different types or error situations should be considered such as: missed or not recognized symbols, confused or wrong recognized symbols and extra or redundantly recognized symbols. In order to evaluate overall recognition's accuracy of such documents formulas similar to (1) may be used:

$$A = \max\left(0, \frac{\sum_{i=1}^{CN} \frac{TR_i - XR_i}{SN_i} \cdot \frac{SN_i}{SN} \cdot w_i}{\sum_{i=1}^{CN} \frac{SN_i}{SN} \cdot w_i}\right) = \max\left(0, \frac{\sum_{i=1}^{CN} (TR_i - XR_i) \cdot w_i}{\sum_{i=1}^{CN} SN_i \cdot w_i}\right) \quad (1)$$

where A is the overall weighted accuracy of document's recognition, CN is the number of symbol classes, SN is the number of all symbols, TR_i, XR_i and SN_i are occurrence numbers of: correct, extra and expected symbols of class i, and w_i are weights determining the importance of class i.

An important problem is to determine the values of weights w_i that would properly represent significance of particular class within the whole document. Some dissertations of that problem for musical documents may be found in [3][4][5]. Another problem arises for documents with a certain logical structure that should be reconstructed from recognized symbols. For such documents the evaluation of recognition's accuracy should also include the proper interpretation of their structure. Unfortunately, though this problem is noticed by many researchers [5] there is no common approach to solve it.

Developing of OMR applications requires repeated evaluation of recognition's accuracy for many whole-page musical scores. Unfortunately, if done by-hand, it takes quite large amount of time and does not guarantee full reliability of this process, so some mechanisms for automation of documents verification has to be created.

Methodology proposed in this paper assumes existence of a database with ground truth score images with corresponding MusicXML files representing their full musical information (Fig. 1). Each MusicXML file can correspond to any number of different versions of the same score allowing for example for testing influence of image quality upon recognition's accuracy. In order to evaluate the OMR application's accuracy, the score image is taken from the database with its corresponding reference MusicXML file. The OMR software generates another MusicXML file representing music information retrieved from the score image. Special 'Comparator' compares this

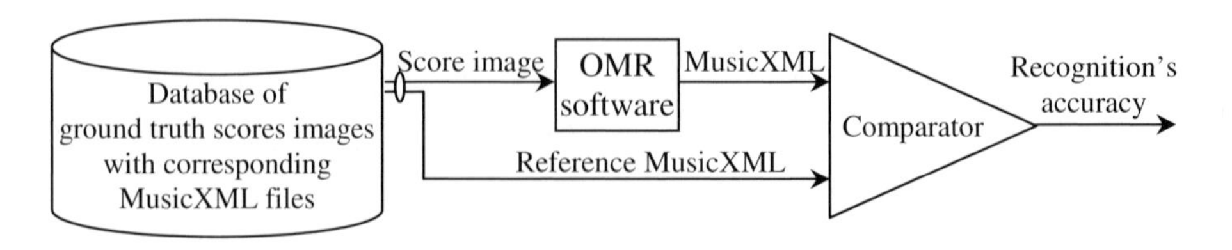

Fig. 1. Evaluation of OMR's accuracy using MusicXML

test MusicXML file with the *reference* one and evaluates the accuracy of the recognition and interpretation process

The 'Comparator' proposed in this attitude may be a standalone application or may be built into the OMR software. Due to a complex nature of the MusicXML format it is not a good idea to directly compare two MusicXML files. Instead they should be both converted into some internal, hierarchical representation conformable with a certain syntax model of musical notation.

The proposed algorithm of comparison realizes top-down attitude, starting the verification process from the root of musical notation's syntax tree (score's page), and moving recursively down towards the leaves, which are symbols of musical notation. On each level of the syntax tree the correspondence between structures in both reference and test score is searched. When no correspondent element is found, the error is counted and the next match is searched. For each matched pair of elements the analysis is recursively passed down to the element's evaluation method at the lower level of the music notation's structure. After completing comparison for a given ele-ment, the control returns to the higher level and the next pair of elements is compared.

This general algorithm may be easily improved by several modifications that would reflect difference in the way of comparing high level structures of the musical notation and low level musical symbols. The first modification use relative information about horizontal location of musical symbols inside each bar. This modification improves results in all cases when lacking or extra symbol destroys the comparison order of other symbols following them (Fig.2 A). The information about horizontal position is stored in 'default-x' parameter of low level symbols' tags in MusicXML 1.0 format. In the future the absolute coordinates of symbols, introduced in MusicXML 2.0, could be used. The second modification is introduced only for bars. It relies on comparing not only corresponding bars but also the following ones and establishing the real correspondence basing on their similarity (Fig. 2 B). This improvement tries to deal with the situation of not recognized or extra bar lines.

3 Experiments

In order to verify the proposed attitude, both comparison algorithm with all improvements and the accuracy evaluation formula (1) were implemented in Guido OMR application [6], developed in KED – Gdansk University of Technology. Also special interface allowing for automatic batch comparison of two sets of scores – test and reference – was introduced. For tests 10 images of full-paper scores were prepared. Images were recognised in Guido and stored in the MusicXML format creating the test set of scores. Then all recognition errors were corrected in a built-in editor and digital scores were saved again creating the reference set of MusicXML scores.

In the experiments, automatic recognition accuracy evaluated automatically by Guido was compared with the accuracy evaluated by human operator. In most cases their values were identical or very similar. The worst case was noticed for lacking or extra bar lines, when some symbols were counted twice as errors: not recognized and extra symbols. The experiments indicate that there is no need to introduce any special measures for the structure misinterpretation [5]. When using the proposed structural top-down approach, any disturbance of high level structure is automatically reflected in overall accuracy by not recognised musical symbols belonging to them.

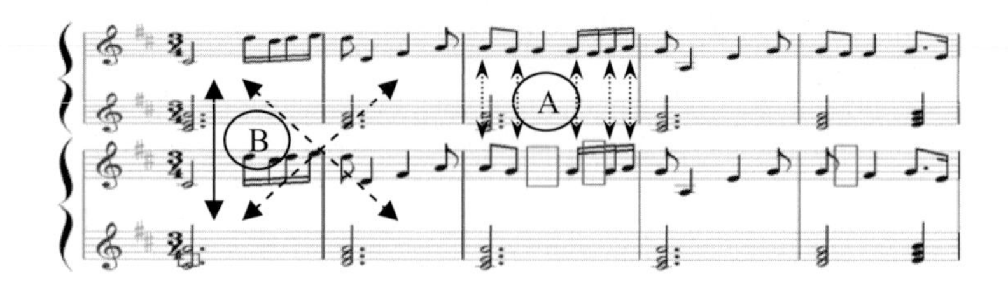

Fig. 2. The excerpt of the reference (top) and test (bottom) score. Automatically evaluated recognition's accuracy A = 92.16% (weights w_i=1) is the same as human's evaluation.

Additionally several experiments were carried on to test the influence of the criterion's weights upon overall accuracy. These experiments indicated general correctness of the formula (1) which binds higher overall recognition's accuracy with higher recognition rate of most important elements in musical notation.

4 Conclusions and Future Work[2]

In this paper the problem of automatic accuracy evaluation in OMR systems was addressed. A proposed, weighted accuracy criterion takes into account importance of musical symbols' classes. Experiments carried on proved that such a criterion, implemented in OMR systems together with automatic mechanisms of MusicXML files' comparison, enables efficient testing of their recognition's accuracy. MusicXML format occurred to be very suitable for scores comparison due to its structural organisation. It is also very convenient in the databases creation of pattern scores.

Further work should be focused on determining proper values of weights in the formula (1), that would best reflect recognition's accuracy according to human evaluation. The algorithm used for comparison of MusicXML scores can also be further improved to better deal with more serious differences in scores' structure.

References

1. Blostein, D., Haken, L.: Using Diagram Generation Software to Improve Diagram Recognition: A Case Study of Music Notation. IEEE Trans. on Pattern Analysis and Machine Intelligence 21(11), 1121–1135 (1999)
2. Blostein, D., Baird, H.S.: A Critical Survey of Music Image Analysis, Structured Document Analysis, pp. 405–434. Springer, Heidelberg (1992)
3. Byrd, D., Shindele, M.: Prospects for Improving OMR with Multiple Recognizers. In: ISMIR 2006, 7th International Conference on Music Information Retrieval (2006)
4. Ng, K., et al.: Coding Images of Music Sheets DE4.7.1, http://www.interactivemusicnetwork.org
5. Bellini, P., Bruni, I., Nesi, P.: Assessing Optical Music Recognition Tools. Comput. Music J. 31(1), 68–93 (2007)
6. Szwoch, M.: Guido: a Musical Score Recognition System. In: ICDAR 2007: Ninth International Conf. on Document Analysis and Recognition, vol. 2. IEEE Computer Society, Los Alamitos (2007)

[2] This paper is partially sponsored by the Polish Government's research funds No 3T11C027 28.

Aestheticization of Flowcharts

Wioleta Szwoch

Gdansk University of Technology, ul.Narutowicza11/12,
80-952 Gdansk, Poland
wszwoch@eti.pg.gda.pl

Abstract. One of the important issues of diagrams is their aesthetics. In this paper a method of its formalization for freehand drawn flowcharts is proposed. In order to formalize the evaluation of flowcharts' aesthetics a criterion consisting of several measures is proposed. Based on this criterion the algorithms for automatic optimization of flowcharts' appearance are proposed.

Keywords: Flowcharts, Aesthetics.

1 Introduction

Aesthetical presentation of information is a very important issue [1-11]. It is especially significant after recognition of freehand drawn flowcharts which are usually done in a careless and imprecise manner. Aestheticization should be done automatically and should allow for optimization of flowcharts' appearance. In order to make it possible a criterion should be formulated, that would measure overall aesthetics of a flowchart. The criterion proposed in this paper is partially based on general features presented in [5], introducing for them special measures suitable for flowcharts. The criterion combines also some newly formulated measures and the density measure from [9].

2 Criterion and Measures of Aesthetics' Features for Flowcharts

The proposed criterion consists of two parts: Kl is responsible for local aesthetics measures whereas a second Kg for the global ones. All measures are defined to produce values ranging from 0 (bad) to 1 (good). The whole criterion is given by formula (1) and its components are explained in Table 1. Greater value of criterion function means that the flowchart is more aesthetic. Because different features have different influence upon the flowchart's aesthetics, additional weights for the measures may be introduced. The weights specify the relative importance of each feature and should be normalized. Because the complexity of flowcharts is not changed essentially after aestheticization, it is not reflected in the criterion (1).

$$K = (w_{S_l} S_l + w_{H_c} H_c + w_E E + w_{VH} VH + w_{SC} SC) + (w_{S_g} S_g + w_D D + w_{H_f} H_f + w_N N) \tag{1}$$

G. Stapleton, J. Howse, and J. Lee (Eds.): Diagrams 2008, LNAI 5223, pp. 423–426, 2008.

where S_l is a local symmetry, H_c - connections' homogeneity, E - equilibrium (alignment), VH - vertical – horizontal alignment of lines, SC - number of all segments in connection, S_g global symmetry, D density, H_f figures' homogeneity, N nodes count, w weights.

Table 1. Features and their measures used in the criterion (1)

Feature	Measure of the feature	Description
Global symmetry	$S_g = \begin{cases} 1 & \text{symmetry exists} \\ 0 & \text{symmetry does not exist} \end{cases}$	Existence of global symmetry
Local symmetry	$S_l = \dfrac{\sum_{i=1}^{N_{fc}} \left(1 - \dfrac{x_{ci}}{x_{oi}}\right)}{N_{fc}}$	N_{fc} - the number of 'sockets' (connections outputs and inputs), x_c - distance between a connection and symmetry axis of figure, x_o - half of the figure's width
Equilibrium (alignment)	$E = \dfrac{\sum_{g=1}^{G} E_g}{G}, E_g = 1 - \dfrac{\overline{x_i}}{x_g} = 1 - \dfrac{\sum_{i=1}^{N_g} x_i}{N_g x_g}$	E_g - alignment in group g, G - number of groups, N_g - number of elements in group g, x_i - distance between the symmetry axis of element and the x center of its group, x_g - half of the width of the group's bounding box
Density	$D = 1 - 2 * \lvert 0{,}5 - \dfrac{\sum_{i=1}^{N_f} P_i}{P_p} \rvert$	N_f - the number of figures, P_i - area of i-th figure, P_p - area of bounding box of the whole flowchart
Homogeneity of figures	$H_f = 1 - \dfrac{\sum_{t=1}^{T} \sum_{i=1}^{N_t} \dfrac{\lvert P_{ti} - \overline{P_t} \rvert}{N_t \overline{P_t}}}{T}, \overline{P_t} = \dfrac{\sum_{i=1}^{N_t} P_{ti}}{N_t}$	T - number of figures' types, N_t - number of figures of a given type t, P_{ti} - area of figure i, $\overline{P_t}$ - the average area of figures of type t
Homogeneity of connections	$H_c = 1 - \dfrac{\sum_{j=1}^{J} \sum_{i=1}^{N_j} \dfrac{\lvert d_{ji} - \overline{d_j} \rvert}{d_S}}{J}$ $\overline{d_j} = \dfrac{\sum_{i=1}^{N_j} d_{ji}}{N_j}, d_S = \sum_{i=1}^{N_j} d_{ji}$	J - the number of connections' sequences, N_j - number of connections in sequence j, d_j - length of the connection j, $\overline{d_j}$ - the average connection's length in sequence j, d_s – summary length of connections in sequence s
Nodes count	$N = \dfrac{N_c - N_{cj}}{N_c}$	N_c - the number of all connections, N_{cj} - number of connections between two nodes
Horizontal – vertical alignment of lines	$VH = \dfrac{\sum_{i=1}^{N_{cs}} (1 - \lvert \tan \alpha \rvert)}{N_{cs}}$	N_{cs} - the number of all connections' segments, α - angle between segment and the horizontal or vertical direction ($\alpha \in$ <$-\pi/4, \pi/4$>). $\alpha = \min(\alpha_X, \alpha_Y)$
Simplicity of connections	$SC = 1 - \dfrac{N_{c4} + N_{c3}}{N_c}$	N_{c4} - number of connections with more than 3 segments, N_{c3} - number of connections with 3 segments between subsequent figures

3 Improving the Flowchart Appearance

The criterion (1) may be used for improving appearance of flowcharts. The optimization of flowcharts' aesthetics can be done by maximization of all measures in the criterion. The order of optimization steps is very important because changing of some features may influence upon others. In general, it is a multi-criteria optimization problem, but in this particular case certain characteristic sequence of operations may be proposed which minimize mutual influences between components of the criterion (1). Also additional feedback from the user may be used to accept or reject changes made on each step.

The proposed sequence starts with local optimization. At first, orthogonalization of all connections is performed maximizing *VH* component. Then, if local symmetry S_l<1, all the sockets are moved to the centre of figure sides. Next, if simplicity of connections *SC*<1, the connections with too many segments are simplified. In the next step, groups of figures are located using FlowGram grammar [12]. For each group their equilibrium and homogeneity of their connections are improved. Next, the values of global components are calculated. If there exists any measure that is less than 1, it means that this feature could be corrected. The algorithm tries to change the features according to appropriate rules. If the value of criterion increases and the values of local measures are not decreased, the changes are accepted.

4 Experiments

To verify proposed solutions the application FCA (Flow Chart Analyzer) [12] was created that allows for calculation of all measures presented in the previous section. In the first test, two sets with 10 corresponding pairs of flowcharts were prepared. The flowcharts in the first set were recognized from hand-drawn images. The second set contained their corrected versions that were intended to look subjectively better. Example pair of flowcharts, calculated measures of aestheticization features and the value of global, local and overall criteria are given in Fig.1. The higher aesthetics of flowcharts in the second set was indicated by greater value of the criterion (nearly 30% in average) and confirmed by the group of 15 students that compared corresponding flowcharts.

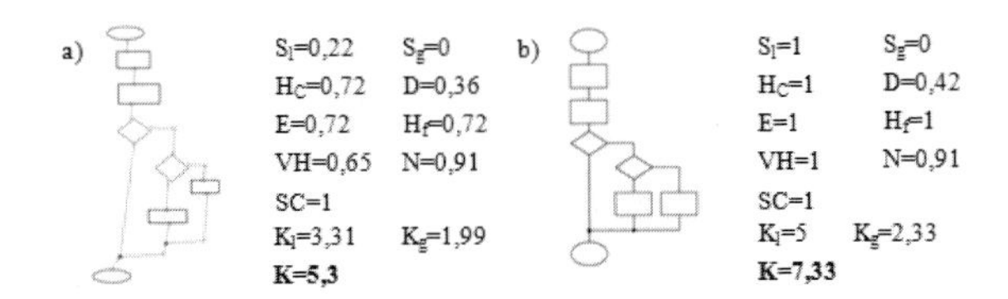

Fig. 1. Examples of the application of the aesthetics measures (near 40% increase of K value)

In the second test, a group of 15 students evaluated for each measure the correlation between the value of criterion and the appearance of proper fragments of flowcharts. In general testers chose flowcharts with greater measure as more aesthetic. Only for *density* and *nodes count* measures some discrepancy occurred.

Thus the first results show that the greater value of criterion indicates the better appearance of flowchart.

5 Conclusions and Future Work

In this paper the problem of aestheticization of flowcharts is presented. The proposed aesthetics criterion allows to calculate the value that shows whether the appearance of flowchart may be improved. The described features represent different aspects of flowcharts appearance like symmetry, alignment and other. The first experiments indicated correlation between the value of criterion and the appearance of flowcharts.

Presented attitude has also some problems and limitations. For instance the geometrical symmetry does not necessarily match the human perception of global symmetry and may only suit as its rough estimation. In some situations connections between nodes that increase criterion value may look better. Also the arbitrary set 50% value of density not always is optimal.

The future works will focus on full implementation of proposed algorithm and automatic improving of flowcharts aesthetics using the proposed criterion. Also more advanced approach, for multi-criteria overall optimization should be examined.

References

1. Birkhoff, G.D.: Aesthetic measure. Harvard University Press, Cambridge (1933)
2. Grabska, E.: Aesthetic measure of polygons and computer evaluation. In: Proceedings of the Third International Conference on Aesthetic, Krakow (1991)
3. Purchase, H.C.: Effective information visualisation: a study of graph drawing aesthetics and algoritms. Interacting with Computers 13, 147–162 (2000)
4. Purchase, H.C.: Metrics for graph drawing aesthetics. Journal of Visual Languages and Computing 13, 501–516 (2002)
5. Ding, C., Mateti, P.: A Framework for the Automated Drawing of Data Structure Diagrams. IEEE Transactions on Software Engineering 16, 543–557 (1990)
6. Ngo, D.C.L.: Measuring the aesthetic elements of screen design. Displays 22, 73–78 (2001)
7. Ngo, D.C.L., Byrne, J.G.: Aesthetics measures for screen design. In: Computer Human Interaction Conference. Australasian, pp. 64–71 (1998)
8. Ngo, D.C.L., Teo, L.S., Byrne, J.G.: Formalising guidelines for the design of screen layouts. Displays 21, 3–15 (2000)
9. Ngo, D.C.L., Teo, L.S., Byrne, J.G.: Modelling interface aesthetics. Information Sciences 152, 25–46 (2003)
10. Hse, H.H., Newton, A.R.: Recognition and beautification of multi-stroke symbols in digital ink. Computers & Graphics 29, 533–546 (2005)
11. Galindo, D., Faure, C.: Perceptually-Based Representation of Network Diagrams. In: Document Analysis and Recognition IV International Conference, pp. 352–356 (1997)
12. Szwoch, W.: Recognition, understanding and aestheticization of freehand drawing flowcharts. In: ICDAR 2007 9th International Conference on Document Analysis and Recognition, pp. 1138–1142. IEEE Computer Society Press, Los Alamitos (2007)

Towards Diagrammatic Patterns

Merete Skjelten Tveit

Faculty of Engineering and Science, University of Agder
Grooseveien 36, N-4876 Grimstad, Norway
merete.s.tveit@uia.no

Abstract. This article presents the idea that the graphical representation (concrete syntax) of a visual language can be specified based on some pre-defined diagrammatic patterns. A diagram from the Specification and Description Language (SDL) is used as illustration.

1 Pattern-Based Language Specification

Visual languages are widely used and have an important role in the area of information technology and software development. Unfortunately, there are not any common agreement and understanding on how these languages are specified. The disagreement is in particular related to how we specify the graphical representation (also known as concrete graphical syntax). This article presents the idea of using a set of pre-defined diagrammatic patterns as a basis for the specification of the graphical representation. A related idea is presented in [3].

The concept of patterns was presented in the area of architecture in the late 1970's, and from 1994 *design patterns* became popular in the fields of software engineering and design [1]. Design patterns describe general and systematic solutions to reoccurring problems in the software development process.

To state it simply, a diagram can be considered as a *collection of diagram components* with special *relations* in between. When specifying the graphical representation, we identify the diagram components and how they are related. There exists an enormous amount of different visual languages, and even if the diagrams from the different languages vary a lot, it is possible to find similarities. In general, graphical languages have more commonalities than differences. An identification of these similarities forms the foundation for the diagrammatic patterns presented in Section 2. The goal is not to cover all possible situations that can occur in a diagram, but to identify the diagrammatic aspects that are most common and to suggest how they should be specified. The patterns could also be seen as guidelines for the definition of new notations, for instance for new domains specific languages.

2 Diagrammatic Patterns

As a starting point for this pattern identification two well-established languages, UML 2.0 [4] and SDL [2], were examined. The diagram components in the languages were categorised based on the *role* they play in the diagrams, and nine

G. Stapleton, J. Howse, and J. Lee (Eds.): Diagrams 2008, LNAI 5223, pp. 427–429, 2008.

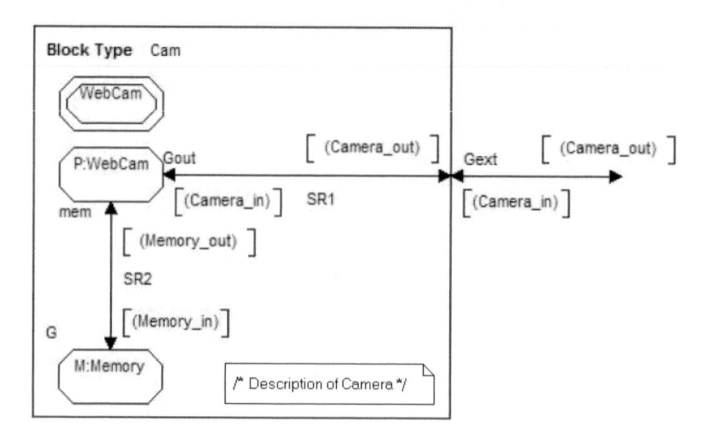

Fig. 1. SDL Block Type Diagram

patterns were identified: *symbol*, *connection*, *compartment*, *structured text*, *free text*, *defined text*, *reference*, *annotation* and *inside*. The patterns can be divided into two main groups: patterns that describe a token in the diagram (the first six), and patterns that describe relations between tokens (the latter three). With the SDL block type diagram in Figure 1 as an illustration, the diagram patterns are briefly described.

- The **symbol pattern** is used to specify graphical tokens in a diagram. A symbol can stand alone, like the process type named "WebCam", or can be connected to other symbols by connections.
- The **connection pattern** is used to specify graphical tokens that connect symbols together. The channels named "SR2" and "SR1" are examples of connections.
- The **compartment pattern** is appropriate for symbols that are divided into different compartments. The diagram example has two compartments, one for the heading and one for the rest of the diagram components.
- The **structured text pattern** is used for more complex and composite text expressions. The expression in the process symbols "P:WebCam" and "M:Memory" are examples as they consist of a name, a colon, and a reference.
- The **free text pattern** is used for text without restrictions on format or syntax. The comment "Description of a Camera" at the bottom of the diagram is an example.
- The **defined text pattern** is used for strings that are pre-defined in the language, sometimes also called keywords. "Block Type" is an example of defined text.
- The **reference pattern** is used in situations where it is necessary to reference another diagram object. "WebCam" is an example of a reference, it actually refers to a process type diagram which is accessed when double-clicking on the reference.

- The **annotation pattern** is useful for describing a relation between two tokens in a language, for instance between some kind of text and a connection. The label "Gext" on the gate is an example of an annotation relationship.
- The **inside pattern** is a pattern for a spatial relationship. It is used when one graphical objects is placed inside another object. In the diagram, the process type "WebCam" is placed inside the block type "Cam".

2.1 Diagrammatic Patterns Versus Syntactic Mapping Patterns

Another interesting aspect with this work is how the graphical representation is related to the structure (also called abstract syntax). The earliest attempts to separate the representation from the structure in visual languages used a one-to-one (mapping) relation between the two syntactic aspects. This view on the relationship is in general too simple. Even if the relationship in many cases is one-to-one, there exist also situations where the relationships are more complex.

- The **duplication pattern** is useful when we have several representations of the same structure element. In the diagram example we have the references to the signal lists, "Camera_in" and "Camera_out", which is both represented twice. This mapping pattern corresponds to the diagrammatic pattern *Reference* described earlier in this section.
- The **merge pattern** is used when tokens are merged. A typical situation is when a declaration of variables and types are merged to a shorten form. This has consequence for the mapping since important information the mapping is based on, is left out. The signal declaration in SDL is an example, where `signal a, signal b;` could be merged to `signal a, b;`.

3 Future Work

This presentation has focused on giving a brief overview of the most common patterns for graphical representation and mapping. The future plan is to focus also on some of the more special cases. Another important part is to document the patterns in a more precise way.

References

1. Gamma, E., Helm, R., Johnson, R., Vlissides, J.: Design Patterns: Elements of Reusable Object-Oriented Software. Addison-Wesley, Reading (1995)
2. ITU-T. SDL - ITU-T Specification and Description Language (SDL-2000). ITU-T Recommendation Z.100 (1999)
3. Kastens, U., Schmidt, C.: Visual patterns associated to abstract trees. Electr. Notes Theor. Comput. Sci. 148(1), 5–18 (2006)
4. OMG: UML 2.0 Superstructure Specification. Object Management Group, ptc/04-10-02 (October 2004)

Visualizing Meaning: Literacy Materials for Dyslexic Children

Myra Thiessen

Department of Typography and Graphic Communication
University of Reading

1 Introduction

Dyslexic children experience a difficulty in learning to read at a level that is expected from their level of intelligence. As dyslexia is commonly characterized by a deficit in phonemic processing, many literacy intervention programs focus on developing skills in this area. However, language is processed by two separate cognitive subsystems: the verbal (language) and the nonverbal (imaging) [3]. Having strengths in visual processing and problem-solving [7], dyslexic learners may be able to use these abilities inherent in the nonverbal subsystem to more easily acquire those skills related to the verbal subsystem, with which they typically struggle. It is suggested here that common teaching strategies that pair concrete nouns with simple and literal visual explanations may be used to teach more complex verbal concepts to children with dyslexia. As verbal information becomes more demanding, it is likely that visual explanations will also become more complex. It is therefore important to understand how visual cues can be used to reduce ambiguity and to create meaning in visual explanations. Current research is exploring how children with reading difficulties use, read and interact with visual information. Through both preference and performance testing, the ways in which visual explanations can be used in the literacy education of dyslexic children will be investigated.

2 Visual Explanations

A diagram can be defined as a visual representation of verbal information. This type of meaningful imagery is described here as *visual explanations*, and may also include illustrations or other forms of pictorial language. It is common to include meaningful imagery of this sort in literacy education materials developed for children today. Research has shown that using visual explanations are an effective approach to the design of literacy materials because visualizations have been proven to aid the learner in recall and in their comprehension of a story's events [4]. Such visualizations can be used to *echo* the verbal information and reinforce its meaning, or they can be used to *supplement* the text and provide additional and perhaps more elaborate meaning [5]. By using visualizations alongside verbal information, the scope of learning strategies available to the reader may be increased [5] and the nonverbal cognitive subsystem may be accessed for meaning and comprehension.

In current literacy materials, developed to teach individual concrete nouns, visual explanations are used to echo textual information and are a direct and literal representation of the word. Concrete nouns, like the word /cat/, are typically easy to visualize

G. Stapleton, J. Howse, and J. Lee (Eds.): Diagrams 2008, LNAI 5223, pp. 430–432, 2008.

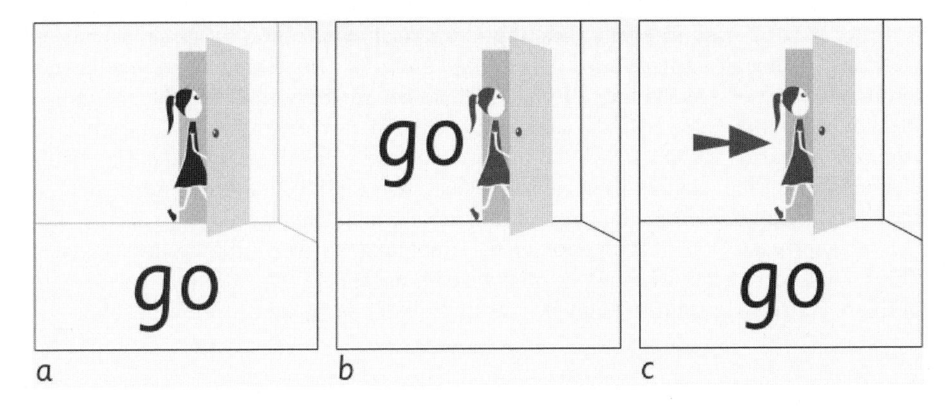

Fig. 1. Visual cues can be used to create visual meaning through a. tints of a color, which in this case is black; b. position; and c. graphic elements

and therefore provide the majority of the content in this type of teaching strategy. This approach can also be utilized to teach words that are typically hard to visualize and that are known to be difficult to learn. Such words are described here as *complex verbal concepts*.

Figure 1 shows a series of conceptual prototypes for the complex verbal concept /go/. These flashcard representations depict alternative visual explanations that show an action in a context: to go from one place to another, in this case through a door. Although the word /go/ may seem relatively simple to acquire, it does appear on a list of 'tricky' or 'sight' words as outlined by popular literacy programs [2, 6]. 'Tricky' words generally occur at a high frequency in written language, but in many cases, due to irregular orthography, they cannot be decoded by traditional grapheme-phoneme mapping and must be memorized. The difficulty that children may experience in reading these words could be the result of more than irregularity in spelling as many of these words are complex verbal concepts. The difficulty with visualizing these words means that they are generally taught without association to a visual meaning, which may limit the learning strategies available to pupils taught in this way.

3 Visual Cues

In visually representing a concrete noun, a visual explanation provided in isolation is likely to be sufficient to communicate the verbal concept. As the complexity of the verbal concept increases it is anticipated that the visual explanation will also become more involved. The visualizations shown in figure 1 may appear relatively simplistic, but the context of a room and an open door has already increased the visual complexity compared to that of an isolated image that may be sufficient to describe a concrete noun. Visual cues can be used to create visual meaning or to clarify visual ambiguity [5]. Visual cues that include the use of color, the positioning of related information and the use of graphic elements may be able to reduce visual ambiguity, show relationships, create visual hierarchies, and draw the reader's attention to highlighted information.

Color may be used to create meaning in a visual explanation by grouping related elements. Although limited to a single color, figure 1a demonstrates this relationship through the use of various tints of black, where the figure performing the action and the word are represented with the same tint. Visual hierarchies or highlighting specific elements can also be shown with the use of color and color tints. By grouping related elements, meaning can be created and visual relationships depicted through their position in relation to one another, an example of which can be seen in figure 1b. The use of graphic elements can reduce visual ambiguity by providing necessary visual information to indicate what might otherwise be unclear [5]. In addition, and as seen in figure 1c, graphic elements may be used to literally point out the relevant action.

4 Conclusion

The use of visual explanations in literacy education has proven to be an effective way to teach simple verbal concepts that are easy to visualize, such as concrete nouns. It is suggested that this established practice could also be used to teach complex verbal concepts to dyslexic children. Traditionally, complex verbal concepts are expected to be learned by sight, but by providing visual meaning the dyslexic child may be able to use their strengths in nonverbal processing to more easily acquire these 'tricky' words. It is likely that as verbal concepts become more involved, the complexity of the associated visual explanation will also increase. Visual information that is misleading, or is not an accurate representation of the concept, may lead to more confusion and misinterpretation than if no image is present [1]. Therefore it is necessary to understand how visual cues may be used to create meaning and reduce ambiguity in a visual explanation.

References

1. Arnheim, R.: Visual thinking. University of California Press, London (1969)
2. Bell, N.: Seeing Stars: Symbol Imagery for Phonemic Awareness, Sight Words and Spelling. Gander Publishing, San Luis Obispo (2001)
3. Bell, N.: Visualizing and Verbalizing for Language Comprehension and Thinking, 2nd edn. Gander Publishing, San Luis Obispo (2007)
4. Goldsmith, E.: Research Into Illustration: An Approach and a Review. Cambridge University Press, Cambridge (1984)
5. Lowe, R., Pramono, H.: Using graphics to support comprehension of dynamic information in texts. Information Design Journal 14(1), 22–34 (2006)
6. Lloyd, S., Stephen, L.: The Phonics Handbook: A Handbook for Teaching Reading, Writing and Spelling, 3rd edn. Jolly Leaning Ltd, Essex (1988)
7. Pollock, J., Waller, E.: Day-to-day dyslexia in the classroom. Routledge, London (1994)

Diagrammatic Interrelationships between Global and Local Algebraic Visual Objects: Communicating the Visual Abstraction

Julie Tolmie

Centre for Computing in the Humanities, King's College London,
26-29 Drury Lane London, UK
Julie.Tolmie@kcl.ac.uk

Abstract. This poster considers mathematical diagrams composed of algebraic visual objects where the role of words and symbols has been explicitly removed with the intention of (re)assigning that role to visual object. Interrelationships between global and local algebraic visual objects become the working abstraction in the space, but challenge text. This transposition is juxtaposed against the historical parallel of early western musical notation, the music as language vs. music as space conflict targeted in the Papal Bull of 1324-25.

Keywords: visual object, visualisation, diagrammatic notation, map, immersion.

1 Compositional Environments

A compositional environment for a discrete abstract visual space has been approached as a non-trivial mathematical choreography problem posed in the visual domain.

In pure mathematics it is generally held that a visual entity of any type is a visualisation and can never be the mathematics itself. In contrast, in mathematical discourse, natural language text and symbols have been appropriated to be the mathematics itself. [1] Diagrammatic argument is viewed as being somewhere in between these semantic extremes. In Category Theory each diagram is composed of objects and arrows and the recognition of parts of a diagram is considered to be valid, well-defined mathematical abstraction. Defining a visual mathematical space in which algebraic and geometric constructions play interrelated and multiple roles as notation is not so surprising in the context of Lie Groups. However conventions do not yet exist for appropriating mathematical form as a structural costume for other parts of itself and mapping these forms together into the 3D space of an empty stage. In part this is because linear text and symbols play a minimal or no role therein.

An abstract visual space composed of visual objects and visual arrows, and in which the abstraction involves the recognition of visual object, itself composed of other visual objects, is referred to as an example. Rational numbers mod1 are notated in two distinct ways; by 2D discrete cyclic configurations of colour coded points; by 3D colour coded curve segments spanning the outer longitude of the torus. [2] The first notation (below) is more visually abstract, the second more visually geometric. Are these visual objects and their interrelationships diagrams? Paradoxically the

G. Stapleton, J. Howse, and J. Lee (Eds.): Diagrams 2008, LNAI 5223, pp. 433–436, 2008.

pairing of the algebraic and the geometric leaves one open to accusations of 'pretty pictures'. However, just as a Lie Group, a mathematical structure commonly found in physics, has both algebraic and geometric properties, these visual objects defy single classification and require abstraction simultaneously with immersive perception. [3]

Similarly to the recognition of parts of a Category Theory diagram, the individual components of the spaces or objects created are recognisable. A major difference is that recognition of individual (and emergent visual groupings of) sub-objects in 3D space involves human perception at human timescales. It will prove useful to introduce an historical parallel: polyphony and the evolution of western musical notation as a spatial form.

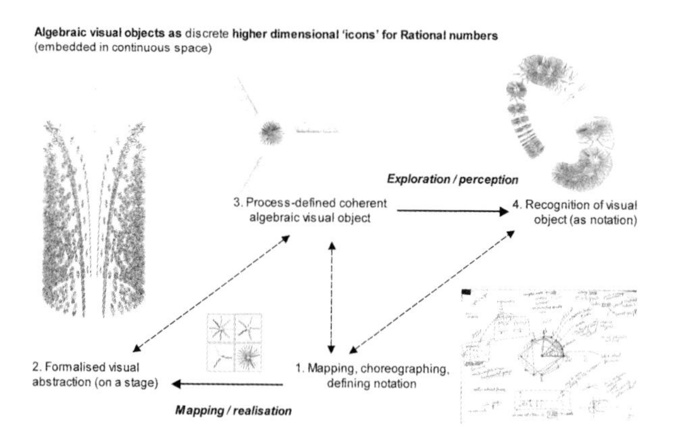

Fig. 1. Static 2D cyclic configurations encoded dynamically using a trace of colour laid down in Rational radial directions by generators of cyclic groups; cyclic point-based 'icons' enable views through the superposition structures; identification of visual sub-objects. [3]

2 Music as Language vs. Music as Space

As polyphony and its associated spatial notation evolved, so did the abstract conceptual frameworks within which musical composition took place, and within which music was understood. A revolutionary shift, the ability to see the whole and the part across time and pitch, enabled composition of polyphonic works using the spatial representational structure to juxtapose sound in the time dimension. [4]

A diastematic musical notation involves discrete pitches represented on lines or in spaces. *The diastematic notation* [5] *[…] assigns a functional place to terms that will determine one another. The portée of Guy d'Arezzo, is at the same time a mediator of operations of transformation, and the distinction of the whole and the part.[…] It implies knowledge of the relationships, the introduction of a system of connections, and the observation of rules appropriated objectively. It announces a history by making it possible.* [6] In both polyphony and visual juxtaposition, discrete objects fuse to create new objects; decomposition into subobjects can be performed in multiple ways, for example in music, by voice or by melody; dynamic abstraction is defined in time and can be perceived in time without language. [4]

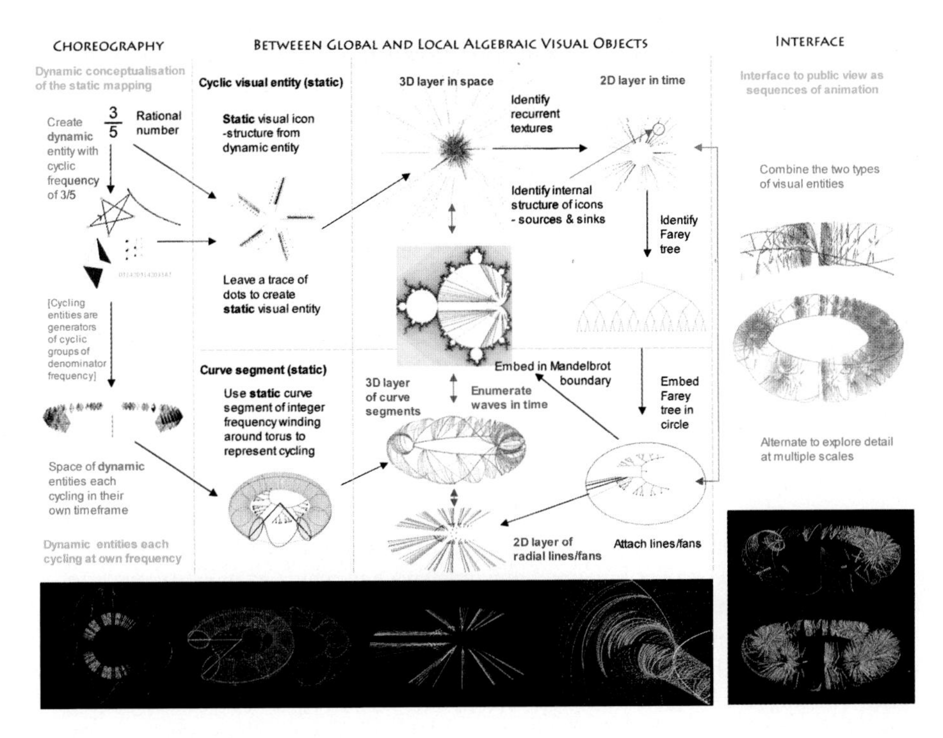

Fig. 2. Global interrelationships between the visual objects: left, the compositional phase; centre the mapping of interrelationships and/or the construction process; right the refined algebraic visual objects within which layers of hierarchy and multiple readings are encoded'; top the cyclic visual entity notation; bottom the continuous curve segment – fan notation. [3]

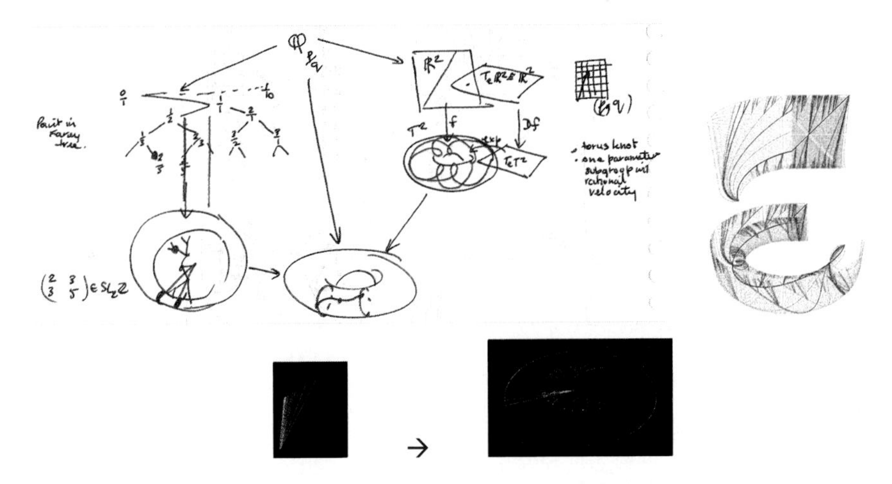

Fig. 3. Notation using continuous curve segments spanning regions of the torus to reveal their interwoven global hierarchy; emergence of local fan structures of curve segments; merging the fan structures and the (Farey) binary tree to create a 2D fan notation for the 3D fan. [3]

3 Communicating the Visual Abstraction

What role did the spatial musical notation play in the history and culture of the time? In 1324-25, Pope Jean XXII issued a Papal Bull condemning polyphony and measured music, insisting that the measure of musical notes is assigned on the basis of the meaning of the phrase sung to God and not on the measured dividing of time. *Primacy of the oral, of linearity, the unequivocal deployment of a syntactic order, and the absolute reference of the music to the phonetic code: that is the injunction of Jean XXII. The famous bull that contemns the modernist tendencies in music is one of the most significant documents of the period, evidence not only of a conflict between two poetics or antagonistic styles, but above all a disagreement between two conceptions of music. One the relation to language, the other the relation to space.* [7]

In music, the spatial notation and the actual sounds are clearly very different types of entities. In visual object, the distinction between the visual object and its notation may not exist; a visual object can also be employed as a notation for a more complex aspect of itself. In figure 2, we see that the concept is a dynamic entity cycling at its own frequency notated into static forms. The superposition of these static forms 'stands for' a very complex imagined dynamic space in which each entity cycles at its own frequency. But the visual superposition also creates the structure of the visualised space. In an environment where the interrelationships between global and local visual objects, including the imagined dynamic ones, become the working abstraction in the space, it is next to impossible to communicate the forming of abstractions and their multiple interpretations without recourse to a global diagram such as figure 2 above.

Certain disciples of the new school, much occupying themselves with the measured dividing of time, display their method in notes which are new to us, preferring to devise ways of their own rather than to continue singing in the old manner; the music, therefore, of the divine offices is now performed with semibreves and minims, and with these notes of small value every composition is pestered. [8]

References

1. Bourbaki, N.: Éléments de mathématiques, théorie des ensembles, Herman Paris: France pp Introduction (1954)
2. Tolmie, J.: Map of curve segment notation (2001), `http://www.tolmie.eu/visual.pdf`
3. Tolmie, J.: Visualisation, Navigation and Mathematical Perception: a visual notation for Rational numbers mod1, Ph.D dissertation (pdf file, embedded animation) (2000)
4. Tolmie, J.: Soundfile (2000), `http://www.tolmie.eu/musicMaths.html`
5. For example see Goldberg The Early Music Portal, The School of Notre Dame, `http://www.goldbergweb.com/en/magazine/essays/2003/09/16246_2.php`
6. Dufourt, H.: Musique, mathesis, et crises de l'antiquité à l'age classique. In: Mathématiques et Art, Colloque, September 2-9, 1991, p. 169. Hermann, Paris (1995) (My translation)
7. Ibid. at pp.171-172 (My translation)
8. 1324-25 Decree of Jean XXII in Munrow David (1976) Music of the Gothic Era, notes accompanying CD Music of the Gothic Era, The Early Music Consort of London, p.18 (1976)

School Curriculum Development to Promote Student Spontaneous Diagram Use in Problem Solving

Yuri Uesaka[1] and Emmanuel Manalo[2]

[1] Tokyo Institute of Technology, Japan
y.uesaka@nm.hum.titech.ac.jp
[2] The University of Auckland, New Zealand
e.manalo@auckland.ac.nz

Abstract. Although previous research has demonstrated that diagrams are powerful tools for problem solving, studies relating to educational practices indicate that students manifest various problems in diagram use. There are studies that have proposed teaching methods to address these problems, but these methods are not fully integrated with the school curriculum. This paper reports on the development of an instruction and support program to promote students' spontaneous use of diagrams as part of a school curriculum. The program was provided to students in the first grade of a public high school in collaboration with the teachers. Findings from a survey and interviews with the students showed that, following the program, their daily learning behaviors involving diagram use improved. These findings suggest the effectiveness of providing graphic literacy education as part of the school curriculum.

1 Introduction

Many previous studies have demonstrated that diagrams are powerful tools for problem solving [e.g., 1, 2]. However, research relating to educational practices indicates that school students do not use diagrams effectively as teachers and researchers do. One of the main problems is lack of spontaneity. Ichikawa [3] pointed out that many students tend to try to work problems out in their head and do not construct diagrams spontaneously when attempting to solve problems – even when they are very much capable of using diagrams correctly. Uesaka, Manalo, and Ichikawa [4] found that lack of spontaneity was particularly prominent amongst Japanese students.

In order to address this problem, researchers have examined factors influencing, and teaching methods to improve, students' use of diagrams. For example, Uesaka [5] reported that adding two manipulations to traditional math classes – teacher explicit instruction about the efficacy of diagram use, and practice in constructing diagrams – promoted the spontaneous use of diagrams as these enhanced perceptions about the usefulness of diagrams, and diagram construction skills, respectively. In addition, Uesaka and Manalo [6] demonstrated that actively comparing diagrams used in class improved abstract conditional knowledge for using diagrams, which in turn promoted the construction of more appropriate diagrams when solving math problems.

Because most of these studies have been conducted in the context of psychology or diagrams research, they have had little or no influence on real day-to-day instructional

G. Stapleton, J. Howse, and J. Lee (Eds.): Diagrams 2008, LNAI 5223, pp. 437–439, 2008.

practices in schools. However, incorporating the effective components of the instructional strategies from these research studies into the school curriculum would appear desirable as that would result in direct benefits to student learning and skills development. The present study, therefore, attempted to do this: by developing a program to promote students' spontaneous use of diagrams in a high school setting, in collaboration with school teachers who share the same awareness and aims about this issue. A questionnaire survey and personal interviews were used to evaluate the effectiveness of the program.

2 Intervention

The program to promote students' spontaneous use of diagrams was provided as part of a course on effective learning strategies, which in turn was one of several courses offered on cross-curricular study in a high school in Kanagawa, Japan. Nine students, aged 15-16 years and all in the first grade of the high school, took this course. The program dealing with diagram use lasted about two and a half hours.

The program was delivered in three parts. Firstly, the participating students were given problem solving tasks to demonstrate the benefits of diagram use: they were asked to use diagrams, and then not to use diagrams. The teacher provided explicit explanations about the efficacy that comes with diagram use. Secondly, the students were asked to select and solve one of two math problems requiring the use of a graph or a table to efficiently solve (from Uesaka & Manalo [6]). These procedures were used based on Uesaka's [5] findings that enhancing 'perceptions about the efficacy of diagram use' and 'skills in constructing diagrams' promoted students' diagram use. Teaching staff from the school and the university were involved in providing instructional support for the students during the program.

Finally, after some of the students had opportunities to present to the class how they solved the problem they chose, the teacher explained how the notion of an 'effective diagram' changed according to the nature of the problem given. Students were asked to make comparisons in order to induce the development of their conditional knowledge about different types of diagrams in relation to different types of problems. This procedure was included based on Uesaka and Manalo's [6] finding that appropriate choice can be promoted via active comparison of diagrams used in class.

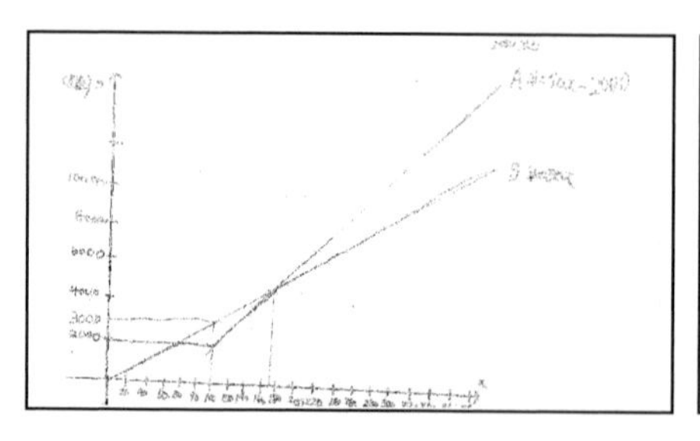

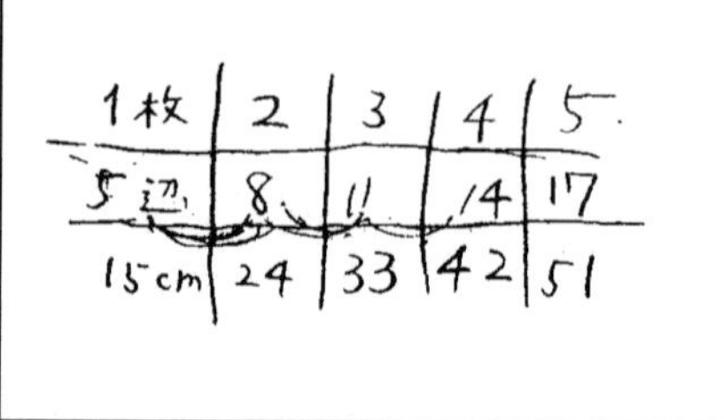

Fig. 1. Examples of Diagrams Constructed by Students During the Program

3 Results and Discussion

A survey in the form of a questionnaire was administered before and after the program. Questions relating to student study/learning behaviors were posed that required responses on a 5-point Likert-type scale (where 1 = Never, and 5 = Always). The improvement that students evidenced on the item about diagram use when attempting to solve problems was statistically significant ($F_{(1, 7)} = 18.00$, $p < .01$) despite the low number of participants.

During the personal interviews, the students were asked to bring the notebook they use in their math class. Upon examination, these revealed improvements in the students' daily learning behaviors, which confirm the finding from the questionnaire. Prior to the program, the students' notebooks contained mainly the teachers' diagrams copied from the board (the case with all of the 8 students; one of the original 9 students was excluded from the analysis for not bringing the notebook). The notebooks rarely contained diagrams that the students constructed by themselves (observed only in 1 out of the 8 students). In contrast, after the program, the students' notebooks contained both types of diagrams: copies of teachers' diagrams (8/8 cases) and ones the students constructed by themselves (7/8 cases).

Evidence was also found for an enhancement of the students' perceptions about the efficiency associated with diagram use. For example, three participants reported that they first realized the efficacy of diagram use through the program, and two participants reported that they previously understood the efficacy of it but their actual perception and appreciation of that efficacy improved through the program.

The findings of this study suggest the effectiveness of providing graphic literacy education as part of the school curriculum. Although the program that was developed and described here was subjected only to a small-scale exploratory evaluation, the results clearly indicate that strategies drawn from psychological/diagrams research can effectively be incorporated into the school curriculum.

References

[1] Larkin, J.H., Simon, H.A.: Why a diagram is (sometimes) worth ten thousand words. Cognitive Science 11, 65–99 (1989)
[2] Schoenfeld, A.H.: Mathematical problem solving. Academic Press, San Diego (1985)
[3] Ichikawa, S.: Benkyouhouga kawaru hon [A book about changing the approach to learning]. Iwanami Press, Tokyo (2000)
[4] Uesaka, Y., Manalo, E., Ichikawa, S.: What kinds of perceptions and daily learning behaviors promote students' use of diagrams in mathematics problem solving? Learning and Instruction 17, 322–335 (2007)
[5] Uesaka, Y.: Suugakutekimondainioite zuhyounoriyouwo unagasu kainyuuhohouno kentou [How to promote the spontaneous use of diagrams when solving math word problems]. In: The Annual Conference of Japanese Educational Psychology, p. 500. Preparatory Committee, Osaka (2003)
[6] Uesaka, Y., Manalo, E.: Active comparison as a means of promoting the development of abstract conditional knowledge and appropriate choice of diagrams in math word problem solving. In: Barker-Plummer, D., Cox, R., Swoboda, N. (eds.) Diagrams 2006. LNCS (LNAI), vol. 4045, pp. 181–195. Springer, Heidelberg (2006)

Author Index

Printing: Mercedes-Druck, Berlin
Binding: Stein+Lehmann, Berlin

Lecture Notes in Artificial Intelligence (LNAI)

Vol. 5009: G. Wang, T. Li, J.W. Grzymala-Busse, D. Miao, A. Skowron, Y. Yao (Eds.), Rough Sets and Knowledge Technology. XVIII, 765 pages. 2008.

Vol. 5003: L. Antunes, M. Paolucci, E. Norling (Eds.), Multi-Agent-Based Simulation VIII. IX, 141 pages. 2008.

Vol. 5001: U. Visser, F. Ribeiro, T. Ohashi, F. Dellaert (Eds.), RoboCup 2007: Robot Soccer World Cup XI. XIV, 566 pages. 2008.

Vol. 4999: L. Maicher, L.M. Garshol (Eds.), Scaling Topic Maps. XI, 253 pages. 2008.

Vol. 4994: A. An, S. Matwin, Z.W. Raś, D. Ślęzak (Eds.), Foundations of Intelligent Systems. XVII, 653 pages. 2008.

Vol. 4953: N.T. Nguyen, G.S. Jo, R.J. Howlett, L.C. Jain (Eds.), Agent and Multi-Agent Systems: Technologies and Applications. XX, 909 pages. 2008.

Vol. 4946: I. Rahwan, S. Parsons, C. Reed (Eds.), Argumentation in Multi-Agent Systems. X, 235 pages. 2008.

Vol. 4944: Z.W. Raś, S. Tsumoto, D.A. Zighed (Eds.), Mining Complex Data. X, 265 pages. 2008.

Vol. 4938: T. Tokunaga, A. Ortega (Eds.), Large-Scale Knowledge Resources. IX, 367 pages. 2008.

Vol. 4933: R. Medina, S. Obiedkov (Eds.), Formal Concept Analysis. XII, 325 pages. 2008.

Vol. 4930: I. Wachsmuth, G. Knoblich (Eds.), Modeling Communication with Robots and Virtual Humans. X, 337 pages. 2008.

Vol. 4929: M. Helmert, Understanding Planning Tasks. XIV, 270 pages. 2008.

Vol. 4924: D. Riaño (Ed.), Knowledge Management for Health Care Procedures. X, 161 pages. 2008.

Vol. 4923: S.B. Yahia, E.M. Nguifo, R. Belohlavek (Eds.), Concept Lattices and Their Applications. XII, 283 pages. 2008.

Vol. 4914: K. Satoh, A. Inokuchi, K. Nagao, T. Kawamura (Eds.), New Frontiers in Artificial Intelligence. X, 404 pages. 2008.

Vol. 4911: L. De Raedt, P. Frasconi, K. Kersting, S. Muggleton (Eds.), Probabilistic Inductive Logic Programming. VIII, 341 pages. 2008.

Vol. 4908: M. Dastani, A. El Fallah Seghrouchni, A. Ricci, M. Winikoff (Eds.), Programming Multi-Agent Systems. XII, 267 pages. 2008.

Vol. 4898: M. Kolp, B. Henderson-Sellers, H. Mouratidis, A. Garcia, A.K. Ghose, P. Bresciani (Eds.), Agent-Oriented Information Systems IV. X, 292 pages. 2008.

Vol. 4897: M. Baldoni, T.C. Son, M.B. van Riemsdijk, M. Winikoff (Eds.), Declarative Agent Languages and Technologies V. X, 245 pages. 2008.

Vol. 4894: H. Blockeel, J. Ramon, J. Shavlik, P. Tadepalli (Eds.), Inductive Logic Programming. XI, 307 pages. 2008.

Vol. 4885: M. Chetouani, A. Hussain, B. Gas, M. Milgram, J.-L. Zarader (Eds.), Advances in Nonlinear Speech Processing. XI, 284 pages. 2007.

Vol. 4884: P. Casanovas, G. Sartor, N. Casellas, R. Rubino (Eds.), Computable Models of the Law. XI, 341 pages. 2008.

Vol. 4874: J. Neves, M.F. Santos, J.M. Machado (Eds.), Progress in Artificial Intelligence. XVIII, 704 pages. 2007.

Vol. 4870: J.S. Sichman, J. Padget, S. Ossowski, P. Noriega (Eds.), Coordination, Organizations, Institutions, and Norms in Agent Systems III. XII, 331 pages. 2008.

Vol. 4869: F. Botana, T. Recio (Eds.), Automated Deduction in Geometry. X, 213 pages. 2007.

Vol. 4865: K. Tuyls, A. Nowe, Z. Guessoum, D. Kudenko (Eds.), Adaptive Agents and Multi-Agent Systems III. VIII, 255 pages. 2008.

Vol. 4850: M. Lungarella, F. Iida, J.C. Bongard, R. Pfeifer (Eds.), 50 Years of Artificial Intelligence. X, 399 pages. 2007.

Vol. 4845: N. Zhong, J. Liu, Y. Yao, J. Wu, S. Lu, K. Li (Eds.), Web Intelligence Meets Brain Informatics. XI, 516 pages. 2007.

Vol. 4840: L. Paletta, E. Rome (Eds.), Attention in Cognitive Systems. XI, 497 pages. 2007.

Vol. 4830: M.A. Orgun, J. Thornton (Eds.), AI 2007: Advances in Artificial Intelligence. XIX, 841 pages. 2007.

Vol. 4828: M. Randall, H.A. Abbass, J. Wiles (Eds.), Progress in Artificial Life. XII, 402 pages. 2007.

Vol. 4827: A. Gelbukh, Á.F. Kuri Morales (Eds.), MICAI 2007: Advances in Artificial Intelligence. XXIV, 1234 pages. 2007.

Vol. 4826: P. Perner, O. Salvetti (Eds.), Advances in Mass Data Analysis of Signals and Images in Medicine, Biotechnology and Chemistry. X, 183 pages. 2007.

Vol. 4819: T. Washio, Z.-H. Zhou, J.Z. Huang, X. Hu, J. Li, C. Xie, J. He, D. Zou, K.-C. Li, M.M. Freire (Eds.), Emerging Technologies in Knowledge Discovery and Data Mining. XIV, 675 pages. 2007.

Vol. 4811: O. Nasraoui, M. Spiliopoulou, J. Srivastava, B. Mobasher, B. Masand (Eds.), Advances in Web Mining and Web Usage Analysis. XII, 247 pages. 2007.

Vol. 4798: Z. Zhang, J.H. Siekmann (Eds.), Knowledge Science, Engineering and Management. XVI, 669 pages. 2007.

Vol. 4795: F. Schilder, G. Katz, J. Pustejovsky (Eds.), Annotating, Extracting and Reasoning about Time and Events. VII, 141 pages. 2007.

Vol. 4790: N. Dershowitz, A. Voronkov (Eds.), Logic for Programming, Artificial Intelligence, and Reasoning. XIII, 562 pages. 2007.

Vol. 4788: D. Borrajo, L. Castillo, J.M. Corchado (Eds.), Current Topics in Artificial Intelligence. XI, 280 pages. 2007.

Vol. 4775: A. Esposito, M. Faundez-Zanuy, E. Keller, M. Marinaro (Eds.), Verbal and Nonverbal Communication Behaviours. XII, 325 pages. 2007.

Vol. 4772: H. Prade, V.S. Subrahmanian (Eds.), Scalable Uncertainty Management. X, 277 pages. 2007.

Vol. 4766: N. Maudet, S. Parsons, I. Rahwan (Eds.), Argumentation in Multi-Agent Systems. XII, 211 pages. 2007.